Microbial Contamination Control in Parenteral Manufacturing

DRUGS AND THE PHARMACEUTICAL SCIENCES

Executive Editor

James Swarbrick

PharmaceuTech, Inc.
Pinehurst, North Carolina

Advisory Board

Larry L. Augsburger
University of Maryland
Baltimore, Maryland

Harry G. Brittain
Center for Pharmaceutical Physics
Milford, New Jersey

Jennifer B. Dressman
Johann Wolfgang Goethe-University
Frankfurt, Germany

Anthony J. Hickey
University of North Carolina School of Pharmacy
Chapel Hill, North Carolina

Jeffrey A. Hughes
University of Florida College of Pharmacy
Gainesville, Florida

Ajaz Hussain
U.S. Food and Drug Administration
Frederick, Maryland

Trevor M. Jones
The Association of the
British Pharmaceutical Industry
London, United Kingdom

Hans E. Junginger
Leiden/Amsterdam Center
for Drug Research
Leiden, The Netherlands

Vincent H. L. Lee
University of Southern California
Los Angeles, California

Stephen G. Schulman
University of Florida
Gainesville, Florida

Jerome P. Skelly
Alexandria, Virginia

Elizabeth M. Topp
University of Kansas School of Pharmacy
Lawrence, Kansas

Geoffrey T. Tucker
University of Sheffield
Royal Hallamshire Hospital
Sheffield, United Kingdom

Peter York
University of Bradford School of Pharmacy
Bradford, United Kingdom

DRUGS AND THE PHARMACEUTICAL SCIENCES

A Series of Textbooks and Monographs

Introduction to the Pharmaceutical Regulatory Process, *edited by Ira R. Berry*

Drug Delivery to the Oral Cavity: Molecules to Market, *edited by Tapash Ghosh and William R. Pfister*

Microbial Contamination Control in Parenteral Manufacturing

edited by
Kevin L. Williams
Eli Lilly and Company
Indianapolis, Indiana, U.S.A.

MARCEL DEKKER, INC.   NEW YORK · BASEL

Library of Congress Cataloging-in-Publication Data
A catalog record for this book is available from the Library of Congress.

ISBN: 0-8247-5320-8

This book is printed on acid-free paper.

Headquarters
Marcel Dekker, Inc., 270 Madison Avenue, New York, NY 10016, U.S.A.
tel: 212-696-9000; fax: 212-685-4540

Distribution and Customer Service
Marcel Dekker, Inc., Cimarron Road, Monticello, New York 12701, U.S.A.
tel: 800-228-1160; fax: 845-796-1772

Eastern Hemisphere Distribution
Marcel Dekker AG, Hutgasse 4, Postfach 812, CH-4001 Basel, Switzerland
tel: 41-61-260-6300; fax: 41-61-260-6333

World Wide Web
http://www.dekker.com

The publisher offers discounts on this book when ordered in bulk quantities. For more information, write to Special Sales/Professional Marketing at the headquarters address above.

Preface

Parenteral drugs require sterility in every case; there is no gray area in that regard. What, then, is the relevance of "contamination control" in parenteral manufacturing? Contamination control occurs at a multitude of sites along the series of processes that lead invariably to the final result that is a sterile product. Any highly specialized, multidisciplinary endeavor must of necessity be viewed from many perspectives. Each chapter in this book represents a facet, an expert perspective, of the control of contaminants in parenteral manufacturing. It is the product of a diverse and international group of experts. It is an effort not merely to describe the multitude of activities involved in the manufacture of parenteral products, but to capture the theory, essence, and pitfalls associated with an endeavor that occurs in an environment in sharp contrast to almost all other earthly activities. Amazingly, perhaps uniquely, this planet teems, festers, even boils with life. To carve a sterile niche free of microbial residue, even briefly, in the ecosphere is no small task. Once a given biomolecule is in hand, thus begins the daunting task of large-scale production, separation, and purification amid a myriad of other byproducts, all the while repelling repeated attempts by microbes to reclaim nature's substance. The breaking of the genetic code to produce complex biomolecules, the dedication of facilities, the active and passive methods used to exclude contaminants from processes, and the delicate twists and turns of engineering necessary to mass-produce a final drug can be counted as among man's most prized achievements.

Part I, Chapters 1–5, lays a foundation, both historical and modern, for microbiology, parenteral therapeutics production, and sterile concepts. Williams describes in overview form the beginnings of microbiological and aseptic knowledge, highlights the major events in the history of regulations governing parenteral drug manufacturing, and surveys emerging trends in the production of biologics. Rusmin describes in graphic detail current theory on the origin of the universe and microbiological evolution and gives us an appreciation of the tasks involved in the "microbiological function" as an oversight capacity in all things microbiological in the production of parenterals. Bansal describes the benefits of parenteral therapy over other forms of drug administration, categorizes the types of parenteral products produced, and overviews the processes used in their manufacture. In Chapter 4 Dabbah and Porter set the framework for current regulatory oversight from the USP perspective that includes detailing new chapters relative to microbiological control to be added to the USP. McCormick, Finocchario, and Kaiser (Chapter 5) introduce the basic concepts of sterility, its probabilistic definition, and the theory and use of Biological Indicators.

Part II, Chapters 6–10, begins with Claerbout's description of the equipment commonly used to isolate product from the outside environment, including operators: biological safety cabinets and isolators. In Chapter 7, Wirchansky describes the multidisciplinary process of planning for the design, construction, and validation of a new parenteral facility. The subsequent chapters in Part II deal with general and specific methods of achieving contamination control over various routes or vectors that would otherwise bring contaminants into contact with product and/or processes (via surfaces, personnel, air, water, etc.). In Chapter 8 Meltzer and Livingston describe emerging concepts in the design of pharmaceutical water systems, garnered from their experiences in both the pharmaceutical and semiconductor industries. Chapter 9, by Gail and Stanischewski, describes the necessity and means of attaining clean air in production environments, including the classification of airborne particulates, cleanroom design and operation, and the concepts governing airflow. Denny and Marsik in Chapter 10 detail extensively the chemistry, classification, and utility of various disinfectants used to sanitize specific environments.

Part III, Chapters 11–17, begins with Jornitz detailing the manifold choices available and the means of assuring the correct use of sterilizing-grade filters, including integrity, pressure, and chemical compatibility testing, to name a few. Sigwarth discusses the use of alternative sterilization methods and provides a detailed example, using hydrogen peroxide, for achieving the validation of one such method in a manner transferable to other prospective alternative methods. Haberer in Chapter 13 speaks of the limitations of end-product testing as a gauge of product quality, particularly in sterility testing, and addresses the basic and advanced concepts necessary to achieve parametric release of drug products. In Chapter 14, Gonzales discusses the necessity of microbiological purity of raw materials. The exclusion of and survey for both contaminants and artifacts or microbial residues in products, namely endotoxin and adventitious agents, are covered in some detail by Williams (Chapter 15), Cooper (Chapter 16), and Aranha (Chapter 17). Aranha's chapter encompasses conventional and nonconventional adventious agents (viral and prion respectively) and their in-process removal.

Part IV describes the important functions of sampling (numerical assessment) and gauging (making quality judgments) the microbiological state of critical environments, in-process materials, and final product beginning with Saranadasa's Chapter 18 on statistical sampling in regard to contaminants. Chapter 19, by Emerson et al., describes the critical function of monitoring clean areas (Environmental Monitoring) for contaminants. Sartain, in Chapter 20, discusses the investigation of environmental excursions from the sterile ideal. Nigel Halls in Chapter 21 describes an analogous activity, process simulation, which is a critical gauge of the ability of an aseptic process to perform at a suitable sterility assurance level (SAL). Finally, in Chapters 22 and 23, Sasser and Jimenez detail, respectively, the classical and rapid methods used to identify organisms isolated from areas critical or adjacent to parenteral manufacturing and/or filling activities.

Thus, in as linear fashion as is practicable, the characterization and control of microbial contaminants as encountered in parenteral manufacturing are presented. From any perspective, the complex environment associated with parenteral manufacturing with an eye toward the control of contaminants is being managed by increasingly sophisticated and specialized means. Parenteral manufacturing achievements will continue to allow doctors to fulfill their oath of providing medicines to patients that first "do no harm" and second, in many cases, are the only means to restore health and preserve life.

Kevin L. Williams

Contents

Contributors

Hazel Aranha   Pall Corporation, New York, U.S.A.

Arvind K. Bansal   National Institute of Pharmaceutical Education and Research, Punjab, India

Mark Claerbout   Eli Lilly and Company, Indianapolis, Indiana, U.S.A.

James F. Cooper   Endotoxin Consulting Services, Greensboro, North Carolina, U.S.A.

Roger Dabbah   United States Pharmacopeia, Rockville, Maryland, U.S.A.

Vivian Denny   Peak to Peak Pharmaceutical Associates, Boulder, Colorado, U.S.A.

John Emerson   Aventis Pharma, Holmes Chapel, United Kingdom

Petra Esswein   Aventis Pharma, Frankfurt/Main, Germany

Catherine J. Finocchario   Bausch & Lomb, Inc., Rochester, New York, U.S.A.

Lothar Gail   Siemens Axiva, Frankfurt/Main, Germany

Lisa Gonzales   Amersham Biosciences, Sunnyvale, California, U.S.A.

Klaus Haberer   Compliance, Advice and Services in Microbiology, Köln, Germany

Nigel A. Halls   Chorleywood, Herts, U.K.

Luis Jimenez   Genomic Profiling Systems, Bedford, Massachusetts, U.S.A.

Maik W. Jornitz   Sartorius Corporation, Edgewood, New York, U.S.A.

James J. Kaiser   Bausch & Lomb, Inc., Rochester, New York, U.S.A.

Robert C. Livingston   Arion Water, Hyannis, Massachusetts, U.S.A.

Frederic Marsik   Microbiology Consultant, New Freedom, Pennsylvania, U.S.A.

Patrick J. McCormick   Bausch & Lomb, Inc., Rochester, New York, U.S.A.

Theodore H. Meltzer   Capitola Consulting Co., Bethesda, Maryland, U.S.A.

Ulrich Pflugmacher   Aventis Pharma, Frankfurt/Main, Germany

David Porter   United States Pharmacopeia, Rockville, Maryland, U.S.A.

Simon Rusmin   New Spring Consulting, Parsippany, New Jersey, U.S.A.

Hewa Saranadasa   Pharmaceutical Group Americas, A Division of Ortho-McNeil Pharmaceuticals, Raritan, New Jersey, U.S.A.

Elaine Kopis Sartain   Steris Corporation, St. Louis, Missouri, U.S.A.

Myron Sasser   MIDI Labs, Newark, Delaware, U.S.A.

Volker Sigwarth   Skan AG, Basel, Switzerland

Dirk Stanischewski   Siemens Axiva, Frankfurt/Main, Germany

Kevin L. Williams   Eli Lilly and Company, Indianapolis, Indiana, U.S.A.

Dimitri Wirchansky   Jacobs Engineering, Conshohocken, Pennsylvania, U.S.A.

Microbial Contamination Control in Parenteral Manufacturing

1

Historical and Emerging Themes in Parenteral Manufacturing Contamination Control

Kevin L. Williams
Eli Lilly and Company, Indianapolis, Indiana, U.S.A.

1.1. INTRODUCTION

Man has been described as an obligate aerobe. Oxygen floods the lungs, dissolves in the bloodstream, and spills into a thousand capillaries as a great waterfall aerates a mighty river. The same blood that brings oxygen is the route of choice for many pharmaceuticals that can only reach the innermost depths of the body via this route to dispense their therapeutic properties. The word *parenteral* is derived from the Greek "para" (beyond) and "enteral" (gut) because it bypasses the digestive system. This route is so effective it necessitates a level of cleanliness that approaches the absolute. A single viable organism, bacteria or virus, thusly introduced into the body evades all but the final mechanism of defense and so the medicine designed to bring life could bring infection, fever, shock, or death. Man's war against microbes is never ultimately won. They are deeply entrenched in the air, water, and soil. The body itself is occupied: 1% or more of the human genome consists of retroviral sequences, and microbes on and in the body outnumber the cells that compose the body by 10- to 20-fold (1). Microbes are legion; ubiquitous, unmerciful, and untiring. On a personal basis, those occupying us now, or their offspring, will decompose us when we die. They can be eradicated only in small places and for a short time.

It is desirable to step back and view, even rudimentarily, the scientific, regulatory, and technological events (historical and contemporary) that contribute to the current state of complexity encompassing the control of contaminants in the manufacture of parenteral drugs. Many of the references chosen here are review articles that will facilitate basic and advanced inquiries into the relevant topics that are presented in this chapter as an overview and that contrast in that regard to the highly specialized chapters to come.

Parenterals require sterility in every case; there is no gray area in that regard. Why then is "contamination control" relevant to parenteral manufacturing? Contamination control occurs at a multitude of sites along the series of processes that lead inevitably to the final result that is a sterile product.

1.2. THE BIRTH OF MICROBIOLOGICAL THEORY

The birth of modern microbiology in the later nineteenth century heralded by Louis Pasteur, Robert Koch, Joseph Lister, and others began the quest to clarify the bacterial causation and mechanisms of infection. Though Anton van Leeuwenhock, the "uneducated" Dutch merchant and amateur microscope maker, made detailed observations of microorganisms, even proposing a role in disease causation in letters to the Royal Society in London between 1675 and 1685, the new paradigm of microscopic life was not generally accepted as fact for at least another 200 years (2). Pasteur's refutation of spontaneous generation, description of fermentation as a by-product of microorganisms, ideas on putrefication, and invention of pasteurization (3) dispelled many of the prevalent myths of the day, sometimes in dramatic fashion (i.e., swan-necked flask). Lister, meanwhile, elaborated his "germ theory" from Glasgow and performed the first successful antiseptic operation using carbolic acid (phenol) to steam-sterilize medical instruments. The work of Pasteur and Lister served to dispel the thought that vapors ("miasma," or bad air as it was called) and other vague forms of suspected "contagion" by gases held any role in disease causation (4). Though Edward Jenner developed the first vaccine using the cowpox virus 100 years before Pasteur, it was Pasteur who knowingly manipulated living microbes to alter the course of disease. He heated anthrax bacilli and dried the spinal cords of rabies-infected rabbits to develop vaccines against anthrax in sheep (1881) and rabies in man (1885), respectively (5).

In the late 1870s, Robert Koch established that individual types of microbes were associated with specific diseases, including anthrax and tuberculosis (6). Koch laid out postulates purporting the conditions that must be met prior to regarding an organism as the cause of a given disease. His postulates were as follows: (a) the organism must be present in every case under conditions explaining the pathological changes and clinical symptoms, (b) the organism must not be associated casually with other diseases, and (c) after isolation from the body and cultivation in pure culture, the organism must be able to produce the disease in animals. Koch refined tools and techniques needed to prove his postulates, including solid agar and a method of isolating singular bacterial colonies by means of a heated inoculating loop. Both tools remain staples of the microbiological trade. Koch's methods led to the rapid identification of the specific bacteria associated with many of the infectious diseases of the late 1800s and early 1900s. The Gram stain, invented by Hans Christian Joachim Gram in 1884 (4), proved to be a most useful tool in the study of fever causation in that it split the newly discovered bacterial world into two distinct groups that, unknown at the time, included those containing endotoxin and those that did not (7). Because the cellular wall contents determined the amount of stain retained in the staining process, subsequent observations were based on cellular morphology and were not merely an arbitrary classification technique. These new theories and methods provided the a priori background for further research into the newly discovered microbial world, established the ubiquity of microorganisms as causative agents of disease, and underscored the rational processes on which to base research into aseptic technology and disease prevention and cure.

1.3. HISTORICAL DEVELOPMENT AND REGULATION OF PARENTERAL DOSAGE FORMS

The manner of origin of most dosage forms is largely unknown. Early humans may have fashioned primitive injections modeled after venomous snakes or insect bites and stings (natural puncture injections). Asians inoculated for the prevention of smallpox by pricking with needles dipped in pus centuries before the technique was used in Western cultures. Jenner performed the same in 1796 using a cowpox sore (8). Sir Christopher Wren was first to inject a drug in 1657, a technique which was later used routinely by the English practitioner Johan Major in 1662. In the early 1800s, Gaspard experimented by injecting putrid extracts into dogs (9). Doctors experimented with injecting some potentially useful compounds and some bizarre and even fatal substances. Stanislas Limousin invented the ampule in 1886, and Charles Pravex of Lyons suggested the hypodermic syringe in 1853. The Royal Medical and Chirurgical Society of London approved hypodermic injections in 1867 concurrently with the first official injection (*Injectio Morphine Hypodermica*) published in a monograph in *British Pharmacopoeia* (8). Early progress in injectable therapy was slowed by fever occurrences and other symptoms associated with the crude state of early parenteral manufacturing. Exceptions existed that allowed progress, notably Ehrlich's use of hypodermic injections of salvarsan for syphilis in 1910 (8). Martindale and Wynn proposed active manufacturing techniques to produce aseptic salvarsan in the same year that Hort and Penfold were describing the active agent in producing fevers (bacterial endotoxin) (10).

It is interesting to note that the very first parenteral applications, vaccines, were in effect contaminated solutions used to trigger the body's immune response (rabies, tetanus, tuberculosis, smallpox). The concept of sterility was introduced at the beginning of parenteral manufacturing and was first required in the ninth revision of the *U.S. Pharmacopeia* in 1916 and was accompanied by an introductory chapter on achieving sterility. The only parenteral solutions included at the time were distilled water, solution of hypophysis, and solution of sodium chloride (11). The fever that accompanied early injections was believed to be due to the route of administration (i.e. the body's response to being pricked by a needle) rather than being viewed as a drug contaminant, and it was therefore referred to as "injection fever." In 1912 Holt and Penfold published several conclusive studies, including "Microorganisms and Their Relation to Fever" (10). The pair demonstrated that (a) the toxic material originated from gram-negative bacteria (GNB), (b) the pyrogenic activity in distilled water correlated to the microbial count, (c) dead bacteria were as pyrogenic as living ones, and (d) a rabbit pyrogen test could be standardized and used to detect occurrence of endotoxin in parenteral drugs.

The work of Hort and Penfold was largely overlooked until 1923 when Florence Seibert in the United States explored the causes of pyrogenicity of distilled water (12). She demonstrated conclusively that bacterial contamination was indeed the cause of "fever shots" (13). She determined that even minuscule, unweighable contaminants were biologically very active (14). During this time it became obvious to numerous investigators that GNB possessed a high-molecular-weight complex as part of their outer cell walls. The complex came to be called the endotoxic complex, which as a whole was thought to be responsible for the toxic, pyrogenic, and immunological response induced by GNB. Rademaker confirmed Seibert's findings and stressed the importance of avoiding bacterial contamination at each stage of pharmaceutical production, pointing out that sterility is no guarantee of apyrogenicity (15). Nevertheless, it would be two decades before the U.S. National Institutes of Health and 14 pharmaceutical manufacturers undertook a collaborative study to establish an animal system to be used to determine the pyrogenicity of parenteral solutions. The first official rabit pyrogen test was incorporated into the 12th edition of the *U.S. Pharmacopeia* (USP) in 1942 (16).

A test for parenteral sterility (to support the 1916 contention that parenteral solutions should be sterile) originated in the *British Pharmacopoeia* in 1932 and in the U.S. Pharmacopoeia in 1936 (17). By 1936 there were 26 parenteral drug monographs in the National Formulary (NF VI), many of which were packaged in ampules (18). The methods of gauging sterility have been modified year in and year out since, but the basic concept of what sterility means has not changed. Halls lists some major limitations of the very first sterility test (17). Limitations associated with the necessity of demonstrating the lack of sterility from a quality perspective still exists in today's test 70 years later:

> (a). The test presumed sterility. Even with the limitations of the sterilization technology of the 1930's, the pharmacopoeia was presuming sterility unless nonsterility could be convincingly and conclusively demonstrated. This is rather unusual because it goes against the grain of scientific criticality to assume that a hypothesis is valid unless it can be proven otherwise. The test was far less a critical test for sterility, as one might suppose it was intended to be, than a test for nonsterility—i.e., false nonsterile results were thought to be more likely than false sterile results (the pharmacopoeia had more faith in the potential of the recommended media to recover microorganisms than it had in the ability of the laboratories to perform successful aseptic manipulations).
> (b). The test did not address total freedom from microorganisms for preparations in 2 mL volumes or greater. For these larger volumes it was really a microbial limit test with a lower sensitivity of detection of one microorganisms per mL.
> (c). The test gave no guidance on interpretation of data from replicate recovery conditions (17).

The dawn of drug manufacturing as a means of disease prevention (vaccination) and treatment (antibiotics, insulin, etc.) brought about the concurrent need to both harness microbes to manufacture cures and to eliminate them from contaminating medicine-producing processes. Concomitant with the medical necessity of providing safe and effective drugs was the political necessity of ensuring that manufacturers would not violate the accumulating regulations of manufacture. The laws governing pharmaceutical manufacturing have come about in stair-step fashion side by side with tragic events. To add insult to injury, commercial opportunists blatantly hawked unproven" cures" thus crowding out the few serious medicines that

TABLE 1 Chronology of U.S. Drug Regulation and Related or Precipitating Tragic Events

Year(s)	Event(s)	Subsequent Regulation
1902	Diphtheria antitoxin contaminated with live tetanus bacilli, resulting in the death of 12 by lockjaw	Biologics control act of 1902 required inspections of biological manufacturing
1906	Upton Sinclair's *The Jungle* railed against the unsanitary practices of the food industry and aided passage of stalled legislation championed for over two decades by Dr. Harvey Watson (Indiana Chemist, Purdue professor, FDA commissioner)	Federal Pure Drug and Food Act created the Bureau of Chemistry, forerunner of FDA
1935	Elixir of sulfanilamide killed 107 due to its formulation in diethylene glycol at toxic concentrations	FDC act of 1938 required proof of safety prior to marketing
1941	Sulfathiazole tainted with phenobarbitol; 300 died due to ineffective recall efforts by Winthrop. One lot (29) contained on average 0 mg of sulfathiazole and 350 mg of phenobarbital (100–150 mg dose being hypnotic)	Manufacturing and quality control requirements precursors to GMPs
1940s	Yellow fever vaccine contaminated with hepatitis virus	
1955	Virus not killed in polio vaccine; ~150 contract polio directly or via those infected	

Year	Event	Regulatory Response
1955–1963	Polio and adenovirus vaccines contaminated with SV40 (simian virus)	
1960	Thalidomide marketed in Europe for morning sickness results in ~10,000 severe birth defects	Kefauver-Harris Act of 1962 strengthened animal toxicity & teratogenic requirement est. by 1938 FDC act and made mention of cGMPs for the first time
Early 1960s	Blood products contaminated with hepatitis virus	
1978	GMPs made final	21 CFR part 210 and 211
1980	Toxic shock syndrome outbreak; 314 cases, 38 died	FDA required tampon package inserts to educate on TSS hazards
1982/83	Tylenol cyanide tampering killed seven	Federal Anti-tampering act
1980s–1990s[a]	Iatrogenic prion disease infections (medically induced): Dura matter (brain) grafts (>60 cases), Human growth hormone (animal sourced, not r-HGH, >90 cases), corneal transplants, and gonadatropin from cadavers—all contaminated with Creutzfeld-Jacobs prion	Industry acts to limit use of animal-sourced raw materials; CBER requires BSE testing of raw materials derived from animal sources.
1998	Gentamicin fever reactions; investigations find the bulk from China borderline failures based on off-label use of 3× daily dose.	Drug Modernization Act allowed for off-label dosing of drugs.

[a] Time from infection to symptoms may be ≥10 years.
Source: Refs. 20 through 23.

were available in the early twentieth century. Few companies at the time limited their sales directly to physicians (so-called ethical drugs) but instead appealed to the hopes of consumers seeking easy and inexpensive cures for every ailment.

> Voluminous newspaper advertisements (sometimes one-fourth of the space), traveling "doctors" and pitch men with or without their side shows, druggists, and general storekeepers proclaimed loudly and constantly the merits of various panaceas. So powerful was the influence that millions of people had come to expect, all in one remedy (at a dollar or two the bottle), certain cure for consumption, cholera morbus, dyspepsia, fevers, ague, indigestion, diseases of the liver, gout, rheumatism, dropsy, St. Vitus's dance, epilepsy, apoplexy, paralysis, greensickness, smallpox, measles, whooping cough, and syphilis (19).

Parenteral manufacturing occurs at an interface of science and regulatory compliance. Change, of necessity, must occur together, in lockstep, to balance the risks of life-saving technological advancements with the safety of traditional methods. Because every risk of process failure, human error, or act of maliciousness could not be precluded and addressed by laws governing the manufacture of drugs, broad and general requirements were enacted initially in the 1906 Pure Food and Drug Act and revised notably in 1938 and 1962. The 1962 amendment to the Food and Drug Act expanded the agency's definition of adulteration to include "conformance with current good manufacturing practice" (18). A drug could be considered adulterated if:

> The methods used in, or the facilities or controls used for its manufacture, processing or holding do not conform to or are not operated or administered in conformity with current good manufacturing practice to assure that such drug meets the requirements of this Act as to safety and has the identity and strength, and meets the quality and purity characteristics which it purports to possess [Section 501(a)(2)(B)] (18).

The "c" in cGMP allowed the law to live and regulatory expectations to grow to meet improvements in technology and/or changing hazards. Regulations still favor the most proficient of manufacturers and act as a failsafe for those who seek to form the lowest denominator of industry practice. Few would argue that without strict regulatory oversight some manufacturers would squeak by in terms of SISPQ (safety, identity, strength, purity, quality) and would instead make manufacturing profit the overriding concern. Table 1 is a sampling of U.S. government regulations governing the drug industry along with the corresponding, often tragic, precipitating events. The use of thalidomide as a prescription for morning sickness is a particularly gruesome example of an adverse event that brought about positive, wholesale change even though the existing FDA regulations prevented the approval of the drug in the U.S. (thanks largely to the efforts of Frances Kelsey, who was assigned the application at the FDA) in 1960 (20).

1.4. FROM ANTIBIOTICS TO BIOLOGICS

Drug discovery began in ancient times with the use of plants as medicinal treatments and centered more recently on the isolation and purification of their bioactive ingredients. Fermentation processes have been used since antiquity to produce a multitude of food products (cheese, yogurt, vinegar, wine, beer, and bread), but the scientific basis of fermentation was unknown and became a topic of contention between chemists and microbiologists as to the underlying cause, chemical or microbial. Pasteur's publication on fermentation in 1857 largely laid the matter to rest by not only associating organisms with fermentation in every case but by describing the specific organisms associated with each (alcohol by yeasts, lactic acid by nonmotile bacteria, and butyric acid by motile rods) (24). Fermentation when combined with microbial strain improvement (via mutant screening) was an important technological platform that served to produce everything from food for human and animals to acetone and butanol (needed for war materials) as prototypes for modern manufacturing processes, which clearly remain analogous.

The extraction and use of animal proteins as roughly human equivalents came next (first bovine insulin from animal pancreases and then growth hormone from animal pituitary glands) to treat diseases of deficiency. The discovery of insulin in 1921 by Dr. Banting of Toronto (for which he shared the Nobel prize) set the stage for the mass production of insulin (Lilly Iletin™), which by the spring of 1923 became available to doctors for general administration (19). The widely visible "miracle cure" that insulin provided to critically ill diabetics solidified the budding disciplines of drug development and parenteral manufacturing. The discovery of penicillin by Alexander Fleming* in the 1920s resulted in the use of microbial fermentation by-products (antibiotics) to treat infection followed by the development of fermentation processes for steroids and was accompanied by the introduction of the concept of "randomization" in drug clinical trials (25). The control of infectious disease is credited with much of the improvement in the human lifespan from the beginning of the 20th century to 1950, rising from below 50 years to the mid 70s where it remains today (26). Interestingly, the chart is

---

* The discovery of penicillin from a random *Penicillium* mold growing in Fleming's staphylococcus culture (agar plate) typified the serendipity associated with the discovery of early drugs.

punctuated with epidemics such as the 1918 outbreak of Spanish influenza that reduced the life expectancy for that year to below 40.

Tools developed in the early 1970s (restriction enzymes and plasmids) allowed the development of recombinant DNA technology, whereby the genes encoding human proteins (insulin and growth hormone) could be inserted into *E. coli* followed by their overexpression via fermentation (i.e., the combination of two previous technologies with the new technology). The first product of biotechnology, in the modern sense of the word, was recombinant human insulin in 1982 (Lilly Humulin ^{TM}). The origin of the term biotechnology is said to have been Karl Ereky's 1917 to 1919 publications, in which he dealt with the concept of the "animal-machine" that he envisioned could help supply foodstuff for war-torn Europe (27).

The new recombinant drugs not only replaced the need for using animal-sourced proteins and all the associated contamination problems (i.e., viruses and prions) but also resulted in very efficient and economical manufacturing processes.

It became clear that recombinant DNA technology yielded purer proteins and was much more economical than conventional techniques. As a result, a large number of mammalian peptide genes were cloned and expressed in *E. coli, B. subtilis* and other bacilli, *Saccharomyces cerevisiae* and other yeasts, *Aspergillus niger*, insect cells and mammalian cells. The benefits of *E. coli* as a recombinant host included 1) ease of quickly and precisely modifying the genome, 2) rapid growth, 3) ease of fermentation, 4) ease of reduction of protease activity, 5) ease of avoidance of incorporation of amino acid analogs, 6) ease of promoter control, 7) ease of alteration of plasmid copy number, 8) ease of alteration of metabolic carbon flow, 9) ease of formation of intracellular disulfide bonds, 10) growth to very high cell densities, 11) accumulation of heterologous proteins up to 50% of dry cell weight, 12) survival in a wide variety of environmental conditions, 13) inexpensive medium ingredients, 14) reproducible performance especially with computer control, and 15) high product yields (Swartz, 1996). ... Many benefits to society have resulted from proteins made in *E. coli* (Swartz, 1996). 1) Diabetics do not have to fear producing antibodies to animal insulin. 2) children deficient in growth hormone no longer have to suffer from dwarfism or fear the risk of contracting Kreutzfeld-Jacob syndrome. 3) Children who have chronic granulomatous disease can have a normal life by taking interferon gamma therapy, 4) Patients undergoing cancer chemotherapy or radiation therapy can recover more quickly with fewer infections when they use G-CSF (28).

The 1984 Nobel prize in medicine was awarded to two scientists, Koehler in Germany and Milstein in England, for their efforts to develop a method for producing monoclonal antibodies (Mabs) (29). With this discovery, highly specific antibodies, products of individual lymphocytes, could be generated against specific antigens. Initially, Mabs were made by inoculating mice with an antigen and isolating and purifying the resulting antibodies or antibody producing cells. For large-scale production the utilization of recombinant methods and fermentation with specialized cell lines have been employed to produce human antibodies to various antigenic disease targets (tumor cells by Genentech's Herceptin™, etc.).

Current methods of drug discovery and production are invariably based on genomics and proteomics, which have flowed from the sequencing of over 60 microbial genomes* (30) and more recently the entire human genome (31). Thomas Roderick is credited with coining the word genomics in 1986, defined as the "scientific discipline of mapping, sequencing, and analyzing genomes" and used as a tag for a new journal (32). There are two areas of genomics, functional and structural. Structural genomics is the construction of high-resolution genetic maps for specific organisms; functional genomics involves mining the data generated in the structural genome to explore how it functions, particularly from a disease causation vantage. The term Proteomics was first used in 1995 to describe the characterization of all the proteins of a cell or organism, referred to as the proteome (33).

Biologics are macromolecular (>500 kd) substances either composed of, or extracted from, a living organism†. Biologics bring with them increasing complexity, including glycosylation‡ that often cannot be manufactured by older technologies employing single-celled organisms as expression systems (i.e., bacteria and yeasts). Biologics, considered by USP 26 as predominately recombinants and monoclonals, tend to be less well defined analytically (34). While biologics as a group are not new (the biologics control act of 1902 covered vaccines, antitoxins, blood, and blood derivatives), the ones that are new are often derived from new technologies, are very different in their method of manufacture, and are susceptible to nontraditional contaminants. As the complexity of the biomolecules (biologics) manufactured has increased, so too

---

* See the TIGR microbial database for prokaryotic and eukaryotic (completed and ongoing) sequencing projects: www.tigr.org/tdb/mdb/mdb.html.

† "For Pharmacopeial purposes, the term "biologics" refers to those products that must be licensed under the Act (1944 Public Health Service Act) and comply with Food and Drug Regulations-Code of Federal Regulations, Title 21 Parts 600–680, as administered by the Center for Biologics Evaluation and Research..." USP 26, <1041> (34).

‡ The attachment of carbohydrates that affects the configuration of the molecule (usually a protein) to which they are attached.

have the processes producing them. Changes in manufacturing processes that have the capability to affect the control of contaminants include:

1. Use of new expression systems
2. Use of new media, cell culture, and transgenics that do not subscribe to previous limitations of potential contaminant types
3. Development of altogether new classes of drugs and drug excipients to be parenterally administered.

A simple diagram of the process flow typical of manufacturing a biologic is shown in Fig. 1. Perhaps of greatest relevance from a contamination control perspective associated with the manufacture of biologics and biotechnology-derived articles is the careful analytical monitoring required

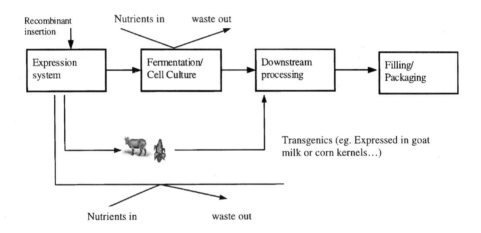

| Addition of gene encoding therapeutic molecule (protein, antibody) into the genome of the desired expression organism: *E. coli*, *S. cerevisieae*, Chinese Hamster Ovary cells (CHO), transgenic animal or plant. | Suitable environment for host organism's growth and production of desired molecule. Suspended in a medium or anchored to a base. Stirred container, continuous growth, batch fermentation, etc. Culture is fed and growth is monitored. | Separation and purification: removal of host cells, by-products that are not the desired drug, and virus or other adventitious contaminants. | Purified product is aseptically filled as per traditional drugs, packaged and sampled for analytical testing. |

FIGURE 1 Simplified process flow diagram for a biological molecule.

(in addition to that already required for injectable drugs), including expression system genetic content, viral particles, expression of endogenous retrovirus genes, and other adventitious microbial agents (mycoplasma as per 21 CFR 610.12), to name a few. The regulations governing biologics are necessarily stringent, requiring the review of every lot manufactured. The types of concerns involved in the manufacturing processes can be surveyed by observing the CBER issued "Points to Consider" documents (http://www.fda.gov/cber/guidelines.htm). Consider those referenced below contained within the "Points to Consider on the Manufacture and Testing of Monoclonal Antibody Products for Human Use" (35):

> Sponsors are encouraged to consult the most recent available versions of the Points to Consider in the Characterization of Cell Lines Used to Produce Biologicals, the Points to Consider in the Production and Testing of New Drugs and Biologicals produced by Recombinant DNA Technology or the Points to Consider in the Manufacture and Testing of Therapeutic Products for Human Use Derived From Transgenic Animals (1, 2, 5), the 1996 CBER/CDER Guidance Document on the Submission of Chemistry, Manufacturing and Controls Information for a Therapeutic Recombinant DNA-derived Product or a Monoclonal Antibody Product for In Vivo Use (4), as well as relevant International Conference on Harmonization (ICH) documents (e.g. 6, 7), if applicable to their expression systems. Sponsors considering novel expression systems not specifically covered by guidance documents are encouraged to consult with CBER.*

---

*1. Points to Consider in the Characterization of Cell Lines Used to Produce Biologicals, CBER, FDA, 1993.

2. Points to Consider in the Production and Testing of New Drugs and Biologicals Produced by Recombinant DNA Technology, FDA, CBER, April 10, 1985 and Supplement to the Points to Consider in the Production and Testing of New Drugs and Biologicals Produced by Recombinant DNA Technology: Nucleic Acid Characterization and Genetic Stability, CBER, FDA, April 6, 1992.

3. No number 3 in this referenced section.

4. Guidance for Industry for the Submission of Chemistry, Manufacturing and Controls Information for a Therapeutic Recombinant DNA-derived Product or a Monoclonal Antibody Product for In Vivo Use, CBER/CDER, FDA, August, 1996.

5. Points to Consider in the Production and Testing of Therapeutic Products for Human Use Derived from Transgenic Animals, CBER, FDA, 1995.

6. International Conference on Harmonisation; Final Guideline on Quality of Biotechnological Products: Analysis of the Expression Construct in Cells Used for Production of r-DNA Derived Protein Products, (February 23, 1996; 61 Federal Register 7006).

7. International Conference on Harmonisation; Final Guideline on Stability Testing of Biotechnological/Biological Products, (July 10, 1996; 61 Federal Register 36466).

The task of excluding microbes from drug manufacturing processes contrasts sharply with their ever-expanding utility in producing medicines. Expression systems for generating unique biochemical entities include single-celled organisms such as *E. coli* and *S. cerevisiae*, mammalian cells, insect cells and hybridomas (immortalized cell lines). The fermentation and cell culture step of biologic manufacture are distinguished in that fermentation refers to the process that utilizes single-cell organisms and cell cultures utilize cells derived from higher, multicellular, organisms. A third type is both an expression system and a "bioreactor" and is referred to as transgenic. The increasing costs associated with greater product complexity* is driving the use of transgenics including (a) transgenic mammals that produce proteins in their milk (cows and goats), (b) transgenic hens that lay eggs containing recombinant proteins, (c) a slime mold (*Dictyostelium discoideum*) that secretes recombinant proteins, (d) the use of plants such as corn or tobacco[†] (e) the use of insect cells inoculated with viruses specific to insects (Baculovirus) that have been genetically modified to encode for therapeutic proteins instead of viral proteins (36), and (f) the recent similar utilization of the silkworm (37). Each drug discovered eventually presents a preferred method of production, and each production method has an associated set of benefits as well as constraints.

Parenteral drug presentations include new classes, some of which are still being defined.[‡] Biologics may bring with them unknowns as to their ability to contain and/or mask microbial contaminants and/or associated artifacts and the possible effects of unique therapeutic activities on potential contaminant interactions with host systems [i.e., endotoxin liberation was found to be associated with the use of some antibiotics (38)]. Some medications contain ingredients that are typically associated with parenterals, many of them of natural origin, that are not readily soluble or contain ingredients not historically used, and, therefore, that may bring with them new potential contaminants (39). Some, such as sustained-release, liposome-contained, and bone-paste drugs may contain polymers, plastics, or adhesives intended to delay degradation or support their therapeutic function (40). "Furthermore, many emerging delivery systems use a drug or gene covalently linked to the molecules, polymers, antibody, or chimera responsible for drug

---

* Consider the initial average annual cost of treating Gaucher disease using human glucocerebrosidase: $160,000 (36).

[†] CropTech introduced the gene that produced glucocerebrosidase into the tobacco plant.

[‡] "The novel approaches permitted by biotechnology can make it difficult to apply classic definitions of these [drug biologic or diagnostic] categories and FDA has advised manufacturers to seek clarification in the early stages of development for how a product will be regulated when classification is not obvious."–USP 26 <1045> (34).

targeting, internalization, or transfection" (39). To demonstrate the uncertainty and novelty of some presentations, the FDA has recently reclassified a paste injected into bone from a parenteral to a device due to its relative inertness. There is also research into delivering toxic substances directly to diseased or infected tissues to bypass toxicity associated with systemic administration and ongoing attempts to combine the convenience of non-parenteral administration with the benefits associated with parenteral drugs such as those designed for inhalation or those presented as oral tablets intended to pass directly from the small intestine to the bloodstream. The concern for the characterization of biologics revolves around more than the finished product:

> Given the importance of biopharmaceuticals, regulatory authorities are emphasizing the requirements of well characterized biologicals* (WCB). This has focused particular attention on the composition not only of the final biological product, but also of the processing materials in contact with that product. In the downstream processing system, contaminants may be carried through to the final product either through the feedstock or through components of the processing media (41).

Indeed, the beginning of concern for product contamination begins before downstream processing with the characterization of the cell bank used:

> The production of biologics requires the use of raw materials derived from human or animal sources, and that poses a threat of pathogen transmission. Potential contamination may arise from the source material (such as cell bank of animal origin) or as adventitious agents introduced by the manufacturing process (such as murine monoclonal antibodies used in affinity chromatography). Evaluating the safety of a biological product begins at the level of the source material, such as the manufacturer's working cell bank (WCB) or, in our case, the master cell bank (MCB) (42).

One can speculate that for every niche man finds to manufacture and administer drugs, microbes (or their genetic insertions, by-products, artifacts, toxins, etc.) will seek to hitch a ride [i.e., invasins, adhesins, etc. (43)] via that particular endeavor, thus providing unique challenges in contamination control.

---

* Embedded references: (2) Little, L. "Current Trends: Impact of the Well-Characterized Regulatory Paradigm," BioPharm 10(8), 8–12 (1997). (3) R.J. Seely et al., "Defining Critical Variables in Well-Characterized Biotechnology Processes," BioPharm 12(4), 33–36 (1999).

1.5. CHANGING PERSPECTIVES ON CONTAMINANTS

Microbiology and genetics are in the midst of unprecedented historical change. Changes occurring that are affecting the way that microbial "contaminants" are viewed include (a) the recent explosion of knowledge in microbial genetics that has brought about the wholesale change in microbial classification* as evidenced by the ongoing retooling of classification from historical phenotypic to genotypic based approaches, (b) the realization that most organisms remain unculturable by standard methods (c) the discovery of emerging microbial pathogens in the form of genetic insertions (free of associated microbes) into human and animal genomes, and (d) the discovery of a previously inconceivable form of infectious disease causing agent: the prion.

Historically, bacteria and other microbes have been classified by "what they do" (i.e., ferment various sugars, retain crystal violet in the Gram stain, etc.) but are now being reclassified by "what they are" or "who they are" (i.e., their genetic relatedness). The most widely used genetic classification system as proposed by Woese (44–46) can be briefly described as centered around the similarity or dissimilarity in ribosomal RNA (rRNA) sequences, which are conserved genetically across species barriers (and significantly in all life forms). The use of the rRNA avoids a caveat that exists in the characterization of genomes in that they contain errant or wandering sequences (horizontal transmission) associated with insertions from plasmids, phage, pieces of phage, and so forth that may confound attempts toward classification, whereas rRNA is not shared. As an example, the 16S rRNAs for *E. coli* and *P. aeruginosa* are members of Proteobacteria and differ by about 15%, whereas *E. coli* differs from *B. subtilis* by about 23% (47).

The realization that most of nature's microbes, by some estimates 99%, cannot be cultivated by standard methods has supported genomics-based reclassification efforts. Genetic methods have allowed for the classification of unknown organisms that cannot be cultured and have the ability to place such organisms between known species within a genetic-based continuum. Amann et al. point out that the 5000 known species of Bacteria and Archaea must represent a tiny fraction of species existing in nature (48). They note that there are 800,000 species of insects and each insect harbors millions to billions of bacteria and "thus, consideration of insect symbionts alone could increase the number of extant bacterial species by several orders of magnitude" (48). While the relevance to parenteral manufacturing is not known, it supports the contention that there are most certainly forms of contamination that are invisible to current methods of detection, particularly in water, air and naturally sourced raw materials and culture media. Methods used for

---

* Note that classification does not equal identification (44).

bacterial identification and genetic sequence determination is discussed in Chapters 22 and 23. Fredricks and Relman maintain that:

> The petri dish and traditional tissue stains have been supplanted by nucleic acid amplification technology and in situ oligonucleotide hybridization for "growing" and "seeing" some microorganisms. The power of these techniques has opened a new window on the diversity of environmental and human-associated microorganisms. It has also led to an explosion of amplified sequences purportedly derived from uncultivated or fastidious microbes that are associated with pathology or disease. The quest to find relevance in these sequences demands a reassessment of our analysis of disease causation (6).

The lines of disease causation have become blurred at the genetic level by the discovery of microbe-induced disease processes not originally associated with microbial causes and only recently identified by genotypic approaches. The latter include viral-induced cancers* (49–51) [even schizophrenia (52) and diabetes mellitus (1) have been implicated], *Borrelia burgdorferi* DNA incorporated in the genome of arthritic mice (53) [and detection in humans (54)], and a list of organisms referenced by Relman (6) that have been found using genotypic approaches to detect microbial genes inserted into the genome of man and animals and therefore associated with specific diseases. These include *Helicobacter pylori* (peptic ulcer disease), Hepatitis C virus (non-A, non-B hepatitis), *Bartonella henselae* (bacillary angiomatosis), *Tropheryma whippelii* (Whipple disease), sin nombre virus (Hantavirus pulmonary syndrome), and Kaposi sarcoma–associated herpes virus (Kaposi sarcoma). In this context, Fredricks and Relman have called for the modernization of Koch's postulates of disease causation.

The discovery of emerging pathogens brings with it the implication of precluding organisms that may be only vaguely associated with disease and that are very difficult to detect and cultivate. Relman maintains that the human intestinal tract harbors Archaea but there are no known pathogens from this group: "in vitro cultivation methods for many Archaea are unavailable, so how would we know if Archael pathogens existed?" (1). Archaea represent an entire domain as defined by Woese (the other two being Bacteria and Eucarya). The limitations of microbial sampling have not been

---

* The discovery of SV40 and subsequent detection in polio vaccines administered to an estimated 100 million people (1953–1960) is an interesting detective story. The vaccines were made from viruses grown in Rhesus monkey kidney cells that harbored SV40 and researchers now wonder if SV40 infection in man originated from those early polio inoculations to now cause specific cancers (brain, bone, lymphomas, and mesotheliomas) that mirror those occurring in hamsters infected with SV40 (see "Simian Virus 40 Infection of Humans", Jour. Virology, May 2003, vol. 77, no. 9, p. 5039–5045).

lost on some in the parenteral industry: "our industry has conventionally defined sterility in aseptic processing only in terms of bacteria, yeasts, and molds because of technical limitations in detection, growth and measurement rather than scientific realities" (55).

The change in microbial classification and new microbial-host disease associations come at a unique time in microbiological history concurrent with a new type of infectious agent that is being elucidated: the prion. Dr. Prusiner proposed the existence of prions, or proteinecious infectious agents, in 1997, for which he received the Nobel prize in Medicine (23). These agents of disease are not alive; indeed they do not contain DNA or RNA, but propagate within living hosts (with resulting neurological damage) by a domino effect of altering the three-dimensional protein conformation of the normal prion protein (PrP^c) in the neurological systems of several mammals including humans, sheep, cattle*, mink, deer, elk, and cats[†] (56). The body can break down the normal form of PrP^c but not the abnormal form (PrP^{Sc})[‡] (58). The prion concept as elucidated by Prusiner, demonstrates how prion-generated disease may be manifested by spontaneous mutation, heredity, as well as infection (by ingestion, injection, transfusion and transplantation[§]) (59). The existence of prions has affected the parenteral manufacturing industry by necessitating the exclusion of certain animal-sourced raw materials and requiring additional testing for those that cannot be replaced. Furthermore, traditional methods of detection do not work and decontamination has little or no effect on prions, which have been described as virtually indestructible[¶] by heat, chemical treatment, or desiccation. Iatrogenic (medically induced) passage of prions has been documented in several instances and point to the tenacity of the prion molecule:

> An electrode that had been inserted into the cortex of an unrecognized
> CJD patient was subjected to a decontamination procedure involving

---

* And other ruminants in UK zoos between 1986–1992: bison, nyala, gemsbok, oryx, greater kudu, and eland (56).

[†] Including puma, cheetah, ocelot, and a tiger in the same zoos and period noted above (56).

[‡] Interesting as an analogy is the comparison of amyloid plaques to spider silk which is the same protein (spidroin) in liquid (glandular) and soluble form (spider's web) (57).

[§] The normal prion protein is coded by mammalian genomes and occurs predominately in white blood cells and brain cells.

[¶] "…sheep were imported from Belgium and the Netherlands and may have consumed tainted feed. The sheep were euthanized and their carcasses dissolved in boiling lye. Barn surfaces and implements were disinfected with sodium hypochlorite or incinerated, and the pastures have been put off limits for five years to allow residual infectivity to diminish." –(59).

treatment with benzene, 70% ethanol, and formaldehyde vapor. It was then used in succession on two young patients and cleaned as above after each use. Within 2 yr, both patients came down with CJD. After these events, the tip of the electrode was implanted into the brain of a chimpanzee where it too caused lethal spongiform encephalopathy, proving that the electrode had retained infectious prions over several years and despite repeated attempts at sterilization (60).

Lastly, relevant to paradigm changes in the view of contaminants, consider current speculation that the prion concept of infection may apply to other disease processes:

Ongoing research may also help determine whether prions consisting of other proteins play a part in more common neurodegenerative conditions, including Alzheimer's disease, Parkinson's disease and amyotrophic lateral sclerosis. There are some marked similarities in all these disorders. As is true of the known prion diseases, the more widespread ills mostly occur sporadically but sometimes "run" in families. All are also usually diseases of middle to later life and are marked by similar pathology: neurons degenerate, protein deposits can accumulate as plaques, and glial cells (which support and nourish nerve cells) grow larger in reaction to damage to neurons. Strikingly, in none of these disorders do white blood cells—those ever present warriors of the immune system—infiltrate the brain. If a virus were involved in these illnesses, white cells would be expected to appear (23).

This frightening proposal begs the question: will discoveries follow of additional infectious proteins and, if so, how might this be relevant to the use of transgenics given that the crossover of pathogenic contaminants has at times gone unrecognized*? The degree of similarity or dissimilarity in the mammalian gene that encodes the PrP has been proposed to explain the mechanism of barrier between animals that can and cannot contract the disease in terms of protein conformation similarity (and susceptibility to being converted) relative to the gene that encodes it (the PrP gene of cows, sheep, and humans are very similar). It is not known at what levels of concentration prions are infective or the cause(s) of variability in the time of onset of symptoms. Governments around the world have enacted precautions in food, medical (including blood collection and handling), and drug regulation to contain the spread of known prion diseases (61–63).

---

* Somewhat analogous to the issues facing xenotransplantation (see Infectious Disease Issues in Xenotransplantation, Clin. Micro. Rev. Jan. 2001, p. 1–14).

1.6. EMERGING APPROACHES TO FINDING AND IDENTIFYING CONTAMINANTS

From a less theoretical vantage, the processes used to manufacture paren-teral drugs can be separated into two broad categories: those that manu-facture, fill, package, and end in terminal sterilization and those that manufacture and fill aseptically without terminal sterilization. The former category is possible only for those drugs capable of withstanding the pro-tracted heating cycle associated with steam sterilization (or alternative treat-ment), whereas the aseptically filled category encompasses a greater variety and more problematic route of production from a contamination control perspective. The "problems" associated with aseptic manufacturing have multifaceted aspects but the FDA has noticed a common theme as summa-rized in this PDA Letter excerpt:

> The Agency (FDA) has also looked at 10-year non-sterility trends. Non-sterility in the recall context means the distributed drug was found to be non-sterile by FDA or another government laboratory, or by the manufacturer's own laboratory. When the FDA looked closely at these data trends, they distilled one overwhelming fact from it: all drugs recalled due to non-sterility over the last 10 years were produced by aseptic processing (64).

The numbers associated with such recalls include 135 drugs (in some cases multiple lots) in three recent years (1999, 2000, and 2001) (64). Many of the tasks associated with contamination are brought about by the interaction of humans with the drug material during aseptic processing. The causes in order of occurrence listed by survey respondents (64) include:

Personnel-borne contaminants
Human error
Non-routine operations
Assembly of sterile equipment prior to use
Mechanical failure
Inadequate or improper sanitization
Transfer of materials within APA
Routine operations
Airborne contaminants
Surface contaminants
Failure of sterilizing filter
Failure of HEPA filter
Inadequate or improper sterilization

Halls (17) (see Chapter 22) condenses the sources of contamination into five overarching routes: (a) environmental air, (b) manufacturing equipment facilities and services, (c) dosage form with product containers and closures, (d) personnel operating the manufacturing equipment, and (e) water and drainage.

Issues in parenteral manufacturing contamination control often revolve around the implausibility (from a statistical vantage) of finding microbiological contamination by way of quality testing without exhaustive sampling schemes. Since the absence of contamination (sterility) is only a statistical likelihood of occurrences and can never be proven absolutely without consuming (testing) an entire manufactured drug lot, the industry-regulatory tension always exists to prove the unprovable (i.e., that a given lot is in fact sterile by a number of criteria including, but not limited to, end-product testing). Furthermore, because the likelihood of the occurrence of artifacts (false-positives) arising during analytical testing is not negligible, it creates an additional layer of tension between manufacturers and their own quality processes.

The 1978 case of Northern District of New York vs. Morton-Norwich Products, Inc. involved the sterility of gauze pads containing an antibacterial dressing (18). Sterility testing by the FDA determined that units of the gauze were adulterated. The defendants argued that sterility is a probabilistic and not an absolute concept and that by passing the in-house sterility test the article was in fact sterile by definition.

> The importance of the court's finding to persons involved in the manufacturing and testing of injectable drug products is immense. The court, knowing that an absolute cannot be measured, insisted that the absolute situation must prevail. Every single unit in every single manufacturing batch is required by the Act to be sterile if the product purports to be sterile or is represented in its labeling to be sterile (18).

The advance of genetic-based identification [DNA fingerprinting (65)] may come to aid the resolution of sterility test failure ambiguity in that genomic characterization makes it possible, in theory, to determine the origin of contaminants (i.e., a true product contaminant or an artifact of testing) based on an organism's genetic relatedness to environmental isolates, either of production or lab origin). Genetic methods are being developed for analogous epidemiological purposes in other disciplines, including diagnosing, identifying, and tracking the origin and progress of infectious agents (66) and foodborne disease without the concomitant need for microbial enrichment (67), in some cases supplanting traditional, culture-dependent serotyping (68), tracking antibiotic resistance genes (69), and tracking the origin of organisms used for bioterrorism (i.e., anthrax) (70,71). This later field has been referred to as "microbial forensics" (72).

The PDA Journal of Pharmaceutical Science and Technology technical report No. 33 (73) describes three broad categories of microbiological testing technologies including (a) viability-based, (b) artifact-based, and (c) nucleic acid-based technologies. Clearly, the latter category is primed to have a profound effect on pharmaceutical analytical testing for contaminants given the genesis of microarrays* (oligonucleotide arrays) (74–76), instrumental biosensors (77,78), and DNA probes (PCR) (79,80) that are capable of detecting femtogram levels (10^{-15}) of DNA or mRNA (or ribosomal RNA); some have noted a paradigm shift from the detection of gene products [such as proteins and contaminating antigens (endotoxin)] to genome fragments especially given the sequencing of the whole genomes of numerous organisms (79). DiPaolo et al. (80) describe the importance of monitoring for potential host cell DNA contamination in the production of drugs using recombinant methods:

> The use of recombinant DNA technology and continuous cell lines in the manufacture of biopharmaceuticals has raised the possibility of introducing potentially oncogenic or transforming DNA into the product as an impurity. Although the actual risk of incorporating tumorigenic sequences into the recipient's DNA is negligible, the FDA continues to require lot-to-lot testing for residual host cell DNA, recommending that the final product should contain no more than 100 pg cellular DNA per dose, as determined by a method with a sensitivity of 10 pg (35). These recommendations have resulted in a significant scientific challenge to develop sensitive and robust assays that can meet the criteria with samples typically containing milligram amounts of biotherapeutic protein.

As seen from both historical and emerging perspectives the increasingly complex environment associated with parenteral manufacturing and the control of contaminants is being met by increasingly sophisticated means utilizing, in many cases, tools derived from microbial contaminants themselves. Parenteral manufacturing achievements will continue to allow doctors to fulfill their oath of providing medicines to patients that first "do no harm" and secondly, in many cases, are the only means to restore health and preserve life.

---

* "When gene sequence information is available, oligonucleotides can be synthesized to hybridize specifically to each gene. Oligonucleotides can be synthesized in situ, directly on the surface of a chip, or can be pre-synthesized and then deposited on to the chip" (74).

REFERENCES

1. Relman, D.A. Detection and identification of previously unrecognized microbial pathogens. Emerg. Infect. Dis. 1998, *4* (3), 382–389 (http://www.cdc.gov/ncidod/eid/vol4no3/relman.htm).
2. De Kruif, P. *Microbe Hunters*, 4th Ed.; Harcourt, Brace and Company: New York, 1940; 350.
3. Demain, A.L.; Solomon, N.A. Industrial Microbiology. Sci. Am. 1981, *245* (3), 67–75.
4. Westphal, O.; Westphal, U.; Sommer, T. *History of Pyrogen Research in Microbiology*; Schlessinger, D., Ed.; American Society for Microbiology: Washington, D.C., 1977; 221–238.
5. Roitt, I.; Brostoff, J.; Male, D. *Immunology*, 4th Ed.; Mosby: London, 1996.
6. Fredricks, D.N.; Relman, D.A. Sequence-based identification of microbial pathogens: reconsideration of Koch's postulates. Clin. Microbiol. Rev. 1996, *9* (1), 18–33.
7. Talaro, K.P.; Talaro, A. *Foundations in Microbiology*, 3rd Ed.; McGraw-Hill: New York, 1999; 873.
8. Buerki, R.A.; Higby, G.J. History of dosage forms and basic preparations. In *Encyclopedia of Pharmaceutical Technology*, 2nd Ed.; Swarbrick, J., Boylan, J.C., Eds.; Marcel Dekker: New York, 1997; Vol. 2, 1447–1478.
9. Gaspard, B. Memoire physiologique sur les maladies purulentes et putrides, sur la vaccine etc. J. Physiol. (Paris) 1822, *2* (1).
10. Holt, E.; Penfold, W.J. Microorganisms and their Relation to Fever, 1912.
11. *U.S. Pharmacopoeia* IX, 1916; 58.
12. Seibert, F.B. Fever-producing substance found in some distilled waters. Am. J. Physiol. 1923, *87*, 90–104.
13. Seibert, F.B. The cause of many febrile reactions following intravenous injections. Am. J. Physiol. 1925, *71*, 621–652.
14. Seibert, F.B. Proceedings of the Research Conference on Activities of Bacterial Pyrogens, Philadelphia; 1951; 58.
15. Rademaker, L. The cause and elimination of reactions after intravenous infusions. Surg. Gynecol. Obstet. 1933, *56* (956).
16. *USP Pharmacopoeia* XII, 1942.
17. Halls, N.A. Sterility and sterility assurance. In *Achieving Sterility in Medical and Pharmaceutical Products*; Swarbrick, J., Eds.; Marcel Dekker: New York, 1994; Vol. 64, 17–54.
18. Munson, T.E., et al. Federal regulation of parenterals. In *Pharmaceutical Dosage Forms: Parent Medications*, 2nd Ed.; Avis, K.E., Lieberman, H.A., Lachman, L., Eds.; Marcel Dekker: New York, 1993; Vol. 3, 201.
19. Clark, R.C. Threescore Years and Ten: A Narrative of the First Seventy Years of Eli Lilly and Company 1876–1946, privately printed in 1946.
20. Immel, B.K. A brief history of the GMPs for pharmaceuticals. *Pharmaceutical Technology*; July 2001, 44–52.
21. Swann, J.P. The 1941 sulfathiazole disaster and the birth of Good Manufacturing Practices, PDA. J. Pharm. Sci. Technol. 1999, *53* (3), 148–153.

22. Swann, J.P. History of the FDA, FDA History Office. In *The Historical Guide to American Government*; Kurian, G., Ed.; Oxford University Press: New York, 1998; http://www.fda.gov/oc/history/historyoffda/fulltext.html.

23. Prusiner, Stanley B. Prions. Proc. Natl. Acad. Sci. USA 1998, *95*, 13363–13383. Nobel lecture.

24. Rosenberg, E. Biotechnology and applied microbiology. In *The Prokaryotes: An Evolving Electronic Resource for the Microbiological Community*, 3rd Ed.; Dworkin, M., Ed.; Springer-Verlag: New York, 2001; http://link.springer-ny.com/link/service/books/10125/.

25. Fisher, L.F. Advances in clinical trials in the twentieth century. Annu. Rev. Public Health 1999, *20*, 109–124.

26. Lederberg, J. Infectious disease as an evolutionary paradigm. Emerg. Infect. Dis. 1997, *3* (4), 1997.

27. Bud, R. History of biotechnology. In *Encyclopedia of Life Sciences*; Nature Publishing Group: London, 2001; (http://www.els.net).

28. Demain, A.L.; Lancini, G. Bacterial pharmaceutical products. *The Prokaryotes: An Evolving Electronic Resource for the Microbiological Community*, 3rd Ed.; Springer-Verlag: New York, 2001; http://link.springer-ny.com/link/service/books/10125/.

29. *The BioPharm Guide to Biopharmaceutical Development, Supplement to BioPharm*, 2nd Ed.; March 2002.

30. Zhour, J.; Miller, J.H. Meeting Review: Microbial genetics-challenges and opportunities: The 9th International Conference on Microbial Genomes. J. Bacteriol. 2002, *184* (16), 4327–4333.

31. Venter, J.C., et. al. The sequence of the human genome. Science 2001, *291*, 1304–1351.

32. Hieter, P.; Boguski, M. Functional genomics: it's all how you read it. Science 1997, *278*, 601–602.

33. Graves, P.R.; Haystead, T. Molecular biologist's guide to proteomics. Microbiol. Mol. Biol. Rev. 2002 March; 39–63.

34. *U.S. Pharmacopoeia* 2002, *26*.

35. Points to Consider in the Manufacture and Testing of Monoclonal Antibody Products for Human Use, U.S. Department of Health and Human Services, Food and Drug Administration, Center for Biologics Evaluation and Research, February 28, 1997.

36. The BioPharm International Guide to Fermentation and Cell Culture, Supplement to BioPharm, 2nd Ed.

37. Tomita, M., et al. Transgenic silkworms produce recombinant human type III procollagen in cocoons. Nature Jan 2003, *21* (1), 52–56.

38. Kirikae, T., et al. Biological characterization of endotoxins released from antibiotic-treated *Pseudomonas aeruginosa* and *Escherichia coli*, Antimicrob. Agents Chemother. 1998, *42*, 1015–1021.

39. Apte, S.P.; Ugwu, S.O. A review and classification of emerging excipients in parenteral medications. Pharm. Tech. March 2003, 46–60.

40. Kannan, V., et al. Optimization techniques for the design and development of novel drug delivery systems, Part II. Pharm. Tech. March 2003, 102–118.

41. Behizad, M.; Curling, J.M. Comparing the safety of synthetic and biological ligands used for purification of therapeutic proteins. BioPharm, 2000; 42–46.

42. del Rosario, M., et. al. Some methods for contamination testing of a master cell bank. BioPharm, 2000; 48–52.

43. Finlay, B.B.; Falkow, S. Common themes in microbial pathogenicity revisited. Microbiol. Mol. Biol. Rev. 1997; 136–169.

44. Woese, C.R. Prokaryote systematics: The evolution of a science. In *The Prokaryotes: An Evolving Electronic Resource for the Microbiological Community*, 3rd Ed.; Dworkin, M., et al. Eds.; Springer-Verlag: New York, 2001; http://link.springer-ny.com/link/service/books/10125/.

45. Woese, C.R., et. al. Towards a natural system of organisms: Proposal for the domains Archaea, Bacteria, and Eucarya. Proc. Natl. Acad. Sci. USA 1990, *87*, 4576–4579.

46. Woese, C.R. Interpreting the universal phylogenetic tree. Proc. Natl. Acad. Sci. 2000, *97* (15), 8392–8396.

47. Hugenholtz, P., et al. Impact of culture-independent studies on the emerging phylogenetic view of diversity. J. Bacteriol. 1998, *180*, 4765–4774.

48. Amann, R.I., et al. Phylogenetic Identification and in situ detection of individual microbial cels without cultivation. Microbiol. Mol. Biol. Rev. 1995, *59*, 143–169.

49. Garcea, R.L.; Imperiale, M.J. Simion virus 40 infection of humans. J. Virol. 2003, *77* (9), 5039–5045.

50. Urnovitz, HB.; Murphy, W.H. Human endogenous retroviruses: nature, occurrence, and clinical implications in human disease. Clin. Microbiol. Rev. 1996, *9*, 72–99.

51. Lower, R., et al. The viruses in all of us: characteristics and biological significance of human endogenous retroviral sequences. Proc. Natl. Acad. Sci. USA 1996, *93*, 5177–5184.

52. Karlsson, H.; Bachmann, S.; Schroder, J.; McArthur, J.; Torrey, E.F.; Yolken, R.H. Retroviral RNA identified in the cerebrospinal fluids and brains of individuals with schizophrenia. Proc. Natl. Acad. Sci. U. S. A. 2001, *98*, 4634–4639.

53. Yang, L., et al. Heritable susceptibility to severe Borrelia burgdorferi-induced arthritis is dominant and is associated with persistence of large numbers of spirochetes in tissues. Infect. Immun. 1994, *64*, 492–500.

54. Schmidt, B.L. PCR in laboratory diagnosis of human *Borrelia burgdorferi* infections. Clin. Microbiol. Rev. 1997, *10*, 185–201.

55. Akers, J.; Agalloco, J. Sterility and sterility assurance. PDA Jour. Pharm. Sci. Tech. 1997, *51* (2), 72–77.

56. Pattison, J. The emergence of bovine spongiform encephalopathy and related diseases. Emerg. Infect. Dis. 1998, *4* (3), 390–394.

57. Kenney, J.M., et al. Amyloidogenic nature of Spider Silk. Eur. J. Biochem. 2002, *269* (16), 4159–4163.

58. Aranha, H.; Larson, R. Prions: Part I, General Considerations in mayhem and management. BioPharm, 2002; 11–17 (suppl).

59. Yam, P. Keeping Mad Cows at Bay. Sci. Am. July 2002, (59).

60. Weissmann, C., et al. Transmission of prions, PnAS Early Edition, www.pnas.org/cgi/doi/10.1073/pnas.172403799, pg. 1–6.
61. ASM Comments on Docket No. 02D-0266, FDA Draft Guidance Document for Industry: Preventive Measures to Reduce the Possible Risk of Transmission of Creutzfeldt-Jakob Disease (CJD) and Variant Creutzfeldt-Jakob Disease (vCJD) by Human Cells, Tissues, and Cellular and Tissue-Based Products (HCT/Ps) (http://dev.asmusa.org/pasrc/CJDvCJD.htm), last modified January 21, 2003, (ASM).
62. Coulthart, M.B.; Cashman, N.R. Variant Creutzfeld-Jakob disease: a summary of current scientific knowledge in relation to public health. CMAJ 2001, *165* (1), 51–58.
63. Aranha, H.; Larson, R. Due diligence in the manufacture of biologicals. BioPharm. 2002, (suppl) 36–40.
64. PDA Letter, Regulatory News: Aseptic processing: How good science and good manufacturing Practices can prevent contamination. 2002; 10–11.
65. van Belkum, A. DNA fingerprinting of medically important microorganisms by use of PCR. Clin. Microbiol. Rev. 1994, *7*, 174–184.
66. Cunningham, M.W. Pathogenesis of group A streptococcal infections. Clin. Microbiol. Rev. 2000, *13*, 470–511.
67. Cocolin, L., et. al. Direct identification in food samples of *Listeria* spp. and *Listeria monocytogenes* by molecular methods. Appl. Environ. Microbiol. 2002, *68*, 6273–6282.
68. Muir, P., et. al. Molecular typing of enteroviruses: current status and future requirements. Clin. Microbiol. Rev. 1998, *11*, 202–227.
69. Chopra, I.; Roberts, M. Tetracycline antibiotics: mode of action, applications, molecular biology, and epidemiology of bacterial resistance. Microbiol. Mol. Biol. Rev. 2001, *65*, 232–260.
70. Sacchi, C.T., et al. Sequencing of 16S rRNA gene: a rapid tool for identification of *Bacillus anthracis*. Emerg. Infect. Dis. [serial online] 2002 Oct [*date cited*];8. Available from: URL: http://www.cdc.gov/ncidod/EID/vol8no10/02-0391.htm
71. Hoffmaster, A.R., et al. Molecular subtyping of *Bacillus anthracis* and the 2001 bioterrorism-associated anthrax outbreak, United States. Emerg Infect. Dis. [serial online] 2002 Oct [*date cited*];8. Available from: URL: http://www.cdc.gov/ncidod/EID/vol8no10/02-0394.htm
72. Cummings, C.A.; Relman, D.A. Microbial forensics—'Cross-Examining Pathogens.' Science 2002, *296*, 1976–1978.
73. PDA. Evaluation, validation and implementation of new microbiological testing methods. PDA Journal of Pharmaceutical Science and Technology 2000, *54* (May/June), 1–39.
74. Watson, A., et. al. Technology for microarray analysis of gene expression. Curr. Opin. Biotech. 1998, *9*, 609–614.
75. Braxton, S.; Bedilion, T. The integration of microarray information in the drug development process. Curr. Opin. Biotech. 1998, *9*, 643–649.
76. Gingeras, T.R. Studying microbial genomes with high-density oligonucleotide Ar-rays. ASM News 2000, *66* (8), 463–469.

77. Nice, E.C.; Catimel, B. Instrumental biosensors: new perspective for the analysis of biomolecular interactions. BioEssays 1999, *21*, 339–352.
78. Briggs, J. Sensor-based system for rapid and Sensitive measurement of contaminating DNA and other analytes in biopharmaceutical development and manufacturing. J. Parenteral Science and Technology 1991, *45* (1), 7–12.
79. Swarbrick, J. DNA probes for the identification of microbes. In *Encyclopedia of Pharmaceutical Technology*; Boyan, J.C., Ed.; Marcel Dekker, 2002; Vol. 19.
80. DiPaolo, B., et al. Monitoring impurities in biopharmaceuticals produced by recombinant technology. Pharm. Sci. & Tech. Today 1999, *2* (2), 70–82.

2

Microbial Origins, Microbiological Function, and Contamination Hazard Analysis in Sterile Product Manufacturing

Simon Rusmin

New Spring Consulting, Parsippany, New Jersey, U.S.A.

2.1. AN OVERVIEW OF THE EVOLUTION AND DIVERSIFICATION OF MICROORGANISMS

Before we start with the discussion of analyzing microbial contamination hazards in sterile drug product manufacturing, it is worth briefly reviewing what we know about the origin and evolution of microbes to better understand and appreciate the magnitude of efforts required to keep drug products sterile.

The story of microbes and, as a matter of fact of all things, may have started with the Big Bang, an estimated 14 billion years ago. At time zero, it is theorized, all matter in the universe was concentrated as a point source of infinite mass, which exploded and very rapidly expanded to an ever-enlarging volume, spewing intense radiation. Within seconds after the explosion, the temperature had decreased enough to allow fundamental particles to coalesce, forming the nuclei of light elements like hydrogen, helium, and lithium. These nuclei captured electrons and formed the first atoms. At 300,000 years, the composition of the universe was mostly clouds of hydrogen and helium.

Gravitational attraction brought these light elements closer and they coalesced violently to form stars and galaxies of stars. Within the stars, heavier atoms were formed via fusion reactions. Some of the stars disintegrated as novas and supernovas, generating clouds of interstellar matter, from which our solar system was formed 5 billion years ago. The birth of Earth set the stage for the emergence of LIFE—organized complex organic macromolecules performing metabolization and reproduction. The emergence of life was estimated to have happened about 3.8 billion years ago. To better comprehend the formation and progression of life, a review of the past 5 billion years of Earth's history is in order.

2.1.1. The First Billion Years: Earth's Cooling, Formation of Solid Crust, The First Atmosphere and Liquid Water

As the Earth further cooled down, a solid crust formed. This crust was made up of oxides of light elements, mainly of silicon and some aluminum and magnesium. At the same time the first atmosphere formed; it was mildly oxidizing, consisting mostly of water, carbon monoxide, and carbon dioxide, with some nitrogen, sulfides, methane, and ammonia, but little oxygen. As Earth continued to cool, the first liquid water was formed in the atmosphere and came down to earth as rainfall, which was instantaneously vaporized, creating a continuous cycle of rain. The acidic water dissolved the minerals on land and eventually accumulated in ponds, lakes and oceans. The atmospheric gases, exposed to lightning and intense UV light, created organic molecules such as fatty acids, sugars, amino acids, purines, primitive nucleotides, and polymers, accumulating into pools of organic soup. Through still unresolved processes emerged the first living things, metabolizng and reproducing macromolecules packaged in defined structures.

2.1.2. The Second Billion Years: First Living Organisms, the Bacteria

It is postulated that the first living organism was an organization of molecules of ribonucleic acid (RNA), performing both enzymatic and genetic functions, enclosed within membranes made of fatty acid in a high-temperature environment. This assembly was facilitated by the surfaces of pyrite or clay minerals. The first metabolic pathways were thought to be thermophilic anaerobic fermentation (obtaining both carbon and energy from organic compounds) or thermophilic anaerobic lithotrophs (obtaining energy from inorganic compounds and carbon from organic compounds). Changes in the genetic elements (mutation) and selection by nature of individuals that were more adaptive to the environment due to these changes (natural selection) resulted in increasing the adaptability and complexity of the living species—a process called evolution. The process began the instant the first living molecules arose and has been in continuous operation since. The first identifiable life forms on earth were bacteria, which were very likely similar to the present Archeobacteria. Today's Archeobacteria inhabit extreme living conditions such as extremely hot volcanic springs, lava vents at mid-ocean ridges or in hot springs, very acidic or very salty waters, or environments devoid of oxygen, conditions similar to those of Earth when living molecules first arose. The now predominant bacteria, the Eubacteria, subsequently arose, filling niches that were not inhabited by the Archeobacteria (some Eubacteria are thermophilic, retaining their ancient origin).

Halfway through the second billion years, the anoxygenic photosynthetic pathway, or Photosystem I (using light as the source of energy without reducing molecules of water to produce oxygen) evolved in the Eubacteria. A representative of such bacteria is the present day Cyanobacterium. As a matter of fact, the cyanobacteria became the oldest fossil discovered—in the form of *stromatolites*, which consist of laminated layers of sediments trapping chains of cyanobacteria cells. The age of the fossils was determined to be 3.2 billion years.

2.1.3. The Third Billion Years: Age of Prokaryotes and Oxygenic Photosynthesis

Bacteria were the sole inhabitants of Earth for close to 2 billion years (the second and third billion years of our story). As we can see, the structure of bacterial cells has hardly changed up to the present time. However, within the limits of their cell structure, bacteria invented diversified biochemical pathways to exploit the resources of the Earth. This fact will not be missed by those who use metabolic characteristics as the main method of differentiation and identification of bacteria, aside from the genomic analysis. By the end of this long period as the sole custodians of Earth, bacteria established and accumulated a large repertoire of metabolic pathways, which became the foundation of which subsets were utilized by all subsequent organisms. One crucial biochemical invention of the Eubacteria was the oxygenic photosynthetic pathway, or Photosystem II (the Archeobacteria never evolved photosystems except for the halobacterium, which possesses a simple light energy transfer mechanism). Unlike Photosystem I, Photosystem II reduces water molecules, resulting in free oxygen as a by-product. The invention of Photosystem II, which occurred midway through the third billion years, crucially impacted the earth's atmosphere by gradually increasing oxygen concentration from 0.1% in the beginning to the present content of 21% over a period of 1.5 billion years. The increase of oxygen in the atmosphere brought the possibility of another bacterial invention, aerobic metabolism, by which organic molecules are oxidized to carbon dioxide and water yielding much higher energy compared to fermentative metabolism.

2.1.4. The Fourth Billion Years: The Age of Eukaryotes and Sexual Reproduction

The bacteria (Archeobacteria and Eubacteria) belong to the Prokaryote group, and the rest of the organisms on Earth today fall under the Eukaryote group (e.g., amoebas, insects, fish, elephants, algae, fungi, beans, and orchids, to name a few). These two big groups of organisms differ fundamentally in size and complexity of their cell structure. Prokaryotic cells are small (0.3–1.5 μm) with simple structures, consisting of ringlike DNA chains and ribosomes within their cytoplasm which is enclosed by a cell membrane and cell wall. In contrast, eukaryotic cells are much larger (20–100 μm) with complex organelles within the cytoplasm such as the nucleus, mitochondria, cytoskeletons, and plastids (i.e., the chloroplast of plants). Prokaryotic cells have a mature structural design (like the bicycle), to which not much more complexity can be added. Therefore, the emergence of eukaryotic cell design was a breakthrough that made it possible for the evolution of diversified unicellular organisms, and later, of the exponential diversification and complexity of multicellular organisms.

Where did the eukaryotes (or *eukarya*, a parallel term to *bacteria*) come from? Today, the general consensus is that the emergence of eukarya was through *endosymbiosis*, a theory advanced by Lynn Margulis of the University of Massachusetts in the 1960s. It was postulated that certain archeobacteria evolved nucleation, the process in which genetic materials are enclosed within a membrane structure. Such organisms adopted a mode of living of ingesting other bacteria. Some of the foods resisted digestion and symbiotically lived within the host cell, contributing to the organelles of the eukaryotic cells—for example, the mitochondria were symbiotic purple bacteria, the plastids symbiotic cyanobacteria, the flagella a spirochete. Other eukaryotic features, such as the cytoskeleton, may also be of symbiotic bacterial origin.

Fossils of eukarya have been found that dated at the beginning of the fourth billion years (1.8 million years ago). Certainly, the processes of nucleation and endosymbiotic acquisitions were more ancient than the fossil evidence shows. For the next one and a half billion years, single-cell eukarya, while coexisting with the bacteria, diversified into different modes of living, including photosynthetic (the algae), phagocyte (the protozoa), or saprophytic (the fungi), and many built cell coverings made of calcium or silicate. Genome exchange between cells enhanced distribution of genes resulting from mutation. Bacteria exchange genes by transferring pieces of DNA from one cell to the other. Eukarya invented genome shuffling (i.e., sexual reproduction), in which two individuals with double sets of chromosomes shuffle the sets and create sex cells (eggs and sperms) with a single set of chromosomes. When the two sex cells merge into genetically distinct new individuals, a variety of progeny is created within one generation. Thus, the invention of sexual reproduction sped up evolution.

2.1.5. The Past Billion Years: Multicellular Organisms and the Grand Parade

At the beginning of the past billion years of earth history, the oxygen in the atmosphere gradually increased to the current composition of 21%. Life on earth did not exhibit any drastic change and was seen as a continuation of the previous time until something amazing happened 0.53 billion years ago, the Cambrian Explosion. Within a very short 100 million years, a pageant of multicellular animals with myriad body structures appeared all at once in the ocean. They were then weeded out just as quickly, resulting in the surviving animal phyla of the present time. Some scientists correlated this phenomenon with the increase in oxygen level. This drama was followed by the diversification of sea animals, which was soon followed by the invasion of land 400 million years ago. Plants developed into mosses, ferns, gymnosperms (the evergreens), and angiosperms (the flowering plants). Two phyla of animals dominated land life, the arthropods, represented by the insects, and the vertebrates, represented by the reptiles, birds, and mammals (see Fig. 1). After the slow and long process of life represented by the bacteria and the unicellular eukarya, the past half billion years of Earth history can be likened to a procession of a grand parade, with more and more complex body designs of plants and animals sequentially appearing on Earth (see Fig. 2). During this half-billion-year drama, the microbes (bacteria, protozoa, algae, and fungi) played crucial roles as pioneers during land invasion, setting the stage for the development of plant and animals on land. The microbes also played decisive roles in maintaining the ecosystem, recycling organic materials for the benefit of the rest of the organisms. Microbes have enmeshed themselves into other creatures through coevolution, resulting in mutually beneficial exo- or endo-symbioticism, or destructive Parasitism. The microbes have used the newcomers as a place to live or as food while alive or at their eventual death.

2.2. A VIEW OF MICROBES AND THEIR RELATIONSHIP TO STERILE DRUG PRODUCT MANUFACTURING

The most recent view in biology holds that all successful creatures surviving today are the result of continuous adaptation to their own special ways of

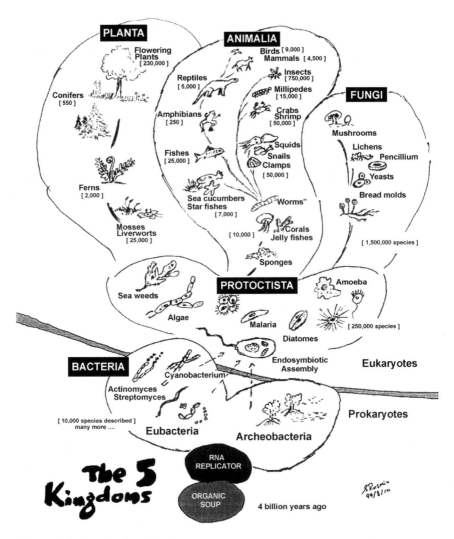

FIGURE 1 The diversity of life.

living. Humans are a special creature with a highly developed brain; the bats are a special creature with highly developed hearing for echo-location; and the bacteria are a special creature with the ability to quickly adapt metabolically to its living environment. Words such as advanced vs. primitive and higher vs. lower are used less and less in the current biological literature. When intelligence is defined as the ability of an organism to solve problems of living (rather than the common definition of having the most complex brain),

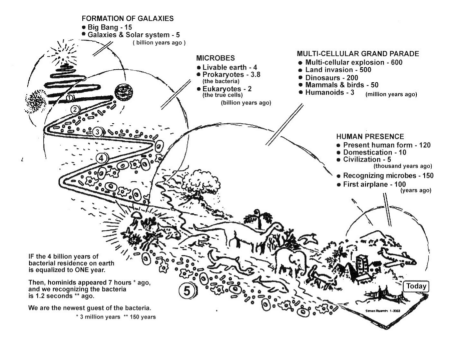

FORMATION OF GALAXIES
● Big Bang - 15
● Galaxies & Solar system - 5
(billion years ago)

MICROBES
● Livable earth - 4
● Prokaryotes - 3.8
(the bacteria)
● Eukaryotes - 2
(the true cells)
(billion years ago)

MULTI-CELLULAR GRAND PARADE
● Multi-cellular explosion - 600
● Land invasion - 500
● Dinosaurs - 200
● Mammals & birds - 50
● Humanoids - 3   (million years ago)

HUMAN PRESENCE
● Present human form - 120
● Domestication - 10
● Civilization - 5
(thousand years ago)
● Recognizing microbes - 150
● First airplane - 100
(years ago)

IF the 4 billion years of
bacterial residence on earth
is equalized to ONE year.

Then, hominids appeared 7 hours * ago,
and we recognizing the bacteria
is 1.2 seconds ** ago.

We are the newest guest of the bacteria.
* 3 million years ** 150 years

Today

FIGURE 2   Earth's last five billion years.

bacteria are not less intelligent than humans. Human intelligence has discovered and modified antibiotics to kill the bacteria. Yet our equally intelligent opponent, the bacteria, within a short 50-year time frame has been circumventing the effectiveness of antibiotics. The late Stephen J. Gould, a naturalist, said of the bacteria: "The most salient feature of life has been the stability of its bacterial mode from the beginning of the fossil record until today and, with little doubt, into all future time so long as the earth endures. This is truly the 'age of bacteria'—as it was in the beginning, is now and ever shall be."

Thus, there is a need in drug companies to remember the following points in dealing with microorganisms in sterile drug product manufacturing:

A.   Bacteria have lived on Earth for the past 3.8 billion years, and single-cell eukarya, the past 2.0 billion years. Hominids came into being only 3 million years ago. We modern humans did not realize that microbes existed until only about 100 years ago. We are latecomers on Earth, which has been hosted by microbes who have been keeping our home (Earth's ecosystem) in an orderly balance. In spite of our wishing for the disappearance of pathogenic microbes, they are an indispensable part of the ecosystem.

B. Within an ecosystem ubiquitously populated by microbes, all multicellular plants and animals carve out a space and defend their space from microbial invasion. In us, the border of this space is defended by our skin and gut lining, and the inside of this space is defended by our sophisticated immune systems. Making drug products is akin to creating a clean space defended against microbes by applying sophisticated technologies and control systems. (see Fig. 3).

C. Our skin, gut lining, and lumens are optimal habitats of warmth and richness of organic nutrient for microorganisms. Therefore, microbes are concentrated at the border of our microbial free space, and will invade that space at any given opportunity. Being a concentrator of microbes, we are the largest bioburden carrier into drug manufacturing facilities.

2.3. THE MICROBIOLOGICAL FUNCTION IN DRUG PRODUCT MANUFACTURING

What is the Microbiological Function? The term Microbiological Function, or MF, was coined during the development of PDA Technical Report #35, A

FIGURE 3 Bug-free zones.

Proposed Training Model for the Microbiological Function in the Pharmaceutical Industry, PDA, 2002, the writer being a member of the Task Group. Microbiological Function refers to all coordinated and integrated microbiological activities that are needed to assure that harmful microbes do not contaminate nonsterile drug products and that all microbes are excluded from sterile drug products. From a company's point of view, Microbiological Function is the ultimate expertise in building and maintaining microbial controls in product design, process design, product manufacturing, and storage and distribution, so as to (a) protect drug users from harmful microorganisms, (b) be in compliance with government regulations in microbial issues, and (c) be in step with industry trends in microbial control policies and technologies.

2.3.1. Why Should There Be a Distinct Microbiological Function?

In traditional drug product manufacturing (as apposed to biotechnology), the technologies applied are based on the sciences of physics (engineering) and chemistry (drug product formulation). Engineers and chemists are trained to deal with the behaviors of atoms and small molecules, which are endowed with high degree of precision, repeatability, and predictability. In contrast, biology deals with living entities composed of interacting networks of complex macromolecules within structural units called cells. From one moment to the next, cells are in a continuous dynamic of change, adapting to the changing environment to survive and propagate. Therefore, the sciences of biology are imprecise, highly variable, and time dependent, baffling most engineers and chemists. Thus, the need in drug companies for the Microbiological Function to be managed by biologists/microbiologists.

2.3.2. What Does the Microbiological Function do in a Pharmaceutical Company?

Table 1 is a list of expertise and services the Microbiological Function provides the different areas of a pharmaceutical company. The list is presented as a series of questions and divided into two parts (expertise and services) and separated as well according to area supported by the MF.

In many companies, these subfunctions are performed and supervised by individual units without communicating the results with each other. Ideally, a centralized Microbiological Function integrates and oversees all microbiological activities in the company, so that all data collected can be analyzed and evaluated to provide a total view of the microbiological heath of the company and of the corrections and improvements needed to be made (Table 2).

TABLE 1 Expertise/Services Provided Various Pharma Manufacturing Areas by the Microbiological Function

Expertise	Services
1. Product Development	

1. Product Development
   - Does MF know the formulation of each product and its bioburden/susceptibility to microbial contamination?
   - Does MF know the preservatives in or the preservation of the formulation?
   - Does MF know the designs of compounding, filling and packaging, viewed from the perspective of susceptibility to microbial contamination?
   - Does MF know the bioburden of raw materials and their susceptibility to microbial growth?
   - Does MF know the primary packaging, viewed from perspective of microbial integrity during storage and transportation?
   - Does MF know the microbial test specifications of in-process and finished products?

   - Is MF involved in microbiology evaluation during product development?
   - Does MF provide expertise to develop products that will consistently meet microbiological quality requirements?
   - Does MF advise on setting microbial specifications for materials and finished products?

2. Process Development—Microbial Contamination Hazard Analysis & Critical Control Points (HACCP)
   - Does MF know the manufacturing process flow of each product?
   - Does MF know the facility design, equipment layout, raw materials and their microbial quality, and processing from compounding all the way to final packaging and distribution of the finished product?
   - Does MF have knowledge of and skills to conduct microbial contamination hazard analysis and the setting of critical control points (HACCP)?

   - Does MF take part in the design and development of manufacturing processes?
   - Does MF advise on facility design, equipment selection, and equipment placement for minimizing microbial contamination and for ease of cleaning and sanitation?
   - Does MF advise on the designs of material/personnel flow during manufacturing to prevent microbial contamination?
   - Does MF advise on the designs of operators' activities in handling of materials/equipment to prevent microbial contamination?
   - Does MF conduct microbial contamination HACCP?
   - Does MF advise on the designs of process and environmental monitoring based on analysis?
   - Does MF provide input in writing operational SOPs to include measures based on analysis?
   - Does MF provide input in operator training to include measures based on analysis?
   - For old processes, does MF retrospectively conduct microbial contamination HACCP?

TABLE 1 Continued

Expertise	Services
3. Engineering: Manufacturing Support Systems • Does MF know the systems of HVAC, Water Purification, Compressed Gases, Steam Generators, Water Chiller and other pertinent support systems supplying or related to product manufacturing? • Does MF know the control, monitoring, and maintenance of these systems? • Does MF know the generally accepted microbial standards, levels, guidelines, regulations, and industry trends for the systems?	• Is MF involved in or does it have knowledge of the IQ/OQ/PQ related to microbiology for the support systems? • Is MF involved in the initial setting of microbial specifications, alert & actions levels, and sampling plans of the systems? • Does MF conduct or audit the implementation of microbial control and monitoring of the systems? • Does MF test the samples for the control and monitoring of the systems? • Does MF have a documentation system for prompt notification to Engineering and Manufacturing of any deviations? • Is MF involved in investigating the causes of deviations and in the subsequent corrective actions? • Does MF periodically review the long-term microbial quality and trends of the systems, and give feedback to Engineering and Manufacturing for decisions on improving the systems?
4. Manufacturing: Facility Cleaning & Sanitation • Does MF know the manufacturing facility layout? • Does MF know the cleanliness requirements at each stage of manufacturing operations? Are the requirements derived from HACCP? • Does MF know the general principles and the practices of cleaning and sanitation, including selection of antimicrobial agents and equipment, cleaning & sanitation methods, and method qualification? • Does MF know the generally accepted microbial standards, levels, guidelines, regulations, and industry trends for facility cleaning and sanitation?	• Is MF involved in the initial design of facility cleaning and sanitation procedures, specifications, and alert/action levels? • Is MF involved in the qualification of cleaning & sanitation procedures? • Does MF conduct or audit the implementation of cleanliness monitoring systems? • Does MF test the samples from control and monitoring of the systems? • Does MF have a documentation system for prompt notification to Manufacturing of any deviations? • Is MF involved in investigating the causes of deviations and in subsequent corrective actions? • Does MF periodically review the long-term microbial quality and trends of the facility, and provide timely feedback to Manufacturing for making improvement decisions?

TABLE 1 Continued

Expertise	Services

5. Manufacturing: Equipment Cleaning, Disinfection/Sterilization
   - Does MF know the equipment used in manufacturing, and the cleanliness/sterility requirements of the equipment?
   - Does MF know the methods of cleaning, disinfection/sterilization of the equipment —e.g., manual, dishwasher, wash bath, autoclave, oven, clean-in-place/steam-in-place?
   - Does MF know the principles, practices, and validation of equipment cleaning and disinfection/ sterilization?
   - Does MF know the generally accepted microbial standards, levels, guidelines, regulations, and industry trends for equipment cleaning and disinfection/sterilization?

   - Is MF involved in the initial design of the cleaning and disinfection/sterilization procedures and in the setting of specifications and alert/action levels of cleanliness/sterility of the manufacturing equipment?
   - Is MF involved in the qualification/ validation of the procedures for cleaning and disinfection/sterilization of the manufacturing equipment?
   - Is MF involved in the cleanliness/sterility monitoring of equipment, the testing of samples, the prompt notification of deviations, the investigation of deviations, and the periodic review and trending of equipment cleanliness/sterility?

6. Manufacturing: Incoming Material Evaluation & Control
   - Does MF know all the incoming materials[a] that may harbor undesired bioburden or are susceptible to microbial contamination? The prevention of microbial contamination during storage and use? The impact on product quality of using microbiologically failed materials?
   - Does MF know the microbial specifications of materials and the inspection/testing procedures?
   - Does MF know the generally accepted microbial standards, guidelines, regulations, and industry trends for incoming materials?

   - Is MF involved in the selection of raw materials and in the setting of specifications for microbiologically risky materials?
   - Is MF involved in vendor evaluation of these materials?
   - Does MF periodically review the microbial quality of raw materials for trending information provided to Manufacturing?

7. Manufacturing: In-Process Control and Environmental Monitoring
   - Does MF know the systems for controlling and monitoring manufacturing processes and environment, including sampling plans, sampling methods, sample processing, data generation and reporting? Are these controls consistent with the HACCP?
   - Does MF know the standards, alert/action levels, guidelines, regulations, and industry practices and trends in process control and environmental monitoring?

   - Is MF involved in designing process control and environmental monitoring during process development? If not, has MF conducted an HACCP on the current processes and reviewed the current control and monitoring systems to judge their adequacy based on analysis?
   - Are samples for environmental monitoring promptly processed and reported to Manufacturing?
   - Are deviations immediately reported to Manufacturing and discussed with Manufacturing for investigation and corrective actions?

TABLE 1   Continued

Expertise	Services
	• Are monitoring results periodically reviewed to reval trends and potential problems? Are the results of review and trending reported in a timely fashion to Manufacturing? • Does MF implement computerized data handling systems to cope with the large amount of data generated that usually hinder prompt feedback to Manufacturing?

8. Validation of Sterilization Processes and Microbial Integrity of Packaging

• Is MF well versed in microbial cell death kinetics (F, D, Z values) and sterile filtration dynamics (LRV), as well as the applications in validation of sterilization processes? • Is MF well versed in the principles, technologies, and statistical analysis of growth media filling for validation of aseptic filling processes? • Is MF well versed in bacterial endotoxin characteristics, detection & measurement, destruction kinetics, and methods and processes of depyrogenation? • Does MF know the principles and practices of bacterial ingress study to validate the integrity of product primary packaging? • Does MF know the standards, specifications, limits, sterility assurance levels, guidelines, regulations, and industry practices and trends in validation of sterilization, media fill simulation, and primary packaging integrity evaluation?	• Does MF provide expertise in validation of sterilization processes, growth media aseptic filling simulation, sterile filtration, primary packaging microbial integrity, depyrogenation, and validation of other processes involving microbiology? • Does MF provide services in the preparation and qualification of growth media, microbial inocula, bio-indicators, and microbial endotoxin challenge to support validation? • Does MF provide services in the collection of microbial/endotoxin samples, testing of samples, interpretation of data, and reporting of results to support validation? • Does MF provide services in the investigation of failed sterilization/ integrity validation and recommendations for corrective actions?

9. Manufacturing: Deviation Investigation, Microbial Troubleshooting, Annual Review & Trending of Microbial Cleanliness

• Is MF well versed in environmental microbiology (versus clinical microbiology), including microbial species found in the habitats of air, ground, and wet places typical of a drug manufacturing facility? • Is MF well versed in the microbial species found on human body surfaces, in the nose and mouth, and in the intestines? • Is MF well versed in the microbial species found in incoming materials, especially packaging cartons, brought into the manufacturing facilities?	• Does MF compile a historical catalog of the microbial burden of the air, manufacturing surfaces, personnel working garments, materials coming into the manufacturing area, and microbiological quality of the supporting system (water, compressed gases, etc.) to be used as a reference for investigation? • Does MF compile and trend all deviations related to microbial contamination or out-of-specifications to make judgments for product/process improvement?

TABLE 1   Continued

Expertise	Services
• Is MF well versed in the classification of microbes, and methods of bacterial and fungal identification? • Does MF know all the microbial contamination prevention measures and procedures installed and practiced in the manufacturing facilities? • Is MF well versed in the approaches and methodologies of investigating microbial failures, in making judgment as to the cause or source of contamination, and in recommending corrective actions and prevention of future occurrences? • Does MF have systems of periodic review and trending of overall plant cleanliness and microbial deviations, including timely feedback to Manufacturing?	• Is MF actively involved in microbial deviation investigation, microbial troubleshooting, and recommendation of corrective and preventative actions? • Does MF periodically review microbial contamination status and provide feedback to Manufacturing for process improvement?

10. Training of Personnel in the Kowledge of Microbiology and Aseptic Techniques

• Does MF establish the requirements of minimum knowledge and skills for each job description of personnel performing the Microbiological Function? • Does MF have a training program to fulfill the qualification/certification of the personnel and for their advancement in the profession? • Does MF have instructors and instructional methods to provide basic microbiology knowledge training for non-microbial-oriented personnel and for manufacturing operators?	• Internally, does MF conduct the personnel qualification and training program? • Is MF involved in the design of training in basic microbial knowledge for non-microbial personnel and manufacturing operators, especially those working in aseptic areas? • Does MF provide instructors for such training?

[a] Materials include raw materials for compounding, packaging materials (bottle, vials, syringes), and manufacturing aids materials (e.g., equipment lubricants).

2.4.   UNDERSTANDING PROCESS ANALYSIS

From the above discussions, the Microbiological Function of a pharmaceutical company is the custodian who assures that the ubiquitous microbes surrounding manufacturing processing are under control in such a way that they do not affect the final product quality. To accomplish this task the Microbiological Function needs to have full knowledge of microbiology as well as full knowledge of manufacturing processes and their support systems. Thanks to the quality movement in the manufacturing industries during the

TABLE 2   Core Activities of Microbiological Function

Managing Microbiology Laboratories and Microbial Services

Liason & Communication
- Does MF have the persons and the communication channels to link with Product Development, Process Development, Facility and Utility Supports, Validation, Manufacturing, Manufacturing Controls & Monitoring, QC-chemistry, QA, RA, Complaint Handling, Marketing, and Training?

Materials & Products Specifications
- Does MF have complete microbial specifications, monographs, standards, and alert-action levels for incoming materials, water and other utility systems, in-process controls, environment monitoring, and finished product testing?
- Are the above periodically reviewed for consistency with compendial monographs, guidelines, regulations, and industry standards and trends?

Sampling, Testing, Data Handling, & Result Reporting
- Does MF have systems for sampling, testing, data handling, and result reporting that provide accurate and timely information, crucial for decision making by Manufacturing and other departments?
- Are the systems being evaluated for their effectiveness and efficiency, leading to system upgrading and improvements?

Qualification & Validation of Analytical Methodology
- Are the testing and other analytical methods qualified/validated before use?
- Are the methods consistent with those filed in the NDA and NDA supplements?
- Are compendial methods consistent and in accordance with the most recent revision of the compendia?

Maintenance of Cultures, Media, Reagents
- Are the in-use reference microbial cultures not more than five generations away from the original standard culture?
- Are the cultures, media, and reagents prepared according to written procedures and qualified before/during use?
- Are the cultures, media, and reagents clearly identified, stored and used within their expiration date, and the efficacy verified before use when necessary?

Microbial Identification
- Has MF defined policies and procedures for identification of microorganisms found contaminating materials, processes or products? Is the policy based on product risk for reject/release decision, and on benefit for pinpointing sources of contamination?
- Are the results of identification promptly used for manufacturing process corrections in the short run, and for review and trending of manufacturing cleanliness in the long run?

Facility, Equipment, and Instrumentation
- Are the facilities, equipment, and instruments suitable and adequate for the purpose?
- Have the facilities, equipment, and instruments been qualified before used?

TABLE 2 Continued

Managing Microbiology Laboratories and Microbial Services

- Are the facilities, equipment and instruments correctly cleaned, maintained, and used?
- Are critical measuring instruments identified and calibrated?
- Have processes of cleaning/sanitation/sterilization been qualified/validated?

Lab Method & Management Advancement

- Is MF highly aware of rapid advances in the sciences of microbiology and the application of technologies in pharmaceutical manufacturing controls and laboratory techniques?
- Does MF have a policy of anticipating the timely use of new technologies?
- Does MF apply time-saving and data-speeding computerized technologies to promptly feed back the necessary information for Manufacturing's decision-making, correction, and improvement?

Personnel Training & Advancement

- Are the managers and supervisors of MF equipped with adequate knowledge, skills, and experience in the Pharmaceutical Microbiology? Are they given the opportunity to attend training, seminars, and professional conventions, and to take part in association committees for standard setting and preparation of advanced technology application?
- Does MF have updated books and references on pharmaceutical microbiology centralized and shared by all MF personnel?
- Does MF design and implement qualification, certification, and training programs for analytical and managerial personnel? Are analytical and managerial personnel given the opportunity to communicate and share knowledge and experience between sub-functions and among plants, as well as with fellow professionals in the industry?

Function Integration

- Will the Microbiological Function of the company be facilitated by a central intelligence, so that through communication and experience-sharing, all microbial function personnel within and between sites will see the same forest and understand their contribution in tending their own individual trees? (see Figure 4)

1980s and 1990s, concepts and tools for process analysis have been well established. A brief summary will be beneficial to the discussion on practical methods of microbial hazard analysis.

2.4.1. What Is a Process?

In manufacturing, a process is the changing of a low value input (raw materials) into a higher value output (products). A process requires certain parameters to run. Parameters can be classified as hardware (facility, utilities, equipment), software (procedures), and wetware (operators). *Process validation* consists of proving that the input used and the parameters set result in an output of the desired specification. *Process controls* are measures to ensure that the parameters remain unchanged with time; and *process monitoring* is checking the constancy of critical parameters (see Fig. 5). Early in the past century, manufacturing processes were analyzed using the science of statistics. The result was the application of Quality Control Statistics and Statistical Process Control in the manufacturing industries today. During the quality movement, process analysis was expanded into processes in the service sectors and ultimately to mapping and analyzing the whole business process.

2.4.2. What Are Process Flows?

A manufacturing company consists of flows of processes in which the output of one process becomes the input of the next process. The main process flow is from a vendor's raw materials to the manufacturing of a product, and to the marketing of the product to final customers. It is also called the core process. Other process flows are tributaries of the core process. For example, QC processes samples as input and produces data as output. This process flow feeds into the manufacturing process as part of parameter monitoring. Consider also the Human Resources department processing employees' information in order to provide welfare to the wetware parameters of all other processes, including its own (see Fig. 6).

2.4.3. What Are the Principles of Improving Quality?

In the pharmaceutical industry, quality is driven mainly by government regulations (cGMP), whereas in other industries, it is driven by stiff competition. Competition drives the industry to provide consumers with the highest value of products (goods and services) with the lowest production cost. The key to staying in competition is to reengineer all aspects of business processes and to enhance the effectiveness of the people who run the processes. The key to effectiveness is for each process owner to ask the questions: (a) Does my output satisfy my internal customers (the owner of the next process)? and (b) How can I improve my process to be more value adding? These principles underlie the familiar buzzwords heard during the decades of the quality movement: Total Quality Management, Process Reengineering, Continuous Improvement, Six Sigma, Team Empowerment, Learning Organization, Lean Manufacturing, and so forth.

FIGURE 4 Knowing the forest.

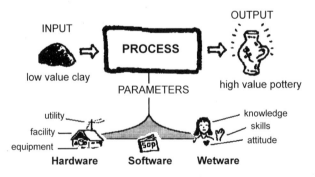

FIGURE 5 Process concept.

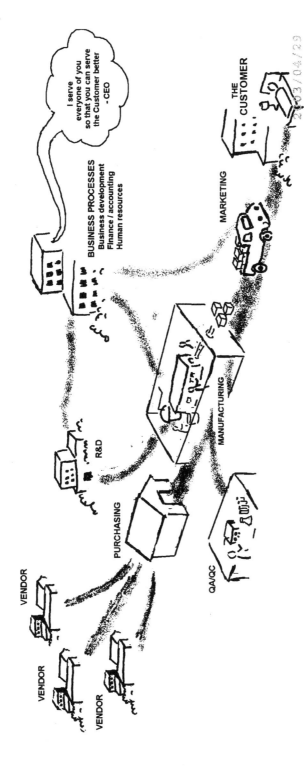

FIGURE 6 Process flow.

2.5. MICROBIAL CONTAMINATION HAZARD ANALYSIS
STEP-BY-STEP—AN EXAMPLE

The microbial hazard analysis to be described below was performed for a filling process of a newly transferred product using a brand-new filling machine that had just been installed and qualified. The product was a liquid that is aseptically filled in 5 mL plastic bottles. The processes being analyzed were the production steps in which human operation and human intervention prevailed. The following is the step-by-step approach to the analysis.

2.5.1. Initiation and Team Organization

The decision to conduct microbial hazard analysis came from the manager who was responsible for the manufacturing of the new product. The exercise of analysis was meant to be used as a model for later application to other sterile manufacturing processes. A team consisting of an experienced supervisor for sterile product manufacturing, a microbiologist from quality control, and the author as the senior microbiologist was established. A simple charter was written to define the project as follows:

Purpose—to establish a scientific base for microbial process controls and operator training

Scope—limited to the new process and to production steps where operators interfaced with the process

Responsibilities—which defined who should complete the tasks described in the work plan

Work Plan—which described who should accomplish which sequential tasks and at what timeline.

2.5.2. Study of Manufacturing Plan & Related SOPs

Guided by the experienced supervisor, team members visited the manufacturing site during nonoperating time to thoroughly study the following processes: (a) compounding and sterile filtration, (b) assembly of surge tanks to the product line, (c) assembling the sterile filling heads, (d) general sanitization of the filling machines before the start of filling, (e) operating the filler and capper, and (f) machine interruption and manipulation for weight and torque adjustments and for unjamming the machine (see Fig. 7). Guided by the production manager, the team studied the production target, filling rate, number of operators per shift, classification of operators, operator task description and skills requirement for each position. Finally, the team studied standard operating procedures (SOPs), for sterile product manufacturing in general, and the newly written SOPs in particular for the

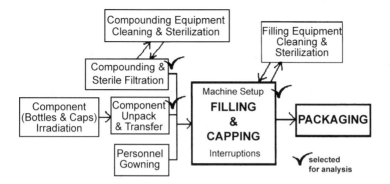

FIGURE 7   Manufacturing flow chart.

new product. With this information the team was ready for the following step.

2.5.3. Observation of Real Operations

During the Performance Qualification of the filling machine, nonaseptic water fillings were conducted. The manufacturing manager took advantage of this period to verify production output and other logistics; the supervisor, to dry-run the new SOPs; and the operators, to practice the new operational manipulations. The team also made good use of the opportunity to observe and take detailed notes of operator task manipulation. The team's focus was observing and asking the question, "What could go wrong with this manipulation that could cause microbial contamination?" Equipped with the knowledge of actual manipulation, the team was now ready to conduct analysis of the tasks performed by each operator in the next step.

2.5.4. Task Analysis for Microbial Contamination Hazards
& Control Point Setup

Task analysis was the core of this project's exercises. It was conducted using the following steps:

1. Describe the Objective of a process-step and the Work Environment in which the process takes place (e.g., the classification of the room, the attire of the personnel, and the number of persons in the room at one time).
2. Hypothesize all possible microbial contamination hazards—e.g., possible sources of contaminants and their route of entry into the finished product.

3. Evaluate the probability and severity of such contamination and give a hazard level value.

4. Design controlling and monitoring measures for high hazard processes.

As an example, the manufacturing flow chart of Fig. 7 was further magnified into Fig. 8, where each block represents a task of unpacking of components, transfer, and hopper feeding process. The resulting Task Analysis is shown in Table 3.

2.5.5. Translation of Control Measures into Operator Skills Through Training

The results of the analysis are only meaningful when Control Points and Monitoring Points are implemented in the actual operations especially those related to operators' manipulation. To accomplish this, Task Guides were written for high hazard tasks. The Task Guides would be used to modify SOPs

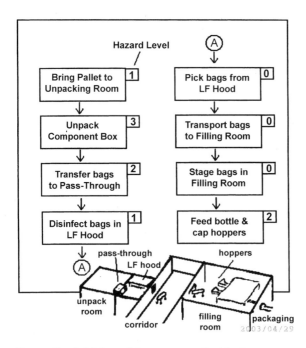

FIGURE 8   Addition of items to sterile block.

TABLE 3   Task Analysis of Unpacking Components

Purpose of Process Step. To take out the bags that contain bottles or caps irradiated in triple bags from the packaging carton boxes and to place the bags into the pass-through built through the wall in a manner so as not to contaminate the sterilized bags.

Process Environment. The room is Class 100,000 with HEPA filtered air and limited access. The attire of the personnel consists of a clean but not sterilized work coat over the area work uniform, hair (moustache) net, shoe covers, gloved hands, and face mask.

Number of Operators and Operation in Room. Two operators are working in the room at the same time. Operator 1 gets the carton boxes from the pallet, examines the condition of the packaging, places them near the pass-through, disinfects the box seams to be cut and cuts the seams with a sanitized box cutter blade. Operator 1 also cleans up the empty boxes and records the box lot number and quantity in the documentation. Operator 2 takes the bags out of the boxes and places them inside the pass-through. This operator performs cleaner operation compared to Operator 1

Microbial Contamination Hazards. Notation: 0 very low, 1 low, 2 high, 3 very high
For *probability of occurrence/severity of impact on product quality*
  1. Unsterilized box released by mistake—0/3
  2. Damaged box causing contamination of the bags inside—1/3
  3. Damaged bag during vendor's packaging—1/2
  4. Damaged bag during cutting of the box—2/2
  5. Bag contamination by body and hands during opening of the box—2/3
  6. Bag contamination by body and hands during transport to pass-through—1/3
  7. Unsanitized pass-through—1/3
  8. Ineffective room sanitizing procedures (including pass-through)—1/3
  9. Ineffective box and bag manipulation procedures (including attire)—2/3
  10. Inattentive operators—1/3
Hazard Level = 30-very hazardous

Rationale:
  1. Material (carton boxes) has been exposed to street environment without any subsequent treatment before coming into the room
  2. Carton boxes are known to carry *Bacillus* spores
  3. Operation is at the borderline between Class 100,000 and Class 10,000.
  4. Prevention of contamination depends on the skills and attentiveness of operators

Control Points
  1. Validation of irradiation sterilization
  2. Box inspection during incoming release
  3. Qualify procedure for material transport
  4. Vendor audit and certification program

TABLE 3   Continued

 5. Validation of room-cleaning procedures
 6. Validation of pass-through cleaning procedure
 7. Validation of box sanitation & cutting procedures
 8. Validation of bag retrieval and transport to pass-through procedures
 9. Operator training / retraining program
10. Inspection of box and bags by operators

Monitoring Points
 1. Room differential pressure, temperature, relative humidity monitoring program
 2. Room air monitoring program (viable, nonviable particles)
 3. Room surfaces monitoring program
 4. Operators' coats and gloves microbial monitoring program
 5. Bag cleanliness auditing program
 6. Audit of operators' procedure implementation

and to conduct on-the-job training of the operators. At the end of writing Task Guides, the manufacturing manager and the team designed a training program for the operators consisting of classroom courses on basic microbiology and sterile product manufacturing, followed by on-the-job training consistent with the Task Guides. Table 4 shows Task Guide for unpacking as an example. The above Task Guide is accompanied by the illustration shown in Fig. 9.

2.6. HAZARD ANALYSIS DISCUSSION AND CONCLUSION

Two methods of manufacturing process hazard analysis have been introduced into pharmaceutical manufacturing (1,2,4): Hazard Analysis and Critical Control Point (HACCP) and Failure Mode Effect Analysis (FMEA).

HACCP was developed as a discipline to safeguard against microbial (e.g., pathogen), chemical (e.g., pesticides), and undesired particulates (e.g., metal shreds) in the processing of food products. The system has been well established and in its systematic implementation bears the following scope:

1.   Conduct hazard analysis (HA).
2.   Determine the critical control points (CCPs).
3.   Establish critical limits.
4.   Establish monitoring procedures.
5.   Establish corrective actions.
6.   Establish verification procedures (e.g., auditing).
7.   Establish record-keeping and documentation procedures. The U.S.A. Food and Drug Administration has incorporated HACCP

TABLE 4 Task Guide of Unpacking Components

Description of task
  - The OBJECTIVE of the task is to take out the irradiated sterilized bags that contained bottles/caps from the box and transfer them to the pass-through.
  - The WORK AREA is class 100,000 with HEPA filtered air and limited access.
  - The ATTIRE of the operators consists of a work coat, hair net, shoe covers, gloves, and face mask.
  - The NUMBER OF PERSONS performing the task is 2 (two): Operator 1 cuts the seams of the boxes, records documentation, and cleans up work area. Operator 2 takes the bags out of the box and places the bags into the pass-through box.

Microbial contamination hazards:
  The outside of the boxes is the major source of contamination. The cleanliness of the room, the pass-through, and the personnel attire not being sterile.
Hazard level: 3
Critical control points:
  - Do not use a box that looks dirty or damaged.
  - Alcohol-wipe doors of pass-through during process.
  - Alcohol wipe the cut areas on the box.
  - Alcohol gloves before cutting or handling bags.
  - Do not allow bags to touch anything (e.g. body, floor or wall) on the way to pass-through.

Unpacking procedure
Operator 1: Box cutter
  1. Alcohol-wipe gloves and cutting areas of box.
  2. Alcohol-wipe blade of cutting knife.
  3. Cut box along top seams.
  4. Crack open box, but keep it closed.
Operator 2: Bag handler
  1. Alcohol-wipe gloves.
  2. Pry open box by touching ONLY inner (sterile) sides of box.
  3. Pick bag by neck one bag at a time.
  4. Carry bag to pass-through by not touching objects.
  5. Open doors of pass-through.
  6. Place bag inside pass-through; arrange bags inside pass-through when necessary.
  7. Close doors of pass through.

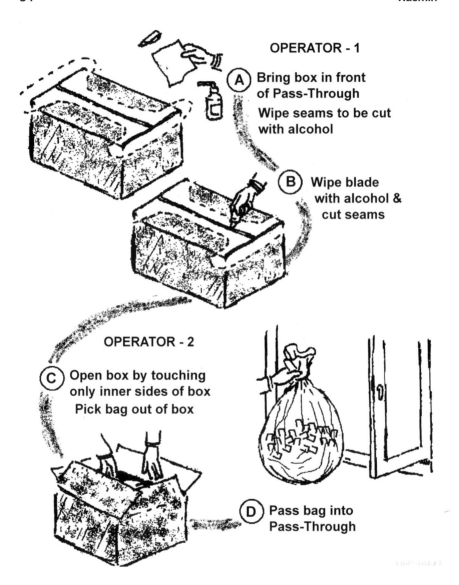

FIGURE 9 Unpacking training diagram.

into U.S. federal regulations to control food manufacturing and its inherently high microbial contamination risks.

FMEA is a process failure risk (hazard) analysis system used by general manufacturing industries. Manufacturing process steps are analyzed based on:

Failure mode (how failures can occur)
Failure effect (the consequences of failure)
Failure cause (the possible causes of failure)
The probability (P), severity (S), and likelihood of detection (D) of failures. These three criteria are given values from 1 to 5, where 5 denotes most unfavorable. A risk priority number (RPN) is calculated from the three criteria:

$$RPN = P \times S \times D \text{ (where the least risk is } 1 = 1 \times 1 \times 1 \text{ and the}$$

$$\text{highest risk is } 125 = 5 \times 5 \times 5)$$

Preventive and corrective actions are taken to prevent, reduce, or eliminate high-risk process failures.

The method of analysis described here was based on principles common to both HACCP and FMEA,. The principles were then adapted to the uniqueness of sterile product manufacturing. The rationales and circumstances of the adaptation are as follows:

1. HACCP is applied to the whole chain of manufacturing processes because in food processing, the raw materials and final product are excellent media for microbial growth, and a large part of the processing is conducted in open processes. Pharmaceutical products are processed in closed containers resulting in less exposure to the environment. For these reasons the analysis described was limited to the stages where the bulk drug and packaging components had been rendered sterile and were transported to and processed in an aseptic area.

2. FMEA model of analysis in which microbial hazard is analyzed at the same time with other hazards (e.g., mechanical, chemical, safety) was not considered appropriate. Microbes, being a living system, render microbial hazards conceptually different from the other hazards. Therefore, the microbial hazard analysis is best done independently. Moreover, FMEA analysis is best incorporated into process development and validation to assure the performance of the manufacturing hardware.

3. The hazard analysis described was only limited to process steps where operators interfaced with the process. It is recognized that humans are the major bearer of microbial load into a manufacturing area; therefore, processes with human interfaces are most likely to introduce microbial contaminants into the product. Process steps with no human participation

belong to the overall machine performance and the contamination hazard is dependent on the cleanliness of the machine and filling room, for which cleaning and disinfecting procedures and monitoring of cleanliness would have been qualified and established.

4. Measuring microbial hazard levels as described involved much more subjectivity compared to the decision tree for determining Critical Control Points in HACCP or the calculation of Risk Priority Number in FMEA. Assigning values to "probability of contamination occurrence" and "severity of impact to product quality" was used as an aid in assigning Hazard Level to a process step. Such subjective judgments could be made more objective by conducting extensive microbial burden testing with contact plates and swabs on the manipulated objects, workplace surfaces, and operators' gloves and gowns, as well as air sampling (3). Such empirical data would enhance the accuracy of hypothetical routes of contamination.

5. The described analysis was deliberately designed to be simple and straightforward for the reason that in this company, as is true with most companies, resources are limited. Yet, as simple as it was, the impact on manufacturing was significant. It was the first time this company had written information on the critical points for microbial hazards in the manufacturing processes, with which rational control measures were incorporated into SOPs and into the training program of the operators. The author believes that the first order of business of the Microbiological Function of a pharmaceutical company is to conduct microbial contamination hazard analysis on all sterile manufacturing processes, giving the rationales for their controls and monitoring.

REFERENCES

1. Janke, M. Use of HACCP Concepts for the risk analysis of pharmaceutical manufacturing processes. Eur. J. Parent. Sci. 1997, 2 (4), 113–117.
2. Kieffer, R., et al. *Applications of failure mode effect analysis in the pharmaceutical industry*; Pharmaceutical Technology: Europe, 9/1997.
3. Ljunggvist, B., et al. Hazard analysis of airborne contamination in cleanroom application of a method for limitation of risks. PDA J. Pharm. Sci. Technol. 1995, 49 (5), 239–243.
4. Sahni, A. *Using Failure Mode and Effect Analysis to Improve Manufacturing Processes*; Medical Device and Diagnostic Ind., 1993; 47–51.

GENERAL READING

1. Corlett, Donald A. Jr. *HACCP User's Manual,* Kluwer Academic Publishers, 1998.
2. Damelio, Robert. *The Basics of Process Mapping,* Productivity Inc., 1996.

3. Lurquin, P.F. *The Origin of Life in the Universe*, 2003.
4. Margulis, Lynn. *Five Kingdoms: an Illustrated Guide to the Phyla of Life on Earth*, W H Freeman & Co., 1998.
5. Stamatis, D.H. *Failure Mode and Effect Analysis, FMEA from Theory to Execution*, 1995.

3

Overview of Modern Parenteral Products and Processes

Arvind K. Bansal

National Institute of Pharmaceutical Education and Research,
Punjab, India

3.1. PARENTERALS—GENERAL INTRODUCTION AND PRODUCT TYPES

The term parenteral is derived from Greek words *para* and *enteron*; meaning "to avoid the intestines," and would broadly include all routes of administration other than oral. However, in the healthcare fraternity the term is restricted for the injectable route wherein the drug is directly introduced in the body tissues, blood vessels, or body compartments. Unique advantages offered by parenteral products, as listed below, have earned them a special place in therapeutics:

1. Fast onset of drug action, due to direct introduction of parenteral product into the biological system, for critical patient care coupled with rapid termination of action offers a unique combination to the clinician

2. Highly predictable and accurate response of therapeutic agent as either all or most of the barriers preventing a drug from reaching the site of action are by-passed by the route of delivery

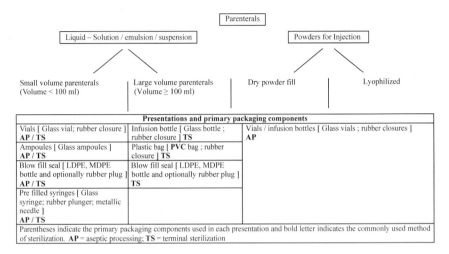

FIGURE 1   Classification of parenteral formulations.

3. Ease of dose titration and individualized therapy based on disease state
4. Preferred route of delivery for drugs having poor permeability, high first-pass metabolism, which are poorly absorbed by oral route
5. Allows administration of drug to unconscious and uncooperative patients

The major disadvantages offered by the parenteral route are the physical/psychological discomfort associated with injection and difficulty in self-administration. However, rapid strides are being made in the area of parenteral delivery devices, which will surmount these barriers and increase the role of parenteral delivery in therapeutics. Introduction of needleless injections and home infusion programs signal toward this emerging trend, which will gain momentum in the near future.

There are a large variety of parenteral products available in the market developed to address (a) physicochemical needs of the drug molecules and (b) specific clinical indications. Figure 1 captures parenteral product classification based on the formulation type, presentation, and manufacturing process (terminally sterilized or aseptically processed). Refer to section 3.3 for details.

3.2. DEMANDS ON PARENTERAL PRODUCTS AND MANUFACTURING PROCESSES

All the pharmaceutical products administered by various routes of administration need to meet the criteria of safety and efficacy. Parenteral products pose health hazards as they are injected through the skin or mucous membrane into the biological system, bypassing the body's natural defense mechanisms. The latter, on one hand, confers unique delivery characteristics to parenterals but at the same time puts stringent demands on product quality and safety. The level of risk involved is enormous and leaves no scope for even minor deviations from the desired specifications.

Parenteral products are radically different from other dosage forms in terms of standards of purity and safety. Apart from complying with standards of potency and stability, parenterals have to meet exacting standards of microbial (sterility and pyrogens), physical (particulate matter), and chemical (isotonicity, buffering capacity, etc.) parameters. Achievement of these standards requires concerted efforts at the formulation and manufacturing level.

The formulator has to overcome many challenges, including (a) achievement of desired solubility profile using parenterally acceptable solvents, (b) osmolarity to the biological fluids, (c) avoiding extremes of pH, (d) minimal use of preservatives due to their inherent toxicity profile, and (e) control of particulate matter. Simultaneous efforts are required from the manufacturer, aimed primarily at achieving microbial standards of sterility and apyrogenicity. A holistic approach addressing all the aspects of manufacturing such a quality of raw materials / packaging materials / manufacturing area environment, and stringent control of process parameters has to be followed, and involves laying down raw material (RM) specifications and introducing

TABLE 1 Parenteral Manufacturing Steps and Their Control Points

Manufacturing step	Possible hazards/control points
RM dispensing	Weighing error
Solubilization of API and excipients in the vehicle	Quality of RMs and API
Aseptic filtration through membrane filter	Sterility of filter assembly
	Filter integrity
Cleaning and sterilization of rubber closures and glass vials	Microbial load
	Particulate load
	Cleaning and sterilization of rubber closures
Filling and sealing	Environmental contamination
	Cleaning and sterilization of filling equipment
Terminal sterilization by autoclaving	Conditions during sterilization e.g. temperature and pressure during autoclaving
Visual inspection	Human error
Batch release	

standard operating procedures. The greatest challenge, however, is posed by the manufacturing process due to its dynamic nature and multiplicity of control parameters of varying hazard potential. Table 1 lists steps of a typical liquid parenteral manufacturing process and highlights the multiplicity of parameters involved therein.

A great deal of stress has been laid by the regulatory agencies on controlling the various process steps of parenteral manufacturing. This, coupled with technological advancements in manufacturing and control machinery, has helpd in improving the control over the processes. An increasing knowledge base of microbiology, bacteriology, filtration, clean room design, and manufacturing technologies over the past few decades has helped in developing sound scientific manufacturing practices. The following sections discuss the advances made over the years in the area of parenteral manufacturing processing.

3.3. PARENTERAL MANUFACTURING PROCESSES

Parenteral manufacturing, like any other manufacturing process, involves a number of integrated process steps. Figure 2 depicts the various parameters contributing toward the success of parenteral manufacturing processes.

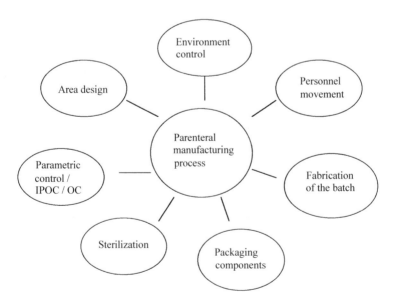

FIGURE 2 Contributors towards the success of parenteral manufacturing processing.

3.4. AREA DESIGN AND ENVIRONMENT CONTROL

Achievement of desired product parameters of a parenteral product demands that the manufacturing be done in a meticulously designed facility, having areas designed to match the criticality of the process to be performed there. A clean production area providing freedom from particulates and microbial contaminants, to the maximum possible extent, is necessary for parenteral manufacturing. A complex system consisting of buffer zones, barriers, environmental control and restricted personnel movement is required to reduce the contaminants. Areas providing varying degree of cleanliness are designed according to the hazard potential of a particular process step. Figure 3 shows various layers of production area designed for specific tasks, with the most critical step filling and sealing being performed in the most protected zone. This concept, with minor modifications, forms the basis for designing of any parenteral facility.

Thus, selection of the manufacturing site, design of the facility, materials and features of construction, and movement of people and materials are critical contributors toward the success of a parenteral manufacturing facility. Based on the number of products to be manufactured, type of manufacturing—continuous versus batch processing and formulation-specific require-

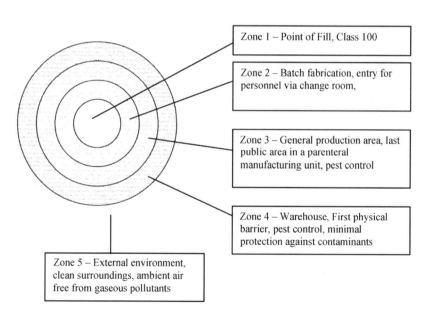

FIGURE 3 Various layers of protection in a typical parenteral manufacturing process.

ments—a balance has to be struck between the process desirables and optimum utilization of space and utilities. The reader can refer to excellent detailed coverage of these parameters elsewhere (1,2).

3.5. FABRICATION OF BATCH

Parenteral products can broadly fall into any of four categories—solutions, suspensions, emulsions, and powders for injection. The typical steps in manufacture or batch production are similar to those followed for nonsterile products. Figure 4 captures the typical manufacturing steps involved in the batch fabrication of various types of parenteral products.

Additional steps in the batch fabrication may be required, especially in cases wherein (a) the dissolved oxygen content (DOC) in the vehicle needs to be reduced before fabrication of the batch and (b) products require post-filling nitrogen purging in the headspace, with inert gases like nitrogen and argon. These steps need to be validated to ensure uniform treatment in all the fabricated batches.

3.6. STERILITY ASSURANCE IN PARENTERAL
MANUFACTURING

One of the most critical requirements to be met by a parenteral product is that of sterility, which is defined as the complete absence of living or potentially living organisms. The effectiveness of an industrial sterilization manufacturing process is conventionally described in terms of probability of a nonsterile unit (PNSU), which is described as a negative exponent of 10. A more commonly used, though less precise term is sterility assurance level (SAL). PNSU is a more accurate descriptor of design parameters and measured performance of physical sterilization processes that involve rendering microbes nonviable, whereas SAL is more suited to processes such as aseptic processes intended to exclude organisms from manufacturing environment and product stream (3). SAL is the probability of a supposedly sterile item being contaminated by one or more microorganisms. All major pharmacopeias require assurance of less than 1 chance in 100,000 that viable microorganisms are present in a sterile dosage, which means a 10^{-6} probability of nonsterility.

Sterility in parenteral products can be achieved either by terminal sterilization or aseptic processing. The former involves filling of formulation in primary packaging containers followed by thermal, ionizing, or chemical modes of sterilization. Certain products which cannot withstand the rigors of terminal sterilization are aseptically processed, which involves (a) sterilizing all the primary packaging components, (b) sterilizing the formulation before

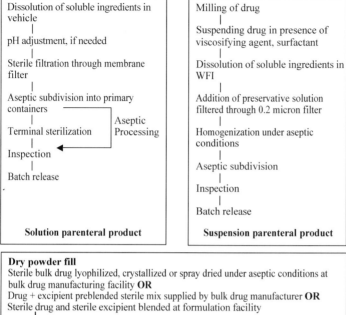

FIGURE 4 Manufacturing steps of various types of parenteral products.

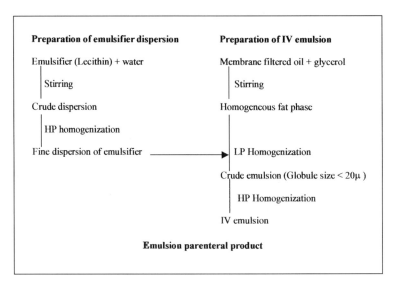

FIGURE 4 Continued.

filling, and (c) carrying out filling in a sterile or as near sterile as possible environment.

3.6.1. Terminal Sterilization

In 1991 USFDA proposed that all sterile products should be terminally sterilized, unless data were available to prove its adverse effects on product stability (4). This was primarily due to the fact that all product recalls during 1981 to 1991 involved aseptically processed products (5). Certain drug classes, such as biologics, multidose ophthalmic products, and dispersed systems, are exempt from this.

The most commonly used technique for terminal sterilization is autoclaving, which makes use of saturated steam. Compendial cycles for autoclaving in USP/EP/BP prescribe a 15-minute exposure at 121°C. Terminal sterilization is based on an "overkill" approach wherein the product is treated to provide a lethality input of 12 D. D is defined as the exposure time required to reduce the microbial population by 90% and is specific for a particular microorganism. The most heat resistant microorganism actually contaminating the formulation—or *Bacillus stearothermophillus* because of its high heat resistance—is used as a standard microbe for development of autoclaving cycles. USP quotes a D121 value of 1.5 minutes for *B. stearothermophillus* and hence an overkill cycle of 18 minutes exposure at 121°C should be used

(1.5×12). D value is affected to a variable degree by the formulation parameters such as pH, antimicrobial activity of drug, preservatives, and excipients such as polymers and sugars. These determinations are typically carried out in specially developed equipment such as BIER (Biological Indicator Evaluator Resistometer), as specified by the Association for Advancement of Medical Instrumentation (AAMI).

Overkill cycles do not take into account the number of microbes actually contaminating the product, which influences the probability of microbe survival after a lethality input of 12D. Bioburden-based sterilization cycles are a rational alternative to ensure acceptable SALs without undue excessive thermal exposure (and potential) degradation. This method of using highly resistant spores of *B. stearothermophillus* involves identification and quantification of actual bioburden in the product, based on which sterilization cycles are developed. The autoclaving of IV emulsions cannot be carried out in traditional autoclaves due to the possibility of emulsion breakage caused by localized heating of the product. Rotating autoclaves with provision for spraying iced water provide the answer. They help by reducing the lag time of the post-autoclaving cooling phase and reduce chances of physical instability of the emulsion during autoclaving.

3.6.2. New Techniques of Terminal Sterilization

There is a strong need to develop terminal sterilization techniques that can help in achieving acceptable SALs without causing damage to the product. PurePulse® Technologies USA has developed a technique called Pure Bright sterilization. It uses broad-spectrum pulsed light (BSPL) to effectively inactivate bacterial organisms and spores in static and flowing solutions, as well as on dry surfaces. Additionally, viruses contaminating the blood-derived products can be inactivated. BSPL generated from xenon lamps contain visible, infrared, and UV wavelengths in ratios similar to sunlight. The major difference from sunlight is that the UV wavelengths are removed due to filtration by Earth's atmosphere. Rapid intense pulses of BSPL are used for inactivation of pathogens. SALs of 10^{-6} have been achieved in liquid materials and dry surfaces using a variety of biological indicators such as bacillus spores, enveloped and noneveloped virus particles, and yeast. The technique is also useful in case of therapeutic proteins, wherein dsDNA, ssDNA, and RNA viruses and bacteria were inactivated by BSPL with economically significant recoveries of the therapeutic proteins. This is possible as only minor changes in temperature are observed at the energy levels used for sterilization, allowing recovery of more that 98% of therapeutic activity.

Energy levels necessary to achieve desired sterility assurance levels are dependent on the organism to be eliminated, the type and thickness of

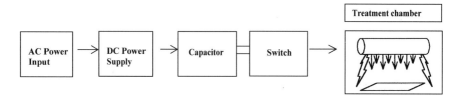

FIGURE 5   Broad-spectrum pulsed light generator.

packaging material, the type and viscosity of the solution and effects on the material being treated. BSPL inactivates both prokaryotic and eukaryotic cells by causing dimer formation and strand breakage of RNA and DNA at magnitudes beyond repair. Similarly exposure of purified RNA, ssDNA, and dsDNA to BSPL results in degradation of nucleic acids into low-molecular-weight species with thymine dimer formation in the surviving fragments. ssDNA and RNA are more susceptible than dsDNA viruses.

The outline of the BSPL system is given in Fig. 5. It consists of xenon lamps, an electric power generator called a capacitor which stores the electrical energy from the DC power supply; the stored energy is discharged to the lamp by switch module, then the lamp flashes intense broad-spectrum white light. Each flash of light is very intense and lasts only for a few microseconds, which is sufficient to kill microbes.

3.7. ASEPTIC PROCESSING

Despite its capacity to provide high SALs, an overkill approach (terminal sterilization) cannot be universally used, as the drastic conditions of temperature can cause (a) degradation of active molecules, especially proteins and peptides, (b) changes in physical attributes of the product, (c) adverse effects on integrity of container closure systems, and (d) generation of leachables/particulate matter. This brings aseptic processing into the picture, which involves filling of presterilized formulations into cleaned and presterilized primary packaging components. In absence of any post-filling sterilization step, the SAL of finished product is a direct function of SAL of individual components.

Products commonly manufactured by aseptic processing include liquid and solid sterile parenteral and ophthalmic dosage forms. They may be single dose or multiple dose. Many aseptically manufactured presentations, especially multidose products, contain preservatives that serve a dual purpose of (a) providing antimicrobial activity and (b) preventing proliferation of any microbe that might contaminate the product during repeated use. However, it needs to be emphasized that standards of protection and controls for aseptically manufactured preserved and nonpreserved formulations should be the same, and preservatives should not be used for providing protection during manufacturing.

Aseptic processing has long faced the skepticism of regulatory agencies because it involves a passive process of protecting against microbial contaminants, as opposed to the active killing of microbes in terminal sterilization (6). The degree of sterility assurance cannot be predicted for aseptic manufacture, as it can be for terminal sterilization. A value of 10^{-3} SAL is frequently stated for aseptic processing and determinations for which are carried out by performing media fill trials (MFTs). The concept of SAL has set a common goal of 10^{-6} for all parenteral manufacturing processes, which places high demands on contamination control during critical processes of aseptic manufacturing. The potential sources of contamination and strategies for controlling them are given in Table 2. An integrated contamination control system addressing all these areas is necessary for the success of aseptic manufacturing. Many of these strategies are also used for products manufactured by terminal sterilization to ensure low initial product bioburden.

The contribution of personnel is the most critical operation within the aseptic filling stage because it is the most difficult to control. A healthy human sheds about 10 million skin scales daily and many of them carry microbes. Disease conditions such as infections and wounds can further add microbial load. Despite controls such as training and medical screening, the human factor remains one of the most variable contributors and improved aseptic controls can be achieved by avoiding contact between product and personnel. Traditionally, this has been addressed by the use of HEPA filtered laminar air-flow. Significant contributions have been made by advances in aseptic processes, in minimizing contact between personnel and sterile products/ surfaces, during aseptic manufacturing.

3.7.1. Advanced Aseptic Processes and Isolator Barrier Technology

The SALs of human-scale clean rooms can be considerably improved by using advanced aseptic processes that reduce or eliminate personnel contact with critical zones. Isolator barrier and blow-fill-seal technologies have contributed significantly toward this. After being successfully used in the area of sterility testing, isolator systems are now revolutionizing parenteral production operations (7) and helping to achieve SALs of 10^{-6}. An isolator is defined as "a device creating a small enclosed controlled or clean classified environment in which a process or activity can be placed with a high degree of assurance that effective segregation will be maintained between the enclosed

TABLE 2 Sources of Contamination and Control Strategy During Aseptic Manufacture

Contributing parameter	Control strategy
Environmental air	Passing air supply through High Efficiency Particle Air (HEPA) filters
	Laminar air flow (90 feet/min) is used to "sweep away" particles and microbes from the sensitive areas.
	Pressure differentials to protect areas of critical operations
Manufacturing equipment	For fixed equipment
	Vacuum cleaner equipped with HEPA filtered exhaust
	Wet wiping with disinfectant solution
	For demountable equipments
	Cleaning and autoclaving
Formulation and primary packaging components	Powders for injection are supplied as sterile by bulk drug manufacturers
	Liquid products are filtered through sterile 0.22 μm membrane filters
	Glass vials are cleaned and dry heat sterilized
	Rubber stoppers are cleaned and sterilized by autoclaving
Personnel	Medical examination to screen personnel working in aseptic area
	Entry of personnel to aseptic area should be through changing rooms
	Containment of personnel microbial flora by protective clothing
	Localized barriers between personnel and areas of filling operations, by means of laminar airflow or by using isolator barriers
Water and drainage	Purification of water by distillation or reverse osmosis
	Storage of WFI at temperatures >80°C and in vessels fitted with continuous circulation loop
	Efficient drainage at the manufacturing shop floor to prevent accumulation of water

environment and the surroundings" (8). The following are the salient features of the isolator barriers used for aseptic manufacturing:

An enclosed controlled environment of minimum volume, installed in an area of Class 100,000 or 10,000.

People segregated from the process

Access for personnel for performing manipulations may be through glove ports or "half suits" made of latex, neoprene, nitrile, PVC, urethanes, or laminated polymers.

Isolator walls may be rigid (stainless steel/glass/Perspex®) or flexible (PVC).

Internally pressurized with turbulent or unidirectional air/inert gas flow filtered through HEPA or ULPA filters, to work at positive pressure relative to general environment.

Under pressurized isolators are used to prevent spreading of potent toxic drugs to the environment.

Internal surfaces are sterilized by gaseous sterilants such as vapor phase hydrogen peroxide, ozone, chlorine dioxide.

Barriers are integrated into manufacturing lines to provide an effective locally controlled environment (7).

Barrier isolator systems are being routinely used for applications such as sterility testing, manufacturing of powders for injection, hospital pharmacies, and large-scale aseptic production. Containment isolator systems are used for subdivision/dispensing of potent, hazardous, or biologically active compounds. There have been notable advances made in this area, and successful implementation of this technology to pharmaceutical operations should address the following issues:

Removal of residual solvent used for gaseous sterilization

Integration of isolator to main production line without compromising the integrity of the isolator

Sound validation strategy justifying installation in class 10,000 or 100,000 areas

Detection of leaks in the glove ports

Reliability of transfer ports

One of the most critical causes of loss of integrity is the transfer of material into and out of an isolator barrier system. Various devices with increasing efficiency and technological sophistication have been used. The "jam-pot" or single doors have little ability to separate internal and external environment. Double-door pass-through hatches with mechanical or electro-mechanical interlocking provides better control. The best assurance can be obtained by using interlocked docking port systems, also called alpha-beta systems, and airflow protected tunnels for continuous discharge. Interesting case studies can be found involving system design, installation and operation of barrier isolator technology as applied to areas of liquid SVP manufacturing, aseptic potent powder filling, and lyophilization (9).

A recent development in the area of improving transportation in and out of the isolator barrier systems has been its integration with BSPL (refer to section 3.6.2). This provides rapid and cost-effective sterilization of the product while transferring it from a nonsterile area into a sterile environment. The sterilization process is reduced to a matter of seconds, compared to earlier processes, which required considerably longer duration of treatment. Such technological innovations provide manufacturing solutions that are effective, fast, and safe.

3.7.2. Blow-Fill-Seal Technology (BFS)

Isolator technology has offered means of improving confidence in the aseptic processing of pharmaceuticals, primarily due to minimal intervention required to run such assemblies using a dedicated air flow, sterilization and depyrogenation. This allows achievement of a finely controlled micro-environment. A similar concept is that of BFS, which instead of using glass containers uses plastic containers formed by blowing within a clean environment. BFS technology involves a fully automated process in which the primary container for the formulation is (a) formed from a thermoplastic, (b) aseptically filled with filtered solution, and (c) sealed, in a single operation in a controlled environment. BFS uses an automated process requiring minimal human intervention once the machine settings have been set. The most critical fillings are carried out in an enclosed compartment, protected by laminar HEPA filtered air. The formed plastic container is filled with sterile product and instantly sealed, thus avoiding contamination. Additionally, the BFS machine is designed for clean-in-place and sterilization-in-place, which also eliminates human intervention. A hermetically sealed bottle formed during the process helps avoid the use of sealing devices like rubber closures and seals. Although optional, rubber closures can be incorporated into the pack to help in integrating with delivery devices during drug administration. Figure 6 shows the steps involved in the BFS machine and Figure 7 shows the flow of materials during manufacturing on a BFS machine.

BFS has been in use in the food industry since the 1960s and a number of European pharmaceutical companies are now using it. The machine is suitable for aseptic processing as has been proved by media fills and challenge tests (9,10). However, their area of installation has been a source of debate, with opinion divided on their installation in Class 10,000 area or Class 100 area. A study conducted by Bradley et al. indicated that the SAL obtained with BFS technique is a complex function of the microbiological quality of

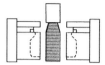

a) Extruding: The plastic parison, extruded from polymer, is received in the opened blow mould and cut below the die of the parison head.

b) Moulding: The main mould closes and simultaneously seals the bottom. The special mandrel unit settles onto the neck area and forms the parison into a container, using compressed air. Small containers are formed by vacuum.

c) Filling: By way of the special mandrel unit, the precisely measured product, is filled into the container.

d) Sealing: After the special mandrel unit retracts, the head mould closes and forms the required seal by vacuum.

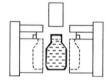

e) Mould opening: With the opening of the blow mould, the container exits from the machine and the cycle repeats itself.

FIGURE 6   Formation of blow-fill-seal pack.

the environment in which the machine is installed (11). Another study investigated the routes of airborne contamination into BFS containers, using sulfur hexafluoride tracer gas (12). A known concentration of gas was released in the clean room housing the BFS machine and later the gas was measured in BFS units. The study concluded that the container was effectively protected by the localized air shower. However, it is now recommended to install the BFS machine in well-controlled environment.

Various thermoplastics like polyethylene (PE), polypropylene (PP), various copolymers and polyalomers are commercially available. Quality and regulatory issues related to contamination by plastic granules used for formation of primary pack have largely been resolved by use of high quality virgin polymer granules having low bioburden. A study using a challenge of *Bacillus subtilis* var. *niger* in polymer granules has demonstrated spore inactivation on granules with strong evidence of lethality associated with the extrusion process (13). PP offers distinct advantages by allowing exposure

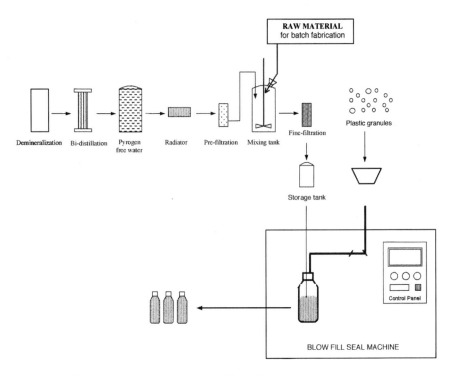

FIGURE 7   Flow of materials in a blow-fill-seal manufacturing unit.

to 121°C for autoclaving. Bottles produced on the BFS machine can be individual or strip-dose formats, in sizes from 0.1 ml to 2000 ml, and outputs as high as 30,000 units/hour can be achieved (14). BFS technology has offered a cost-effective means of introducing high-quality aseptically processed products with additional advantage of reduced breakage, reduced hazard of accidental injury, and reduced pack volume over glass containers.

3.7.3.   Terminal Sterilization of BFS Containers by Autoclaving

Pharmacopoeias recommend standard sterilization conditions of 121°C for 15 minutes equaling a Fo of 15. Most commonly used BFS containers are made up of PE, which can only withstand temperatures of 105–106°C. The manufacturers of BFS machines argue that a Fo of 2 to 4.5 is acceptable for BFS containers because the filled product is already sterilized as it is passed through one or two 0.22 μ membrane filters. Exposures at these temperatures would be termed 'sublethal' from the viewpoint of regulatory agencies. The standard microbe for validation of moist heat sterilization is *B. stearother-*

mophillus, which will not be destroyed at this temperature. Hence, this sterilization exposure is unacceptable and this has prevented introduction of injectable products in BFS packs in the United States. The argument that "if bio-burden of solutions filled in BFS is kept low by using sterilizing filters which maintain integrity throughout the manufacturing process" has found takers in some of the Asian and South American countries and many infusion, SVPs, and ophthalmic products have been introduced into the market.

Apart from introduction of PP as a packaging material, introduction of BSPL sterilization technique is improving the SALs achievable in BFS process. The suitability of BSPL introduced by PurePulse® Technologies, a division of Maxwell Technologies™, was investigated for its suitability of terminally sterilizing the PE containers. The BSPL system produces flashes of intense light of broad wavelength (200 to 1100 nm) and one to three flashes are sufficient for sterilization of PE vials. With one flash of 5 mJ/cm^2 at a wavelength of 260 nm, a SAL of 10^{-6} was achieved for a challenge of 12 different microbes including *B. stearothremophillus*. This technique provides flexibility as it allows in-line/on-line sterilization of the filled vials (15).

3.8. RECENT ISSUES IN STERILIZATION BY FILTRATION

Sterilization by filtration has traditionally involved use of 0.2 / 0.22 μm rated filters and *Pseudomonas diminuta* (now called *Brevundimonas diminuta*) as the standard challenge organism. USFDA defines minimum qualifying area of $10^7/cm^2$ of filter area (i.e., filter must be able to retain the microbes at a level of $10^7/cm^2$ of the filter area). Over the years there have been reports that in response to stressful conditions, as may be encountered in a pharmaceutical formulation, microbes can change their size and morphology and can pass through filter membranes that would have normally retained them (16). In this study, 40% reduction in size of *Burkholderia pickettii* was observed. Thus, the FDA expects that pharmaceutical manufacturers test the microbial retentivity of the filter with the microbial challenge in the actual drug product (17). These developments are causing a shift from 0.2 / 0.22 rated filters to 0.1 filters. However, the transition will only be possible after issues including revalidation, reduced filtration throughputs, and increased process times are addressed satisfactorily.

3.9. STERILE PREFILLED SYRINGES (PFS)—MANUFACTURING AND TERMINAL STERILIZATION PERSPECTIVES

PFS have become a popular packaging system for parenteral products due to their advantages of ease of administration, dosing accuracy, and increased assurance of sterility. PFS consists of a barrel, a plunger rod with rubber fitting, and a luer-lock tip/stainless steel needle. The manufacturing process for PFS may involve (a) filling of formulation in previously cleaned and sterilized PFS or (b) cleaning, sterilization, depyrogenation of nonsterile syringes, followed by filling (18). The filling is carried out in Class 100 area, and other operations can be carried out in Class 10,000/100,000 area.

Terminal sterilization of PFS by autoclaving poses a unique challenge due to the possibility of rubber plunger migration during the process. This "pop-off " of rubber plunger can be prevented by using autoclaves with a counter-pressure feature. The counter pressure should be calculated all along the cycle during the heat-up phase, forming the sterilization plateau, and also during the cooling phase. The pressure inside the syringe is dependent on the drug product temperature, which varies during different process steps and also among autoclave load. It is important to maintain a positive pressure throughout the cycle and these autoclaves achieve this by accurately linking any change in pressure closely to the product temperature.

3.10. PROCESS VALIDATION, HAZARD ANALYSIS AND CRITICAL CONTROL POINTS (HACCP)

The significance of validating any pharmaceutical process and especially parenteral manufacturing process is immense. Validation of a process is the demonstration that controlling the critical steps of a process results in products of repeatable attributes or causes a reproducible event.

A new technique being employed, which fits into validation needs very well is HACCP, which although not approved by USFDA for validation can help in immaculate control for the manufacturing process. Most of the principles of HACCP are concordant with the main provisions of GMP. HACCP is a technique used to analyze a process, determine the high-risk steps, and control or monitor those steps to ensure that a process yields quality product (19). It has become an accepted practice by USFDA in both the food and medical device industries. Soon this may become a "current" requirement in process validation (20). Owing to the extremely sensitive nature of parenteral manufacturing to some of the process parameters, HACCP can improve the reliability of the overall process. The success of this technique depends on successful identification of critical parameters and focusing efforts and resources where they are required. Validating noncritical parameters, apart from wasting time and money, dilutes the validation effort.

HACCP involves seven principles: (a) analyzing each step for hazard, (b) identifying all critical control points (CCP), (c) verifying the limit for each CCP, (d) verifying monitoring and testing of limits, (e) verifying corrective actions, (f) verifying operational procedures for CCPs, and (g) verifying that

TABLE 3 Critical Control Points in a Typical Parenteral Product Manufacturing

Parameter/process	Possible risk	Level of risk
Liquid Injection		
Dissolved oxygen content in the vehicle	Oxidative degradation of active drug	Major for drugs prone to oxidative degradation
pH adjustment using acid or alkali	Changes in the isotonicity of the product	Variable depending on the quantity of acid or alkali added
Buffer concentration	Changes in isotonicity and post-injection pH changes in blood	Variable depending on the quantity of buffer added
Evaporation of volatile preservative	Loss of preservative efficacy	Extreme
Batch holding of aseptically processed product before filling into final pack	Loss of sterility	Extreme
Powders for Injection		
Moisture deposition in glass vials during cool-down phase after dry heat sterilization	Degradation of drug by hydrolysis	Extreme for powders for injection, Nuisance for liquid products
Residual moisture in rubber plugs	Degradation of drug by hydrolysis	Major for powders for injection, Nuisance for liquid products
Relative humidity of filling area	Degradation of drug by hydrolysis	Major or Minor depending on stability profile of the drug

records of each CCP are documented in the batch record (19). After a parameter gets identified as a hazard, a risk assessment should be carried out to classify it as extreme, major, minor, or nuisance, based on its overall impact on the product quality.

Focusing on the active ingredient is obvious for any drug product but parenteral products require scrutiny of various excipients and processes that are vital to the quality of the final product. The hazard in a parenteral product could be physical, chemical, biological, or environmental. Although HACCP can be applied in all areas of product development and manufacturing, it is better to focus on the batch manufacturing process. Other processes and parameters can be effectively controlled using existing Standard Operating Procedures and quality systems. Table 3 lists some of the areas that owing to their criticality would require close scrutiny using HACCP. As is evident, the possible 'hazard' needs to be assessed vis-à-vis stability and performance characteristics of the specific formulation. The strength of HACCP lies in the fact that it integrates drug quality control into the design of the manufacturing process rather than depending on end product testing.

3.11. CONTINUOUS PROCESSING OF PARENTERAL PRODUCTS

Pharmaceutical manufacturing processes, including parenteral manufacturing, traditionally have been batch processes, due to issues related to process, batch size, and quality control. This is reflected in the ubiquitous reference to "Batch Production Records" for all issues related to processes and quality compliance. Apart from the fact that certain processes, such as dissolving drug in vehicle and autoclaving, cannot be made continuous, batch operations offer additional advantage of step-by-step controls at end of each unit process. However, they are labor intensive, time consuming, and prone to contamination due to 'difficult-to-validate' multiple transfer steps between various unit operations. Additionally, high human intervention is the most common reason for introduction of viable and nonviable particles into the product. Despite the practical and psychological barriers, attempts have been made to introduce continuity in the parenteral manufacturing processes.

Various steps involved in a typical parenteral manufacturing process are as follows: (a) weighing, (b) charging, (c) formulation preparation, (d) filtration, (e) filling, (f) sterilization, and (g) inspection. Continuity in the entire manufacturing process from weighing to finished product inspection has not been possible but significant advances have been made in automating a single or a number of adjacent processes. In a continuous process, (a) the incoming material from the previous step is automatically fed, (b) processed, and (c) delivered to the next step. Some of the unit processes frequently performed automatically are sterilization of glass vials, filling, filtration, packaging and visual inspection, whereas some of the processes that require lag times and hence impede continuity include weighing, mixing, dissolving, terminal sterilization, and freeze drying. It is common to have islands of continuous processing in a typical parenteral manufacturing process, interspersed with discontinuous batch processes (Fig. 8).

The trend of the future will be the introduction of continuous processes, to harness the benefits of shorter process times and reduced environmental exposure, leading to improved quality and productivity. This will be coupled with shifts in quality control concept from "batch production record" to "in line" testing. The emerging concepts of computer data processing merge well with these objectives and will be increasingly used in parenteral manufacturing. Newer technologies like 'Purebright' sterilization will help in transforming batch process intensive stages such as terminal sterilization into continuous processes. Modular conveyor technology can also be beneficially utilized for automation in several steps and can facilitate the use of same equipment for many different product lines. A mixture of automatic, semi-automatic, and manual operations for low to medium volume production can

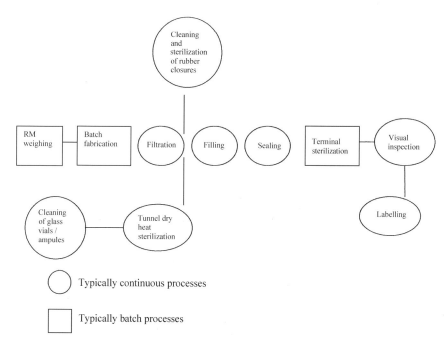

FIGURE 8 Continuous and batch processes in a parenteral process.

be handled using these systems. They also allow mixed sequential and parallel flow of goods simultaneously.

3.12. COMPUTER-CONTROLLED AUTOMATION OF UNIT PROCESSES

Although complete continuity of parenteral manufacturing process still remains far-fetched, significant strides have been made in automation of unit processes. Development of programmable logic controllers (PLCs) and sensor technology now allows precise control of process parameters. Two types—open loop and closed loop process controls—are utilized. In the former, the sensor records the parameters, compares against desired standards, and notifies a human operator for corrective actions. A closed loop system works without human intervention and directly communicates with regulating devices for corrective actions.

Freeze drying of powders for injection has now been largely automated, using a micro-computer and PLC. The microcomputer initiates/monitors the process and archives the data, whereas PLC controls freeze-drying, sterilization, and cleaning process by means of instructions downloaded from a computer (21). Similarly detection of foreign matter in ampoules and vials has been automated by using principles of light transmission, absorption, and reflection (22). Using the same principle, instruments have been developed to detect the fill volume in ampoules, and vials, by determining the meniscus of the filled solution.

The core advantages of automated and continuous processing, in terms of minimal human intervention and precise control, are ideally suited for manufacturing of parenteral products. Application of "in line" controls to automated processes and utilization of automated inspection instruments is already greatly improving productivity and quality. Greater application of parametric release over end-point testing in parenteral manufacturing, and its regulatory acceptance, encourages automation. Simultaneous developments of user-friendly computer languages has made programming easy even for nonexperts and would help development of expert systems, addressing intricate details of the parenteral manufacturing process.

3.13. PROCESSING OF BIOPHARMACEUTICAL PRODUCTS

The drugs obtained from pharmaceutical biotechnology could be antisense compounds, complex carbohydrates, or proteins, the latter being the major current contributors. Introduction of proteins and polypeptides as drugs is going to experience a quantum jump in years to come. Preparation of proteins as medicinal agents has become an integral part of the pharmaceutical industry (23). Proteins are highly sensitive to environmental factors such as temperature and electrolyte concentration because apart from their primary chemical structure, maintenance of their secondary, tertiary, and quaternary three-dimensional structure is necessary for biological activity. This presents unique challenges in their purification, separation, formulation, storage, and delivery. Preservation of protein structural conformation throughout the processing and shelf life is a critical issue for their successful formulation. High molecular weight and low permeability through biological membranes causes poor absorption from routes requiring crossing of biological membranes, and this makes parenteral route of delivery promising for these molecular entities. The requirements of physical, chemical, and microbiological parameters for proteins and polypeptide parenteral formulations are the same as for any other parenteral product.

Stability of proteins can be classified as chemical and physical stability. Chemical instability involves bond formation or cleavage leading to changes in primary chemical structure and may involve deamidation, oxidation, proteolysis, disulfide exchange, and racemization. Physical instability involves changes in secondary and higher structures. This can involve denaturation, aggregation, adsorption, and precipitation. This phenomenon is rarely observed in the case of small drug molecules. These physical changes in protein molecules occur because of their polymeric nature and higher-order structures and are independent of chemical modifications in their primary structure.

Protein formulations have been introduced in solution as well as freeze-dried solid form, and various strategies available to improve protein stability during processing and storage include addition of excipients, site-directed mutagenesis (24,25), and chemical modification (26,27). Additives used for stabilization include (a) ionic salts, which increase thermal stability by ion binding (28,29), (b) polyalcohols like glycerol and sugars, which stabilize the proteins with respect to denaturation, through selective solvation of protein (30,31), (c) surfactants belonging to nonionic (Tweens, Pluronics) and anionic classes (sodium lauryl sulfate) (32,33), (d) amino acids, and (e) chelating agents. Many protein formulations involve a number of formation and processing steps in liquid and solid state and more than one stabilizer may be required. A stable liquid formation is usually a prerequisite even for freeze-dried or spray-dried formulations, and it is the starting material for subsequent processing (34).

3.13.1. Stabilization During Freeze-Drying of Protein Formulations

One of the main causes of denaturation of proteins during freeze-drying is the loss of the hydration shell, which is required for maintaining the structure of protein (i.e., hydrophobic regions inside and the hydrophilic parts outside). The removal of the hydration shell decreases the free energy barrier of unfolding of the protein and thus makes denaturation easier. This can be prevented using polyols like sugars and dextrans whose hydroxyl groups act as a substitute for the water lost and also mechanically help to maintain the protein structure.

Ions are added in the formulation to maintain osmolarity. Changes in ion concentration in the solution due to crystallization of water during freezing is another factor that influences the stability of proteins. Use of certain buffers (e.g., phosphate buffers) will also cause problems as one of the components of the buffer system crystallizes out preferentially depending on the rate of cooling utilized, leading to a drastic change in the pH of the system.

The temperatures reached during the freezing and heating cycles involved in freeze-drying have to be closely monitored as both are capable of causing denaturation of proteins. Higher temperatures provide the energy necessary to cross the energy barrier to unfolding and the lower temperatures lead to increased solubilization of the hydrophobic parts. Glass transition temperature is another factor to be considered to maintain the stability of proteins because crossing this temperature during secondary drying will cause greater mobility of the lyophilized constituents, making it easier for degradative reactions to take place. The excipients chosen in formulating a lyophilized protein should be such that they do not themselves cause degradation—for example, reducing sugars like glucose react with proteins via Maillard reaction and PEGs produce peroxides on aging that oxidize the proteins.

3.13.2. Spray-Drying of Protein Formulations

Spray-drying is another method used to obtain proteins in solid state. There are many factors that affect protein stability during spray-drying, most of which are similar to the factors encountered during lyophilization. The other factors observed to have an influence on the stability of the proteins being spray-dried are outlet temperature, feed rate, pH, atomization rate/air-water interface, and shear stress. Proteins are usually spray-dried in a cocurrent manner to preserve their stability because the protein droplets are most sensitive at the outlet where its moisture content is very low and it becomes important to ensure that the dry product is in contact with only the coolest air. The methods used to stabilize proteins involve addition of sugars, use of a surfactant, altering feed solution pH, and keeping the residual moisture to a minimum. The crystallinity of the sugars in the final product is of importance because only amorphous form of sugars have the beneficial effect on the stability of proteins.

3.13.3. Assuring Sterility of Protein Formulations

The greatest challenge offered by the processing of protein formulations is during sterilization. Thermal instability of proteins completely rules out the use of thermal methods like autoclaving for sterilization and use of radiation. Protein formulations are most commonly manufactured by aseptic processing, wherein isolator barrier technology will increasingly play its role in improving the SALs. Similarly, BSPL has shown potential in inactivation of viruses and vegetative bacteria from formulations of therapeutic proteins. Future advancements in BSPL technology will have far-reaching impact on sterilization of these unique therapeutic moieties.

3.13.4. Sterile Filtration of Protein Formulations

Filtration of protein formulations through 0.45 or 0.22 μm micron membrane filters poses an additional challenge, because the solvent sometimes passes through the filter surface at a somewhat faster rate than the solute itself. This could mean a higher solute concentration on the upstream side of the filter and in extreme cases could lead to formation of gel-like structure. This gel polarization can significantly slow down or block the filtration process. Having a fibrous prefilter structure immediately adjacent to the membrane can prevent this phenomenon, as the gel forms around the fibers and not on the surface of the membrane filter (35). Generation of vigorous turbulent flow of the protein solution can cause significant precipitation and loss of biological activity. Specially designed filtration systems can prevent extremes of flow pressures and provide greater product stability.

3.13.5. Future Trends

An increasing product pipeline of pharmaceutical biotechnology-based products is going to pose numerous challenges to formulation and process scientists. Introduction of isolator barrier systems for aseptic processing, automated freeze-drying processes, and newer sterilization techniques have already made an impact on manufacturing of these specialized molecules. Future parenteral manufacturing trends are increasingly going to be targeted to meet the functionality and quality needs of these highly promising biological response modifiers.

3.14. PROCESSING OF PACKAGING COMPONENTS

3.14.1. Rubber Closures

Rubber plugs are a critical packaging component of liquid and powders for injection, wherein they critically affect the product stability. A careful selection of the rubber closure is necessary, as it can be a potential source of microbial, particulate, and nonparticulate (leachables) contamination. The processing of the rubber closure before being incorporated into the final pack also affects its performance, and a validated processing treatment goes a long way in ensuring optimum performance. Any failure at different stages of processing can lead to problems such as (a) microbial contamination, (b) particulate contamination—especially difficult to remove fibers from the surface of the rubber closures can dislodge in the formulation, (c) tacky surface leading to interrupted movement in machine channels, and (d) failure of container closure integrity during product shelf life.

Powders for injection normally contain moisture-sensitive substance and the moisture vapor transmission ratio (MVTR) of the rubber closure significantly affects product stability. Here also, processing determines the residual moisture and other performance characteristics of the rubber closure. Their processing essentially remains same for liquid and powders for injection except that in the latter, a post-sterilization drying step is critical for the product stability.

3.14.2. Washing

Machines for washing the rubber closures are based on overflow rinse cycle, wherein the floating debris and fibers are removed by overflow rather than bottom draining. This ensures that the floating particles do not get deposited on the rubber plugs during draining. The bed of the rubber closures is agitated with the help of filtered compressed air, released at the bottom of the bed. The washing cycle can be carried out with distilled water alone or with detergent solution, the most common being Teepol® or Polysorbate 80. The use of the detergent sometimes cannot be avoided because of the stubbornly sticking fibers which are difficult to remove because of the electrostatic charges. However, rinse cycles should be properly validated to ensure complete removal of detergent traces. Suitable analytical techniques such as UV, HPLC can be used to control the process.

3.14.3. Siliconization

The dry and sticky nature of the rubber closures can cause "jamming" in the traveling chutes of the filling/sealing machine leading to line shut-downs. Siliconization imparts surface lubrication and aids in the flow of rubber closures. It also helps in easy insertion of the closures on filled vials and reduces powder sticking in the case of powders for injection. Siliconization traditionally was carried out using silicon oil or silicon oil emulsions available commercially. Rubber closure manufacturers now supply presiliconized rubber closures. There have been reports of product contamination by silicon oil droplets used for siliconization. This has encouraged innovations in the formulations of high-performance rubber closures, which do not require siliconization for optimum functioning. These polymer-coated closures have low friction coefficient and enhanced machineability (Table 4). Additionally, these stoppers offer increased compatibility as well as low-visible and subvisible particulate matter extraction.

3.14.4. Cleaning and Sterilization of Elastomeric Closures

Elastomeric closures can significantly contribute to particulate and endotoxin contamination in sterile pharmaceuticals, thus compromising overall quality objectives (36). These safety concerns and strict compendial limits of particulate matter of parenterals have forced improvements in the processing of rubber plugs aimed at reducing particulate contamination. This acquires greater importance in the case of aseptic processing, wherein the final quality of the product is directly dependent on the quality of the individual components, including the packaging components.

TABLE 4 High Performance Rubber Closures

Rubber closure	Manufacturer	Rubber closure formulation
Omniflex Plus®	Helvoet Pharma, Belgium	Inert, flexible, fluorinated polymer coating for pharmaceutical rubber closures, which covers entire closure surface
FluoroTec® Plus	West Pharmaceutical Services, USA	ETFE copolymer of ethylene and tetrafluoroethylene
Daikyo Fluorotec film	West Pharmaceutical Services, USA	ETFE copolymer of ethylene and tetrafluoroethylene
B2	West Pharmaceutical Services, USA	A mixture of silicones that is polymerized on the surface of rubber by UV radiation. Chemical structure is similar to polydimethylsiloxane
Teflon®	West Pharmaceutical Services, USA	Fluorinated ethylene-propylene
Unishield™	Abbott, USA	Applied barrier coating
Soloshield™	Abbott, USA	Deposition-coated stopper

A number of methods and machines have been developed to effectively remove surface contaminants, leachable materials, and other debris adhering to the closures (37–40). Traditionally, many manufacturers used a non-continuous batch process wherein the rubber plugs were washed, siliconized, sterilized, and dried equipment, individually housed under laminar flow units. This process has a number of "potentially contaminating" manufacturing steps, thus reducing its reliability. These have now been replaced with automated microprocessor-based washers capable of carrying out all the processing steps, thus reducing the chances of contamination. Particles dislodged from the surface of rubber plugs during washing are removed by overflow rinse to eliminate possibility of their re-deposition on the closures.

A noteworthy trend, that drastically reduces the resources necessary to prepare stoppers has been the introduction of (a) ready-to-use radiation sterilized rubber closures (41) and (b) ready-to-sterilize rubber closures. These prewashed, presiliconized, WFI-rinsed closures are supplied in breathable-steam-sterilizable bags. These rubber closures help in streamlining the manufacturing process by (a) reducing likelihood of rejects, (b) simplifying pretreatment, (c) reducing validation efforts, (d) saving time and cost, and (e) providing product of low bioburden and endotoxin levels. Some of the products available in this category are Westar® RS from West Pharmaceutical Services and Ultraclean 6 from Stelmi, France.

3.15. FUTURE TRENDS IN PARENTERAL PROCESSING

The role of parenteral delivery will increase in years to come, due to the biotechnology boom and developments in the field of parenteral delivery devices. The traditional clinical and psychological barriers to injectables will diminish as more patient-friendly devices hit the market. The manufacturing of parenteral products has undergone a sea change over the years with introduction of technologies aimed at improving their SALs and overall product quality. Isolator barrier technology will play an increasing role in ensuring the highest quality of aseptically processed products. Terminal sterilization of BFS packs and therapeutic proteins are already experiencing technology-driven changes, with BSPL overcoming the chronic shortcomings of moist heat sterilization. A booming pipeline of biotechnology-based products is continuously challenging the prevalent parenteral manufacturing practices. These specialized molecules, because of their poor permeability, are most suited to parenteral delivery. Their tendency to physical and chemical instability puts demands on current formulation and processing methodologies. Special delivery and stability requirements of these unique therapeutic moieties will fuel further refinement of aseptic processing and new terminal sterilization techniques.

REFERENCES

1. Keller, A.M.; Hoffman, G.L. Design consideration for a parenteral production facility. In *Pharmaceutical Dosage Forms: Parenteral Medications*; Avis, K.E., Lieberman, H.A., Lachman, L., Eds.; Marcel Dekker: New York, 1993; Vol. 2, 235–316.
2. Halls, N.A. Aseptic manufacture. In *Achieving Sterility in Medical and Pharmaceutical Products*; Hall, N.A., Ed.; Marcel Dekker: New York, 1994; 179–239.
3. Akers, J.; Agolloco, J. Sterility and sterility assurance. PDA J. Pharm. Sci. Technol. 1997, *51* (2), 72–77.
4. FDA. Use of aseptic processing and terminal sterilization in the preparation of sterile pharmaceuticals for human and veterinary use. Federal Register, Docket No. 91N-0074, October 1991.
5. Akers, M.J.; Nail, S.L.; Groves, M.J. Top 10 current technical issues in parenteral science revisited. Pharm. Tech. 1997, *21*, 126–140.
6. Halls, N.A. Aseptic manufacture. In *Achieving Sterilization in Medical and Pharmaceutical Products*; Halls, N.A., Ed.; Marcel Dekker: New York, 1994; 179–180.
7. Farquharson, G.J. Isolators for pharmaceutical applications. In *Encyclopedia of Pharmaceutical Technology*; Swarbrick, J., Boylan, J.C., Eds.; Marcel Dekker: New York, 1996; 1211–36.
8. Mackler, S.E. Barrier isolation technology: facilities update. Pharm. Tech. 2000, *24* (2), 40–48.
9. Leo, F. Blow-fill-seal packaging technology. In *Aseptic Pharmaceutical Manufacturing: Technology for 1990s*; Olson, W.P., Grove, M.J., Eds.; Interpharm Press: BuffaloGrove, IL, 1989; 195–218.
10. Sharp, J.R. Validation of a new form-fill-seal installation. Manuf. Chemist 1988, *239*, 22–23, 27, 55.
11. Bradley, A.; Probert, S.P.; Sinclair, C.S.; Tallentire, A. Airborne microbial challenges of blow-fill-seal equipment: a case study. J. Pharm. Sci. Technol. 1991, *45* (4), 187–192.
12. Whyte, W.; Matheis, W.; Dean-Netcher, M.; Edwards, A. Airborne contamination during Blow-Fill-Seal pharmaceutical production. PDA J. Pharm. Sci. Technol. 1998, *52* (3), 89–99.
13. Birch, C.J.; Sinclair, C.S. Blow-fill seal extrusion of spore contaminated polymer: an exploratory study. BFS News, BFS Operators Association, September 1988.
14. Gander, P. Debating the case for sterile filling. Manufac. Chemist 1998, *69*, 28, 30.
15. Furukawa, M.; Ento, N.; Kawamata, T. Brand new pulsed light sterilization technology can sterilize both injectable solutions and its 20 ml polyethylene container. PDA International Congress, Tokyo, Japan, Feb 22–26, 1999.
16. Le, F. et al. Applications of 0.1 µm filtration for enhanced sterilization assurance in pharmaceutical filling operations, I BFS News, August 1997.
17. Human drug CGMP notes Vol 2 No. 4: 35 FDA CDER office of compliance 1994.

18. Karras, L.; Wright, L.; Abram, D.; Cox, T.; Kouns, D.; Akers, M. Sterile prefilled syringes: current issues in manufacturing and control. Pharm. Tech. Oct. 2000, *8*, 188, 190, 192, 194, 196.
19. Armburster, D.; Feldsien, T. Applying HACCP to pharmaceutical process validation. Pharm. Technol. 2000, *24* (10), 170, 172, 174, 176, 178.
20. www.fda.gov/cdrh/gmp/haccp.html.
21. Basavapathruni, O. Computer in pharmaceutical technology. In *Encyclopedia of Pharmaceutical Technology*; Swarbrick, J., Boylan, J.C., Eds.; Marcel Dekker: New York, 1990; 201–226.
22. Akers, M.J. Parenteral Quality Control; Marcel Dekker: New York, 1985; 159–170.
23. Manning, M.C.; Patel, K.; Borchardt, R.T. Stability of protein pharmaceuticals. Pharm. Res. 1989, *6* (11), 903–918.
24. Shortle, D. Probing the determination of protein folding and stability with amino acid substitutes. J. Biol. Chem. 1989, *264*, 5315–5318.
25. Querol, E.; Parrilla, A. Tentative rules for increasing the thermostability of enzymes by protein engineering. Enzyme Microbiol. Technol. 1987, *9*, 238–244.
26. Buchowski, A.A. Cell. J. Biochem. 1987, *11A*, 174.
27. Ribiero, A.A.; Saltman, R.P.; Goodman, M.; Mutter, M. Biopolymers 1982, *21*, 2225–2239.
28. Arakawa, T.; Timasheff, S.N. Mechanism of protein salting in and salting out by divalent cation salts: balance between hydration and salt binding. Biochemistry 1984, *23*, 5912–5923.
29. Almog, R. Effects of neutral salts on the circular dichorism spectra of ribonuclease A and ribonulease S. Biophys. Chem. 1983, *17*, 111–118.
30. Lee, J.C.; Timasheff, S.N. Biochemistry 1975, *14*, 5183–5187.
31. Lee, J.C.; Timasheff, S.N. The reconstitution of microtubules from purified calf brain Tubulin. Biochemistry 1975, *14*, 5183–5187.
32. Twardowski, Z.J.; Nolph, K.D.; McGary, T.J.; Moore, H.L. Influence of temperature and time on insulin adsorption to plastic bags. Am. J. Hosp. Pharm. 1983, *40*, 583–586.
33. Tandon, S.; Horowitz, P.M. Detergent assisted refolding of Gunnidinium chloride denatured rhodanase. The effect of the concentration and types of detergent. J. Biol. Chem. 1987, *262*, 4486–4491.
34. Carpenter, J.F.; Crowe, J.H. An infra-red spectroscopic study of the interactions of carbohydrate with dried proteins. Biochemistry 1989, *28*, 3916–3922.
35. Groves, M.J.; Alkam, M.H.; Hickey, A.J. The formulation of proteins and peptides. In *Pharmacetical Biotechnology—Fundamentals and Essentials*; Klergerman, M.E., Groves, M.J., Eds.; Interpharm Press: 1992; 226–228.
36. Borchert, S.J.; Abe, A.; Aldrich, D.S.; Fox, L.E.; Freman, J.E.; White, R.D. Particulate matter in parenterals, a review. J. Parenter. Sci. Technol. 1986, *40* (5), 212–241.
37. Kapoor, J.; Murty, R. Pre-treatment of rubber closures for parenteral containers. Pharm. Technol. Colorado, U.S.A., 1977, *1* (6), 53, 80, 83.

38. Smith, G.G. New process for treatment of parenteral closures. Bull. Parenter. Drug Assoc. 1976, *30*, 53–63.
39. Hopkins, G.H. Improved machine design for stopper washing. Bull. Parenter. Drug Assoc. 1973, *27*, 114–125.
40. Nishimura, T.; Kishimoto, J.; Nishida, Y.; Noguchi, Y.; Imai, S. A novel system for washing parenteral rubber closures individually. J Parenter. Drug Assoc. 1979, *33* (2), 96–103.
41. Marceille, J.P.; LeGall, P. Radio-sterilization of rubber stoppers for injectable preparations, American Stelmi Corporation, technical article. 1988.

4

The Role of USP in the Microbiological Assessment of Parenteral Manufacturing

Roger Dabbah and David Porter
United States Pharmacopeia, Rockville, Maryland, U.S.A.

4.1. INTRODUCTION

The role of the United States Pharmacopeia (USP) in the microbiological assessment of parenteral manufacturing is comprehensive as well as being very diversified. It is part of the microbiological continuum assessment that starts in the development of the product and continues through the stability of the product once it has reached the distributor and the patient. The microbiological assessment continuum starts with raw ingredients, excipients, drug substances, manufacturing, aseptic processing or terminal sterilization, and preservation of products during storage as well as maintenance of sterility. At every level in the continuum, USP has a monograph, several general chapters, and some information chapters that are to be used critically to ensure the microbiological quality of the final product.

The role of USP as well as the existence of USP and its relationship to the Food and Drug Administration (FDA) is not very well understood by industry and by regulators, including FDA compliance inspectors and FDA reviewers. In an attempt to put the record straight, it is beneficial to examine in some detail the USP organization, its mandate, and its objectives as well as the composition of the USP that establishes the framework for the role of USP in the microbiological assessment of parenteral products manufacturing. Following this framework we will follow the role of USP at every major step in the continuum, referencing specific monographs and chapters. We will also discuss another dimension of the pharmacopeia that is the harmonization of compendial requirements among the three major pharmacopeias, the European Pharmacopoeia, the Japanese Pharmacopoeia, and the United States Pharmacopeia.

4.2. THE USP ORGANIZATION

USP was founded in 1820 by physicians that wanted to have a compendium of best drugs in use in the United States. USP 1 was a compendium of such drugs, a total of 217 drugs, describing compounding methods for these products. It evolved during the year to become a compendium of manufactured drugs. It is ironic that in the year 2000, USP 24, to service the practitioners and benefit the patients, has returned to the development of monographs for compounding in addition to manufactured drugs.

The USP mission is to "promote the Public Health by establishing and disseminating officially recognized standards of quality and authoritative information for the use of medicines and other health care technologies by health care professionals, patients, and consumers." USP is the only non-government pharmacopeia in the world, yet its standards are enforceable by FDA under the provisions of the Food, Drug and Cosmetic Act (FD&C Act). These provisions are located under the Adulteration (501) (b) and the Misbranding (502) (g) sections of the Act.

USP is a not-for-profit organization that develops and revises public standards for drugs. These public standards include identification, purity, quality, strength, packaging, and labeling standards. In order to accomplish its mission, the USP holds a convention every 5 years. That convention, around 400 delegates, represents a very diversified group from schools of pharmacy, schools of medicine, state medical and pharmaceutical societies and associations, U.S. government representatives, and some foreign government representatives, manufacturers, distributors, trade and affiliated associations, consumer organizations, and persons representing the public interest, and national and state professional and scientific associations.

The convention considers strategic directions, elects officers, the board of trustees, and the Council of Experts Committee chairs. It also proposes and debates resolutions and issues that set the agenda for the next 5 years. About 700 volunteers serve in Expert Committees, distributed in 62 Expert Committees—31 in the development of standards and 31 in drug information. The composition of the Expert Committee is balanced: 36% from industry, 48% from academia, and 16% from government.

The Expert Committee responsible for microbiological assessment is the Analytical Microbiology Expert Committee, composed of 10 experts from industry, government, and academia. The members of the Committee represent themselves and not the organization that employs them. This is an important stipulation since it will ensure scientific judgment and decisions that are as unbiased as possible. The members of the committees are elected and every 5 years a call for candidates is made to ensure up-to-date technological expertise in new technologies. When the expertise for a given technology is not represented in the Expert Committee, Advisory Panels are nominated to help the Committee review these new areas. In addition, project teams are formed that are asked to specifically address some issues and advise the Committee. Both advisory panels and project team members are non-voting members but help the Committee in its decision-making process.

The USP has been in continuous revision since 1820. Currently we are at the USP 27 revision (2004). In addition to USP, the National Formulary (NF) has been part of USP since 1975. NF is a compendium of excipients. We are at the 22nd revision of NF. The continuous revision feature of USP-NF allows it to keep up with the technological advances that affect the analysis of drugs, including the microbiological analysis of materials and finished products.

The USP-NF revision process that includes the addition of new monographs and general chapters, as well as revision of current monographs and general chapters, is a simple and open process. It includes publication of proposals in the Pharmacopeial Forum (PF) to ensure public comments from all interested parties. Monographs and general chapters become official after all interested parties are able to comment. These comments are reviewed and considered by the Committee that accepts or reject them, giving rationales for its decisions. Additional feedback is obtained through USP Open Conferences, workshops, Regional Compendial Groups, Stakeholders Forums, Industry Forums, and International Communication Groups.

4.3. USP LEGAL RECOGNITION AND USP RELATIONSHIP TO FDA

4.3.1. USP Legal Recognition

USP's legal recognition is based on the 1848 Drug Import Act in which USP was legislatively mandated, followed in 1906 by the Federal Pure Food & Drugs Act whereby USP and NF standards were recognized, and in 1938 in the Federal Food, Drug and Cosmetic ACT, whereby the USP and NF standards were made enforceable by FDA. In 1990, the OBRA legislation recognized USP–DI (Drug Information) for off-label uses reimbursement. And in 1994 the DSHEA recognized USP-NF as the official compendium for conformity for Dietary Supplements (voluntary).

4.3.2. USP and Federal Statutes

USP and Federal Statutes vary according to the FDA centers. CDER and CVM follow the FD&C Act using the sections on Adulteration and Mislabeling; CDRH follows the FD&C Act for the definition of medical devices. In addition a number of USP standards have been accepted by CDRH for device submissions. Following the 1997 FDA Modernization Act, CFSAN includes USP under DSHEA legislation while CBER functions under the Public Health Service Act that does not mention USP. However, submission to CBER include references to USP general chapters such as Sterility Test ⟨71⟩.

USP does not have access to confidential government data submitted to FDA by pharmaceutical companies, thus it must obtain its information from manufacturers. USP does not deal with the safety or effectiveness of drugs but with the quality parameters.

A program, the Ad-Hoc reviewers program, started some years ago, involves FDA representatives that attend Expert Committee meetings and represent the FDA viewpoint. The Analytical Microbiology Expert Committee has a number of FDA ad hoc reviewers, one from each center. Members of the Expert Committees that are employed by FDA represent their own views as experts.

4.4. MICROBIOLOGICAL ASSESSMENT CONTINUUM

4.4.1. Step 1—Assessment of Raw Materials

The microbiological assessment of raw materials, excipients, drug and biological substance is governed by a number of chapters in USP. The fact that parenterals will be eventually aseptically processed or terminally processed does not mean that the materials used for the preparation of the parenteral products need not have a certain microbiological quality that would ensure that the final product will be sterile after aseptic processing or terminal sterilization. Using good microbiological quality ingredients, in addition to being a GMP requirement, makes a lot of sense since high microbiological counts are a signal that manufacture of these ingredients is not of a quality that is acceptable. Microbial limits for these ingredients are generally indicated in specified monographs. However, a number of monographs do not have microbial limits, but this does not relieve the manufacturers of ensuring that the quality ingredients are microbiologically sound. General guidelines are given under chapter ⟨1111⟩ Microbial Attributes of Pharmacopeial Articles.

4.4.1.1. Microbial Limit Tests ⟨61⟩:

This chapter provides tests for the estimation of aerobic microbial count of an ingredient and for the determination of the absence of designated microbial species that are either objectionable or that signal the presence of objectionable microorganisms. Chapter ⟨61⟩ is being modified under the Harmonization initiative and will be discussed later in this chapter. It is interesting to note that this chapter indicates that an automated method can be used provided that it is validated to give results that are equivalent (or better) than those obtained by USP testing. Discussions on how to show that a microbiological method is equivalent to the USP method will occur later in this chapter.

The crucial and important characteristic of the USP methods for microbial limits is that it is essential that a preparatory test be conducted prior to analyzing the ingredient sample. The principle of the preparatory test is simple, yet often questionable results are obtained if it is not done. The principle of the preparatory test is that you have to ensure that the product itself does not inhibit the detection or quantitative determination of the microbiological counts or the detection of the presence/absence of objectionable microorganisms. The result of a microbial limit test is valid only if the preparatory test has passed or that removal or inhibition of the inhibitory factors are validated. The preparatory test uses the inoculation of small quantity of microorganisms, namely, *Staphylococcus aureus*, *Salmonella*, *Pseudomonas aeruginosa*, and *Escherichia coli*. If you have prior knowledge that your ingredients might contain other species, it is the responsibility of the analyst to use additional microorganisms in the preparatory test. The same applies to the determination of the absence or presence of specified microorganisms. Other microorganisms that are objectionable can be present, and restricting the analysis to the USP microorganisms is not microbiologically sound.

When a requirement for Total Combined Mold and Yeast count is indicated, the same preparatory test using common yeast and mold is indicated, even if USP does not require it. Another characteristic of microbial limit tests is that you want to ensure that the media used for counts or for determination of presence or absence of specified microorganisms are appropriate and are conducive to the growth and detection of the microorganisms. In addition you also want to ensure that the media used are not contaminated. Microbial limit Tests for Nutritional and Dietary Supplements are indicated in chapter ⟨2021⟩. Again, as for the USP-NF drugs the Dietary Supplements, such as botanicals, have specialized testing depending on their composition.

4.4.1.2. Microbial Attributes of Non-Sterile Pharmaceutical Products ⟨1111⟩:

This is an information chapter that presents some principles that a manufacturer should use in determining the microbial quality of the ingredients used for parenteral products. It does also deal in general terms with nonsterile finished products, which we will not discuss in this chapter on parenteral products.

The principles indicated in this chapter are that sometimes ingredients, especially from natural sources, have inherently high microbial counts. Special treatment might be necessary to make them microbiologically acceptable for parenteral products. This is done through treatment, including sterilization with dry heat, ethylene oxide, moist heat sterilization, or irradiation. Regardless of the treatment given, unacceptable residues, depending on the treatment agent used, must be determined and eliminated. The frequency of testing for ingredients depends on the track record of that ingredient and its supplier. Skip-lot testing for microbiological assessment is not endorsed by USP; however, a body of data would be helpful to convince the regulatory agency that skip-lot testing for that particular ingredient from a specific supplier is appropriate.

4.4.2. Step 2—Microbiological Assessment Continuum

Once the raw materials, excipients, drug substances have been microbiologically assessed and deemed appropriate for use in parenteral products, meaning that their microbiological quality fulfills the requirements of the monographs, then the product has to be formulated and manufactured. Close to 80% of pharmaceutical products are aseptically processed, with the remaining being terminally sterilized. USP addresses these two modes of manufacture in a combination of general chapters and information chapter. USP-NF is not a book of "how to manufacture products," but it does discuss the principles of manufacture from a microbiological perspective.

4.4.2.1. Sterility and Sterility Assurance of Compendial Articles ⟨1211⟩:

This information chapter sets the stage for the microbiological control of parenteral products as well as other sterile products by discussing various methods of sterilization. This chapter will be updated in the near future. It reviews the concepts and principles involved in the microbiological quality control of sterile products. The most important information relates to the limitation of the Sterility Test ⟨71⟩ in assuring that a batch is sterile and puts the assurance of sterility in the context of a total quality assurance approach that includes validation of sterilization cycles as well as sterility tests. It discusses the validation of sterilization cycles.

A number of methods of sterilization are reviewed including steam sterilization, dry-heat sterilization, gas sterilization, sterilization by ionizing radiation, sterilization by filtration, and aseptic processing. The role of USP in this context is educational. It applies both for the manufacturers and the regulatory agencies and provides a leveled field platform in terms of information.

4.4.2.2. Biological Indicators for Sterilization ⟨1035⟩:

This information chapter continues the educational role of USP by reviewing in some detail the types of biological indicators, the selection of specific biological indicator depending on the mode of sterilization used. It also defines the responsibility of the manufacturers of biological indicators and the responsibility of the users of biological indicators in quality assurance and control.

4.4.2.3. Biological Indicators—Resistance Performance Tests ⟨55⟩:

This general chapter, a chapter that is enforceable by FDA, describes in considerable detail the determination of D values in a variety of sterilization modes. The importance of the determination of a D value of a biological indicator in the development and monitoring of sterilization cycles cannot be overstated. The D value determination is explained step by step, including the apparatuses used, the procedures, and the mathematical calculations. Although the chapter describes the use of the Limited Spearman-Karber Method, it does also recognize the use of the Survival Curve Method and the Stumbo-Murphy-Cochran procedures. Other performance tests such as Survival Time and Kill Time are briefly discussed, as well as the measure of the spore count for biological indicators. This chapter also references a number of technical publications, especially from AAMI, including the BIER vessels where the determination of D values are actually performed.

4.4.2.4. Biological Indicators Monographs:

The standard requirements for biological indicators are described in monographs in USP. A monograph defines a product, provides standard requirements for packaging and storage, expiration date, labeling, identification, resistance performance tests, purity, and even disposal. A biological indicator labeled USP must fulfill the requirements of the monograph. If it does not, it is mislabeled and/or adulterated according to the FD&C Act. USP has monographs for Biological Indicator For Dry-Heat Sterilization, Paper Carrier; Biological Indicator for Ethylene Oxide, Paper Carrier; Biological Indicator for Steam Sterilization, Paper Carrier; and Biological Indicator for Steam Sterilization, Self-Contained.

4.4.2.5. Microbiological Evaluation of Clean Rooms and Other Controlled Environments ⟨1116⟩:

This general information chapter reviews the various issues relative to aseptic processing and the establishment and maintenance and control of the microbiological quality of controlled environments. This is particularly important for aseptic processing because terminal sterilization is not generally done after aseptic processing. It is important that some basic principles be used in the construction, maintenance, and operation of clean rooms. Microbial evaluation programs of clean rooms or controlled environments assess the effectiveness of cleaning and sanitization practices by and of personnel that could have an impact on the microbiological quality of the product being aseptically manufactured.

The critical factors involved in the design of a microbiological environmental control program are the establishment of a sampling plan, sampling sites, and frequency of sampling. It does also include the establishment of alert and action levels. The methodologies and instruments and equipment used for sampling these environments are critical. The identification of each and every microbial isolate from the environmental microbiological control program is problematic, although under certain circumstances it might be helpful in troubleshooting. The use and utility of a media-fill to assess the adequacy of aseptic processing is discussed. The number of units filled, the frequency and types of interventions during the media fill, the temperature of incubation of the samples, and the acceptable rate of positives are all in state of flux and rather controversial. Of interest to all is the section on glossary that attempts to standardize the definition of words related to aseptic processing and microbiological control of environments. This is another role of USP: standardization of nomenclature.

4.4.3. Step 3—Microbiological Assessment Continuum

The role of USP in Step 3 is critical because it references the finished product, the product that will be administered to the patient. A parenteral product has to be sterile and nonpyrogenic, and if it is a multidose formulation it has to contain antimicrobial preservatives. Each requirement is addressed by a USP general chapter that is enforceable by FDA.

4.4.3.1. Antimicrobial Effectiveness Testing ⟨51⟩:

Chapter ⟨1⟩ Injectables require the use of antimicrobial preservatives for multidose containers. The General Notices section of USP that is applicable to all the monographs in USP defines an antimicrobial preservative as an "Added Substance." These added substances are prohibited unless "(a) they are harmless in the amounts used, (b) they do not exceed the minimum quantity required to provide their intended effect, (c) their presence does not impair the bioavailability or the therapeutic efficacy or safety of the official preparation, and (d) they do not interfere with the assays and tests prescribed for determining compliance with Pharmacopeial standards." These four characteristics are fully applicable to antimicrobial preservatives.

The microbiological test that determines the minimum amount of antimicrobial preservative is Chapter ⟨51⟩: Antimicrobial Effectiveness Testing. This chapter defines different categories of products and, based on the category, defines the criteria for effectiveness. Parenterals are in Category 1. For these products, the test is to be conducted in original containers if sufficient amount of product is available per container; if not, combine several container in a sterile container to obtain the required volume. Stock culture preparations of *Escherichia coli* (ATCC No. 8739), *Pseudomonas aeruginosa* (ATCC No. 9027), *Staphylococcus aureus* (ATCC No. 6538), *Candida albicans* (ATCC No. 10231), and *Aspergillus niger* (ATCC No. 16404) are standardized and an appropriate volume of inocula are transferred to each container to obtain a concentration of microorganisms between 10^5 and 10^6 cfu per mL in the samples. The antimicrobial effectiveness is appropriate for bacteria when there are not less than 1.0 log reduction from the initial calculated concentration per mL at 7 days of incubation, not less than 3.0 log reduction from the initial concentration per mL at 14 days of incubation, and no increase from the 14 days count at 28 days. For yeast or molds there is no increase in count from the initial calculated concentration per mL at 7, 14, and 28 days. Antimicrobial preservatives are added to multidose containers to inhibit the growth of microorganisms that may be introduced from repeatedly withdrawing individual doses.

An issue often encountered is that manufacturers prepare one batch of product that they dispense into single-dose containers as well as multidose containers. There is no requirement for single-dose containers to have antimicrobial preservatives; however, if you have a single dose container with product containing an antimicrobial preservative, the antimicrobial preservative should be effective according to the conditions indicated in chapter ⟨51⟩. Another issue is in the stability of the parenteral when at regular intervals the product is tested for stability. Repeating effectiveness testing at every stability testing period is not appropriate. However the concentration of preservative in the parenteral product can be tested by chemical means, without having to repeat an antimicrobial preservative effectiveness test. Once the product has been formulated to contain a preservative at the minimum amount determined, then there is no need to test each batch of product for antimicrobial effectiveness as a release test. As for stability samples, the concentration of antimicrobial preservative is monitored.

4.4.3.2. Sterility Tests ⟨71⟩:

A USP Sterility Test is "applicable for determining whether a Pharmacopeial article purporting to be sterile complies with the requirements set forth in the individual monograph with respect to the test for sterility." Failure of an article to meet the sterility requirement occurs when microbiological growth is evident following the procedure indicated in this chapter.

The pharmaceutical literature has a number of articles showing the limitations of the test to ensure that a batch is sterile. The USP sterility test is not intended to show if a batch is sterile or not, it is used to show compliance with the specific requirement of sterility in a monograph. Limitations that are generally related in the literature to the small sample size, the inadequacy of the media used to detect all potential contaminants or surviving micro-organisms, and the preordained problem of potential contamination by analysts. These limitations are well taken but are not relevant for the purpose to which the USP sterility test is intended. USP tests and assays are not release tests but are compendial tests that ensure compliance to the monograph. There are far better methods to determine that a batch is sterile, including validation of physical parameters of sterilization cycles and monitoring of batches using biological indicators, and for aseptic processing, media-fills. One can answer the limitations of the tests indicated above by using a statistical sampling, but increasing the number and type of media used, and by removing the analysts from the contamination equation by the use of isolators. But, as indicated above, a compendial test is only designed to ensure compliance with the sterility requirement of the monograph, not to declare that a batch is sterile or not sterile.

Since the sterility test is a compendial test it needs to be standardized, and this is provided in this general chapter. Starting in USP 27 (2004), this test will be harmonized with the sterility tests of the European Pharmacopoeia and the Japanese Pharmacopoeia.

Standardization of the sterility test requires that media used be standardized, that the sample size be standardized, that the temperature and time of incubation be standardized, that the growth promotion of the media used be standardized, that the test for bacteriostasis and fungistasis of samples to be tested is established, and that validation of the procedure be standardized. All USP tests are validated but manufacturers have to qualify the test for their own products using the procedures and principles indicated in this chapter.

Critical parameters include sample preparation that varies with the type and nature of the product to be tested. Sample preparations to be used with membrane filtration are shown for Liquid Miscible with Aqueous Vehicles;

Liquid immiscible with Aqueous Vehicles; Ointments and Oils Soluble in Isopropyl Myristate; Prefilled Syringes; solids for Injection other than Antibiotics; Antibiotic Solids for Injection including Pharmacy Bulk Packages; Antibiotics Solids, Bulk and Blend; Sterile Aerosols Products, and Devices with Pathways Labeled Sterile. For Direct Transfer Method, sample preparations are indicated for Non-filetrable Liquids; for Ointments, Oils, and non-filterable Liquid Insoluble in Isopropyl Myristate; for Purified Cotton, Gauze, Surgical Dressings, Sutures, and Related Articles; and for Sterile Devices.

Two methods are indicated, the Membrane Filtration Method and the Direct Transfer Method. Unless indicated in a monograph, or when the nature of the product is such that it cannot be tested using the Membrane Filtration Method, this method is the preferred method. Facilities where sterility tests are done include clean rooms, clean zones (see ⟨1116⟩ for additional detail) or isolators. Isolators will be discussed further later in this chapter. Another issue is the retest of samples. There is no retest allowed unless one can demonstrate conclusively that growth in a sterility test is due to factors independent from the batch, such as documented contamination at the laboratory level, nonsterile media, or other critical parameters.

4.4.3.3. Sterility Testing- Validation of Isolator Systems ⟨1208⟩:

This information chapter provides guidelines for the validation of isolator systems for use in sterility testing. Since the analyst is shown to be a major source of external contamination to sterility testing of products, isolating the analyst from the samples under test has been a common occurrence in the microbiology laboratories. This chapter describes the principles of validation used, from construction to Installation Qualification, followed by Operational Qualification, and Performance Qualification. A critical issue is the maintenance of asepsis within the isolator environment, and the transport of materials and samples from the outside to the inside of the isolator.

4.4.3.4. Bacterial Endotoxins Test ⟨85⟩:

This general chapter provides a test to detect or quantify bacterial endotoxins that may be present in parenteral products. There are a number of techniques that can be used, but they all use Limulus Amebocyte Lyzate (LAL) reagents. These reagents have been formulated for use in Gel-clot procedure as well as for Turbidimetric or Colorimetric (photometric) tests. This test has been fully harmonized with the European Pharmacopoeia and the Japanese Pharmacopeia. There are over 650 monographs in USP that have a bacterial endotoxins requirement. A reference standard, the USP Endotoxin RS, is needed to complete the test and is available in the USP RS catalog. This RS is harmonized with the EP and The International (WHO) standard for endo-toxin; the Japanese RS is indexed to the International standard.

4.4.3.5. Pyrogen Test ⟨151⟩:

The Pyrogen Test is used for products that cannot be tested by the Bacterial Endotoxins Tests ⟨85⟩ due to the nature of the sample or that may have a specific regulatory requirement (such as vaccines). It involves the measure-ment of the rise in temperature of rabbits following standardized intravenous injection of the test solution in a dose not to exceed 10 mL per kg. The Pyrogen test is also used in conjunction with the BET during development of new product. There are some pyrogenic reactions that are due to constituents other than bacterial endotoxins. It is the responsibility of the manufacturer to determine if the BET is appropriate or not for its products.

4.4.4. Step 4—Microbiological Assessment Continuum

Issues of validation of microbiological methods in USP have been raised numerous times in this chapter. Chapter ⟨1225⟩, "Validation of Compendial Methods," does not apply to microbiological methods. USP has developed chapters on validations that give guidelines to the microbiologists as well as to manufacturers of instruments, so-called Rapid Methods for Microbiological Analysis.

4.4.4.1. Definition of Validation as Applied to Microbiological Methods:

With apologies to FDA, ICH, and USP various guidances & guidelines and draft guidances & guidelines for analytical methodology validation, we have adapted them to microbiological methods.

    a. Microbiological Methods Validation is the process of demonstrating that microbiological procedures are suitable for their intended use.

    b. Microbiological method validation includes all the procedures recommended to demonstrate that a particular method for the quantitative measurement of a count in a given biological matrix is reliable and reproducible.

    c. Establishing documented evidence that the microbiological method will consistently evaluate the microbiological quality of the product.

    d. The main objective of validation of a microbiological procedure is to demonstrate that the procedure is suitable for its intended purpose.

    e. Validation of a microbiological method is the process by which it is established, by laboratory studies, that the performance characteristics of the method meets the requirements for the intended analytical applications.

A summary of the validation requirements indicates that validation of a microbiological method should include the following characteristics:

Suitable for intended use
Reliable
Reproducible
Documented evidence
Consistency
Meets the requirements for the intended analytical applications.

4.4.4.2. Validation of Microbial Recovery from Pharmacopeial Articles ⟨1227⟩:

This information chapter provides guidelines for the validation of micro-biological methods for the estimation of the number of viable microorganisms, for the detection of indicators or objectionable microorganisms, for counts done in the antimicrobial effectiveness test, and for the sterility testing of articles.

For products that have intrinsic inhibitory properties, or to which antimicrobial preservatives are added, the recovery procedure must inactivate, neutralize, or remove the inhibitory factors. This is why this chapter describes some common neutralizers for different types of antimicrobial preservatives. Recovery comparisons among control group, inoculated group, and inoculated group with product under test will establish if the product is inhibitory to the recovery of the inoculum. At least three independent replicate experiments are performed, and each should demonstrate that the average number of microorganisms recovered from the challenged product and that of the inoculated control should not be less than 70% of each other. As the CFU count on a plate decreases, say from 30 to 7 per plate, the standard error decreases and the error as a percentage of the mean increases.

4.4.4.3. Validation of Alternative Microbiological Method ⟨1223⟩:

This information chapter is not yet official and was published as a proposal in the Jan–Feb. 2002 issue of *Pharmacopeial Forum*. The purpose is to provide guidance for validating methods for use as alternatives to the official compendial microbiological tests. It stems from the General Notices that indicate that for compliance purposes a manufacturer can use in-house tests provided that they are equivalent to the USP referee tests. Alternative analytical procedures can be validated using the guidelines indicated in

Validation of Compendial Methods ⟨1225⟩. Chapter ⟨1225⟩ defines characteristics such as accuracy, precision, specificity, detection limits, linearity, range, ruggedness, and robustness. However, this chapter cannot be directly applied to microbiological methods or procedures because one has to show at least equivalency of the alternative method to the USP method.

It is necessary to take into account large degrees of variability within each microbiological method before attempting to show that two methods are equivalent. Variability in microbiological methods can be due to sampling errors, dilution errors, plating errors, and operator errors. The allure of a new method must be demonstrated in a comparison study to the compendial method. A critical issue is whether or not an alternative microbiological procedure will yield data equal to, or surpassing in quality the data generated by the compendial method.

The types of microbiological assay will determine the approach that one will use in the validation of microbiological method alternatives. These methods are qualitative, quantitative, or reference identification. The role of USP is to facilitate the use by manufacturers of rapid microbiological methods as well as alternative methods. This chapter provides guidelines for determining equivalency of qualitative microbiological tests using the accuracy, precision, specificity, ruggedness, and robustness characteristics. For quantitative microbiological tests one can use the accuracy, precision, specificity, limit of quantification, linearity and range, ruggedness, and robustness characteristics to test the equivalence of alternative methods to compendial methods. For microbial identification tests, it is sufficient to consider accuracy, precision, ruggedness, and robustness characteristics to show equivalency.

4.4.5. Step 5—Microbiological Assessment Continuum

The Analytical Microbiology Expert Committee activities include the development and proposal of a number of new information chapters designed to provide to industry and regulators extensive guidelines on a number of subjects.

4.4.5.1. Terminally Sterilized Pharmaceutical Products-Parametric Release ⟨1222⟩:

This information chapter is in the In-process section of Pharmacopeial Forum 29(1) [Jan.–Feb. 2003]. This version of the chapter incorporates a number of changes from the Pharmacopeial Preview version published in PF 23(6) [Nov.–Dec. 1997]. References to sterility assurance levels (SAL) are given using positive exponents to clarify the SAL concept. A new section discusses three general categories of terminal sterilization (bioburden-based processes, biological indicator/bioburden combined processes, and overkill processes). The discussion pertaining to critical aspects of biological indicators used for sterilization validation, as well as the discussion regarding the Sterilization Microbiology Control Program has been expanded and clarified. The section, Physicochemical Indicators and Integrators, that was a portion of the earlier version of this chapter has been removed and placed into the proposed chapter ⟨1209⟩ Sterilization-Chemical and Physicochemical Indicators and Integrators [also published in PF 29(1)].

The chapter begins with a discussion of general issues related to parametric release, regardless of the specific mode of sterilization, and then discusses some specific modes of sterilization. Chapter ⟨1222⟩ in part addresses the limitations expressed in chapter ⟨71⟩ Sterility Tests pertaning to the usage of the test ("These Pharmacopeial procedures are not by themselves designed to ensure that a batch of product is sterile or has been sterilized. This is accomplished primarily by validation of the sterilization process or of the aseptic processing procedures"). Chapter ⟨1222⟩ indicates that once a sterilization process is fully validated and operates consistently, "... a combination of physical sterilization data such as accumulated lethality or dosimetry in combination with other methods such as biological indicators or physicochemical integrators, can provide more accurate information than the sterility test regarding the release of terminally sterilized product to the marketplace."

4.4.5.2. Disinfectants and Antiseptics ⟨1072⟩:

This information chapter was first published in Pharmacopeial Forum 28(1) [Jan.–Feb. 2002] and subsequently as an In-Process Revision in PF 29(3) [May–June 2003]. Its purpose is to provide guidance pertaining to disinfectants and antiseptics. It includes sections on definitions, the role of disinfectants in aseptic processing, types of disinfectants, disinfectant practices in the pharmaceutical industry, disinfectant effectiveness validation, environmental monitoring, and operator training.

The USP Analytical Microbiology Expert Committee is planning to prepare and publish chapters on the following subjects (⟨xxx⟩ indicates a mandatory chapter, ⟨xxxx⟩ an informational chapter):

Process Simulation (⟨1224⟩)
Mycoplasma Determination (⟨xxx⟩)
TSE/BSE (⟨xxxx⟩, with input from a Project Team)
Isolators for Aseptic Processing Applications (⟨xxxx⟩)
Process Simulation testing for Aseptically Manufactured Active Pharmaceutical Ingredients and Dosage Forms (⟨xxxx⟩)

Ethylene Oxide Gas Sterilization (expansion of section in current chapter ⟨1211⟩, ⟨xxxx⟩)

Dry heat Sterilization and Depyrogenation (expansion of section in current chapter ⟨1211⟩, ⟨xxxx⟩)

Vapor Hydrogen Peroxide Decontamination & Sterilization (expansion of section in current chapter ⟨1211⟩, ⟨xxxx⟩)

Moist Heat Sterilization (expansion of section in current chapter ⟨1211⟩, ⟨xxxx⟩)

4.4.6. Step 6—Microbiological Assessment Continuum

The five steps of the Microbiological Assessment Continuum indicated above cover the USP role in the assurance of the microbiological quality of parenteral products. The scope of the USP involvement in microbiological assessment is very wide and ranges from raw materials, excipients, drug substances, dosage forms, manufacturing microbiological control via clean rooms or other controlled environments, to aseptic processing, terminal sterilization, and preservation using antimicrobial preservatives. In addition to providing referee methodologies for microbiological assessment, it fulfills its educational role through the development of information chapters that can be kept current because USP is in continuous revision.

The participation of industry and regulatory agencies in the development of microbiological standards and methodologies amplifies the role of USP as a neutral scientific body with the objective of assuring that patients get the best quality product available in a consistent manner. USP cannot cover every detail of the microbiological methods or cover all eventualities; it is up to the users of USP to use sound microbiological principles, in addition to the sound microbiological principles indicated in USP.

5

Sterility and Bioindicators

Patrick J. McCormick,
Catherine J. Finocchario,
and James J. Kaiser

Bausch & Lomb, Inc., Rochester, New York, U.S.A.

5.1. INTRODUCTION

The first use of biological indicators in industrial sterilization is attributed to Kilmer, who inoculated spores of *Bacillus anthracis* in the center of packages of gauze dressings to verify the efficacy of sterilization (1). Today biological indicators are available in a variety of different formats ranging from spore suspensions employed to inoculate product according to the method pioneered by Kilmer to rapid readout biological indicators that provide an indication of the efficacy of a sterilization process within hours of processing (2). Despite the progress that has been made in the field of biological indicator technology, biological indicators are often a source of frustration to those who are not familiar with their characteristics and idiosyncrasies. As noted by Spicher (3):

> We use monitors to obtain objective information on the efficacy of microbiocidal procedures. The judgement of the microorganisms is

infallible. In this case the responsibility for any errors rests on the maker of the biological indicators, the employer of the biological indicators, the tester for surviving organisms, namely man, alone! There are numerous opportunities for making errors.

The biological indicator is the only measurement system that is capable of directly integrating the lethality of a sterilization process with regard to its ability to inactivate viable microorganisms, making it an essential tool for the development and validaton of sterilization processes. As noted by Pflug and Odlaug, however, biological indicators themselves cannot be used as primary measurement standards but must instead be calibrated against other accepted measurement standards (4). As no one sterilization indicator by itself can demonstrate that a given product or load is sterile, biological indicators should always be employed in combination with physical and/or chemical measurements to demonstrate the efficacy of a given sterilization process (5). The effective use of biological indicators requires an understanding of their role in the development, validation, and routine monitoring of sterilization processes as well as knowledge of those factors that impact their performance.

5.2. STERILITY ASSURANCE

Fundamental to the application of biologial indicators in the development and validation of sterilization processes is the concept of log-linear or semi-logarithmic inactivation kinetics. Although sterile is defined in absolute terms as the absence of viable microorganisms, sterility itself is a function of probability. When the logarithm of the number of surviving microorganisms versus exposure time or dose is plotted, a linear relationship is evident (Fig. 1). This relationship is fundamental to all sterilization processes, as it is the basis for determining the exposure conditions necessary to achieve the desired level of sterility assurance based on the initial population of microorganisms and their resistance to sterilization. A Sterility Assurance Level (SAL) of 10^{-6}, or less than one chance in a million that viable microorganisms will be present in the sterilized article or dosage form, is typically regarded as the minimum acceptable SAL for terminally sterilized injectable articles or critical devices purporting to be sterile (6). For noncritical devices that do not contact compromised tissue, a SAL of 10^{-3} may be acceptable (7). Biological indicators provide a convenient means to directly demonstrate attainment of a minimum SAL. The use of biological indicators during routine sterilization processing also serves to alert the user to conditions that may compromise the ability of the process to attain the desired SAL.

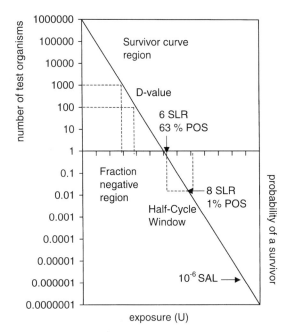

FIGURE 1 Survivor curve illustrating semi-logarithmic nature of microbial inactivation and attainment of 10^{-6} SAL (sterility assurance level). SLR = Spore Log Reduction. POS = Positive.

5.3. RESISTANCE PERFORMANCE

The potency of a biological indicator is largely a function of its population of test organisms and their inherent resistance to sterilization. Although the construction of the biological indicator may affect its overall response, these factors are generally secondary to the number of test organisms present on the biological indicators and their resistance to sterilization. Other factors that may affect the potency of biological indicator include storage, handling, processing, and recovery conditions.

5.3.1. Population

The initial population of test organisms on a biological indicator may vary depending on whether the biological indicator is being employed for cycle development and validation studies or as a routine monitor of the sterilization process. Populations ranging from 10^4 to 10^6 test organisms per biological indicator are typical, although other population levels are available. The USP and other guidelines describe methods for assessing the populaton of test organisms on biological indicators and information on the acceptable range of recovery (8). These methods typically include blending or pulping of the inoculated carrier to yield a suspension of test organisms. Other means, including sonication or vortexing with sterile glass beads, may be necessary to ensure the optimum recovery of test organisms. The resulting suspension of test organisms is then diluted and plated in microbiological media. An aliquot of the test organism suspension is also placed in a heated water bath and heat-shocked as appropriate both to activate any dormant test organisms and to reduce the presence of any contaminating microorganisms. Care should be taken to ensure that the heating characteristics of the test organism suspension when placed in the water bath are consistent with those described in the pharmacopeia. The manufacturer of the biological indicator should be consulted with regard to the need for any specialized recovery methods or conditions for test organism enumeration.

5.3.2. D-Value

The resistance of a biological indicator to sterilization may be described by the decimal reduction time, or D-value. The D-value may be defined as the exposure time or dose required to achieve a one logarithm (90%) reduction of the initial population of microorganisms under the stated exposure conditions and is typically referenced to those conditions (D_{121C}, $D_{600mg/L\ EO}$, D_{kGy}, etc.). As can be seen from Fig. 1 and Eq. (1) (see Appendix), the D-value may also be defined as the negative reciprocal of the slope of the survivor curve. Either the direct enumeration method or the fraction negative method may be used to determine the D-value of biological indicators. The direct enumeration method requires that the surviving test organisms be removed from their carrier following sterilization processing, diluted as appropriate, and plated onto microbiological growth medium in a manner similar to that used to determine the initial population. The plates are then incubated under the appropriate conditions, and the number of colonies enumerated by standard techniques. This method is generally acceptable for enumerating survivors in excess of 5×10^1. The direct enumeration method requires that a plot of the number of surviving microorganisms versus exposure conditions (time or dose) exhibit a linear response over a range of at least three logarithms with a correlation coefficient of 0.8 or greater (5). When the number of survivors is anticipated to be $<5 \times 10^0$ fraction-negative methods may be employed. Under these conditions the intact carrier is incubated in the presence of liquid growth medium following sterilization processing and the number of samples negative for growth determined. For the data to be acceptable there must be a progressive increase in the number of samples negative for growth with increasing exposure or dose (i.e., no "skips"). The D-value may then be determined based on an estimation of the number of surviving test organisms using Most Probable Number (MPN) analysis (Stumbo-Murphy-Cochran method—equation 2) (see Appendix) or based on an estimation of the Mean Time Until Sterility (Spearman-Karber method—equation 3) (see Appendix) (5). It must be emphasized that analysis of the D-values of biological indicators using either direct enumeration or fraction-negative techniques must be performed under "square wave" exposure conditions as provided by Biological Indicator Evaluator Resistometer (BIER) vessels (9,10). Performing D-value analysis in a standard laboratory or production sterilizers should be discouraged as these vessels are generally not capable of maintaining exposure conditions within the necessary tolerances nor are they capable of achieving the desired ramp rates (11,12).

5.3.3. Survival/Kill Time

The concept of variable survival/kill times for biological indicators based on the D-value and population of the indicator was first introduced in USP 21. Earlier editions of the pharmacopeia included either fixed survival/kill times (USP XIV–XIX) or recommended D-value ranges (USP XX). The labeling of some commercially available biological indicators may still reference earlier fixed survival/kill times, although this is becoming less common as most biological indicator manufacturers have revised their labeling to reflect the current approach where:

Survival Time = not less than (labeled D-value)

$\times$ (log labeled test organism count per carrier $-$ 2)

Kill Time = not greater than (labeled D-value)

$\times$ (log labeled test organism count per carrier $+$ 4)

To verify labeled survival/kill times, biological indicator samples are processed in a BIER vessel in a manner similar to that employed for determination of the D-value and the number of samples positive for growth at each parameter are determined. Although it has been suggested that survival/kill testing may be performed in a laboratory sterilizer, the use of a BIER vessel is strongly recommended. The minimum number of biological indicators tested at each parameter and provisions for retest varies with the specific biologic indicator performance standard. USP procedures require a minimum of 20 samples tested at each parameter with a provision for additional testing should one sample fail to meet the required criteria at either parameter (8). Biological indicator performance standards based on ISO methodology require the testing of 50 samples at each parameter with no provision for additional testing should one sample fail to meet the acceptance criteria at either parameter (5).

5.4. SELECTION

5.4.1. Test Organism

The selection of a biological indicator appropriate for use with a particular sterilization process requires the consideration of a number of factors. First is identification of the appropriate test organism. The test organisms indicated in Table 1 are generally recognized to exhibit greater resistance to the indicated sterilization processes than typical bioburden. Exceptions include the use of *Bacillus pumilus* as a test organism with radiation sterilization, as some bioburden organisms have been found to exhibit greater resistance to radiation (13). For this reason, validation methods for radiation sterilization processes typically employ a bioburden approach (14), with the use of biological indicators being indicated only in certain circumstances (15). Other exceptions include nonconventional transmissible agents (NCTAs), as the resistance of these agents to sterilization is believed to be greater than that of the test organisms typically employed with most sterilization processes (16). When developing or validating novel sterilization processes, extensive screening may be necessary to select an appropriate test organism in the absence of supporting documentation or literature review (17).

5.4.2. Format

Once the appropriate test organism and population have been determined, one must then select a biological indicator format suitable to the application in question. Biological indicators are available in a variety of different configurations including suspensions of test organisms for direct inoculation onto products, inoculated carriers, inoculated carriers packaged in glassine or Tyvek envelopes, self-contained biological indicator systems that incorporate both the test organisms and growth medium in the same unit, and others (Fig. 2). The type of cycle development studies and validation work being performed will in many instances determine the biological indicator configuration to be employed. In some situations it may be necessary to prepare a specialized or in-house biological indicator for use with a particular application. Caution should be exercised when preparing in-house biological indicators for, as noted by Spicher, "there are numerous opportunities for making errors (3)."

TABLE 1 Biological Indicator Systems

Sterilization mode	Test organism	Conditions	D-value (minutes)	Performance Standards
Moist heat	Geobacillus stearothermophilus[a]	121°C	1.5–3.0	Yes[b-f]
	Clostridium sporogenes[g]	103°C	43	None
	Bacillus coagulans[h]	115°C	6–7	
	Bacillus atrophaes[g,i]	121°C	0.5	
Ethylene oxide	Bacillus atrophaes[i]	600 ± 30 mg/L EO	2.5–5.8	Yes[b-f]
		54 ± 1°C		
		60 ± 10% RH		
Dry heat	Bacillus atrophaes[i]	160°C	1.0–3.0	Yes[b-f]
Irradiation	Bacillus pumilus	e-beam or gamma	≥1.9 kGy	Yes[f]
Steam-formaldehyde	Geobacillus stearothermophilus[a]	10 ± 2 mg/L HCHO	≥3	Yes[f]
Gas plasma[j]	Geobacillus stearothermophilus[a]	Contact vendor	Contact vendor	None
Hydrogen peroxide vapor	Geobacillus stearothermophilus[a,b]	30°C, 200 ppm[k]	1.2	None

[a] Formerly Bacillus stearothermophilus.

[b] USP 26. Official monographs.

[c] ANSI/AAMI ST59 (General requirements), ST21-ethylene oxide; ST19-moist heat.

[d] ISO 11138-1 (General requirements), -2 (ethylene oxide), -3 (moist heat).

[e] European Pharmacopeia.

[f] EN866 series.

[g] AD Russell. Destruction of Bacterial Spores by Thermal Methods. In: AD Russel, WB Hugo, GAJ Ayliffe, eds. Disinfection, Preservation and Sterilization. 3rd Ed. London: Blackwell Science, 1999, pp. 640–656.

[h] AT Jones, IJ Pflug. Bacillus coagulans FRR B66, as a Potential Biological Indicator Organism J Parenteral Sci Technol 35(3):82–87, 1981.

[i] Formerly Bacillus subtilis var. niger.

[j] PT Jacobs, SM Lin. Sterilization Processes Utilizing Low-Temperature Plasma. In: SS Block, ed. Disinfection, Sterilization, and Preservation. 5th Ed. Philadelphia: Lippincott, Williams & Wilkins, 2001, pp. 695–728.

[k] I Taizo, A Sinchi, K Kawamura. Application of a newly Developed Hydrogen Peroxide Vapor Sensor to HPV Sterilizer. PDA J Pharmaceutical Science & Technology 52(1):13–18, 1998.

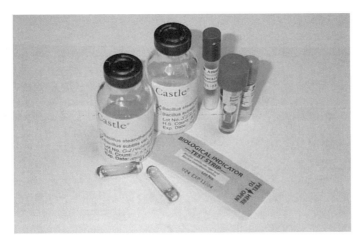

FIGURE 2 An example of different biological indicator formats, including spore suspensions, inoculated carriers, paper strip biological indicators, and self-contained biological indicators.

5.4.2.1. Suspensions and Inoculated Carriers

Although simple in concept, the use of a suspension of test organisms for the development and validation of sterilization processes can also be one of the most problematic. Suspensions of test organisms are typically employed to prepare inoculated product or inoculated simulated product for closure validations and similar studies. Aliquots of the test organism are either inoculated directly onto product or onto suitable carriers that are then placed in those locations of the product considered to be the most difficult to sterilize. Following sterilization processing the number of surviving test organisms is determined by either direct transfer to growth medium (fraction-negative analysis) or removal of the test organisms from the product or carrier for direct enumeration. Note that the labeled D-value of the test organism suspension may not accurately reflect the actual D-value when the test organisms are inoculated directly onto the product. Biological indicator manufacturers typically determine the D-value of a suspension of test organisms by inoculating the test organisms onto paper strips or other carriers. The substrate onto which the test organisms are inoculated will affect their resistance to sterilization and the effective D-value when inoculated onto product or different carriers and packaging may be higher or lower than the labeled D-value provided by the manufacturer with the test organism suspension (18). Other complications include the potential for uneven application of the test organisms onto the product surface, leading to clumping of the test organisms that may result in a nonuniform response or excess resistance to sterilization processing conditions. Although this effect may be mitigated to some extent by the use of appropriate suspending media (aqueous/ethanol mixtures or aqueous mixtures with surfactants) and proper drying techniques, it is difficult to consistently eliminate clumping of the test organisms when using direct inoculation methods (19). Likewise, the use of suspending media with high salt contents should be avoided, as encapsulation of the test organisms within salt crystals may lead to extraordinary levels of resistance (20). Where possible, the use of inoculated carriers in the form of paper strips or other carriers is preferred for applications such as closure validations where clumping of the test organisms with the direct inoculation method could lead to extreme resistance to sterilization and overprocessing of the product. When performing closure validations in this manner, care must be taken that the inoculated carrier employed does not create a path for sterilant entry to the closure site that would not otherwise be present on the product (21).

5.4.2.2. Paper Strip Biological Indicators

One of the best-characterized forms of biological indicators is the paper strip biological indicator. This type of biological indicator, consisting of a paper strip carrier inoculated with a suspension of test organisms and packaged in a glassine or Tyvek outer envelope has seen little change since its commercialization during the 1960s. Despite this, paper strip biological indicators are still widely employed in a variety of industrial and health care applications. The outer envelope serves to protect the inoculated carrier from external contamination, while allowing sterilant access to the test organisms. The small size and mass of the paper strip biological indicator provide for ease of placement within the product or load with minimal perturbation to the local sterilization environment. Aseptic technique is required when removing the inoculated carrier from the outer envelope and transferring it to growth medium, however, which may lead to false-positive cultures if appropriate caution is not exercised during the transfer process.

5.4.2.3. Self-Contained Biological Indicators

Self-contained biological indicators are widely used in many applications, as they do not require the user to aseptically transfer the inoculated carrier to growth medium. Self-contained biological indicators incorporate both the test organisms and the growth medium within the same unit and are typically of two distinct types. The simplest form of self-contained biological indicators consist of a hermetically sealed glass ampule or vial containing spores of *Geobacillus stearothermophilus* suspended in growth medium with a pH indicator dye. Following sterilization processing the ampule is incubated and growth of the test organisms detected as a change in the color of the growth medium. Although this form of self-contained biological indicator is simple to use, it is sensitive to temperature only and is best suited to the validation of the moist heat sterilization of solutions.

Perhaps the most widely used form of a self-contained biological indicator consists of an inoculated carrier packaged with a frangible [i.e., easily broken] ampule of growth medium within an outer vial with a filter or tortuous path providing access of the sterilant to the test organism while precluding adventitious contamination. Following sterilization processing the inner ampule containing growth medium is ruptured in a controlled manner to bring the growth medium into contact with the test organisms. In some circumstances the response of this type of self-contained biological indicator design to moist heat sterilization may lead to atypical results due to their greater thermal mass as compared to paper strip biological indicators (22). Although convenient to use, self-contained biological indicators may be difficult to place within the product or load. Many self-contained biological indicators contain cautions or warnings regarding processing with moist heat sterilization due to the potential for the ampule of growth medium contained within to burst prematurely and cause personal injury if adequate cooling is not allowed prior to activation of the ampule.

5.5. PROCESS CHALLENGE DEVICES

In addition to the biological indicators formats described above, biological indicators may also be employed within the context of process challenge devices. A process challenge device is a device into which a biological indicator or inoculated carrier is placed in order to provide a "worst case" challenge to the sterilization process. A number of such devices (also known as "test packs") are available commercially from various sources (Fig. 3). Process challenge devices are frequently employed in situations where it may be inconvenient or impossible to place a biological indicator in the desired location within the product or load (5). Under these circumstances the ability of the Process Challenge Device to present a suitable challenge to the sterilization process as compared to the product must be documented. This may be performed by conducting fractional sterilization cycles with biological indicators placed within the Process Challenge Device and within product and comparing their recovery.

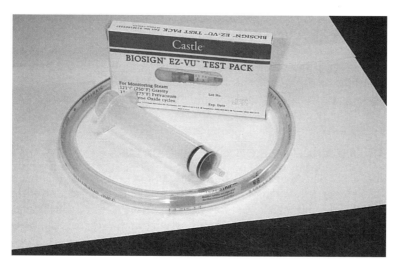

FIGURE 3   An example of different process challenge devices (PCDs).

5.6. REGULATORY STATUS

Over 30 different performance standards for biological indicators have been promulgated by various agencies, including USP (8), ANSI/AAMI (23), CEN (24), and ISO (25). This has led to some confusion regarding application of these standards, although there is a general trend toward harmonization. As there may be significant differences between biological indicator standards with regard to test procedures and acceptance criteria, the reader is advised to review the relevant biological indicator standard(s) when selecting a bio-logical indicator for use with the sterilization of products distributed in a particular market. Consideration should also be given to the regulatory classification of biological indicators. Within the United States, biological indicators employed for use in healthcare applications are considered to be accessories to medical devices and are therefore regulated as class II medical devices and subject to the appropriate controls and enforcement activity (26). Biological indicators employed in industrial applications are treated not as accessories to medical devices but as components of the manufacturing process. This may lead to significant differences with regard to the labeling of the biological indicator and FDA enforcement activities. An additional consideration is that the labeling of biological indicators employed in health-care applications typically reflects that cleared by the FDA at the time of submission of the biological indicator for premarket 510([k]) clearance and may not reflect current biological indicator performance standards. Depending on the nature of the process with which the biological indicator is employed, this may not be a critical distinction, but the reader should be aware that some biological indicator products may not comply with the requirements of current biological indicator standards. The labeling provided with the biological indicator should be carefully reviewed in this regard.

5.7. QUALIFICATION OF BIOLOGICAL INDICATORS

In addition to the selection of the appropriate test organism and biological indicator format, a system must be in place to both qualify the vendor of the biological indicator and each lot of biological indicators received. The former is typically addressed by means of a quality systems audit, whereas the latter requires actual testing. Guidance pertaining to quality systems audits of biological indicator manufacturers can be found in ANSI/AAMI/ISO 14161 (5). During audits of the biological indicator manufacturer, particular attention should be given to the test equipment including the BIER vessel employed to qualify biological indicators. There may be considerable variation between vendors with regard to the construction and extent of validation of the BIER vessels employed to qualify biological indicators. The biological indicator manufacturer should provide documentation of conformance to current BIER vessel standards along with documentation of the routine calibration and validation of the unit (10). Other areas to be examined include the methods employed to propagate and purify the test organisms, traceability and control of the test organisms, assay procedures, raw material qualification procedures, stability studies, etc. Whenever possible, it is prudent to qualify more than one vendor for each type of biological indicator employed.

Once suitable vendors have been identified, an in-house biological indicator qualification program must be established. This program must encompass, as a minimum, determination of the population of each lot of biological indicators received and the purity of the biological indicator test organisms to at least the genus level. Although the determination of the D-value for each lot of biological indicators is not required, it should be verified on a periodic basis either internally or at an appropriately qualified external laboratory. Internal acceptance criteria should be established for each type of biological indicator employed and conformance of each lot of biological indicators received to the certificate of analysis provided with the biological indicator verified. Should the user elect to propagate test organisms and prepare their own biological indicators in-house they will also be required to verify the D-value of each lot produced in addition to its population and purity. Records will also need to be maintained regarding the traceability of the test organisms, the media and reagents employed, culturing conditions, and so forth as for a commercial manufacturer (6).

When qualifying biological indicators for use in the development, validation, or routine monitoring of sterilization processes, it is important to recognize that, just as biological indicators are sensitive indicators of the efficacy of a sterilization process, they are also sensitive to the processing and recovery conditions employed. Therefore, caution should be exercised in the storage, handling, processing, and recovery of biological indicators during qualification testing to adhere to the recommendations of the manufacturer of the biological indicator. This is particularly important with regard to use of the proper growth medium and recovery conditions because these may significantly impact the overall performance of the biological indicator (27).

5.8. APPLICATION

5.8.1. Cycle Development

The application of biological indicators to the development and validation of sterilization processes generally falls into one of two categories, the "overkill" sterilization method or the combined biological indicator/bioburden method. The overkill method requires that the resistance of the biological indicator to the sterilization process exceed that of the bioburden present on the product by a significant margin (at least two- to threefold). The overkill sterilization process is designed to achieve a minimum 12-logarithm reduction of the biological indicator, corresponding to a 10^{-6} SAL of the biological indicator as opposed to product. This is typically demonstrated by placing biological indicators inoculated with a minimum population of 10^6 test organisms in the most difficult to sterilize location within the product and distributing them throughout the sterilization load. The biological indicator challenge system employed may consist of biological indicators or inoculated carriers placed within the product, test organism suspensions inoculated directly onto the product, or previously qualified process challenge devices. The load is then processed for one-half the sterilization exposure time anticipated under routine production conditions and a six logarithm reduction in the population of biological indicator test organisms demonstrated. Biological indicators with initial populations less than 10^6 test organisms per carrier may also be employed provided that the multiple of the logarithm of the population and D-value is equivalent to or greater than the multiple of logarithm 10^6 and the minimum D-value requirement for the biological indicator (5). Intuitively, one would anticipate that demonstrating a six-logarithm reduction in the population of biological indicator test organisms at one-half the normal sterilization exposure time would result in total kill of the test organism population. In theory, however, it is possible to recover as many as 63% biological indicators positive for growth consistent with a six-logarithm reduction based on MPN analysis (Fig. 1). This "Half-Cycle Window" concept is discussed in various publications and guidance documents (5,28,29). It should be emphasized that this concept is applicable only to those sterilization processes with a documented log-linear relationship between exposure time or dose and inactivation of the biological indicator test organism such as moist heat, dry heat, or ethylene oxide sterilization. Application of the "Half-Cycle Window" concept with novel sterilization processes may require extensive testing to establish the range of sterilization conditions consistent with log-linear inactivation kinetics.

Overkill processes intended to achieve a 10^{-6} SAL of the biological indicator may result in a very high product SAL as the bioburden population and resistance on many products is often significantly less than that of the biological indicator. In situations where such extended sterilization processing may have a deleterious effect on the product, the use of a combined biological indicator/bioburden method may be appropriate. The combined biological indicator/bioburden method requires that both the population and resistance of the product bioburden be determined, but provides for reduced processing conditions as compared with the overkill method. This method assumes that the resistance of the bioburden may be equivalent to that of the biological indicator. Based on the known bioburden population and the resistance of the biological indicator, conditions are selected which provide for the appropriate SAL of the bioburden on the product, but less than a 12-logarithm reduction of the biological indicator. For example, for products with a maximum average bioburden of 100 CFU, demonstration of an eight-logarithm reduction of the appropriate biological indicator under full cycle conditions (≤ 1 biological indicator positive for growth per 100 biological indicators processed) would be equivalent to an SAL of 10^{-6} or greater for the product bioburden. Further information on the biological indicator/bioburden method can be found in AAMI TIR 16—Process development and performance qualification for ethylene oxide sterilization—Microbiological aspects (28).

5.8.2. Cycle Validation

Once the appropriate sterilization cycle conditions have been established and a suitable biological indicator system has been qualified, routine cycle validation procedures may be implemented according to the appropriate regulatory or compendial guidelines. These guidelines generally require replicate testing under reduced cycle and full cycle exposure conditions to demonstrate attainment of the desired SAL and the integrity of the product and packaging. The complexity of the validation procedures will vary depending on the nature of the sterilization process, type of product, and load sizes involved. In conjunction with validation of the sterilization process, consideration should be given to validating any particular restrictions concerning the handling or processing of biological indicators employed with the sterilization process. Such items as retrieval of the biological indicators from the sterilization load, the interval of time and conditions between retrieval of the biological indicators and culturing the biological indicators, and shipping conditions if the biological indicators are forwarded to a culturing laboratory should be validated. Although it is desirable to retrieve the biological indicators from the sterilization load as soon as possible, this may not be practical due to potential worker safety issues or other considerations. Current guidelines for the qualification of biological indicators stipulate that they should be cultured within two (25) to four (8) hours of sterilization processing. For large-volume loads processed by ethylene oxide sterilization, attempting to retrieve biological indicators within 2 hours of processing may present undue risk of exposure to ethylene oxide. Under these circumstances biological indicators should be retrieved as soon as possible without compromising the safety of the personnel responsible for their retrieval. Written guidelines should be established, and any post-sterilization processing shipping or storage conditions validated as appropriate. The biological indicator manufacturer should be consulted with regard to any special considerations that may apply with regard to the handling of the biological indicator post-sterilization processing.

In conjunction with validating post-sterilization processing conditions, it is essential to maintain continuity between the biological indicator challenge system employed for initial cycle development and that subsequently employed for routine validation and monitoring of the sterilization process. It is not unusual to encounter situations where a biological indicator manufacturer cannot supply a particular biological indicator or Process Challenge Device due to production problems or raw material issues. In these circumstances it may be possible to prepare a written rationalization comparing the relevant properties and labeling of the current biological indicator system and substitute product if both systems are of similar construction and design. Factors to be considered include the extent of equivalence between the design and construction of both systems, use of the same test organism and population, instructions for use, and the labeled performance characteristics of each system. Where the construction of the biological indicator systems or Process Challenge Device or performance claims are significantly different however, it will be necessary to perform validation testing to demonstrate acceptable performance under the desired sterilization conditions. Such testing may include processing samples of both systems concurrently in a BIER vessel or production sterilizer for a series of fractional exposures. Note that the performance of each system may not be equivalent due to inherent variability of biological systems not only on a lot-to-lot basis but across different designs qualified to the same performance standards. In these circumstances a reasonable correlation between the performance of the current biological indicator and the new biological indicator should be demonstrated within the desired sterilization parameters. For process challenge devices employed with moist heat sterilization processes, qualification may include heat penetration studies with statistical analysis to demonstrate no significant differences in the physical lethality (F_o) at the site of the biological indicator challenge.

5.8.3. Routine Processing

Written procedures should be established for the storage of biological indicators, their placement within the sterilization load, and post-sterilization processing and culturing conditions consistent with the procedures employed during cycle development and validation and the recommendations of the biological indicator manufacturer. When developing these procedures, allowance should be made for the potential loss of biological indicators during sterilization processing and subsequent handling. For example, the medium ampules of many self-contained biological indicators are constructed of frangible glass that may rupture on an infrequent basis during processing, leading to a loss of integrity of the biological indicator. Other potential problems include evaporation of the growth medium with prolonged incubation over weekends or holidays, improper placement of the biological indicators within the sterilization load, failure of the operator to activate self-contained biological indicators prior to incubation, etc. These problems may not be apparent during initial cycle development and validation, but may present themselves during routine sterilization processing when larger quantities and multiple lots of biological indicators are processed. As this may lead to the loss of otherwise acceptable product, consideration should be given to the outcome of such failures. Development of a Failure Mode and Effects Analysis for the sterilization process with particular emphasis on the biological indicator system and the institution of effective control measures may protect against both significant financial loss and the potential release of nonsterile product (30).

During routine sterilization processing it is possible to occasionally encounter a biological indicator positive for growth subsequent to sterilization processing for no apparent reason. This requires that the sterilization load is quarantined and a review of the sterilization processing conditions performed to determine probable cause. As noted previously, biological indicators such as paper strip biological indicators or inoculated product or carriers are subject to potential contamination during the process of aseptic transfer to growth medium. An investigation similar to that performed for investigating sterility positives with product sterility testing should be performed and the results of the investigation documented (31). The positive culture should be identified to at least the genus level and compared against the genus of the test organism normally present on the biological indicator. If differences are noted, further microbial ID testing may identify the positive culture as a contaminant. Factors that may lead to contamination and false positive results with self-contained biological indicators include cracks in the outer vial that may not be visible to the naked eye, breaching of the filter or tortuous path intended to preclude microorganisms, collection of excess condensate within the vial, or other factors. If no cause can be determined for growth of the biological indicator subsequent to routine sterilization processing, the load should be either rejected or reprocessed as appropriate and a review of both the sterilization process and biological indicator conducted. This review should be documented and corrective actions taken as necessary.

Of particular concern with self-contained biological indicators employed with ethylene oxide or other chemical sterilization processes is the potential for residual sterilant remaining within the vial with subsequent inhibition or delay of the outgrowth of the test organisms upon incubation (32). This may lead to a worst-case false negative situation with the subsequent release of a nonsterile load as sterile. Under these circumstances the directions of the manufacturer of the biological indicator with regard to the need to allow for post-process aeration of the biological indicator prior to activation should be followed. Likewise, overprocessing of the growth medium of self-contained biological indicators employed with moist heat sterilization may affect its coloration and growth-promoting properties, leading to potential false-negative results. This potential may be evaluated by performing growth promotion testing on the medium ampule separate from the biological indicator subsequent to sterilization processing (8). Some manufacturers of self-contained biological indicators also provide separate negative controls that are processed in conjunction with the test indicators to guard against this potential.

5.8.4. Incubation Times

By their nature, biological indicators are a retrospective test, as they require a defined period of incubation to allow for the outgrowth of any test organisms that may have survived sterilization processing. Depending on the particular application, this delay in the release of product may or may not be of consequence. The FDA has provided a strategy to validate the reduction of biological indicator incubation time (33). This strategy requires processing a minimum of three sets of one hundred biological indicators from three separate lots under sublethal sterilization conditions intended to yield between 30 and 80% of the biological indicators positive for growth. After processing, the biological indicators are then incubated for 7 days at the appropriate incubation conditions. The minimum incubation time allowed is the maximum time required to attain 97% or greater recovery of the number of biological indicators positive for growth at 7 days for any one lot. Although this procedure provides only a 19% probability of acceptance at true growth readout of 97%, it has been successfully employed by many manufacturers to validate reduced biological indicator incubation times. Reduced incubation times established by this method should be periodically verified to ensure that they are not impacted by changes in the design of the biological indicator or changes in manufacturing materials or methods employed with the biological indicator.

5.9. RAPID READOUT BIOLOGICAL INDICATORS

In applications where the rapid release of product subsequent to sterilization processing is desired and parametric release or validation of a reduced incubation time is not practical or possible, the use of rapid readout biological indicators may be considered. These indicators represent a significant advance in biological indicator technology and provide an indication of the efficacy of a sterilization process within hours of processing as opposed to days. Rapid readout biological indicators rely on the use of a heat-resistant enzyme present in the spore coat to provide an early indication of the efficacy of the sterilization process (34). Enzyme surviving processing converts a nonfluorescent substrate in the growth medium to a fluorescent form that is detected by a separate instrument (Fig. 4). The use of heat-resistant enzyme technology provides for the early detection of surviving test organisms as compared to conventional means that rely on a pH color change or visible turbidity in the growth medium. Various studies have found the response of the enzyme system to be equivalent to that provided by growth response of the test organisms (2,35). Rapid readout biological indicators may also exhibit a greater potential for identifying both sterile and nonsterile conditions over a range of sterilization conditions as compared to conventional biological indicators that rely on outgrowth of the test organism (Tables 2, 3). In the absence of published performance standards for this type of technology, the use of rapid readout biological indicators in industrial applications requires prior clearance from the FDA. As a minimum, validation studies should include concurrent testing under in-use conditions performed with both the

FIGURE 4   An example of a fluorometric reader for detecting the presence of active alpha-glucosidase enzyme.

rapid readout biological indicator and the biological indicator system currently in use (Table 4). Although suitable for the release of product under routine processing conditions with appropriate FDA clearance, the rapid readout component should not be employed to validate sterilization processes.

5.10. CONCLUSION

The role of biological indicators in the development, validation, and routine monitoring of sterilization processes is occasionally questioned within the context of established and well-defined sterilization processes and the development of new physical and chemical sterilization indicator technology. Despite this, biological indicators remain unsurpassed in their ability to demonstrate directly the ability of a given sterilization process to attain the desired sterilization conditions. As noted by Agalloco et al., "One must understand that the biological indicator responds directly to the conditions present and is able to discern more precisely than any physical measurement what has occurred" (36). It can therefore be assured that biological indicators will continue to play an important role in the development and control of many sterilization processes for the near future. Although biological indicator

TABLE 2 Rapid Readout Biological Indicator Moist Heat Temperature Range Study-Summary Data

Temperature	Exposure[c]	Conditions	Attest 1262 Spore[a]	Attest 1292 rapid readout	
				Spore[a]	Enzyme[b]
116.1°C	15.8 minutes	All survival	19/40	40/40	40/40
	47.4 minutes	All kill	0/40	0/40	0/40
121.1°C	5 minutes	All survival	40/40	40/40	40/40
	15 minutes	All kill	0/40	0/40	0/40
126.1°C	1.6 minutes	All survival	40/40	40/40	40/40
	4.8 minutes	All kill	40/40	40/40	40/40

Data presented as the number of non-sterile determinations per number of samples tested.

[a] Spore = Growth positive. Seven days incubation in a humidified incubator (56 ± 2°C Attest 1262.60 ± 2°C Attest 1292) followed by determination of media color change.

[b] Enzyme = Fluorescent response. Three hours incubation at 60 ± 2°C followed by determination of fluorescence in a 3M Auto-reader.

[c] Exposure = For 116.1°C and 126.1°C exposure conditions equivalent to that at 121.1°C were calculated based on a z-value of 10°C.

TABLE 3 Rapid Readout Biological Indicator Moist Heat Temperature Range Study-Index of Efficiency

	Attest 1262		Attest 1292 rapid readout			
	Spore[a]		Spore[a]		Enzyme[b]	
Conditions[c]	Sterile	Non-sterile	Sterile	Non-sterile	Sterile	Non-sterile
Number of correct sterile and non-sterile determinations						
Non-Sterile	21	99	0	120	0	120
Sterile	80	40	80	40	80	40
Index of Efficiency[d]						
Sensitivity[e]	0.83		1.00		1.00	
Specificity[f]	0.67		0.67		0.67	
Efficiency[g]	0.75		0.83		0.83	

[a] Spore = Growth positive. Seven days incubation in a humidified incubator ($56 \pm 2°C$ Attest 1262, $60 \pm 2°C$ Attest 1292) followed by determination of media color change.

[b] Enzyme = Fluorescent response. Three hours incubation at $60 \pm 2°C$ followed by determination of fluorescence in a 3M Auto-reader.

[c] Conditions = Actual conditions under which samples were tested. Non-sterile conditions include all samples processed at $116.1°C$ for 15.8 minutes, $121.1°C$ for 5 minutes, and $126.1°C$ for 1.6 minutes. Sterile conditions include all samples processed at $116.1°C$ for 47.4 minutes, $121.1°C$ for 15 minutes, and $126.1°C$ for 4.8 minutes.

[d] Reich R. and Fitzpatrick G. 1985. Flash Sterilization. J. HSPD. May/June, pp. 60–63.

[e] Sensitivity = probability of correctly identifying non-sterile conditions

$$= \frac{\text{no. correct (non-sterile responses)}}{\text{no. correct (non-sterile responses)} + \text{no. incorrect (non-sterile) responses}}$$

[f] Specificity = probability of correctly identifying sterile responses

$$= \frac{\text{no. correct (sterile responses)}}{\text{no. correct (sterile responses)} + \text{no. incorrect (sterile) responses}}$$

[g] Efficiency = probability of correctly identifying both sterile and non-sterile conditions

$$= \frac{\text{no. correct non-sterile responses} + \text{no. correct responses}}{\text{total no. correct and incorrect responses}}$$

TABLE 4 Air Overpressure Sterilization

Cycle Type		$F_O < 5$		F_O 8-12		Half-Cycle		Full Cycle	
		Sub-lethal		Marginal		Lethal		Lethal	
Exposure Temperature		115°C		≥121.1°C		≥121.1°C		≥121.1°C	
Monitor	Packaging	Spore[a]	Enzyme[b]	Spore[a]	Enzyme[b]	Spore[a]	Enzyme[b]	Spore[a]	Enzyme[b]
Attest 1262	Test pack	30/30	N/A	0/30	N/A	0/30	N/A	0/30	N/A
Attest 1292	None	30/30	30/30	30/30	30/30	0/30	0/30	0/30	0/30
Attest 1292	Pouch	30/30	30/30	30/30	0/30	0/30	0/30	0/30	0/30
Attest 1292	Test pack	30/30	30/30	30/30	0/30	0/30	0/30	0/30	0/30

Data presented as the number of non-sterile determinations per number of samples tested. Total of three cycles processed under Air Overpressure conditions with a full load.

N/A = not applicable.

[a] Spore = Growth positive. Seven days incubation in a humidified incubator (56 ± 2°C Attest 1262, 60 ± 2°C Attest 1292) followed by determination of media color change.

[b] Enzyme = Fluorescent response. Three hours incubation at 60 ± 2°C followed by determination of fluorescence in a 3M Auto-reader.

technology has advanced considerably since the time of Kilmer, the effective use of biological indicators requires both knowledge of their limitations and the implementation of an effective system of controls and procedures for their qualification and application.

APPENDIX 5.1.   D-VALUE CALCULATIONS

5.1.1.   General Survivor Curve Equation

The semi-logarithmic inactivation of microorganisms follows the general equation of a straight line:

$$y = mx + b \qquad \text{(straight line equation)} \qquad (1)$$

where: y = y coordinate
$\quad$ m = slope of line
$\quad$ x = x coordinate
$\quad$ b = y intercept

This equation can also be expressed as the general survivor curve equation:

$$\text{Log } N = -U/D + \log N_o \quad \text{(general survivor curve equation)} \qquad (2)$$

where: N = the number of microorganisms present at time U ($=y$)
$\quad$ D = D-value ($m = 1/D$).
$\quad$ U = exposure time ($=x$)
$\quad$ N_o = the initial number of microorganisms ($=b$)

Note that mathematically the D-value is actually the negative reciprocal of the slope of the general survivor curve. Equation (2) can be arranged as the general D-value equation:

$$D = \frac{U}{\log N_o - \log N} \qquad \text{(general D-value equation)} \qquad (3)$$

This Eq. (3) forms the basis for determining the D-value by either the survivor curve or fraction-negative method using the MPN procedure.

Note that in reality the survivor curve may exhibit shoulders or tailing and that only the straight line portion of the curve should be employed for estimation of the D-value.

5.1.2.   Stumbo-Murphy-Cochran MPN Method

For determination of the D-value using fraction-negative data, the number of surviving microorganisms can be estimated by the means of Most Probable

Number (MPN) analysis (Halvorson and Zeigler in 1938 J. Bact. Vol. 25, pp. 101–102) as follows:

$$N = \ln(n/r) \qquad \text{(MPN estimate)} \qquad (4)$$

where: N = the number of microorganisms present at time U (= y)
      ln = natural logarithm
      n = the number of samples tested.
      r = the number of samples negative for growth

Converting the natural logarithm to the base10 logarithm and substituting for N in the general survivor curve Eq. (3) above:

$$D = \frac{U}{\log N_o - \log [\ln(n/r)]} \qquad \text{(Stumbo-Murphy-Cochran equation)}$$
$$(5)$$

Equation (5) is known as the Stumbo-Murphy-Cochran equation for calculating D-values. The Stumbo-Murphy-Cochran equation provides an estimate of the number of surviving microorganisms (N) at a given exposure time based on the Halvorson and Zeigler MPN equation.

    Example. Calculate the D value from the following resistance performance data for a biological indicator with an average spore count per carrier of 1.5×10^5 test organisms and a labeled D-value of 2 min.

Exposure	# Samples	# Sterile
7 min	20	0
8 min	20	0
9 min	20	2
10 min	20	8
11 min	20	20
12 min	20	20

Calculate the D-value using the Stumbo-Murphy Cochran Method:

$$D_1 = \frac{t}{\log N_o - \log[2.303 \ \log(n/r)]}$$

$$D_1 = \frac{9 \ \text{min}}{\log(1.5 \times 10^5) - \log[2.303 \ \log(20/2)]}$$

$$D_1 = \frac{9 \ \text{min}}{5.1761 - 0.3623}$$

$$D_1 = 1.87 \text{ min}$$

$$D_2 = \frac{t}{\log N_o - \log[2.303 \ \log(n/r)]}$$

$$D_2 = \frac{10 \text{ min}}{\log(1.5 \times 10^5) - \log[2.303 \ \log(20/8)]}$$

$$D_2 = \frac{10 \text{ min}}{5.1761 - (-0.0379)}$$

$$D_2 = 1.92 \text{ min.}$$

$$D_{avg.} = \frac{1.87 \text{ min.} + 1.92 \text{ min.}}{2}$$

$$D_{avg.} = 1.89 \text{ min}$$

5.1.3. Limited Spearman-Karber Mean Time Until Sterility

In contrast to the Stumbo-Murphy-Cochran model, the Spearman-Karber model provides an estimate of the mean time until sterility (U_{SK}), or the estimated time when the samples are all sterile. Returning to the general survivor curve Eq. (2), the Spearman-Karber model can be expressed as follows:

$$D = \frac{U_{SK}}{\log N_o + 0.2507} \qquad \text{(Spearman-Karber equation)} \qquad (6)$$

It can be shown by the Poisson distribution that the probability of a sterile sample at the mean time until sterility (U_{SK}) is 0.57 and that the corresponding value for log N is equal to -0.2507 (Pflug, I.J., Holcomb, R.G., and Gomez, M.M. Principles of the Thermal Destruction of Microorganisms in Disinfection, Sterilization, and Preservation. Block, S.S., Ed. Lipincott Williams & Wilkins. Philadelphia, p. 121). A benefit of the Spearman-Karber method is that it allows one to calculate confidence limits for the D-value estimate. The D-value can be calculated by the Limited Spearman-Karber Method as follows:

Designate the number of specimens taken for each group (i.e., 10) by n, and the difference between adjacent times (in minutes) by δ. Designate for each group of the series the number of specimens showing no growth by:

$$f_1, f_2, \ldots \ldots f_k$$

in which f_1 is the response of all 10 specimens showing growth (0/10 inactivated) in the group held for the shortest time for such result which is adjacent to an intermediate mortality, and f_k is the response of all 10 specimens of the group showing no growth (10/10 inactivated) in the group held for the longest time for such result which is adjacent to an intermediate mortality. Do not use for the calculation observations for groups beyond the ends of the series, f_1 and f_k, giving results that are not adjacent to an intermediate mortality. The test is valid if there is available a valid result (0/10) from a group held for a shorter time than that for the selected shortest time result (f_1) and there is available a valid result (10/10) from a group held for a longer time than that for the selected longest time result (f_k). Calculate the mean heating time (T) for achieving complete kill by the equation:

$$T = T_k - \delta/2 - \left(\delta/n \times \sum_{i=1}^{k-1} f_i \right)$$

in which T_k = time for achieving the result f_k. Calculate the D value by the equation:

$$D = \frac{T}{\log N_o + 0.2507}$$

in which N_o is the average spore count per carrier determined at the time of the test.

Calculate the variance of T (i.e., V_T) by the equation:

$$V_T = \frac{(\delta)^2}{n^2(n-1)} \times \sum_{i=1}^{k-1} f_i(n - f_i)$$

The standard deviation (s_T) is the square root of the variance:

$$s_T = \sqrt{V_T}$$

Calculate the lower and upper 95% confidence limits (approximate CL) for the D value by the equation:

$$\text{upper CL for D} = \frac{T + 2s_T}{\log N_o + 0.2507}$$

$$\text{lower CL for D} = \frac{T - 2s_T}{\log N_o + 0.2507}$$

If not more than one specimen from a group and not more than two specimens from all of the groups giving the results f_1 through f_k are missing, replace each missing value by adding 0 to the number showing no growth, if the number showing no growth in the remaining 9 specimens of that group is 4 or less, and adding 1 if the number showing no growth in the remaining 9 specimens of that group is 5 or more.

Example: Calculate the D-value by the Spearman-Karber Method using the same date employed to calculate the D-value by the Stumbo-Murphy Cochran method above.

	Exposure	(n) # samples	(f_i) # sterile	$(n - f_i)$	$f_i(n - f_i)$
f_1	7 min	20	0	20	0
	8 min	20	0	20	0
	9 min	20	2	18	36
f_{k-1}	10 min	20	8	12	96
f_k	11 min	20	20	0	0
	12 min	20	20	0	0

$$\sum_{i=1}^{k-1} f_i = 10$$

$$\sum_{i=1}^{k-1} f_i(n - f_1) = 132$$

a) Calculate the mean heating time until sterility (T) as follows:

$$T = T_k - \delta/2 - \left(\delta/n \times \sum_{i=1}^{k-1} f_i \right)$$

$$T = 11 \text{ min} - \frac{1 \text{ min}}{2} - \left(\frac{1 \text{ min}}{20} \times 10 \right)$$

$$T = 11 \text{ min} - 0.5 \text{ min} - (0.5 \text{ min})$$

$$T = 10 \text{ min}$$

b) Calculate the D-value as follows:

$$D = \frac{T}{\log N_o + 0.2507}$$

$$D = \frac{10 \text{ min}}{5.1761 + 0.2507}$$

$$D = 1.84 \text{ min}$$

c) Calculate the variance of T (V_T) as follows:

$$V_T = \frac{\delta^2}{n^2(n-1)} \times \sum_{i=1}^{k-1} f_1(n - f_1)$$

$$V_T = \frac{1^2}{400(20-1)} \times (132)$$

$$V_T = \frac{1}{7600} \times (132)$$

$$V_T = 0.0174$$

d) Calculate the standard deviation:

$$s_T = \sqrt{Vt}$$

$$s_T = 0.1318$$

e) Calculate the lower and upper 95% confidence limits (CL) for D:

$$CL_{lower} = \frac{T - 2s_T}{\log N_o + 0.2507}$$

$$= \frac{10 - 0.2636}{5.1761 + 0.2507}$$

$$= \frac{9.7364}{5.4268}$$

$$= 1.79 \text{ min}$$

$$CL_{upper} = \frac{T + 2s_T}{\log N_o + 0.2507}$$

$$= \frac{10 + 0.2636}{5.1761 + 0.2507}$$

$$= \frac{10.2636}{5.4268}$$

$$= 1.89 \text{ min}$$

Further discussion of the D-value determination of biological indicators may be found in ANSI/AAMI/ISO 14161, Sterilization of health care products—Biological indicators—Guidance for the selection, use, and interpretation of results, Pflug (Pflug, I.J., Holcomb, R.G., and Gomez, M.M. *Principles of the Thermal Destruction of Microorganisms in Disinfection, Sterilization, and*

Preservation, Block, S.S., Ed.; Lipincott Williams & Wilkins, Philadelphia, 2001) and Shintani (Biomedical Instrumentation & Technology, March/April 1995, pp. 113–124).

REFERENCES

1. Gaughran, E.R.L.; Fred, B. Kilmer—Pioneer in microbiological control. In *Sterilization of Medical Products*; Gaughran, E.R.L., Kereluk, K., Eds.; Johnson & Johnson: Somerville, 1977; 1–16.
2. Vesley, D.; Langholz, A.C.; Rohlfing, S.R.; Foltz, W.E. Fluorometric detection of a Bacillus stearothermophilus spore-bound enzyme, α-D- glucosidase, for rapid indication of flash sterilization failure. Appl. Environ. Microbiol. 1992, *58* (2), 717–719.
3. Spicher, G. Biological indicators and monitoring systems for validation and cycle control of sterilization processes. Zkl. Bakt. Hyg. A 1988, *267*, 463–484.
4. Pflug, I.J.; Odlaug, T.E. Biological indicators in the pharmaceutical and the medical device industry. J. Parent. Sci. Technol. 1986, *40* (5), 242–248.
5. Association for the Advancement of Medical Instrumentation (AAMI). ANSI/AAMI/ISO 14161: Sterilization of health care products—Biological Indicators—Guidance for the selection, use, and interpretation of results. Arlington, VA, 2000.
6. United States Pharmacopeial Convention. United States Pharmacopeia 27 < 1211 > Sterilization and Sterility Assurance of Compendial Articles. Rockville, MD, 2004; 2616–2620.
7. Association for the Advancement of Medical Instrumentation (AAMI). AAMI/CDV-3 ST67: Sterilization of health care products—Requirements for products labeled "sterile". Arlington, VA, 2003.
8. United States Pharmacopeial Convention. United States Pharmacopeia 27 < 55 > Biological Indicators—Resistance Performance Tests. Rockville, MD, 2004; 2150–2152.
9. Stumbo, C.R. *Thermobacteriology in Food Processing*; Academic Press: New York, 1965; 79–104.
10. Association for the Advancement of Medical Instrumentation (AAMI). ANSI/AAMI ST44: Resistometers used for characterizing the performance of biological and chemical indicators. Arlington, VA, 1992.
11. Mosley, G.A. Estimating the effects of EtO BIER vessel operating precision on D-value calculations. MD&DI 2002, *24* (4), 45–56.
12. Mosley, G.A.; Gillis, J.R. Operating precision of steam BIER vessels and their interactive effects of varying Z values on the Reproducibility of listed D-values. PDA J. Pharm. Sci. Technol. 2002, *56* (6), 318–331.
13. Cooper, M.S. Sterilization and sterilization indicators. The Microbiological Update 2001, *18* (11).
14. Association for the Advancement of Medical Instrumentation (AAMI). ANSI/AAMI/ISO 11137: Sterilization of health care products—Requirements for validation and routine control—Radiation sterilization. Arlington, VA, 1994.

15. EEC Directive 75/318/EEC. The use of Ionising Radiation in the Manufacture of Medicinal Products, 1992.
16. Taylor, D.M.; Fraser, H.; McConnell, I.; Brown, D.A.; Brown, K.L.; Lamza, K.A.; Smith, G.R.A. Decontamination studies with the agents of bovine spongiform encephalopathy and scrapie. Arc. Virol. 1994, *139*, 313–326.
17. Association for the Advancement of Medical Instrumentation (AAMI). ANSI/ AAMI/ISO 14937: Sterilization of health care products—General requirements for characterization of a sterilizing agent, and the development, validation, and routine control of a sterilization process for medical devices. Arlington, VA, 2000.
18. Shintani, H.; Akers, J.E. On the cause of performance variation of biological indicator used for sterility assurance. PDA J. Pharm. Sci. Technol. 2000, *54* (4), 332–342.
19. Gillis, J.R.; Schmidt, W.C. Scanning Electron microscopy of spores on inoculated product surfaces. MD&DI 1983, *5*, 33–38.
20. Ernst, R.R.; Doyle, J.E. Sterilization with gaseous ethylene oxide: a review of chemical and physical factors. Biotechnol. Bioeng. 1968, *10*, 1–31.
21. Owens, J.E. Sterilization of LVPs and SVPs. In *Sterilization Technology. A Practical Guide for Manufacturers and Users of Health Care Products*; Morrissey, R.F., Phillips, G.B., Eds.; Van Nostrand Rheinhold: New York, 1993; 254–285.
22. Joslyn, L.J. Sterilization by heat. In *Disinfection, Sterilization, and Preservation*, 5th Ed.; Lippincott, Williams & Wilkins: Philadelphia, 2001; 695–728.
23. Association for the Advancement of Medical Instrumentation (AAMI). ANSI/ AAMI/ISO ST59: Sterilization of health care products—Biological Indicators—Part 1: General. Arlington, VA, 1999.
24. EN 866-1: Biological systems for testing sterilizers and sterilization processes—Part 1: General requirements.
25. ISO 11138-1: Sterilization of health care products—Biological indicators—Part 1: General requirements, 2d Ed.
26. 21CFR 880.2800.
27. Graham, G.S.; Boris, C.A. Chemical and biological indicators. In *Sterilization Technology. A Practical Guide for Manufacturers and Users of Health Care Products*; Morrissey, R.F., Phillips, G.B., Eds.; Van Nostrand Rheinhold: New York, 1993; 36–69.
28. Association for the Advancement of Medical Instrumentation (AAMI).TIR16: Process development and performance qualification for ethylene oxide sterilization—Microbiological aspects. Arlington, VA, 2000.
29. Spicher, G. Biological indicators and monitoring systems for validation and control of sterilization processes. Zbl. Bakt. Hyg. A 1988, *267*, 463–484.
30. Sherratt, D. Taking a risk-based approach to medical device design. MD&DI 1999, *9*, 84.
31. Sordellini, P.J.; Lang, M. Investigating and preventing BI sterility failures. MD&DI 1996, *18* (8), 66–69.
32. Winckels, H.W.; Boumans, P.G.F. Validation studies for the 48 hour release of products exposed to an ethylene oxide sterilization process. Eucomed Conference Proceedings, London, 1984.

33. Office of Device Evaluation, FDA. Premarket Notifications [510(k)] for Biological Indicators Intended to Monitor Sterilizer Used in Health Care Facilities; Draft Guidance for Industry and FDA Reviewers, 2001.
34. Albert, H.; Davies, D.J.; Woodson, L.P.; Soper, C.J. Biological indicators for steam sterilization: characterization of a rapid biological indicator utilizing Bacillus stearothermophilus spore-associated alpha-glucosidase enzyme. J. Appl. Microbiol. 1998, *85* (5), 865–874.
35. McCormick, P.; Finocchario, C.; Manchester, R.; Glasgow, L.; Costanzo, S. Qualification of a rapid readout biological indicator with moist heat sterilization. PDA J. Pharm. Sci. Technol. 2003, *57* (1), 25–31.
36. Agalloco, J.P.; Akers, J.E.; Madsen, R.E. Moist heat sterilization—myths and realities. PDA J. Pharm. Sci. Technol. 1998, *52* (6), 346–350.

6

Biological Safety Cabinets and Isolators Used in Pharmaceutical Processing

Mark Claerbout

Eli Lilly & Company, Indianapolis, Indiana, U.S.A.

6.1. INTRODUCTION

The pharmaceutical industry has historically employed various means to provide a controlled local environment at critical points in the manufacturing and testing of pharmaceutical products. The design of manufacturing facilities must ensure personnel are protected from potent and/or potentially hazardous active pharmaceutical ingredients and excipients. Equally important is the need to protect pharmaceutical products, especially aseptically produced sterile products, from particulate and microbiological contamination from people and the surrounding environment. Drug manufacturing facilities are typically designed to have the highest environmental quality in the areas where the product is directly exposed to the environment. In this context, high quality means the air and product contact surfaces have low levels of nonviable particulate and microbiological contamination. In the case of sterile products produced by aseptic processing, the expectation is the air contacting the sterilized product and sterile product contact surfaces should be also be sterile. Air quality is achieved using terminal high-efficiency particulate air (HEPA) filtration and monitored through integrity testing of the filter media prior to initial use and periodically thereafter. Similarly, the solutions used to cleanse surfaces are filtered through microbial retentive filters to minimize particulate contamination (1).

The pharmaceutical testing laboratories must also protect the test samples from environmental contamination when conducting microbiological tests and testing for foreign particulate matter. The United States Pharmacopeia (USP) addresses these requirements in various chapters, including Chapter ⟨1211⟩, which in reference to the sterility testing laboratory, states "The facility for sterility testing should be such as to offer no greater a microbial challenge to the articles being tested than that of an aseptic processing production facility."

Chapter ⟨788⟩, which describes the testing for subvisible particulate contamination, provides the following guidance: "Perform the test in an environment that does not contribute any significant amount of particulate matter. Preferably, the test specimen, glassware, closures and other required equipment are prepared in an environment protected by high efficiency particulate air (HEPA) filters, and non-shedding garments and powder-free gloves are worn throughout the preparation of samples."

The USP and other pharmacopeial compendia require rigid environmental controls for these tests because the test results may be critically affected by environmental contamination. Furthermore, because lab personnel require protection from potentially hazardous properties of some samples, it is important to assure the design of lab facilities and containment equipment satisfactorily address both of these issues. This chapter will focus on design aspects of partial barrier enclosure (chemical fume hoods and biosafety cabinets) and full barrier enclosures (isolators). Validation of the use of isolators for sterility testing will also be covered, as this application has become commonplace in the pharmaceutical microbiology lab over the past decade.

6.2. DESIGN CONSIDERATIONS

6.2.1. Partial Barrier Enclosures

Partial barrier enclosures designed to remove chemical and/or potentially biohazardous aerosols away from personnel have been in use in pharmaceutical laboratories for decades. Design variations on these enclosures also allow product protection from contamination due to personnel and the surrounding environment. These enclosures have the advantage of allowing easy access to the work area through openings directly in front of the operator, but offer less containment than full-barrier enclosures such as isolators. Laminar airflow air benches are the simplest solution to create a low particulate environment for the testing of pharmaceuticals. Inlet air is passed through HEPA filters at the back of the enclosure and flows toward the opening and the operator (horizontal air flow hood): or in the vertical airflow design it flows downward toward the work surface. These air benches provide no protection to the operator and thus are not recommended for working with potent, allergenic, or hazardous products. The U.S. Centers for Disease Control and Prevention/National Institutes of Health (CDC/NIH) provide guidelines for primary containment equipment helpful in selecting enclosures for appropriate uses. The CDC/NIH enclosure classification system for biological safety cabinets (BSCs) consists of three classes. Classes 1 and 2 are partial barrier enclosures while class 3 is a full barrier cabinet (or isolator). Class 1 BSCs are designed to provide personnel and environmental protection but not product protection. Air is pulled into the enclosure below the front sash and along the floor to the back of the enclosure, where it enters the exhaust plenum and exhausted after passing through a HEPA filter. The incoming air along the front of the enclosure acts as a barrier to prevent product aerosol-contaminated air within the enclosure from escaping into the lab environment. The Class I BSC may be connected via ducts to the building exhaust system, whereby the building fan provides the negative pressure to draw room air through the cabinet, or it may contain an exhaust fan interlocked to the building exhaust fan system. HEPA filters should be installed on the inlet side and additional HEPA filters may also be installed in the duct system to provide a higher degree of safety (Fig. 1).

A minimum airflow rate of 75 linear feet per minute (lfpm) provides protection to the personnel. The inlet area on the hood front may be restricted with movable panels to increase the flow rate. Class 1 BSCs are often employed to enclose equipment such as centrifuges, mixers, or homogenizers that generate aerosols.

Class 2 BSCs afford protection to both the samples being manipulated in the cabinet and the operator. They use the principle of "laminar" airflow

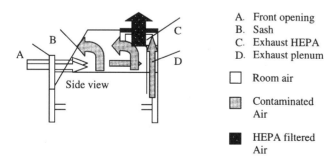

A. Front opening
B. Sash
C. Exhaust HEPA
D. Exhaust plenum

☐ Room air

▨ Contaminated Air

■ HEPA filtered Air

FIGURE 1 Class 1 biosafety cabinet.

(unidirectional air moving at a uniform speed) coupled with HEPA filtration to provide a curtain of essentially sterile air to protect the test materials. Airflow is pulled around the operator into the front grille of the cabinet and then downward under the work surface along with HEPA filtered air. When running properly, the incoming airflow from the surrounding lab area and the operator is not supposed to enter the work area where the product and testing materials are located. There are several variations of the Class 2 BSC (type A, type B1, type B2, and type B3) (Fig. 2).

In the Class 2 Type A BSC, HEPA-filtered air moving downward in the cabinet keeps the sample materials bathed in filtered air. The air column splits as it reaches the cabinet floor with the air nearest the front opening mixing with incoming room air. The mixture passes under the floor and is joined by the air coming from the portion of the air column closer to the back wall of the cabinet. This contaminated air enters the supply blower inlet and climbs through the rear plenum where it is either exhausted through an exit HEPA filter or is passed though the inlet HEPA filter back into the cabinet interior. The exhaust/inlet air ratio is approximately 30%/70%. Unducted Class 2 Type A BSCs cannot be used for work involving volatile or toxic compounds, as the fumes will be released back into the laboratory. The Type A cabinet can be ducted to vent outside, but it must be done without altering the balance of the cabinet exhaust/inlet airflow by use of a canopy hood thimble around the cabinet exhaust filter housing. Generally, the Type A cabinets are not vented to the outside because fluctuations in air pressure and volume that often occur in building exhaust systems make it difficult to match the airflow requirements of the cabinet.

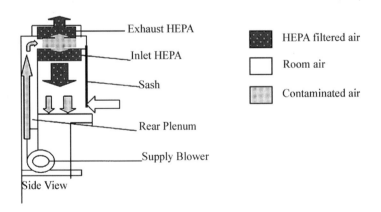

FIGURE 2 Class 2 Type A biosafety cabinet.

The Class 2 Type B1 BSC provide both biological and chemical containment and are often used in cell culture and microbial cultures where hazardous compounds or radionucleotides are involved. Room air is pulled in the front opening and blended with the downward moving column of air near the front opening. This mixture passes through a HEPA filter under the floor work surface into blowers that push the filtered air up along channels within the sidewalls of the cabinet. The air enters the plenum above the cabinet and is either passed out of the exit HEPA filter along with air from the rear half of the cabinet, or it is recirculated through inlet HEPA filters (Fig. 3).

The Class 2 Type B2 BSC has a 100% exhaust design that eliminates the recirculation of air within the cabinet. It provides both biological and chemical containment. The supply blower draws in air (either room air or outside air) at the top of the cabinet, passes it though a HEPA filter and down through the work area of the cabinet. The facility exhaust system and/or the cabinet exhaust system pulls air through the rear and front grilles at the workbench base in the cabinet and circulates it upward though a negative pressure exhaust plenum through the exhaust HEPA. The inflow face velocity is approximately 100 lfpm. This hood type can be expensive to operate because each cabinet exhausts as much as 1200 cubic ft per minute of conditioned room air.

The Class 2 Type B3 BSC is a hard-ducted Type A cabinet surrounded by a negative air pressure plenum. The cabinet exhaust must be connected to the building exhaust system and the design also requires a minimum inward

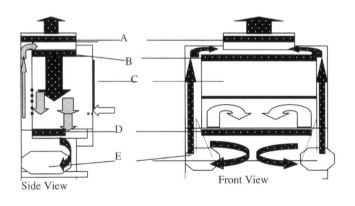

A. Exit HEPA B.Inlet HEPA C.Sash D. Additional HEPA E. Blowers
  Room Air   HEPA filtered air   Contaminated air

FIGURE 3 Class 2 Type B1 biosafety cabinet.

airflow of 100 lfpm. The cabinet is also recommended for both biological and chemical containment.

Class 3 BSCs are full-barrier isolators designed for work with biosafety level 4 microbiological agents and, therefore, are not as common in pharmaceutical laboratories as Class 2 BSCs, although their design has several features in common with the isolators commonly used in pharmaceutical testing laboratories. They are gas-tight enclosures, with HEPA filters on both the inlet and exit ducts. They may be connected to an autoclave or a chamber containing a chemical dunk tank pass-through, which allows the passage of biohazardous material in and out of the enclosure. Unlike sterility testing isolators, the cabinet is maintained under negative pressure using a dedicated independent exhaust system. Long heavy-duty gloves are attached to ports in the cabinet to allow handling of the materials inside. The gloves restrict movement and dexterity but maximize personnel protection.

It is critical that BSCs be properly installed and tested to assure proper functioning prior to certification for use. Retesting the equipment should be performed at least annually and after maintenance. Testing includes verifying the HEPA filter integrity via aerosol challenge and scanning for downstream aerosol leaks, velocity checks of the inlet airflow, and checking alarm systems for proper function. Training of lab personnel in the use of the BSCs is also necessary to assure the equipment is used properly. Rapid and repeated insertion and removal of worker's arms into and out of the enclosure, or fast walking by the enclosure can cause aerosols to escape from the enclosure. The BSCs should be located away from entry doors, hallways, and high-traffic areas. BSCs used for handling biohazardous material should be decontaminated using an appropriate chemical agent prior to preventive or corrective maintenance being performed on the HEPA filter system. Due to the open design that relies on air flowing in defined directions to separate the work area from the surrounding environment, Class 1 and Class 2 BSCs allow some air exchange between the two areas and are not as effective as Class 3 BSCs or other full-barrier isolators for assurance of containment (3).

6.2.2. Full-Barrier Enclosures (Isolators)

PDA Technical Report No.34 proposes the following opening statement for a definition for isolators used in worldwide health care industry- "An isolator is sealed or is supplied with air through a microbially retentive filtration system (HEPA minimum) and may be reproducibly decontaminated." The PDA report differentiates between closed and open isolators and further divides each of these categories based on use: aseptic applications, containment applications or a combination of both. Closed isolators intended for strictly containment applications (e.g., potent and toxic bulk drug dispensing) usually operate under a negative pressure relative to the surroundings. They are designed to not exchange air with the adjacent areas unless the air is purified of the hazardous aerosols generated in the isolator via filtration or "scrubbers" such as activated charcoal canisters. Materials exiting the isolator are either moved into an antechamber and cleaned of the hazardous material residues prior to removal or the entire isolator is cleaned before opening so personnel are not exposed to significant levels of the hazardous material. The cleaning process must be validated to demonstrate adequate hazard removal. Closed isolators for aseptic operations (e.g., sterility testing isolators) normally operate under positive pressure and undergo a validated biodecontamination procedure prior to use. All air entering and leaving the isolator must pass through microbial retentive filters, and materials entering the isolators must be decontaminated or sterilized prior to entry. Closed isolators used for both aseptic operations and product containment are often designed to have positive pressure internally with the air from the surrounding area at a slight negative pressure to adjacent work suites. The workers may also wear additional protective equipment for protection in the event of an isolator leak.

Open isolators are designed for the passage of materials into and out of the isolator during operation while maintaining the integrity of the internal environment. Open isolators are often used for the aseptic filling/sealing of product containers and are designed to not exchange unfiltered air with their surroundings. They typically are biodecontaminated while closed and then opened prior to the start of operations. For the containment of toxic product residues, a cleaning process can be incorporated after the filling/sealing process prior to exit from the isolator(s) to also provide containment. The airflow from the isolator and the surrounding room may be ducted away from workers, who may also wear special protective equipment, depending on the nature of the product. The room containing this type of isolator may be kept under negative pressure relative to the surrounding rooms.

6.2.3. Isolator Materials of Construction

Materials used for isolator construction must have suitable strength to resist puncture damage from normal operations and be compatible with the products, cleaning agents, and biodecontamination agents planned for use. Soft or flexible wall isolators are often used for both sterility testing and drug manufacturing aseptic processing applications. Rigid wall isolators are more likely to be used for product containment applications that may involve cleaning processes using aggressive chemical solvents. Polyvinyl chloride (PVC) sheeting is the most common flexible wall isolator material, as it is easily formed and joined at seams, is transparent, and is generally chemically compatible with common chemical biodecontamination agents like hydrogen peroxide or peracetic acid; and PVC is available in various thicknesses. PVC isolator "bags" can be supported using either an internal or external framework, often made of stainless steel. Acrylic panels bonded with cement have also been used for construction of some rigid wall isolators. Stainless steel is often used for flooring material in both flexible and rigid wall isolators, and is also a common material for internal and external structural members (e.g., door frames, shelving). For maximum compatibility with cleaning and decontamination agents, 300 series stainless steel with low carbon content (304L, 316L) is recommended for use in isolator interiors. Where surface cleanability is a major concern, such as hazardous product containment, electropolishing is recommended. Smooth wall/floor junctions on the isolator interior are desirable from a cleaning and biodecontamination efficacy perspective, and sloped floors with a drain port improve the isolator cleanability. Glazing materials used in windows for rigid wall isolators normally use polycarbonate or acrylic plastics or tempered glass. The design of the window seal and gasket materials must take into account the cleanability and compatibility of the window with the planned cleaning/sanitizing agents— ledges or crevices around the glazing material should be avoided to minimize the possibility of product residue or bioburden retention around the window seals. Silicone rubber is often used for gasket material and Neoprene® or Hypalon® rubber compounds are commonly used for glove construction. In general, any materials may be used in isolator construction as long as they are compatible with product materials, cleaning/sanitizing agents, have adequate structural integrity, and can be cleaned to the degree necessary. It is best to minimize the presence of exposed electrical equipment in isolators due to incompatibility associated with exposure to aggressive cleaning and sanitizing agents. Lights, fan and pump motors, circuit boards and switches will last longer and function more reliably when mounted on the isolator exterior.

6.2.4. Isolator/Operator Interface

The design of the operator interface must consider several challenges resulting in a compromise between optimal isolator integrity and ease of conducting operations within the isolator. Operators need to interface with isolators via sleeve ports with gloves, or through the use of half suit or full suits, or less commonly, automanipulators. These sleeves and suits can create irregular protrusions into the isolator that may have folds or crevices that result in areas difficult to clean and sanitize. They can also impede the airflow within the isolator and may require internal isolator fans to assure adequate airflow for the circulation of biodecontamination agents. The suit, sleeve, and gloves need to be thin and flexible enough to permit adequate operator movement while thick enough to minimize cracks and leaks developing which compromise the isolator integrity and risk product quality and/or operator safety. In general, successful isolator designs will minimize the need for operator activity in the isolator, as the more the interfaces are used, the greater the likelihood for integrity failures. The interface design must also consider ergonomic and operator safety issues.

Glove/sleeve designs are generally either a one-piece or two-piece design. The one-piece design consists of a glove and sleeve in a continuous integral assembly, which may be available in various lengths or glove sizes. The base of the sleeve connects to a port on the isolator. Two-piece units have some type of sealed connection of the glove to the sleeve, which allows for the change-out of gloves without having to remove the sleeve from the isolator port. Isolator gloves tend to be the weakest link in isolator integrity, and the routine exposure of the gloves to cleaning/sanitizing agents will increase the incidences of glove failures. Limiting the use life of the isolator gloves through mandatory changing and/or implementing double gloving practices can reduce the incidence of glove failures jeopardizing isolator operations. Double gloving, in which sterile disposable gloves are placed over the isolator gloves, may reduce tactile ability for delicate manipulations but can reduce wear and tear on the isolator gloves. A second pair of gloves worn by the operator underneath the isolator gloves helps to protect the isolator from the operator's bioburden in the event of a pinhole leak in the isolator gloves. Care in selecting the glove material of construction based on compatibility with the chemical agents used in the isolator can greatly reduce the likelihood of glove failures. Gloves should be inspected and changed on a routine basis. It is a good business practice to implement routine glove integrity testing—for example, pressurizing and immersing gloves in water to look for small leaks, both prior to initial use and periodically during the use life. For critical applications such as sterility testing, these data are useful to help determine whether the isolator meets system suitability criteria.

Half-suits are commonly used to allow the operator to be physically in the isolator environment while being chemically and biologically separated from it. They allow a wider range of motion in the isolator than can be achieved from sleeve/glove ports. The half-suits cover the operator down to the waist and have a face plate to provide a clear undistorted view. The suits are usually made of flexible plastic materials similar to the sleeve materials and may be slightly thicker and more durable. The suits must endure strains from operators reaching and undergo repeated twisting motions, and the seams tend to be weak points. The half-suits should be flexible and long enough to allow operators to turn 180° or more and reach all areas of the isolator that they will need to access. Depending on the height of the operators and the vertical dimension of the isolator, the operators may need to have platforms available to stand on to reach upper areas within the isolators. The half-suits are often equipped with straps to hold them up for ease of entry/exit by the operator and to keep them contained within the isolator. The half-suite construction allows room air supplied by external fans to be circulated within the suits to cool the operator without compromising the isolator environment. For additional personnel protection for containment applications, operators can wear respirators or supplied breathing air equipment under the half-suits if necessary.

6.2.5. Isolator Ergonomics

Ease of access and working in the isolators should be a primary consideration in the design and fabrication of the isolator layout. Sufficient glove and half-suit ports should be provided to permit safe activities and minimal physical extertion by the operator and strain on the interface materials (gloves, sleeves, etc.). To understand the ergonomics of specific isolator uses, modeling or construction of a full-scale mock-up should be considered. Careful attention to these details can reduce operator fatigue and possible ergonomic-related work injuries, wear on interface components, and breaches of the isolator integrity, which could negatively impact product quality.

6.2.6. Isolator Ventilation

Isolators used in the pharmaceutical industry laboratories are generally of the turbulent flow design. Unidirectional (laminar) airflow isolators impose significant limitations on equipment design and placement within the isolator, and sleeves and half suits may also interfere with the airflow. At a minimum, HEPA grade filters are used at the inlet and exhaust flows for the ventilation system. Redundant and/or higher specification filters may also be used. The exhaust system may be designed to enhance the rapid removal of the bio-decontamination agent, since this aeration phase is a major part of time required to complete the biodecontamination process. The filters must be protected from wetting during normal operations, including the isolator cleaning process, and for isolator containment applications the design of the ventilation system must allow for filter changing without releasing hazardous materials in the filters.

The air change rate in isolators can be much less than the minimum 20 changes per hour used for clean rooms due to the smaller size of isolator systems and the exclusion of people-associated particulate contamination. Air velocity only needs to be sufficient to maintain a uniform flow of air for isolators designed for unidirectional flow, or to maintain adequate pressure differentials for turbulent flow isolators. Air classification within the isolator may not be a requirement for isolators used only for containment purposes.

Isolators used for sterile product manufacturing or testing must protect the product from bioburden and in the former case must also minimize nonviable particulate contamination. Where particulate classification is required, generally air is sampled near or on the work surface and testing is conducted under two conditions: static and/or dynamic conditions. For static testing, the cleaned empty isolator with the ventilation system turned on is tested while no activities are being performed. The isolator should meet ISO Class 4 (Federal standard 209E class 10) requirements of NMT 350 particles of 0.5 μm or greater per cubic meter. At least one particulate sample should be taken per 0.5 square meter of isolator floor area. No fewer than three samples should be taken for any unidirectional flow isolator. For dynamic testing, air samples are collected while equipment in the isolator is being operated. Dynamic testing is not required for sterility testing or nonaseptic applications as well as during powder processing operations. Isolators being used for sterile product manufacturing should meet ISO Class 5 (Federal Standard Class 100) during dynamic testing. Area classification for sterility testing isolators is not a requirement but is recommended as a good business practice. Some regulatory authorities expect sterile product manufacturing isolators to be in a classified area (ISO Class 8, FS209E Class 100,000) although this is currently a controversial topic.

Air recirculation is generally not a requirement for isolators, although some ventilation systems are designed to recirculate air through a catalyst or scrubber to facilitate removal of biodecontamination agents. The recirculation process can allow several hundred isolator air volume changes per hour to minimize the time needed to pass materials from the outside environment into the controlled isolator environment.

6.2.7. Isolator Integrity Testing

Isolators used for aseptically processed product manufacture or testing should not allow air ingress from the external environment unless the air passes through microbial retentive filters, to avoid potential product quality issues. Leaks are also a personnel safety concern during the vapor decontamination of isolators or during the handling of hazardous materials in them. Testing for leaks is a critical requirement and every isolator system will leak to some extent. Low leak rates generally are associated with an insignificant ingress of viable/nonviable particulates into isolators. If the isolators are maintained under positive pressure, then any leak should have minimal impact on product quality. An acceptable leak rate should be established based on personnel safety considerations. Leak tightness can be demonstrated in a variety of ways: two commonly used methods are the pressure decay test and tracer gas detection testing. The former involves either positive or negative pressurizing of the isolator to a specified level, then holding the isolator under static conditions (ideally including isolator and room temperature) and measuring the pressure decrease/increase over a time period. The American Glovebox Society has recommended a leak rate of not more than 0.5% of the isolator volume per hour. The latter test uses a tracer gas such as ammonia to be released within the isolator, followed by a detailed scanning of the isolator exterior with a detector to find leaks. It is inevitable and acceptable for isolators to leak to some degree, as long as the user understand the risks to product and personnel and they are deemed acceptable. The use of continual isolator air pressure monitoring and control design features as well as an alarm system can help reduce leak-associated risks. For product critical isolator applications, leak testing should be done on a routine basis, typically prior to daily use of the isolator and before biodecontamination cycles. The leak testing of isolator interface components such as gloves and half-suits should also performed routinely.

6.2.8. Rapid Transfer Ports (RTPs)

These systems allow the connection of transfer isolators, portable waste containers or other devices to stationary isolators to facilitate the movement of materials in and out of the main isolator. These ports rely on tight-fitting gasket systems to seal exterior port surfaces on the components being joined to prevent contamination once the port is opened. RTPs must be manufactured to exacting tolerances, and the vendor should be able to provide quality control specifications and performance testing data for the design. Because the design of these gaskets may not permit optimal exposure of some portion of the ring to the biodecontamination process, the potential for contamination from the RTPs exists. Care should be taken to avoid touching them with gloves or materials used during product processing or testing operations. The RTP gaskets may be swabbed during environmental monitoring activities. When RTPs are used in isolators for containment purposes, care must be taken to clean the gasket rings to protect personnel from product residues.

6.3. COMMON LAB ISOLATOR REQUIREMENTS

Isolators used for testing for microbiological quality attributes of pharmaceuticals such as sterility have some general specifications across the industry. These general specifications and any company-specific specifications are the foundation for acceptance criteria used in isolator qualification testing. These specifications should be documented early in the isolator procurement process, typically as part of the design qualification (DQ) phase. The two most common lab isolator user specifications are sterility assurance and

containment. The requirements for lab isolators used for strictly contain-
ments are the same as those discussed previously in the section on closed
isolator applications.

6.4. LAB ISOLATORS USED FOR STERILITY ASSURANCE

Use of isolators for sterility assurance purposes imposes the need for the
isolator environment to exclude any viable microbial contamination from the
testing process, so that if any microbes are detected in the testing their
presence is considered to have originated in the product. Since product
sterility testing is the most common use for lab isolator systems, the discussion
will focus on requirements for these isolators.

6.5. THE STERILITY TEST

Proving sterility of every container in a product lot and the entire isolator
environment is impossible. Limitations in the sampling process and the test
methodologies limit the sensitivity of sterility tests, but despite its technical
flaws it is still considered to have value in a final check to assure the quality of
pharmaceutical products claimed as sterile. The USP and European Pharmacopoeia sterility test methods typically require 10–40 final product containers
per test (depending on fill weight/fill volume and lot size). If feasible, the
product samples are dissolved in sterile water or other suitable aqueous
diluent, pooled and filtered through microbially retentive filter membrane(s).
The membrane(s) are rinsed to reduce product residuals and half of the filter
membrane material is immersed in soybean casein digest broth and the other
half in Fluid Thioglycollate broth. Products that are not water-soluble are
tested by transferring portions from each sample container to separate
containers of the two types of broth growth media. Incubation for most
products is a minimum of 14 days. A longer (21-day) incubation is required
for test broth containers that are rendered cloudy by the product and must be
subcultured to new media containers to detect growth. The media containers
are periodically examined visually for signs of growth over the incubation
period. Testing of negative controls include samples of the media, diluents,
and other components used during the test session that are also incubated to
verify their sterility. Environmental monitoring of the isolator for microbial
contaminants is routinely done during the test sessions to provide data as to
the microbiological state of the isolator. In addition, the specific test method
must be validated to be capable of recovering low numbers of a variety of
challenge organisms inoculated into the test procedure (bacteriostasis/fungi-
stasis testing) and the test media must also be checked to assure it supports the
growth of low numbers of the challenge microorganisms. A test result for a

product batch tested using a valid method and suitable growth supporting media showing no microbial growth meets the requirements of these compendial tests. A test result in which microbial contamination is detected means the lot fails the test and both lab and manufacturing personnel must rigorously investigate their respective areas for a cause. The lot must be rejected due to lack of adequate sterility assurance unless evidence from the test controls, isolator integrity testing, or environmental monitoring data shows a clear assignable laboratory cause for the contamination. Properly functioning isolators should make sterility test false positives a very low frequency event.

6.6. ISOLATOR ROOM REQUIREMENTS

Sterility testing isolators do not need to be in a classified environment, although the room should have limited access and be clean and well organized. The temperature and humidity of the room should be controlled to ensure operator comfort and to facilitate the conduct of isolator leak testing and the biodecontamination process. The utility requirements are similar to those of a conventional clean room and should be designed and built so as to not be a contamination risk to the isolator system. Point of use filters should be used with compressed air or inert gases and connections (including valves and control instruments) should be free of dead legs. All utility connections need to be leak-tight, and seals and gaskets should be regularly checked for leaks and wear. Vacuum systems should have a means to prevent backflow in the event of a power or system failure.

6.7. BIODECONTAMINATION EFFICACY

Isolators exclude microbes from the testing environment via the physical barriers designed into the system. In addition to maintaining this biological barrier, the isolator must be effectively biodecontaminated to inactivate any bioburden in the isolator prior to beginning the testing. The biodecontamination process is generally of relatively short duration (a few hours), and after it is completed the isolator relies solely on the integrity of the physical barriers and positive pressure of the ventilation system to prevent the reintroduction of bioburden into the isolator. All materials introduced into this isolator at this point must also be biodecontaminated, either by passing through an autoclave interface or via a transfer isolator or airlock system. Many items are sterilized in overwrapped packages using gas or ionizing radiation and are then loaded into the transfer system to undergo a surface biodecontamination cycle. The type of biodecontamination process used for the primary and transfer isolators or airlock systems may differ, but each type must be validated to show effectiveness in killing microorganisms. Another requirement is that materials exposed to a gaseous vapor biodecontamination cycle must be well aerated to bring residual levels down to a safe level from both the perspective of ensuring valid test results and personnel safety. It is known that gas/vapor residuals in exposed growth media can negatively affect its ability to support growth. Recommended safe exposure levels for hydrogen peroxide, a commonly used biodecontamination gas/vapor agent, is 1 ppm or less for a 8-hour time-weighted average. However, this level may not need to be attained to proceed with use of the isolator for testing. The user must conduct studies on the specific types of media and product containers to be used in the testing process to prove the gas/vapor residuals do not negatively affect the ability to recover microorganisms (4). These studies typically include growth promotion testing of media exposed to the worst-case decontamination cycle and may include testing media-filled product containers, or water-filled product containers spiked with low levels of inherently sensitive microorganisms. Residual biodecontaminant levels in the containers can also be evaluated by chemical means, such as indicator test strips (T. Burns, personal communication, 2003). Hydrogen peroxide vapor is the most common agent used for lab isolator biodecontamination in the United States. Peracetic acid, chlorine dioxide, and liquid hydrogen peroxide have also been used (5). It must be shown during qualification studies that the agent is distributed throughout the isolator in sufficient concentration to effectively kill microorganisms. Chemical indicators placed throughout the isolator, including hard-to-reach places, can establish whether the vapor/aerosol reaches various positions within the isolator. Biological indicators (BIs) consisting of a challenge organism (usually a bacterial endospore suspension dried on a coupon in a gas permeable pouch) are placed throughout the isolator, usually 5–10 BIs per cubic meter of volume. The challenge microorganism can be selected using references from the biodecontamination vendor or other sources that show the chosen organism is more resistant than the normal isolator bioflora. A 3-log reduction of a known resistant BI population is often used. This can be determined by a variety of methods:

1. Total kill analysis of BIs containing at least 10^3 spores/indicator, using three consecutive test runs with at least 50 BIs per run.
2. Fraction Negative Studies in which an exposure period calculated to deliver a minimum 3-log kill of resistant spores using the Holcomb, Spearman, Karber Procedures or Stumbo, Murphy, Cochran Procedures.
3. Overkill methods—for example a total kill approach using a resistant BI with a population of 10^5 spores/indicator or more is considered an overkll cycle.

4. An alternative is to test bioburden isolates (bacterial sporeformers are the most resistant) during cycle development studies to determine the most resistant isolate in the representative flora. Spores from the isolates can be placed onto coupons to make BIs. A 6-log reduction of the representative bioburden is sufficient to demonstrate efficacy of the biodecontamination process. It is important to note these approaches are intended to support the efficacy of a biodecontamination process, not a sterilization process (4).

6.8. MICROBIOLOGICAL MONITORING OF STERILITY TESTING ISOLATORS

The monitoring regimen should include a combination of air and surface sampling methods, such as active air sampling, passive air sampling (settle plates or broths) and surface sampling (RODAC plates, swabs, flexible films).

For glove sampling (swabs, immersion in rinse fluids, RODAC or flexible film), the frequency, locations, exposure duration, media types, and sampling equipment must be documented in procedures. The acceptable specification for a properly functioning sterility testing isolator is no organisms detected: an out-of-specification monitoring result should lead to an investigation. Due to the time (several days) that elapses between the sampling and detection of the contamination, the use of daily biodecontamination cycles may in some circumstances limit the impact of the failing result on subsequent isolator operations. Identification of the contaminants to at least the genus level is desirable, and results should be tracked and analyzed for trends.

6.9. ISOLATOR CLEANING

For sterility testing applications, a specification for isolator cleaning of visibly clean is generally adequate. Special cleaning regimens may be needed if antimicrobial powders (e.g., antibiotic products) are tested because these materials could interfere with the testing of other products and the microbiological monitoring of the isolator. The cleaning methods, equipment, frequency, and cleaning agents must be documented (6).

6.10. CONCLUSION

In the past 15 years, isolator technology has gradually joined partial barrier enclosures like biological safety cabinets and fume hoods as common equipment in pharmaceutical manufacturing and testing laboratories. It is important to understand the design features of each type of equipment to optimize their proper use and have an awareness of their limitations.

REFERENCES

1. Del Ciello, R. Designing a pharmaceutical facility—design considerations. In *Microbiology in Pharmaceutical Manufacturing*; Prince, R., Ed.; Davis-Horwood: Surrey, U.K., 2001; 175–202.
2. Chapter ⟨1211⟩ Sterilization and Sterility Assurance 26th Ed.; United States Pharmacopeia: Rockville, MD, 2003.
3. U.S. Department of Health and Human Services. *Biosafety in Microbiological and Biomedical Laboratories*; 3rd Ed.; U.S. Government Printing Office, 1993.
4. Technical Report No. 34, Design and Validation of Isolator Systems for the Manufacturing and Testing of Health Care Products. Davis Horwood, 2001.
5. Technical Report No. 36, Current Practices in the Validation of Aseptic Processing-2001. Davis Horwood, 2002.
6. Chapter ⟨1208⟩ Sterility Testing—Validation of Isolator Systems; 26th Ed.; United States Pharmacopeia: Rockville, MD, 2003.

7

Developing a Process for Aseptic Facility Design and Validation

Dimitri Wirchansky
Jacobs Engineering, Conshohocken, Pennsylvania, U.S.A.

7.1. INTRODUCTION

When preparing for a capital project in aseptic manufacturing, it is important to know what is necessary to carry out the project to completion. While a capital project in aseptic manufacturing is not that different from any other capital project, the aseptic nature of the operations add technical complications to project execution. In the following text, I have attempted to describe the process of performing a project, what resources are needed, and how the various groups interact. If the participants in a project understand the process, they can participate more effectively and increase the project's chance for success.

7.2. DEFINING THE PROJECT

Before a project reaches the implementation stage, the project has to go through a definition phase. There are several terms used for this activity, which generally fall under the heading of preconceptual design activities. Some organizations work with design and engineering companies for this activity, and some organizations keep this activity completely in-house. This depends on the staffing and the level of skills available within the organization. In any case, the initial definition of the project must come from within the organization itself.

7.3. PRELIMINARY PROJECT TEAM

Once a preliminary scope for the project has been defined, a small group of skilled and knowledgeable people are required to complete the process of defining the project scope and facilities impact and to develop a preliminary estimate of the cost. Timing of the project must also be defined leading to a schedule. For the early stages of the work it is important to use knowledgeable, experienced people with the right skill-set for the project. Usually, this involves a process architect and process engineer. The process engineer should know the process involved in the project and the architect should know the people flows and material flows needed for the process. Experienced people are required for this work, because many aspects of the project are undefined at this point. The people involved need to be able to determine what is important to the project and direct their efforts to these items as they work to define the project scope. It is also important for the people involved to have a working knowledge of the cGMP's.

7.4. DELIVERABLES FOR THE PRECONCEPTUAL PHASE

One of the more important tasks of the early work is to adequately define the scope for the project. Defining the scope is to determine what has to be done as part of the project to achieve a desired result. Identifying the objective for the project is an important part of scope definition. Anything that does not contribute to the desired result may be optional to the scope. Yet everything that is required to achieve the desired result must be included as part of the scope for the project to be successful. For example, if the project is to increase the output of SVPs by the installation of a new filling line to an existing facility, the new filling line, the filling area, and supporting utilities are all part of the scope. A new labeler for the packaging line should not be part of the scope unless the existing equipment in the packaging area is unable to label the vials or keep up with the required output. The equipment needs, building needs, and utility needs have to be defined.

Another important part of defining a project is to identify timing needs. Is the new filling line required for a new product launch that must be accomplished quickly, or is it to respond to product forecasts that foresee growth over a several year period? Must we pull out all stops to get this line purchased and installed quickly, paying for overtime and delivery incentives, or must we concentrate on costs proceeding at a more normal pace, devoting efforts to cost reduction as the project progresses? This is another way of asking what the important drivers of the project are. Is the project schedule driven, where cost is sacrificed to speed implementation? Is the project cost driven, where schedule is sacrificed to reduce cost?

Cost is always important to a project. Early in the development of a project, many things are not well defined. That is why it is important to identify items important to the project's success. When critical items are identified, it is less likely that an important cost issue will be overlooked. To return to the example of the new filling line, does formulation have enough capacity to supply the line? Is there enough capacity in inspection? What if the product is a suspension? Have we made suspensions before? Do we know how to prevent uniformity problems? Any of the issues mentioned in the example can affect the project cost. At the early stages of a project, it is likely that the level of information will be incomplete. A historical project database and industry benchmarking are examples of ways to mitigate incomplete information.

In order to document the information that describes the project, the following documents are used. In many cases they are called "deliverables" because they are delivered as documentation at the close of the preconceptual phase. These include:

Preliminary building layout to include equipment
People flow and material flow for this layout
Equipment list (budget priced)
Process block flow diagrams
Preliminary schedule
Preliminary estimate (in the range of $+/- 30-50\%$)
Identify critical issues

At this time it is useful to identify important cGMP issues as they relate to the project. This is typically covered in a cGMP Design Brief, which is prepared with input from Regulatory, QA, and Validation. This brief covers the GMP highlights of the project. Sensitive issues should be identified, and plans made to address them.

If the project deals with potent or toxic materials, a Containment Plan should be prepared. The Containment Plan lists the hazards and limits of exposure as well as a strategy to meet these limits under normal, set-up, cleaning, maintenance, and upset conditions. If the product is a biological, the Containment Plan must incorporate biological guides and standards.

The cGMP Design Brief and the Containment Plan may be revised and updated as the project progresses and additional information becomes available. However, it is useful to identify relevant cGMP and Containment issues early to allow more time to adequately address these potentially critical items. Early recognition and inclusion of the needs of commissioning, qualification, and validation can save time and cost to the project overall.

7.5. CONCEPTUAL DESIGN

The preconceptual design work is general in nature. The purpose is to define the scope and develop a path forward to advance the project. The conceptual design phase builds on the previous work and develops additional details to fill in the specifics of what the project will acomplish and how the project will be executed.

7.6. DEVELOPING THE PROJECT CORE TEAM

Although the preliminary work can be performed by a very small group, a core team comprising key disciplines as well as user representation is required to develop the conceptual design. This core team should be formed from representatives of key disciplines. This usually requires a project manager, a process architect, a process engineer, an HVAC engineer as well as an owner representative for manufacturing and QA-Regulatory. Each team member has a role to fill and deliverables to produce. The number of disciplines involved depends on the requirements of the project. Civil and Structural may be involved if the project is a new building or site. Instrumentation and Controls may be involved for automated systems.

7.6.1. Architect

The architect's job is to identify the building functions required and estimate the space required for these functions. This is accomplished as part of programming the building space. The program is a list of building spaces along with the equipment in each space and an estimate of the area required for each of the spaces. Summing the spaces provides an estimate of the total building space required for the project. Categorizing the spaces into general, controlled-non-classified, grade D, grade C, grade B, grade A, and so forth allows an estimate of building cost to be generated on a dollar per square foot basis. The architect has to understand the process and the flows for each step of the process to develop the requirements for adjacencies or which operations have a process need to be near each other. Once the adjacencies are understood, a building layout may be developed. The layout is then tested for people flows and material flows to verify that cGMP concerns and operational issues have been adequately addressed. This process may lead to revisions in the layout or to the development of several alternatives for the layout that will be evaluated by the team.

7.6.2. Process Engineer

The process engineer is responsible for the configuration of the process. Much the way that the Architect is responsible for programming the building spaces, the process engineer must "program" the process. The process engineer must define what operations are required and how these operations fit together to make the product and meet the objectives of the project. Preliminary Process Flow Diagrams, or PFDs, are developed for each step of the process. The PFDs define what equipment is needed and how the various pieces of equipment fit together and interact with each other. Once the PFDs are developed, an equipment list is made to identify what will need to be purchased to execute the project. This equipment list is reviewed for budget pricing to provide input to the estimate. A process description is also prepared to describe in words how the process works. The PFDs, the equipment list and the process description should all be consistent with each other. A review of one document may lead to revisions to another. This provides a check that all items needed to run the process are provided for.

Sometimes, technical issues requiring further study may be identified as part of this process. These items represent technical risk for the project. An important part of the conceptual design is to identify technical risk and develop a plan to mitigate this risk as the project advances. Items of risk are best flushed out and identified early to avoid unpleasant surprises later in the project. The process engineer must also define the clean utility services required and estimate the demands for these clean utility services. The process engineer must also identify the need for clean utilities such as WFI and Clean Steam, and so forth and estimate the quantities required.

7.6.3. HVAC Engineer

The HVAC engineer has to understand the process and develop plans to support the various parts of the building with the proper quantity and quality of air as defined by the area classifications. An Area Classification Diagram will be prepared in cooperation with the architect and process engineer. This diagram identifies the classes of air required and where they are located. Air flow diagrams are prepared to identify how the various systems will be configured, and zoning diagrams will be prepared to indicate the areas served by each air handler. The HVAC engineer must also identify the need for plant utilities such as plant steam, chilled water, instrument air, and so forth and estimate the quantities required.

7.6.4. Manufacturing Representative

The manufacturing representative is essentially the primary user. This person should develop a clear picture of what is needed for manufacturing and support activities and communicate this to the project team. This requires that a generalized philosophy of operations be developed. Will the operation run over three shifts, two shifts, or less? How much time will be allocated for area shutdown and repairs? Will product be made on a single lot or campaign basis? Manufacturing input is required to properly configure the facility and equipment to produce the product to meet demand and meet the organization's operating parameters.

During this phase of work, it is useful to develop user requirements specifications for critical long lead equipment. Each company may have its own preferred format for the user requirements specification. This document should list important features of the equipment or system as well as critical performance criteria. The document requires input from the manufacturing, as well as QA and validation groups.

7.6.5. QA Representative — Validation

The QA representative is an example of a secondary user. While the primary purpose of the project may be to produce product, this product must be produced in conformance with corporate procedures and guidelines as well as those of the regulatory agencies where the product will be sold. The QA representative must understand the project and how the regulations affect the project. To serve as a guide for the design effort, a cGMP Design Brief is reviewed and modified as required. This document calls out important cGMP issues as they relate to the project and defines how these issues will be addressed as part of the project. As an example, this document will address concerns of product segregation and cross contamination, such as those required for penicillins or live virus vaccines, or other product properties such as uniformity of dose for a suspension.

Once the equipment list is developed, validation can begin to develop the Validation Master Plan. This plan will be updated and require further development as the design progresses.

7.7. DEVELOPING THE BASIS OF DESIGN OR "BOD"

Most companies have a procedure for developing the BOD. Content and level of detail are often described for uniform execution of various projects. Sometimes this work is performed by internal company resources. Many times external support is required because the internal people have other responsibilities and cannot devote sufficient time to perform the project work and document the development of the design.

7.8. LEVERAGING THE EFFORT

7.8.1. Lining Up Internal Support

It is important to determine the level of internal support available for the project, and to determine what level of external assistance is required. An internal company project manager may be assigned to bring in external resources and work with internal resources assigned to the project on a part-time basis. In this case the objective is to leverage the knowledge of the internal people by bringing in external support to understand the needs of the project and develop project documentation. In this case the availability of internal people is key to developing project understanding and direction. Once this is done, the defined work may be performed by outside resources. If the work is to be performed internally, there is less concern for misunderstanding of project issues; however, there is more concern for the amount of time internal people can devote to the project.

7.8.2. External Support

If internal resources are limited, it may be necessary to bring in external resources in the form of an engineering company or consultants. When this is done, it is beneficial to define rolls and expectations. It is also necessary to review the backgrounds of the potential external resources to assure that the people involved have the necessary knowledge and background for the project.

Typical deliverables for a BOD are as follows:

Programming summary
Preliminary layout
People flow diagrams
Material flow diagrams
PFDs
Equipment list
Process description
Containment plan (if required)
Air flow diagrams
Zoning diagrams
Utility flow diagrams
cGMP design brief

Validation master plan
Schedule
Estimate to $+/- 30\%$

7.9. THE DESIGN QUALIFICATION PROCESS

Recently, considerable attention is being given to the process of Design Qualification. The purpose of design qualification is to verify and document that the design meets Quality and User requirements. The design qualification involves the following tasks:

Define user requirements
Perform system classification
Document that systems or equipment address user requirements

Design Qualification is called for by the Q7A Guide to cGMP for API and the Orange Guide Annex 15:

7.9.1. Q7A Guide to cGMP for API

Before starting process validation activities, appropriate qualification of critical equipment and ancillary systems should be completed. Qualification is usually carried out by conducting the following activities, individually or combined:

Design Qualification (DQ): documented verification that the proposed design of the facilities, equipment, or systems is suitable for the intended purpose.
Installation Qualification (IQ): documented verification that the equipment or systems as installed or modified comply with the approved design, the manufacturer's recommendations, and/or user requirements.
Operational Qualification (OQ)
Performance Qualification (PQ)

7.9.2. Orange Guide, Annex 15

The first element of the validation of new facilities, systems or equipment could be design qualification (DQ).
The Compliance of the design with GMP should be demonstrated and documented.

Design Qualification is not mandatory for compliance of manufacturing facilities regulated by the FDA. Design Qualification is not referenced in regulatory publications as regulation rules or guidelines. However, Design Qualification is well referenced in current Good Manufacturing Practices (1).

The code of federal regulations (CFR) title 21 Part 211 Subpart C— Buildings and Facilities, and Subpart D Equipment, make specific reference to "appropriate" design and to "suitable" or "adequate" size, construction and location for cleaning, maintenance and proper operation of facilities, utilities, and equipment (1).

7.10. EXAMPLE OF AN APPROACH TO DESIGN QUALIFICATION

7.10.1. User Requirements

A User Requirements document defines clearly and precisely what the user wants the system to do. It defines the functions to be performed, required system output, and the operating environment. The emphasis should be on the required functions, not the method of implementing those functions. If the User Requirements are not appropriately defined and measurable, the DQ process will be more difficult to perform.

The user requirements must be incorporated into the bid specifications and purchase specifications to assure that these needs are being met. The process of documenting that these user requirements are covered as part of the design is the purpose of Design Qualification.

7.10.2. System Classification

A system may be defined as an organization of engineering components that have a defined operational function (e.g., piping, instrumentation, equipment, facilities, computer hardware, computer software, (2)). To perform system classification, systems are classified as having direct impact, indirect impact, or no impact on product quality. Once the systems have been classified, classification of individual or groups of similar components as critical or noncritical must be completed. This classification must be performed for all components in direct impact and indirect impact systems. Every critical component must be part of a direct impact system. Systems that have direct impact on product quality will undergo DQ, commissioning, IQ, OQ, and PQ. Systems that have indirect impact on product quality will undergo DQ, commissioning, and IQ. Systems that have no impact on product quality will undergo commissioning.

7.10.3. Tying It Together

When the equipment, system, and/or facility is delivered, prior to Process Validation, it must be checked to verify that the final output of the design meets the user requirements and is suitable for its intended purpose. The documentation of this verification step completes Design Qualification.

The description above is one example of how design qualification may be implemented. Each company may have its own procedures and formats for implementing design qualification. The format used will influence the content of equipment and system specifications. This has to be addressed early in the design to assure that the needs of design qualification are integrated into the design process.

It should be noted that this process is somewhat similar to the GAMP model used for process automation. The process is similar but it is not the same, because the needs of process automation and pharmaceutical systems are not the same.

7.11. PRELIMINARY ENGINEERING

Preliminary design builds on the conceptual design work. The project team expands to all disciplines required for execution, and the true "production" work for building the design ramps up. If the BOD has not been finalized earlier, it will be finalized in the early stages of preliminary design. The objective of this phase of work is to finalize the technology and project objectives, and develop engineering and pricing to support a budget cost estimate and funding request.

During this stage of design, the following activities will take place:

7.11.1. Architecture

The architects will perform a code review to check the design for compliance with the relevant building codes. They will also prepare any demolition plans if needed, as well as floor plans and reflected ceiling plans for the new construction. Flow diagrams will be prepared for people, materials, product, and waste.

7.11.2. Environmental Health and Safety (EHS)

EHS wil prepare a hazard analysis and a permitting plan.

7.11.3. Process

Process will continue to expand the process description and prepare P&ID's and utility flow diagrams or UFDs. They will also prepare equipment data sheets for long lead equipment. Process, in conjunction with EHS, will hold a safety review to prepare for the HAZOP.

7.11.4. Process Mechanical

Process Mechanical will prepare specifications and data sheets for long lead equipment. They will also work with Process and Architecture to develop the layouts and facility flows.

7.11.5. Civil

Civil will prepare a geotechnical report, a site plan, a site utility plan, a soil erosion control plan, site zoning plan, a site characteristics plan, and site details. They will also develop specifications for site preparation, earthwork, excavation and backfill, paving, site domestic water system, foundation drainage, storm sewer system, and sanitary sewer system.

7.11.6. Structural

Structural will develop foundation plans and foundation sections, as well as building framing plans and column schedules, concrete details and structural steel details. They will also prepare specifications for cast-in-place concrete and/or structural steel.

7.11.7. Electrical

Electrical will prepare an area classification plan, a grounding plan, single line diagrams, as well as power plans and lighting plans. They will also prepare appropriate specifications.

7.11.8. Instrumentation and Controls (I&C)

Instrumentation will provide input to process P&ID's. They will also develop system architecture block diagrams, typical loop diagrams and installation details. Instrumentation also provides an I/O point list, an instrument list and instrument data sheets. They develop specifications for PLC/SCADA hardware, control panel fabrication, operator control stations, instrumentation and instrument air supply, sequence of operations, testing, documentation, and system training.

7.11.9. Building Automation Systems (BAS)

BAS provides specifications for operator control stations, distributed processing unit (DPU) systems, and an I/O point list. They also prepare a system architecture block diagrams and P&ID's for HVAC units-air side.

7.11.10. HVAC

HVAC systems are important to any aseptic facility. The HVAC engineers will work closely with the architects and the process engineers to develop area classification diagrams, zoning diagrams, and pressurization plans. They will also produce airflow diagrams, sections and P&ID's for HVAC utilities.

7.11.11. Process Piping

Piping produces the equipment arrangements as well as lane studies and pipe rack studies. Process piping works on equipment arrangements to verify that all utilities are routed to the equipment and that the equipment and utilities are able to be serviced and maintained. They also produce general piping specifications. Process piping works primarily on the clean utilities.

7.11.12. Plumbing

Plumbing will develop pipe routing plans, riser diagrams, and a roof drainage plan. They also prepare utility P&ID's.

7.11.13. Manufacturing Representative

The manufacturing representative will develop strategies for operation of the system or facility. These strategies will serve as input for the process and process mechanical engineers to use in finalizing the configuration of the process systems and equipment. This work is also used to confirm adjacency models, which leads to the confirmation of the floor plan. The manufacturing representative will also work with the process mechanical engineers to prepare to place orders for critical and long lead equipment. Procurement will also work to develop the terms and conditions that will be used for placing the orders.

7.11.14. QA Representative — Validation

The QA and Validation representatives will participate in design reviews to assure that company quality standards are being met and cGMP issues are being addressed appropriately for the corporation. The validation master plan has been developed and the validaton representative will assure that proper documentation is being requested from vendors to meet the project's needs for Factory Acceptance Testing (FAT) and site testing.

7.12. THE PROJECT FUNCTION

The project function covers primarily cost, scope, and schedule. When preliminary engineering is completed, the scope should be understood and documented. A schedule for the project will be developed based on primary project drivers, and a cost estimate will be developed that can be measured against as the project progresses. The direction is now set and detailed design may begin.

7.12.1. Detailed Design

The objective of this phase of the project is to manage, coordinate, and perform design activities necessary for the construction and start-up of the facility. This also includes the completion of procurement activities necessary for economic, schedule, or detailed design reasons.

During this phase of design, procedures for change management will be initiated. The scope has been defined and a cost estimate has been prepared to serve as a baseline. Most of the procurement of equipment and materials will occur in this phase.

As the process of detailed design is carried out, it is important to coordinate all the activities and external interfaces. Design reviews will be held to address issues of internal coordination, constructability, cGMP and regulatory compliance, safety (HAZOP), maintainability and operability, and so forth. Items raised during these reviews will be incorporated into the design. It will also be desirable to obtain regulatory authority for the design.

Deliverables for this phase include purchase order documentation, construction documents, control systems programs, appropriate license and permit applications, project schedule, project data books and change control documents.

7.12.2. Construction

During the construction phase, the focus of the project shifts to the construction site. During this phase, equipment will be procured and contracts for construction services will be implemented. A detailed, integrated project schedule will be developed to coordinate the various types of work. Using the example of the filling line, the line cannot be installed before the building is prepared and the floors and most of the walls and ceilings have been installed. Yet the parts of the line have to have a pathway from the outside to their final position. The dry heat tunnel is likely to be the single largest piece of equipment. There has to be a way to take it off the truck, place it into the building, then move it to its final location without damaging the equipment or the building. Large lyophilizers are also a challenge.

If the project involves modifications to an existing aseptic operation, extreme measures are required to perform the construction activities while the facility remains operational. The construction area should be separated from the operating area. The air systems should be separate and the area should be accessed from a different path than that for operations. It is also important to keep operational supervision advised of construction activities as this may impact operations. A system of scheduling construction work and reviewing this plan with operations on a daily basis is one way to avoid construction impacts to ongoing operations.

When pharmaceutical equipment is purchased, this is usually accomplished with technical specifications and data sheets and vendor data requirements forms. We have to tell the vendor what we want in the equipment (the user requirement information is incorporated into the technical specifications), and what documentation is required. If we don't get the proper documentation, design qualification and validation becomes much more difficult. The vendor will make periodic submittals as defined in the vendor data requirements document. Some of the submittals require review and approval and some are for information. This information is reviewed as part of the design qualification. This information is also used by engineering disciplines to finalize their design. The WFI drop and drain line have to be located properly to serve the vial washer. The air supply and power line have to be located properly to serve the dry heat tunnel, the proper utilities have to be available for the parts washer, and so forth.

Before a critical piece of process equipment can be delivered to the site, it undergoes factory acceptance testing or FAT. Many companies will develop FAT protocols that mirror the site acceptance testing (SAT) as well as IQ and OQ as much as possible. This work is documented and becomes part of the qualifications package for the equipment. It is a good idea to perform the same tests at the FAT that will be performed later as part of validation. In this way, difficulties can be found early and rectified. Sometimes this cannot be done. For example, performing a test of vial handling on the filling machine at the factory can produce optimistic results. When the glass is washed and sent through a dry heat tunnel exiting into a clean room, the glass surface acts like it is "sticky" and can result in vial handling problems. Because most filling machine manufacturers don't have clean rooms in their factories, FAT testing may not reveal vial handling problems. The normal environment has dust and humidity that helps to "lubricate" the surface of the glass, enabling the glass to move more easily on turntables and feeding devices. Although not everything can be verified at FAT, this testing is important to the overall task of building the facility. Any problem found and rectified during the FAT is one less problem to face during commissioning and start-up.

7.12.3. Start-Up, Commissioning, and Validation

Once the equipment has been installed, it may be possible to begin the start-up process. A key variable to this process becomes the utilities needed and any other support equipment or structures required. For example, you can't start up an autoclave unless the clean steam and instrument air systems have been started. Many companies are trying to streamline the process of start-up, commissioning, and validation. Much of what is required in IQ is also required for commissioning. Therefore, if the process of commissioning also includes the documentation required for IQ, this part of the work only needs to be performed once. If the checks required for commissioning are not documented, they will have to be performed over again as part of the IQ.

With the equipment installed and capable of running, OQ may begin. As the site acceptance testing is performed, it may become the start of OQ. Testing relevant to the equipment that was not performed as part of the FAT, for whatever reason, should be performed as part of the SAT. Any discrepancies in equipment performance should be investigated and rectified. Draft Operating and Maintenance Procedures should be developed to prepare for PQ. The Validation Master Plan defines the testing required for each system. A schedule should be prepared to integrate the needs of the equipment and the facility. While the initial equipment testing is taking place, testing of the facility and facility systems are also taking place. All systems for facility and equipment must be ready to begin the PQ phase. If the work leading up to Process Validation or Performance Qualification has been well done, there should be few surprises or upsets. The performance qualification work should proceed routinely and lead to an operable facility in a timely manner. If significant items have been missed or not addressed, performance qualification will require additional effort and take additional time. If the work of Design Qualification and IQ and OQ are integrated with the project work, with as much of the work as possible performed and documented early, the overall timing and expense to deliver an operable facility can be reduced.

REFERENCES

1. ISPE Baseline Pharmaceutical Engineering Guide, Volume 5, Commissioning and Qualification, Section 7.2.
2. ISPE Baseline Pharmaceutical Engineering Guide, Volume 5, Commissioning and Qualification, Glossary.

8

Pharmaceutical Water Systems
New Orientations in System Design

Theodore H. Meltzer

Capitola Consulting Co., Bethesda, Maryland, U.S.A.

Robert C. Livingston

Arion Water, Hyannis, Massachusetts, U.S.A.

8.1. INTRODUCTION

The literature contains an ample discussion of the several facets of pharmaceutical water requirements: regulatory constraints; system designs; operational protocols; the microbiological imperatives, biofilms, sanitizations and bioburden analyses; and the various aspects of the validation exercise including that of the individual purification units (1,2). A duplication of such efforts is not intended in this chapter. Indeed, it is assumed that the reader will have some fair knowledge of the principles and operations involved in water treatments such as softening, filtration, total organic carbon (TOC) ion-exchange, reverse osmosis, and so forth. The intention of this writing is to suggest new outlooks for designing pharmaceutical water purification systems, particularly based on practices in other applications.

8.2. WATER PURITY STANDARDS

The degree of water purity suitable for any particular purpose depends, of course, on the applicational requirements. Electronic rinsewaters are used to free semiconductor silicon chips of particles, ions, and undissociated molecules, such as organics, that may detract from their inherent electrical properties. Appropriate standards are set for these rinsewaters by the semiconductor manufacturers (3).

The standards pertinent to the various pharmaceutical waters are, by law, those listed in the Monograph sections of the U.S. Pharmacopeial compendium, as endorsed by the FDA. Because these organizational entities are concerned with the well-being of the American public, and with the purity, safety, and efficacy of drugs, it is sometimes erroneously assumed that the standards for the pharmaceutical waters are set on the basis of their physiological significance.

The early standards set by the USP, which is a standard-setting body, listed substances expected to be found in waters whose removal would result in an untainted liquid, pure water. The freedom from the various listed contaminants was defined in terms of their analytical detection by wet chemistry test procedures. Thus, the absence of chloride ion was attested to by the non-appearance of a white precipitate or cloudiness of silver chloride when an acidified silver nitrate solution was added to a water sample. This was meant to signify zero chloride ion content. If the test indicated the presence of chloride ion, the need to remove it would thereby be signaled. The concern with chloride was that it was a water impurity, whether with or without physiological implications. The same considerations applied to other suspected ionic, and ion-producing contaminants such as calcium, pH or hydronium ion, sulfate, carbon dioxide, and ammonia; each assessed by its wet chemistry test. The presence of organic compounds, largely unionized, was to be detected by the "oxidizable substances test," an analysis wherein a permanganate solution is bleached of its purple color by the reducing action of oxidizable organics. The persistence of the permanganate color upon addition to a water sample would signify the absence of organics.

Subsequently, the qualitative assays indicating the presence or absence of a contaminant were translated into quantitative values defined by the sensitivity of the test detection method. Thus, less than 4 ppm chloride ion in a water sample will not be detected by the silver chloride precipitate method. Therefore, the non-appearance of cloudiness in the qualitative test, signifying zero chloride ion, becomes interpreted as a tolerable chloride ion concentration of less than 4 ppm. From this it is assumed by some that chloride ion concentrations above 4 ppm are a danger to health, and that a water with a higher concentration is more dangerous. In the days of innocence, what was sought was simply an assurance that chloride ion was not a contaminant in the subject water; nothing more, no quantitative measurement was intended and none is of significance. Whatever chloride ion is present requires being removed. The same applies to the other specified impurities (Table 1).

In the case of the organic impurities, the standard set by the USP and enforced by the FDA is a limit of 500 ppb. The intention of this standard is the defining of Purified Water or Water for Injection for manufacturing purposes. Waters with TOC readings in excess of 500 ppb may not be designated by these appellations. The 500 ppb limit is absolutely without physiological meaning. TOC is too general a classification to be described in terms of specific health implications. The injection of many organics can be fatal, whereas polyvinylpyrrolidone has found beneficial use as a blood extender. In this sense waters with TOC levels lower than 500 ppb, while purer, are not necessarily more desirable or less harmful, if at all.

8.3. ELECTRONIC RINSEWATER

Understandably, almost without exception, the focus in pharmaceutical water preparation is upon the goals of the pharmaceutical field as characterized by high standards of purity, especially with regard to microorganisms and their pyrogenic endotoxins. Curiously, the practices of the semiconductor industry in its preparation of high-purity waters is virtually ignored by the pharmaceutical practitioners. Yet, in most respects, if not all, the electronic

TABLE 1 USP Compendial Water Standards

Type	USP purified	Water for injection
Chloride, MG/1	2.0	2.0
Total solids, MG/1	10	10
Microorganisms, maximum per 100 mL	—	—
pH	5.0 – 7.0	5.0 – 7.0
Sulfates, MG/1 as SO_4	4.0	4.0
Ammonia, MG/1 as NH_3	0.3	0.3
Calcium, MG/1	4.0	4.0
CO_2, MG/1 @ 25°C	5.0	5.0
Heavy metals, MG/1 as CU	1.0	1.0
Oxidizable substances as O_2	0.8	0.8
Pyrogens	—	Absent by Rabbit Test 0.25 EU/mL

rinsewaters are more stringent in their requirements. Albeit without the restrain's of governmental oversights, they are, nevertheless, held to the severe technical requirements and the practical realities imposed by the strong competition of the commercial marketplace.

In the preparation of electronic rinsewaters, the goals aimed for are the optimal purity of >18 megohm resistivity, TOC lower by magnitudes, and a striving for organism counts of zero. Bioburden, as a source of particles, while innocent of physiological implications, has ruinous effects on the silicon wafer quality and hence on operational profitability (need more be said!). Therefore, the tolerance for particles is much lower than in pharmaceutical waters. Electronic systems are essentially room-temperature operations. Despite the absence of the elevated temperatures known to discourage microbial growth, disruptive sanitization practices are reduced, and reverse osmosis (RO) cleanings much less exercised in the semiconductor application. An optimization of the RO function results, possibly including a prolongation of the RO unit's life with fewer downtime interruptions. RO unit cleaning and sanitization should be minimized in pharmaceutical water systems as well.

One of the intentions of this chapter is to bring to the pharmaceutical scene advantages that may be forthcoming from the semiconductor practices. In one respect, namely that of TOC analysis, such was accomplished some years ago when the USP approved as replacements for its permanganate-based oxidizable substances test, TOC measurements, already being used by the semiconductor people.

It is interesting that for pharmaceutical waters where as much as 500 ppb of TOC is tolerated, the regulators have strong concerns about extractables from the polymerics. Meanwhile, in the electronics operation where the TOC level attained is 10-fold less than in pharmaceutical waters, the many years' use of polymerics in place of the Austenitic steels continues to be a universal and successful practice (3).

Certain of the high standards of the semiconductor water purification practice are considered not to be required in the pharmaceutical context; very low TOC may be a case in point. For semiconductor rinsewaters the TOC may be less than 5 ppb. Consequently, the organism counts are quite low, fewer than 1 cfu/L, there being a rough correlation between the organism population and that portion of the TOC that serves as its food. It is strongly recommended that TOC levels be kept below 20 ppb for pharmaceutical waters. Other techniques beg for imitative implementations as well. The substitution of polymerics for stainless steels in general, and the avoidance of stills in particular, would eliminate the rouging, utilities support, and metallic content that are often troubling concerns in pharmaceutical water systems and that necessitate the downtimes and costs of intermittent cleanings and passivations.

The elimination of stills, especially for the preparation of Water for Injection, is a most unlikely event given the harmonization effort among the international regulatory authorities. The insistence of the regulatory agencies may well be correct, but their lack of explanation or technical justification can be judged an impediment to the application of new devices and techniques whose development requires and deserves collegial discussions and evaluations within the entire community of industry, academia, and governmental agencies. Encouragingly, the FDA has more recently more deeply involved industry in its technical deliberations.

8.4. EMPHASIS ON PRETREATMENTS

Discussions of pharmaceutical water purification often follow a logical path that traces the train of successive purification units from pretreatment details through principal purifications to storage and distribution operations. In this writing the common bases for the principles governing ion-exchange, reverse osmosis, EDI, and distillation are assumed. Aside from elaborative remarks—usefully informative, it is hoped—the focus will be upon pretreatment units where recent advances have been forthcoming.

At least in principle, any water, regardless of how contaminated, can be purified by any of the principal techniques to a compendial water quality. The question is, how much can be prepared before the process deteriorates and/or the equipment is ruined? Thus, ion-exchange operations can be so managed as to minimize ionic leakage for a time; ROs can be operated suitably until compromised by fouling, whether by mineral depositions, biofilm buildup, or the consequences of concentration polarization; distillations can be operated without contaminative endotoxin or mist entrainments until rendered uneconomical by progressive vitreous silica glazing, or being irrevocably ruined by chloride stress cracking.

The pretreatments are meant to determine, by validation, a documented experimental exercise, how the desired effects of the main purification units can be extended, and for how long, by regenerations, refurbishings, replacements, and by the ameliorative treatments of cleanings, sanitizations, and preventive care. The justification for focusing upon the pretreatment operations arises from the certainty that timely prevention is to be preferred over the cost of item replacements, or as compensation for preventable damage.

The operational requirements of the principal purification units are well known; the proper protocols for the pretreatments apparently less so. As stated, the pretreatments are intended to remove from the feedwaters those contaminants whose contact with the principal purification units would compromise their functions. Thus, the removal of the divalent alkaline earth or hardness elements spares the ion-exchange beds from premature exhaustion; the RO membranes from mineral fouling; the still from developing heat-insulating deposits. The removal of chlorine, chloramine, or ozone initially added to the feedwater for biocidal purposes protects the EDI and/or RO membranes, and the ion-exchange resins from oxidative degradations. Certain pretreatments conduce to the intended final water quality by contributing to TOC removal, and to endotoxin removal through ultrafiltration. Such, too, can be the functions of the several principal purification operations. Their chief purpose is to remove distiller-compromising silica; ionic contents contributory to excesses over conductivity specifications; the reduction of viable organism populations, as by adsorptions to resin surface, by RO or EDI filtrations, or by the thermal killing effects of the distillation process; and as abetted by sanitization practices. In any pharmaceutical water manufacturing installation, the pretreatment and principal purification units that are required become revealed by the analysis of the source waters. It is in this manner that one learns what is present and what needs to be removed.

8.5. SOURCE WATERS

Water is remarkable in its power as a solvent. Its high dipole moment enables it to dissolve salts by the separation of their ionic lattices, and its hydrogen bonding serves to dissolve molecules having partially charged atoms, particularly oxygenated structures. Additionally, waters nurture microorganisms; serve to suspend solid matter such as colloids and soils; dissolve gases; and serve as depositories for vegetative matter that in its subsequent deterioration adds TOC and other impurities. It is questionable whether pure water can be found in nature. Yet it is these waters that must undergo the rigorous purifications required of the compendial waters.

Surface waters such as comprise lakes, reservoirs, streams, and so forth are generally characterized by high TOC, high total suspended solids (TSS), lower clarity, higher organism counts, and higher oxygen contents than groundwaters. Depending on climate and location, they may be subject to seasonal "turnovers" caused by density differences occasioned by the climatic cooling of their top layers. This results in their becoming inverted, slipping to the bottom of the water pool. Their being roiled stirs up sediments and otherwise alters their contamination profile. This upwelling may also introduce anerobes released from the stirred muds. However, although sulfides may be a product of the anaerobic bacteria, they themselves seem not to be a problem if only because of their early demise when in contact with oxygen.

Different groundwaters, as from wells, reflect different geologies. They are usually remarkably constant in composition, and, depending on location, may contain manganese, iron, and even hydrogen sulfide; all three are best avoided. Midwestern limestone deposits contain alkaline earth hardness elements, iron and manganese. It is the presence of these entities that necessitate water softening and precipitative oxidations by permanganate-containing green sand.

In general the water contaminants consist of ionic materials such as dissolved minerals and salts; TOC, usually, organic substances derived from vegetative decay; suspended and colloidal matter including silica in its many forms; organisms, living and dead, and their metabolites and endotoxins; dissolved gases, such as oxygen and carbon dioxide; possibly, industrial pollutants; and farm runoff such as fertilizers, pesticides, and animal excreta (Tables 2,3).

8.5.1. Municipal City Water

An example of the information suggested to profile a difficult municipal water source from which pharmaceutical and semiconductor waters are actually prepared is shown in Table 4.

8.6. PRETREATMENTS

As stated, the principal purification units of distillation, reverse osmosis, ion-exchange, and electrodeionization can purify at least some small quantity of water of any degree of contamination even without pretreatment. The question is how much before the particular purification unit is fouled or possibly irreparably damaged. Chlorine will rapidly and irreversibly degrade polyamide RO membranes. Chloride ions will cause the corrosion of stainless steels. This can, and has been manifest by the pitting, scarring, and eventual stress cracking of stainless steel tanks and distillers. Barium and strontium may cause the fouling of Reverse Osmosis membranes with mineral deposits such as their sulfates, and may influence the inadequate regeneration of ion-exchange resins by the difficulty of the regeneration. Silica will wastefully interfere with the heat transference of stills; and mineral deposits, whether caused by water hardness, or by iron and manganese compounds rendered insoluble by oxidation, can block pipes, membranes, and filter surfaces, and so forth. The pretreatments are necessitated to extend the effective operations of the principal purification units to practical durations.

There are manifold purposes to be served in devising the water purification system. It must be determined what purification units are required, and for how long a time they will perform. Also to be ascertained are how, and how frequently they need to be renewed, refurbished, or replaced to prolong the purification operation. These elucidations result from the documented experimental investigations that define the validation process.

TABLE 2   An Analysis of a Source Water

	Analysis (ppm as such)	Conv. factor (x)	Analysis (ppm as $CaCO_3$)	meq/L
Cations				
Calcium (Ca^{2+})	60.0	2.50	150	3.00
Magnesium (Mg^{2+})	7.3	4.12	30	0.60
Sodium (Na^+)	50.5	2.18	110	2.19
Potassium (K^+)	7.8	1.28	10	0.20
Hydrogen = FMA^a (H^+)		—		
Total Cations			300	5.99
Anions				
Bicarbonate (HCO_3^-)	183.0	0.82	150	3.00
Carbonate (CO_3^{2-})	—	0.5^a	—	—
Hydroxide (OH^-)	—	2.94	—	—
Sulfate (SO_4^{2-})	53.8	1.04	56	1.12
Chloride (Cl^-)	63.8	1.41	90	1.79
Nitrate (NO_3^-)	2.5	0.81	2	0.04
Phosphate(ortho) (PO_4^{3-})	1.3	1.58	2	0.04
Total Anions			300	5.99
Total hardness ($CaCO_3$)	—	—	180	
M.O. alkalinity ($CaCO_3$)	—	—	150	
pH alkalinity ($CaCO_3$)	—	—	—	
Carbon dioxide (CO_2)	211	1.15^a	2.4	
Silica (reactive) (SiO_2)	30	0.83	24.9	
Silica (nonreactive)	5			
Iron (Fe)	2			
Manganese (Mn)	0.1			
Chlorine, free (Cl)	0.5			
Total dissolved solids	360			
COD by permanganate	5			
TOC	8			
Turbidity	2			
Color	—			
Specific conductance (μmho/cm @ 25°C)	660			
Specific resistance (ohm-cm @ 25°C)	1,520			
Temperature -°F	55–68			
pH	8.1			
pH^a	7.62			
Langelier index	0.48			

^{a}Free mineral acidity.
Source: Continental Penfield Water.

TABLE 3 City Water Analysis

Sanger, CA	
Calcium	32.5 mg/L as $CaCO_3$
Magnesium	79.1 mg/L as $CaCO_3$
Sodium	58.7 mg/L as $CaCO_3$
Potassium	5.1 mg/L as $CaCO_3$
Alkalinity	111.0 mg/L as $CaCO_3$
Sulfate	25.9 mg/L as $CaCO_3$
Chloride	38.5 mg/L as $CaCO_3$
TDS	175.4 mg/L as $CaCO_3$
Conductivity	334.0 mg/L as $CaCO_3$
Silica	31.7 mg/L as SiO_2
pH	7.4
Carbon dioxide	8.0 mg/L as CO_2

Results of a Water Analysis

	City Water	RO I Feed	RO I Permeate	RO II Feed	RO II Permeate
Calcium	13.0	12.0	0.2	0.2	0.05
Magnesium	19.0	19.0	0.5	0.5	0.05
Sodium	27.0	22.0	2.8	3.6	0.30
Potassium	7.0	4.0	0.3	0.3	0.00
Sulfate	27.0	71.0	0.4	1.9	0.01
Chloride	27.0	22.0	0.8	1.0	0.01
Alkalinity	111.0	40.0	8.0	8.0	0.80
pH	7.4	5.8	5.1	6.2	5.8
Carbon dioxide	8.0	79.0	78.0	2.0	2.0
Chlorine	0.0	0.4	0.0	0.0	0.00
Silica	31.7	32.0	6.1	6.2	0.035
Conductivity	334.0	335.0	21.0	23.0	1.5

All result are as mg/L, which is expressed as the ion (or molecule), except for alkalinity, which is expressed as $CaCO_3$. Conductivity is expressed as μS/cm.
Source: Comb and Fulford (1991): Courtesy, Ultrapure Water Expo '91 East.

TABLE 4 Municipal City Feed Water Profile

Municipal City Water Department (111) 222–3333

Chemist:	Mr. Proctor
Sources:	Surface feed for approximately two years as the city rebuilds its water treatment plant.
Surface source:	Quagmire reservoir
Transmission:	Water flows through the Quagmire aqueduct to mudflat reservoir.
Chemical addition:	Sodium hypochlorite (bleach) is added. (Target 1.6 ppm)
	At the city Corrosion Control Facility, the following chemicals are added: since 1996
	Soda ash ($NaCO_3$): Target - 28 ppm
	CO_2: target - 4 ppm
	Fluoride: target - 1 ppm
Transmission:	Water flows through the Old Rusty, Old Leaky and Cement aqueducts to the Trafficjam Reservoir.
Chemical addition:	Chloramine formation to transmit active chlorine more effectively.
Chlorine gas:	Target - 2 ppm
Ammonia:	Target - 0.5 ppm (5 min. delay)
T.D.S. (Total dissolved solids):	60 ppm (low)–371 ppm (wide variability)
Total hardness:	9.6 ppm (17% of TDS) (less than 50% hardness in TDS is considered soft water)
PH:	7.8 – 9.5 (elevated pH is now expected)
Temperature:	35 – 86°F (seasonal fluctuations)
Difficult constituents	
Silt density index (S.D.I.):	>25 (High feedwater silt density Index)
Silica:	2.63 ppm
Iron:	0.06 ppm
Total organic carbon (TOC):	2.91 ppm

The Municipal City Water has seen significant changes since the water source switch.

	Well water	Surface water
T.D.S. (Total dissolved solids):	371 ppm	60 ppm (low)
Total Hardness:	63.1 ppm (41% of TDS)	9.6–70 ppm (17–40% of TDS)

TABLE 4 Continued

pH:	9.10	8–9.0
Carbon dioxide	13.0 ppm	6.0 ppm
Difficult constituents		
Silt density index (S.D.I):	4.9 (Low)	>25
Silica:	6.3 ppm	2.63 ppm
Iron:	0.01 ppm	0.06 ppm
Total organic carbon (TOC):	2.38 ppm	2.91 ppm

This is a particularly difficult water source due to alternating sources. Even though the sources of the water appear to be surface water, the Municipal supply is unique in that the actual water drawn from the reservoir far exceeds what inflows the source. Thus the reservoir is acting as a shallow well. A water system is designed to treat its feed water source. The Municipal City Water is a particularly difficult source. From waters containing such types of contamination the pharmaceutical waters are prepared to their requisite standards:
- TOC: maximum of 500 ppb
- Conductivity: Stage I 1.3 ③ S/cm @ 25°C
- Stage II 2.1
- Stage III 4.7–5.8 pH dependent
- Organism alert & action levels

      WFI 10 cfu/100 mL
      PW 100 cfu/mL max. Actual limit depends on application.

Endotoxin: 0.25 EU/mL (WFI Water only)

In terms of ionic content the cations of interest would most likely be ammonia, sodium, the bivalent hardness alkaline earth elements of calcium, magnesium, barium, strontium, iron, manganese, and aluminum, the latter of concern in hemodialysis treatments. The corresponding anions would be: bicarbonate, carbonate, chloride, sulfate, nitrate, and silica.

Depending on the quantities and nature of the contaminants, the pretreatments may consist of any of a number of steps, not necessarily in the following order:

> Chlorination to control organisms
> Adsorption of TOC by activated carbon
> Chlorine removal to protect RO and I-X
> Softening to prevent RO scaling
> Fuoss effect on priority choice between dechlorination and softening
> UV ongoing organism control

8.6.1. Chlorination

Pharmaceutical water systems are obliged by regulations to use feedwaters of potable quality. Where this is accomplished in the pharmaceutical plant, the waters do not have to be segregated. As a first step, raw waters are commonly chlorinated to kill pathogenic microbes. Chlorine concentrations of 1 ppm effect a 97% kill of *E. coli* in 0.6 minutes at 5–25°C, and 0.5 ppm amounts have the same effect in 7 minues at 5°C. *Salmonella* and *Cholera* are killed by 3 ppm.

The chlorine is added to a residual concentration of 0.5 to 2.0 ppm, mostly to less than 1 ppm. This is usually done in (municipal) water treatment plants. Municipally treated waters may fall short. The analysis of incoming water using a chlorine monitor is therefore advised. Chlorine residuals or, increasingly, chloramines formed from the chlorine by reaction with ammonia are deliberately permitted in the water supply existing the water authority's plant, in accord with EPA requirements. These biocides are, therefore, present in the feedwaters entering the premises of the drug manufacturer. Where the feedwaters originate from wells, they are usually similarly chlorinated. The chlorine or chloramines require eventual removal. The means of achieving dechlorination will shortly be considered. When chlorine contacts water it reacts to form hypochlorous acid. HOCl dissociates to yield hydrogen ions, H^+ (or hydronium ions, H_3O^+), and hypochlorous ions, OCl^-. The sum of the hypochlorous acid and the hypochlorite ions is called the "free available chlorine." Hypochlorous acid is about 100 times stronger in its oxidizing potential than is the hypochlorite ion. Therefore, chlorinated waters exhibit stronger oxidizing effects at values below pH 7.4, the pK_a of hypochlorous acid, where it exists in half-dissociated form. Below this pH level, the progressively larger hydrogen ion concentrations increasingly suppress the dissociation of the acid, thereby increasing its concentration. Regrettably, chlorine partakes oxidatively in a free radical chain reaction with TOC present in the water to form the carcinogenic trihalomethanes (THM). The strength of the C-Cl bond, resistant to breakdown by the liver, the body's detoxification organ, is hypothesized to be the initiating cause of cancer. The use of ozone in place of chlorine, creating oxygenated instead of chlorinated molecular structures, would avoid THM formation and its assumed carcinogenic consequences. The EPA, whose responsibility includes drinking water, is studying the use of ozone in certain of its facilities (e.g., Fairfax, VA). Ozone is a more lethal biocide than chlorine and is more effective against viruses; all to the good. However, its very aggressivity against organic materials creates its own problems. It oxidative-degrades elastomeric seals and gaskets, ion-exchange resins, and the majority of polymeric materials including filters. Nevertheless, its usefulness in removing TOC, as will be discussed, is noteworthy.

8.6.2. Removal of Trihalomethanes

In passing, mention is being made of THM removal from water. More detailed accounts exist in the literature (4,5). The trihalomethanes found in feedwaters consist of mixtures of chlorine and bromine atoms substituent on the single carbons created by the free radical chain scission reaction of chlorine on longer carbon-to-carbon TOC chains. Bromine enters the picture when, by one path or another, seawater mingles with the source waters. Monobromo, dichloromethane Br-CH-Cl$_2$; monochloro, dibromomethane Cl-CH-Br$_2$; bromoform HC-Br$_3$; and chloroform HC-Cl$_3$ constitute the trihalomethanes. The THMs, except for chloroform, are destroyed to an 85% extent by 185 nm UV. They are removed by reaction with anion exchange resin in hydroxyl form; chloroform only to the extent of 50%. They are adsorbed by activated carbon in proportion to the surface area of the carbon and increasingly with bromine content; chloroform CHCl$_3$ 10%, bromoform CHBr$_3$ 50%.

8.6.3. Deep Beds and Multimedia Filtration

Deep beds constructed of particles, whether of sand, activated carbon, ion-exchange beads, and so forth serve as filters, the interstices among the granules acting as pores, as conduits that carry the liquid flow. Such beds have nominal porosites of from 10 to 40 μm, depending on the particle size distribution, the newer beds having the lower values. The size of the granules determines the packing density. If such a bed is fluidized by a water backwash prior to being allowed to settle, the particles will arrange themselves in layers according to their size; the smallest particles on top and the largest on the bottom. This inverted V arrangement has the smallest pores forming its top layers, and the larger pores the bottom layer, because the larger particles settle out first; the smaller on top of them. The smaller particles also pack more closely. As a result, the pores are smallest in the topmost layers and become progressively larger in the lower layers of the bed assembly. Particles carried in a water stream will be trapped in the upper layers of this depth filter. The lower layers serve to govern the water's flow; the deeper the bed, the slower the flow. Resuspending the construction-particles by fluidizing (backwashing) the bed frees the trapped contamination. Subsequent settling allows re-constitution of the bed cleansed of the previously trapped contaminating-particles, to nominal porosities of about 40 μm. Deep beds can be constructed of several different materials that differ in their densities. These are the multi-media beds that are employed to create different zones of filtration within the single container, each composed of the V-shaped particle layers. Thus, bituminous coal of density 1.5–1.8 is less finely ground than the sand, density 2.5, which it overlies. Most finely ground is the bottom layer of garnet rock, illminite, of density 3.5–4.5; all rest on a gravel base designed for drainage. Such an arrangement supplies a deep bed with three zones of filtration, each composed of particle sizes whose dispositions of size and density create layers of inverted V-shaped pores of progressively increasing size within each zone; but where each zone, overall, has smaller pores than the one above it. The resulting filter structure increasingly traps smaller particles of suspended contaminants, the uppermost zone, the largest; the bottommost, the smallest (Fig. 1).

Multimedia bed design is versatile, but there is no ready way to match its available constructions to the TSS, to the total suspended solids contents of given waters. Particles too small in size to be retained even by the (bottom) most finely ground, densest medium bed may be present, as also colloidal particles. Coagulation and flocculation techniques are then invoked to agglomerate the ultrafine particles to sizes that can be removed by the deep beds. In the event, rather arbitrary selections are made from among the relatively few multimedia designs that are offered commercially. Adaptations in operations are instituted as needed, principally by the use of "polymer," high-molecular-weight polyelectrolytic coagulants and flocculants, to help agglomerate colloids in order to facilitate their removal by the multimedia bed. If a maximum removal of colloids and other suspended matter is essential in the pretreatment, the filtration process is greatly facilitated via the use of polymer coagulants. "Polymer" is expensive but is typically utilized at less

Multimedia Filtration

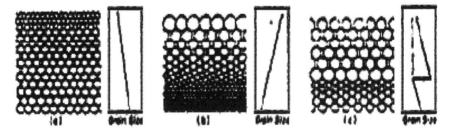

FIGURE 1 Cross sections of representative filter particle gradations. Diagram (a) represents a single-medium bed such as a rapid sand filter. The bottom half of a filter of this type does little or no work. Diagram (b) represents an ideal filter uniformly graded from coarse to fine from top to bottom. Diagram (c) represents a dual-media bed, with coarse coal above fine sand, which approaches the goal of the ideal filter. (Courtesy American Water Works Association.)

than 10 ppm dosage. However, its use can be overdone. Its excess can combine with water contaminants to create hard-to-remove deposits.

The importance of the deep bed operation deserves to be stressed. Its removal of colloids is very important, as otherwise clogging, at least partially, of interstices in the carbon and ion-exchange units may occur. Colloids detract from the effectiveness of UV light by shielding suspended organisms. If not previously removed, they can deposit on RO membranes, reducing their available surface area to such a degree as to necessitate their being cleaned. Cleanup of fouling deposits from the RO units should be kept to a minimum, as in the electronics industry. RO cleanings involve downtimes disruptive and costly to the purification process and taxing to the membrane service life. Sanitizations of RO membranes are also to be avoided because TFC RO membranes have little oxidant tolerance, and most sanitizers are oxidants.

Green sand is a zeolite. Zeolites are crystalline structures of aluminum silicate wherein the negative charge of the silicate anion holds metallic ions so fixedly that they undergo ion-exchanges, or oxidation-reduction reactions without disruption of the crystal structure. Green sand is a naturally occurring zeolite (also synthesized) containing manganic ions. These can oxidize ferrous and manganous ions dissolved in the feedwater to their insoluble hydroxide (oxide) counterparts, which then precipitate from solution to be removed by filtration. The zeolitic manganic ion is reduced thereby to manganous, its lower electron state. This treatment, at a time and place of one's own choosing, eliminates random, inconvenient depositions of iron and manganese oxides occasioned by contact with oxygen of the air. The manganese ion, manganous in its reduced state, can then be regenerated by permanganate oxidation to its manganic state for a repeated application. Green sand, perhaps because of its sharp angular structure, packs more closely than the layers of the multimedia beds. Its use after the multimedia bed, therefore, offers a better removal of the feedwater's TSS, to the betterment of the subsequent RO operation. It should be used before the softening operation to remove iron and its almost inevitable accompaniment, manganese. Otherwise, on encountering the softening-resin, the water will cause precipitation of the oxides to form a coating blocking the resin surfaces. This causes compression of the resin particles and results in slowed water flows and in increased differential pressures. It should be noted that organisms retained within the multimedia and nurtured on its adsorbed or otherwise trapped impurities can grow within its confines. Backwashing of the beds will minimize microbial activity. A pressure drop should not be relied upon to initiate the backwashing, as commonly recommended. Backwashing should be performed on a scheduled biweekly basis.

In extreme cases, sanitization of the multimedia beds is made by use of hyperchlorination. A shock treatment of as much as 50 ppm of chlorine, usually in the form of sodium hypochlorite solution, is used. The exact concentration of chlorine required for control of organisms, in this or any other context, can be learned only by actual trial as assessed by microbiological analysis. Deep beds, whether multimedia, carbon, or ion-exchange, generally utilize flows of 5 to 15 gpm/ft^2, the lower end of the range providing superior service. Slower flows may encourage channeling and result in local overloading; larger flows may attenuate ion exchanges and adsorptive contamination removals. The flow requirements are set by the dimensions of the beds. A 30-in. deep bed will trap the contamination it acquires within its top 6 in. or so. The remaining depth, as stated, serves to moderate the downward rate of flow of the water by imposing the impediment of longer flow paths. The bed performance is described in terms of its square footage, ft^2, the face area seat of its action. (The cubic foot, ft^3, dimension applies when a time dependency is involved.) The deep beds are backwashed at about 15 psi delta pressure to a "quicksand" consistency, thereby releasing the entrapped particulate matter to be flushed away. The backwash operation can be automated to respond to a pressure buildup but this does not result in optimal filtration capability. Depending upon contamination loads and flow rates of 5 to 15 gpm/ft^2, it may have to be done as frequently as every 4 to 6 hours.

8.6.4. Softening and Solubility Product

Water softening is usually accomplished by way of sodium-form ion-exchange wherein the ion-exchange resin in the softening unit removes the hardness-causing elements from the feedwaters by exchanging them for the sodium ions it releases. This forestalls subsequent mineral fouling of the RO by membrane-blocking deposits of alkaline earth salts of limited solubilities, such as the sulfates, carbonates, and fluorides of calcium, barium, or strontium. Such compounds are characterized by low solubility product values. The solubility product of a salt is the maximum product of its cation and anion concentration expressed in moles per liter that can exist in equilibrium with its undissolved phase at any one temperature.

$$C_{Ca}^{+2} \times C_{(SO_4)}^{-2} = K_{a_{CaSO_4}}$$

When such salts exceed their particular solubility product levels, they precipitate from solution. This can happen in the RO operation where ongoing water recovery progressively increases, within the recirculating stream, both the concentrations of the salts and the likelihood of their precipitation through exceeding their solubility products.

The blockage of the RO membrane that would ensue would make less membrane area available, resulting in a diminution in the rate of permeation, as well as a decrease in rejection by the RO. To restore the system, cleaning of the RO is necessitated. This involves downtime, and the expense of RO cleaning. More significantly, the RO membrane, of a delicate structure, is exposed to harsh chemical treatment. Such effects would be catastrophic in the electronics industry and are rigorously guarded against. The consequences involved in RO cleaning in the pharmaceutical setting are not so dramatic, but they are best avoided. It is advised to clean as required, based upon increasing pressure drop across the membranes or as indicated by RO product flow loss. Cleaning schedules should be so written into the SOPs (weekly, monthly) that the cleaning activity is performed only as needed; perhaps on the basis of a 25% decrease in permeate. Cleaning is not necessarily performed optimally, and the RO membrane, being delicate, can be damaged, and its longevity compromised. Perhaps the semiconductor practice of avoiding RO cleanings should be emulated. The incidents of RO cleaning can and should be reduced via improved pretreatment and RO design and operations.

As stated, sodium-form ion-exchange resin is ordinarily used to remove the bivalent, hardness-forming elements from the water. Sodium chloride is used to regenerate the resin upon exhaustion of its exchange capacity. Countercurrent regeneration is estimated to save from 50% to 60% of the salt used. Saturated brine, 26% concentration, is diluted to a 10% strength as attested by conductivity measurements, for use over a contact time of 20 minutes. The hydrogen-form cation exchange resin would be more efficient, but the low pHs resulting from its release of hydrogen ions restricts its use to the absence of carbonates and bicarbonates or to systems where the removal of carbon dioxide is provided for.

Barium and strontium are removed from the resin with difficulty in the regeneration process. Their incomplete removal is very possible and can lead to a progressive fouling of a downstream RO or electrodialysis unit. Ideally, the resin regeneration is performed on a gallonage or totalized water flow basis. Regenerations are also set on a time basis. In such cases, incomplete regenerations result in a diminished efficiency in hardness-removal because of mistimed regeneration cycles. Hardness monitors or test kits should be used to assure that complete resin regeneration is achieved. Continuous monitoring is preferred, but is expensive and is maintenance intensive. It would also be wise to include barium and strontium in the initial feedwater analyses so as to be alerted to possible complications caused by their presence.

As stated, softening of the feedwater is essential in keeping the RO from being blocked by mineral deposits to the point where its cleaning is necessary. Softening failures are to be avoided. Curiously, a relatively inexpensive pump is often relied on to supply the regenerant salt solution. Care should be taken to make sure that the pump is in good operating order. Above all, a hardness reading should be taken at the end of the service run to make sure of its adequacy, as a means of ensuring against softening failures and overruns.

8.6.5. Softening or Dechlorination First

At the pretreatment stage a choice can be made between first softening the feedwater or removing its chlorine. The choice should depend on the application intended for the water but may often be made rather arbitrarily. Indeed, where adsorption onto activated caron is used to remove TOC, the order in which softening is performed may make a difference. This is due to the Fuoss effect, which will be discussed subsequently. When softening precedes dechlorination, some of the ion-exchange resin will be oxidatively degraded by contact with the chlorine. TOC will be generated. Also, some chlorine may permeate the softening unit to create a protective presence against organisms in the softened water. In so doing the chlorine also, to some extent, sanitize the resin bed. The extent of each of these happenings depends on the rate of flow of the feedwater through the softening unit and on the chlorine concentration. The protective chlorine umbrella against organisms is desirable. However, the generation of TOC makes this usage unsuited to situations wherein low TOC levels are of great importance. When chlorine degrades cation exchange resin, the structure's cross-links are severed. This leads to a greater swelling of the resin, but not necessarily to its loss of cation exchange function. TOC is created in the process in the form of smaller, more water soluble molecular species. Where the anion exchange resin is attacked by chlorine, in addition to breaking of the carbon-to-carbon cross-link bonds, the carbon-to-nitrogen bond of the quaternary group is broken. Amines and amine-like molecules result. Unlike the case of the cation exchange resin, exchange functionality is lost. If in twin-bed ion-exchange usage the water flow is rapid enough to carry chlorine into the anion exchange column, a fishy odor may become apparent due to the release of trimethylamine from the oxidized quaternary amine group responsible for the anion exchange function.

8.6.6. Fuoss Effect and TOC Adsorption

There is reason to believe that TOC is better removed by adsorption to activated carbon in the presence of bivalent ions such as are responsible for water hardness. TOC derived from vegetative sources are usually in the form of long-chain molecules that exist in coiled forms. The coils are extended to different degrees in direct accord with the extent of their dilution. Higher ionic strengths, such as are forthcoming from the presence of ions, have the effect of concentrating the solution, as if preempting the available water. The result is a folding of the molecular chains into tighter coils. This folding results in the exposure of the hydrophobic amino acids–namely, tryptophan, phenylalanine, and tyrosine—onto the TOC's surface. The result is an increase in the molecules' overt hydrophobicity (Figs. 2,3). Adsorptions to activated

Extended Protein Molecule

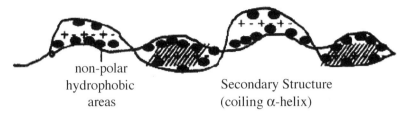

non-polar
hydrophobic
areas

Secondary Structure
(coiling α-helix)

FIGURE 2 Polymer in dilute solutions.

carbon are largely hydrophobic in character. Thus, TOC adsorptions are promoted by the hydronium ion (H_3O^+) concentrations of low pHs, and more so by the bivalent charges of the alkaline earth ions. It is therefore better to dechlorinate using active carbon before softening, so that the simultaneous adsorptive removal of TOC can be optimized.

8.7. CHLORINE REMOVAL

The choice may be made to remove the chlorine before the water is softened. This order of removal avoids the degradation of the resin by chlorine and avoids TOC generation. However, the dechlorinated waters are no longer protected against bacterial infection.

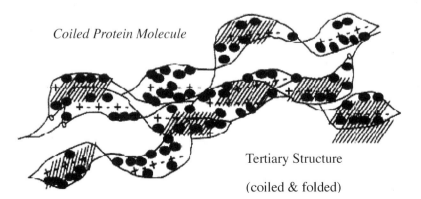

Coiled Protein Molecule

Tertiary Structure

(coiled & folded)

FIGURE 3 Polymer in concentrated solution.

Dechlorination can be managed in a number of ways, some more suited to particular water applications than others:

Adsorption to and reaction with activated carbon
Reduction with bisulfites
Reaction with ion-exchange resins
Reaction with ion-exchange resins within electrodeionization units (this technique, although practiced, is not recommended)
Destruction by ultraviolet light

8.7.1. Activated Carbon

Adsorption to activated carbon (GAC, granulated activated carbon) is a method long in use. It is dependent on the vast surface area within the carbon granules made available by the roasting, in atmospheres of limited oxygen, of carboniferous materials, such as wood, bituminous coal, and various nut shells.

Not too much is known regarding which of the many commercially available activated carbons is most suitable for a particular adsorption. Choices seem to be made mostly on the basis of suitable use by others. Economics, not surprisingly, also plays its part. At least two known properties are desired in a carbon: resistance to abrasion and a regularity of porosity. Other properties include iodine number, molasses number, carbon tetrachloride activity, surface area, pore volume, moisture content, particle size distribution, mean particle size, and soluble ash content. The iodine number is regarded as indicating opportunities for chlorine addition to unsaturated bonds, and the molasses number as being an indicator of TOC adsorptions. The activation process of firing the carbon source in a limited oxygen atmosphere creates metallic oxides of the accompanying minerals. These add leeched alkalinity to carbon-treated waters, raising their pH. Water-extracted and washed activated carbons, "free" of extractables, are available.

The pore characteristics of the various carbons differ. Coconut shell gives a low ash (5%) activated carbon with a high microporous density but a low macroporosity. This is seen as favoring adsorptions of low-molecular-weight TOCs but not of the larger TOCs presumed excluded by size from the smaller pores. It has resistance to abrasion, a property that minimizes carbon dust, fines whose exiting of the carbon bed tends to clog downstream interstices and obfuscate sight glasses.

The activated carbon derived from bituminous coal is less expensive than that prepared from nutshells. It has a higher ash (10%) content, a medium dust content, and good resistance to abrasion. High micropore and medium macropore densities are expected to encourage the adsorptive removal of a wider molecular range of organics. Activated carbons derived from bituminous coal are recommended for use in liquids applications.

In the application, the carbon removes the chlorine by adsorption followed by reaction with it. An efficient removal requires a flow rate of about 2 to 3 gpm/ft^3. TOC adsorption by activated carbon requires a slower flow of from 0.5 to 1 gpm/ft^3; and the chloramines, a still slower flow. (The unit ft^3 applies rather than ft^2 because contact time rather than extent of surface is the issue.) Backwashing of the carbon beds to free them of the fines which may otherwise unhelpfully coat and block downstream surfaces and interstices is a necessary practice. The backwashing will also reduce the microbiological loadings in the beds, which otherwise may be distressingly high. Elemental carbon is not a metabolite preferred by organisms. The prolific microbiological production from the carbon beds is thought to be the result of high suspended solids loads and TOC feeding the bacteria. Bacteria shedding from GAC carbon beds can be significantly reduced by two backwashes per week. The active carbon beds are replaced at intervals of 6 to 12 months. Our data indicate the standard plate count from the carbon effluent should not exceed 300 cfu/mL (TNTC) and is often less than 100 cfu/mL.

As stated, microbes find the ambiences of carbon beds conducive to their growth. The beds require the attentions of sanitizations at a frequency determined by the method of prime reliability; namely, microbiological testing, or on a prophylactic basis as established during the system's validation. Sanitization of the carbon beds is best managed by steam or hot water flows. Hot water, above 65°C, is preferred because the lower viscosity of steam makes it more likely to permeate the bed rapidly only through the wider pores. Hot water, with its higher viscosity, is seen to have a less discriminatory effect. However, the use of stainless steel carbon containers that are corrosion resistant, as compared to those of black iron, or of those fitted with hard rubber liners are, in the interests of (shortsighted) economics, often slighted. The hot water or steam treatments are, therefore, not applicable. Curiously, active carbon is then judged by some as being unsuited for dechlorinations, and substitutes are sought.

The activated carbon filtration step in the pretreatment serves also to remove TOC as well as chlorine. Its omission will result in an increase in TOC, and very likely in the bacteria counts of the product water. It is known from semiconductor system operations that product water TOC can be reduced significantly simply by changing the carbon beds. It is seldom possible to forecast when a carbon bed will exhaust its TOC-adsorbing capacity. This determination necessitates periodic TOC testing of the bed's effluent waters. Some pharmaceutical installations seek to revive the exhausted capacity by steaming the carbon bed in order to desorb whatever adsorbed TOC can be volatilized. This is not done in the electronics industry. In the authors' opinion, this requires too expensive an apparatus and is not sufficiently efficient to make it a profitable undertaking. Carbon bed replacement is recommended instead. Pharmaceutical operations have successfully utilized activated carbon beds for over three decades. Given the dedicated attention it merits, the activated carbon removal of chlorine is a dependable procedure, and the TOC reduction pays dividends in imparting greater microbiological control to final product water quality due to nutrient deprivation.

8.7.2. Reductions with Sodium Sulfite

Chlorine, being an oxidizing agent, is consumed by reactions with reducing agents. Solutions of the sodium salts of sulfites, including bisulfite and metabisulfite, are used. This avoids the organism problems connected with carbon beds. However, bisulfite additions, although successfully employed, have their own limitations. The reaction is not stoichiometric and the interference of oxygen as an oxidizer necessitates reasonably freshly prepared solutions of the reducing reagent. The use of excess sulfite is almost inevitable, and its eventual removal a necessity. Addtionally, there are organisms that can grow in the sulfite solution. There are also sulfate-reducing organisms that live on the product of the sulfite's activity. The use of the sulfites to remove chlorine is not a panacea and excess sulfite is considered conducive to RO membrane fouling, probably due to microbiological biofouling of the membranes. It is accomplished every day, but it too requires care and discipline, qualities sometimes in short supply.

8.7.3. Destruction by Ultraviolet Light

The newest technique for removing chlorine is by ultraviolet destruction. Employing the same wavelengths that are used in UV sterilizations, namely, 254 nm, chlorine is destroyed by ultraviolet rays. The rule of thumb at present requires 10 times the sterilizing UV dosage to accomplish a 1.0 ppm chlorine destruction, and the higher the chlorine content, the greater the UV dosage required, and the more expensive the equipment. Equipment modules are available that are designed to completely remove the 0.5 to 2.0 ppm chlorine or chloramines that is the usual content of a water system. If larger destruct units are required, they can be modularized as becomes necessary. Many such installations have already been placed into operation, prompted, no doubt, by a desire to avoid the protocols essential to good carbon bed usage.

Current expectations of this newer technique seem cautiously to allow for less than total chlorine removal. However, substantial reductions are expected. Interestingly, in two cases where larger chlorine concentrations were present, or where less than a total chlorine destruction was a concern, the single module was augmented in one installation by a carbon bed, and in another by a downstream EDI unit. Given that all systems and devices have their limitations, it is fair to say that this application is still too new for its limitations to have become defined.

8.7.4. Reaction with Ion-Exchange Resins

The perceived inability of the ultraviolet destruct method to totally remove chlorine does not vitiate its practical application. The bulk of the chlorine having been destroyed by the UV, the small remainder may be removed by reaction with the ion-exchange resins it is permitted to encounter by its entrance into the DI beds. (Clearly, where a chlorine-sensitive RO membrane precedes the ion-exchange column, this technique is not an option.) For the same reason the UV destruct method is suitable as a precursor to the use of EDI units, which also contain ion-exchange resins available for interaction with chlorine residuals. As detailed above, TOC is generated thereby, but this occurrence is tolerable in many applications.

8.8. CHLORAMINE REMOVAL

As noted, the UV destruct method is said to be capable of removing completely 2.0 ppm of the chloramines. A chemical equation that can be written for the reaction has nitrate ion and ammonium chloride as products of a first-order reaction. This leaves ammonia as the entity still to be removed following chloramine destructions. The EPA decrees that the potable waters leaving water purification facilities must contain some biocidal residuals to offer antimicrobial protection during the water's distribution to consumers. Chlorine is eschewed because its oxidative powers can turn TOC into the carcinogenic THMs. Therefore, increasingly the chlorine is converted to chloramines, of lesser oxidation potential incapable of THM formation (or less so), as the biocidal residue. Because the chloramines are less reactive chemically than chlorine, they react at lower rates or find fewer molecular species with which to react. As a result, they last longer. In this regard they are more stable, there being a reciprocity between stability and reactivity, usually expressed as a time/concentration relationship. However, over time the oxidative strength of the chloramides is enough to deteriorate RO membranes.

Ammonia added to the chlorinated waters undergoes reaction to form the chloramines, a series of three compounds separately characterized by their degree of substitution by chlorine atoms for the three hydrogens of the ammonia molecule. Monochloramine, $H_2N\text{-}Cl$; dichloramine, $HN\text{-}Cl_2$; and trichloramine, $N\text{-}Cl_3$, constitute the chloramines. The addition of ammonia to the chlorinated water, a condition wherein the chlorine concentration is greatest, favors, by the Law of Mass Action, formation of the most highly chlorinated molecule, trichloramine. It is also the strongest of the three in terms of taste and odor. The biocidal action of monochloramine is such that its 0.3 ppm is equivalent to 0.1 ppm chlorine. It is the monochloramine that is least offensive in terms of taste and odor. The preferred mixture of the chloramines consists of two-thirds monochloramine and one-third dichloramine. This composition adsorbs evenly to carbon, the individual adsorption rates of the components becoming balanced by their 2:1 ratio.

8.8.1. Removal of Chloramines Usually Managed by Adsorption to Activated Carbon

The uptake is quite slow, a flow rate of about 0.5 gpm/ft^3 or 3.785 liters per 0.0283 m^3. Ammonia is a product of the adsorption reaction. The very high solubility of ammonia in water makes its removal problematical, whether by spray ball or hydrophobic membrane contactors. The distillation of ammoniated waters, while resulting in the release of ammonia, is unsuited for its significant removal because of its high water solubility. Reverse osmosis does not remove ammonia whether in its NH_3 or NH_4^+ form. Being the base anhydride of ammonium hydroxide, at higher pHs ammonia is converted to the ammonium ion, the product of the feeble dissociation of ammonium hydroxide. In the form of the ammonium ion, NH_4^+, it can be removed by use of cation exchange, but only with difficulty because it is just above sodium in the displacement hierarchy. It is lower, however, in the series than the bivalent hardness elements. Practical considerations lead to the practice of utilizing dual sodium-form ion-exchange softener units in series to effect its sure removal. Hydrogen-form cation exchanger can also be employed, provided that no carbonate or bicarbonate is available to release carbon dioxide. Were CO_2 to be released, its acidifying influence would counter the alkaline pH necessary to the conversion of the NH_3 to NH_4^+.

The chloramines can also be removed by their oxidative destruction, as by chlorine. The reaction, known as breakpoint chlorination, is rapid; the optimum rate is at pH 7.5. Nitrogen gas and nitrate ion are formed as products in a complex fashion. However, this reaction (attractive and convenient) is little used.

8.9. ORGANIC ENTITIES, TOC

The term TOC is too encompassing to be very meaningful with regard to how organic contaminants are to be removed from waters intended for pharmaceutical purposes. Those of low molecular weights may be removed by volatilization. However, the distillation of waters containing organics, instead of separating the entities, may give rise to azeotropes, combinations of the organics and water that codistill at a set temperature and composition. The universe of organic compounds is too all-embracing to permit a general approach to be applied to the numerous individual compounds. The different TOCs react variously. They may oxidize at different rates with the same oxidizer, and to different extents.

References were earlier made to the removal of TOC by adsorption to activated carbon surfaces. Such adsorptions can take place on almost all surfaces, ion-exchange resins and filters included. For this reason, TOC or organic traps are constructed of activated carbon black mixed with anion-exchange resins; weak base resins are used for easier regeneration. The mixture features an extensive surface area conducive to adsorptive actions. The ion-exchange interactions involve charge attractions, but hydrophobic adsorptions motivated by reductions in free surface energies likely predominate. The extent of the carbon's surface area, electric double layer effects, the molecular weight influences of the adsorbent, and the hydrophobic nature of the adsorption, all play a role in the TOC removal process.

As a generalization, perhaps the best way of removing whatever constitutes the contaminating TOC, most of which originate from vegetative origins, is to subject it to the strong oxidizing action of ozone. The oxidation, also that of nonpolar hydrocarbons, follows a free radical chain reaction wherein the large molecules are lysed into smaller entities and where, incidentally, cross-linking may also take place; but most importantly, where a hierarchy of oxygenated structures stepwise leads from hydroperoxides and peroxides, to alcohols, to aldehydes or ketones, to carboxylic acids—the last characterized by the $-COOH$ group. It is this organic acid moiety that enables removal through interchange with strong anion exchange resins. The organic carboxylic acids are also oxidizable by ozone, ultimately to carbon dioxide and water; albeit at much slower rates. However, it is the carboxylic acid stage of the oxidation chain that is utilized in an anion exchange to effect TOC removal. As with all ion-exchange reactions, the kinetic rate of exchange at the resin surface is high; the rate of diffusion into the resin interior is lower, but the high gel phase of the resin makes it possible. The sulfate ion competes with the carboxylic ion exchange. Elution is performed with sodium chloride. The chloride affinity is 20 times greater than that of the hydroxyl.

8.10. ENDOTOXINS

The endotoxins, derived from the cell walls of gram-negative organisms, are lipopolysaccharidic in character. As present in waters containing neither bivalent ions nor surfactants, they are of a size that permits their removal by ultrafilters of 10,000 to 20,000 dalton ratings (6). Being negatively charged, they can also be removed by adsorption to positive-charged filters, usually composed of charge-modified polyamide microporous membranes. Their sieve retention by ultrafiltration is less conditional, and hence, more reliable. The ultrafiltration activity can be part of the pretreatment train. However, ultrafilters, being fine in their porosities, can consequently trap much suspended matter and may clog relatively rapidly when liquids with high total suspended solids (TSS) are involved.

8.11. ULTRAFILTRATION

Reverse osmosis provides an important technique to pharmaceutical water purifications. The role of RO in front of stills is primarily to remove chloride ion and silica whose presence could compromise still performance. Early applications had misdirected its use to the absolute removal of organisms: a role for which RO units are not suited (RO membranes are not subjected to standard integrity testing; opinions differ regarding whether RO membranes are dependably fabricated without flaws). However, RO treatment does accomplish a considerable reduction in organism levels for treated waters. Unfortunately, the organisms retained by the RO membrane serve as foulants in poorly staged RO's or in systems operated above 60% product conversion (product to reject ratio). By progressively reducing the remaining effective filtration area, they compel cleaning of the RO units. This detracts from the RO's longevity. The use of ultrafilters as prefilters to RO installations would largely spare the RO membranes the burden of fouling. Cleaning ultrafilters is less hazardous to their service lives than cleaning the RO filters; their polymeric structures, usually polysulfones, are far more robust than the polyamides and more resistant to oxidizers and other aggressive reagents. The point being made is that the use of ultrafilters as prefilters for RO devices would help prolong the RO service run between cleanings and improve the RO product water quality, without incurring the expenses of premature RO membrane replacements. A reluctance to utilize ultrafilters may be encountered from those who consider membrane processes as not being secure enough to assure filtrate sterility. Be that as it may, sterility is not the objective of the proposed ultrafiltration use; silt load (suspended solids) and organism count reductions are. Neither Water for Injection nor Purified Water is expected to be a sterile article. That ultrafilters can and do minimize the organism populations of filtered waters has been established. They should be used to bestow this advantage in water purifications.

8.12. PRINCIPAL PURIFICATIONS

The incoming feedwaters, having been pretreated in manners encouraging to the continued long-term operation of the principal purification units, are next purified of their ionic and organic contents and of their silica and other, perhaps less common, contaminants. Appropriate selections are made from among the available principal purification units. The choices are reverse osmosis, ion-exchange, electrodeionization, and distillation. The latter option, being expensive, is employed almost exclusively in preparing Water for Injection; but even in that situation, one or more of the other techniques are also utilized. To convert the raw source waters into the pharmaceutical articles is the intention of the principal purification design. The purification systems are very individualistic. They are site specific, depending on the particular source water. Even where the same water might be used, the treatments could require different purifications depending on the intended water application, and even on different rates of usage. A very real consideration in the choice of the water purification units is the economic burden they entail. This may differ regionally, as also the cost of the source water and the disposal of waste waters. Personal preferences also come into play. Each purification technique offers its own advantages and limitations. As stated, this writing is meant to focus on the pretreatment options available to the pharmaceutical processor. The principal purification operations are plentifully described in the literature (1,2). Elaboration of these will here be confined to modifications and/or practices considered noteworthy.

8.12.1. Distillation

The distillation process utilizes heat to vaporize water and its volatile impurities; thus separating them from their nonvolatile analogues. The subsequent condensation of the vapors separates their volatile contaminants from the liquefied water. The entrainment of the inevitable mist formation, very small water droplets with their contamination, is guarded against by such still (distiller) design features as demisters and centrifugal force promoters. Blowdowns, now largely automated, periodically empty the stillpots of their progressively accumulating nonvolatile contaminants. This potentially helps minimize the consequences of mist entrainments by reducing the contained impurities. The big advantage of distillation is the killing effect of heat on microorganisms. Its chief drawback is the cost of the heat-generating fuel. Therefore, heat conservation is a consideration of still design. As a result, two still-type designs have evolved; namely, the multistage still and the vapor compression type. The multistage still uses the heat acquired by the condenser water during its cooling of the steam of the first stage to heat the feedwater entering the second distillation stage. In this way, heat is conserved and fuel costs are reduced. If the first stage distills at temperatures of 140°C, with the aid of plant steam, then the 11° difference required by a successive stage can enable as many as seven separate stages to be utilized. At some point, the cost of an additional distilling stage outweighs the fuel savings benefits. The water produced by the multistage distillers is distilled only once in whichever stage it was prepared. (Erroneous opinion has the water distilled progressively through the successive stages of the still.) However, the total volume of distilled water produced is the quantity processed by all of the stages combined. Even, as in the case of WFI preparation, where distillation is the principal purification technique ultimately relied upon, certain of the other purification methods are also utilized. The vapor compression distiller causes the water molecules that are at distances from one another in the steam form (vapor or gaseous) to come together close enough to convert them into water in its liquid form. In going from the gaseous state of matter to the liquid state of matter, the latent heat of condensation is released. This is utilized to help elevate the next batch of water to its boiling point. By means of its compressive action, the vapor compression still recovers heat that helped vaporize the previous quantity of feedwater. It should be noted, however, that the vapor compression still may be operated at relatively low temperatures. Where these approach room temperature, the thermal killing action on organisms becomes increasingly attenuated.

The elevated temperatures involved in distillations promote rouging, the product of stainless steel corrosion. Neither corrosion nor the passivation exercise that, at least for some duration, protects against it is a subject of this writing. Interested readers are directed to the literature (7,8).

Distillations, properly conducted, serve to purify waters of their endo- toxic contents. However, the endotoxin removal is of a log reduction nature. It is not absolute. FDA considers the distillative removal of endotoxin to be reliably effective to a 3-log extent. Therefore, the feedwaters to stills must not exceed that capability in their endotoxin contents. As previously stated, ultrafilters of 10 kd should be used to remove endotoxin from the still feedwaters.

8.12.2. Ion-Exchange

Certain aspects of the long-established ion-exchange operation bear discus- sion. These deal with silica removal and countercurrent flow. The hot water sanitization of ion-exchange installations also deserves mention. The possible advisability of conducting barium and strontium analyses on the feedwaters for ion-exchange has already been addressed under "Softening."

8.12.3. Silica

The chemistry of silica is complex, and its removal from waters is therefore complicated, particularly as it exists in equilibrium among its several types— namely, the soluble, the many colloidal or polymeric kinds, and the suspended. Soluble and colloidal silica may be in equilibrium. The soluble ionic type can be detected by the molybdate color (blue) reaction. In its ionic form it is removable by strong base anion-exchange; very low molecular weight colloids especially so; higher weights progressively less so. Colloidal or polymeric silica may be too little ionized to be removed by ion-exchange at all. Also, so-called Giant Silica, silica colloids about 10 µm in size, some 10 to 1000 times larger than the standard silica colloid, is too large to be adsorbed by conventional anion-exchange resins, although it can be by large-pored resins. Some silica colloids containing organic materials and even heavy metals are large enough to be removed by ultrafilters of 100,000 dalton ratings.

Strong base anion-exchange macroreticular resins in hydroxyl form that have 5 to 10 µm pores are depended on to remove colloidal silica, other than the giant type, by ion-exchange. Nevertheless, silica leakage from ion-exchange beds, even from mixed beds, is a common occurrence. Reverse osmosis performs excellently in removing colloidal silica, and thin-film composite (TFC) membranes can remove some 90% of soluble silica. Unless silica is removed from waters prior to their distillation, it will form a vitreous glaze on the stillpot surfaces that will waste fuel by its insulating properties.

8.13. COUNTERCURRENT OPERATIONS

In two-bed ion-exchange practices, cocurrent flow means that both the service flow and the subsequent regeneration flow are in the same direction through the ion-exchange column. Countercurrent flow signifies they are in opposite directions. The countercurrent posture, far more widely used in Europe than in the United States, results in a substantially less expensive process that offers operational benefits as well.

In the operation of an ion-exchange column, the ion-exchange resin becomes exhausted from the top down. The ion-exchange sites further upstream are the first to be utilized. The completeness of their conversion is assured by the initial high ratio of ions to ion-exchange sites. The ion-exchange sites lower down the bed are less depleted, and toward the bottom of the column may remain unspent altogether. To prevent sodium ion leakage, the ion-exchange operation is best left with a margin of safety. The ion-exchange operation is therefore not conducted to complete exhaustion. The beds are operated to a breakthrough point, to a percentage of total exhaustion. In semiconductor usage, the bed operation may be halted before the effluent waters decrease in quality to 15 megohm-cm resistivity.

Consider the regeneration of the cation-exchange column. As it is carried out in cocurrent fashion, the incoming hydrogen ions (or hydronium ions, H_3O^+) preferentially displace the sodium ions, then the magnesium, then the calcium ions, in accordance with the more tenacious hold of the multivalent elements. Also, as the multivalent elements are released, they in turn displace the monocharged ions downstream in a continuing progression. As a result, there may come to be a broad band of resin in the hydrogen form on top of the column, with narrower bands of resin combined with magnesium and calcium below that, and yet another band in the sodium form beneath that. If the regeneration proceeds to a sufficient extent, the sodium band may be eluted by the downflowing calcium ions released by displacement by the regenerating hydrogen ion. There will, in any case, be a heel of some unregenerated resin left at the end of the regeneration cycle because it is not economically feasible to completely convert the resin to the hydrogen form. The unregenerated zone is perhaps 3 to 6 inches deep. This cocurrent regeneration is limited by the law of diminishing returns. More and more acidic regenerant is required to obtain progressively less and less added conversions. Furthermore, it increases the problem of acid disposal. Even the dumping of water can be expensive, costing as much as $10 or more per 1000 gallons.

When the service flow is again resumed, the water first issuing from the column will contain ions eluted from the unconverted heel; sodium and, possibly, other ion leakage will be evident. Eventually, the water quality will improve until, as the bed nears exhaustion, its breakthrough will be reached. If, however, the exhausted bed were to be subject to countercurrent regeneration, the broad band of hydrogen-form resin would become situated at the bottom of the column with overlays of resin in calcium, magnesium, and sodium forms. The resumption of ion-exchange operations would then not occasion any sodium leakage into the descending and exiting waters until the column reached its breakthrough point (Fig. 4).

For the counterflow of liquid with the regenerant chemical solution to be in the upflow configuration, some form of restraint must be imposed on the top of the resin column to maintain its proper degree of compaction in order to prevent its physical movement, the disturbance of the ion-exchange wave front. This may be managed in any of several ways; by blocking the flow of water, or by an imposition of air pressure, or by the hindrance of physical restraints that are porous to permit the passage of regenerant solution but that will not become clogged by the resin beads (Fig. 4).

Backwashing is disruptive of the hydrogen-form resin zone. The backwash frequency is, therefore, reduced to once every ten cycles or so, instead of every cycle. This necessitates filtration of the service water to reduce or eliminate its particle load that would otherwise more frequently clog the column. Also, countercurrent operations require the maintenance of a rather constant flow rate to avoid disruption of the ion-exchange wave front. This is not achieved too easily. Countercurrent installations cost more by 15 to 25%. It is

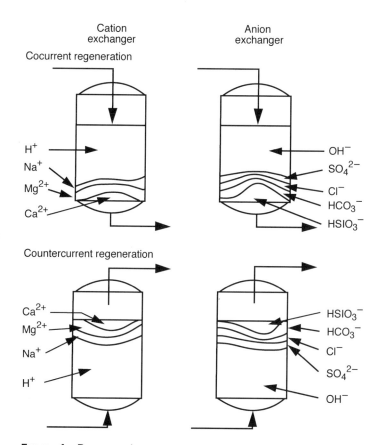

FIGURE 4 Regeneration.

estimated, however, that while cocurrent regeneration uses 300 to 400% the amount of chemicals theoretically required, countercurrent would exceed the actual need by only 5 to 10%. In terms of performance, the effluent quality from separate bed deionizers in cocurrent operations is normally in the range of 1 to 10 ppm total dissolved solids (TDS), with a resistivity of 0.05 to 0.5 megohms-cm. Countercurrent operations could produce an effluent with one-tenth that quantity of TDS, and a resistivity of 0.5 to 10 megohms-cm.

8.13.1. Hot Water Sanitizations

Ion-exchange beds offer conditions conducive to microbial growth. For this reason they require intermittent sanitizations. Regenerations of the resins perform this service, but the TOC, including that of organisms not removed by the relatively brief regeneration process, may then provide suitable nutrients to promote significant regrowth before and during the return to service. There is the belief that virgin resins, free of this type TOC, are in less frequent need of sanitizations. Hot water sanitizations are to be preferred (9). The upper temperature limit is set by the thermolability of the strong base anion-exchange. The quaternary functional group in its hydroxyl form undergoes a Hoffman degradation involving loss of its functionality. The hot water sanitizations should therefore be performed with due regard for the operational temperature and time limits.

8.13.2. Electrodeionization

Perhaps the newest ion-removal technique enjoying present applications is electrodeionization (EDI). A particular brand is called CDI. The process is also referred to as continuous electrodeionization (CEDI). The competitive types operate on the same principle. As shown in Figs. 5 and 6, a series of compartments or cells is created by the alternate spacing of ion-permeable

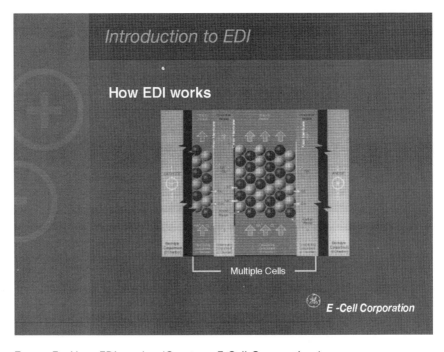

FIGURE 5   How EDI works. (Courtesy E-Cell Corporation.)

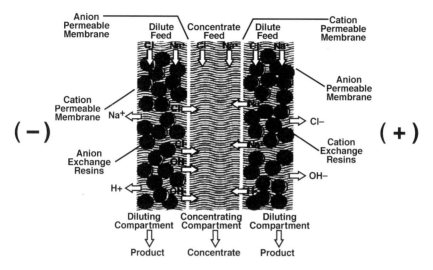

FIGURE 6   The ionpure CDI process. (Courtesy Gary Zoccolante.)

membranes bearing plus or minus charges, the result of their respective anion-exchange and cation-exchange compositions. A basic unit can be thought of as consisting of three cells. The middle compartment contains ion-exchange resin beads. A plus-charged anion-exchange membrane containing quaternary amine groups and therefore permeable to anions, which are minus charged, separates it on one side from the middle cell. The other side of the center compartment is bounded by a negatively charged cation-exchange membrane that is permeable to cations, bearers of positive charges.

   The solution intended for deionization flows in parallel fashion through the three compartments. Under the influence of a direct current, the cations (positive charged) in solution in the middle cell migrate through the cation-exchange membrane (negative charge) toward the cathode. Were they to reach it, they would each gain an electron to become atoms. The anions (negatively charged) in the central migrate through the anion-exchange membrane (positively charged) in the opposite direction toward the anode where, were they to reach it, they would become atoms by surrendering the electron, the loss of which is needed to neutralize their plus charge. While the ions, of whatever charge, leave the purified water cell through the appropriately charged boundary membrane, their oppositely charged counterions are repelled by these same (like-charged) membranes. They remain imprisoned within the waste stream compartment. The contents of these compartments, enriched in salts by the inflowing ionic migrations, are ultimately consigned to waste. Though the central compartment or channel flow the product waters, freed of their ionic contaminants by the ionic emigration, it must be made plain that in the CEDI technique the water is not filtered through the membranes. Instead, it is the electrically driven removal of ions through the cell-bounding membranes that purifies the water.

The modularized equipment, depicted above as three cell units for easy representation, consists of an alignment of many compartments, every other of which forms a purified water channel. Every other alternative compartment receives cations from one side and anions from the other, thus maintaining the electrical neutrality of the saline waste solution. The cells bounding the central purified water compartment are simultaneously parts of other three cell units. Separated from adjourning compartments by membranes bearing charges opposite from those on their other side, they lose cations to the waste water cell which receives anions from its other neighbor. The effluents from the ion-containing compartments are directed to discard. The flows from the several purified water cells or channels are joined to form the product stream.

The direct electric current motivates the ionic migrations in the direction of the appropriate electrode. At the cathode the cations receive electrons; at the anode, they lose them. Thus, one can think of the electrons as being one of the reactants in the chemical transformation. At any voltage, the amperes, or stream of electrons, is quantified by the electrical wattage. Each electron alters an element's cation or anion form into an atom. For every ion carried in the flowing stream to be thus altered, an electron must simultaneously be made available. But this depends on the current density or amperage. It follows, therefore, that for a given wattage over a unit of time, the concentration of the charge-carrying impurities can be excessive. Calculations can be made regarding the electrical input and the amount of deionization that can be expected therefrom. These calculations, however, may be rendered inexact by the presence of carbon dioxide or other weakly ionized compounds. This may limit the efficiency of the CEDI operation.

As the liquid in the cell being purified of its ionic content grows more dilute, the electric current, having fewer ion carriers to rely on, requires a higher driving force, voltage, to sustain it. The presence of ion-exchange resin beads in the purified water cell ameliorates this condition; the contiguous resin beads provide a pathway for the current. In time, the overvoltage characteristic of the dissociation of water into hydronium and hydroxyl ions (electrolysis) is reached. The ongoing regeneratng power of these ions, it is believed, maintain the functionality of the resin beads.

CEDI easily exceeds TOC requirements and gives satisfactory reductions of bacteria, endotoxin and colloids. As with most systems, CEDI has its limitations. When it is working well, it requies attention only about four times per year, whereas ion-exchange installations are much more demanding of servicing in their operation. When, however, the CEDI equipment is "down," it takes a substantial amount of time to reactivate. For whatever reason, the ion-exchange resins are loaded into the alternate compartments in their exhausted state. For the necessary regeneration to take place, time must be allowed. Estimates range from 12 to 24 hours. CEDI is being increasingly installed, but, as with all equipment, it does have its limitations.

The use of an CEDI device necessitates the use of an RO upstream as a pretreatment unit, which itself calls for pretreatments, and it needs to be followed with a "sterilizing grade" (0.2 μm-rated) cartridge membrane to remove organisms. It is essential to remove the hardness elements that would otherwise foul the ion-exchange membranes that form the cell boundaries. It is perhaps some indication of its limitations that the use of CEDI is not advocated separate from its RO "prefilter." In fact, CEDI with its RO unit is suggested as an alternative to a two-pass RO. However, like its ion-exchange counterpart, CEDI is more efficient than RO in its deionization activities. Usually it is installed to avoid the use of the ion-exchange technique and of the frequent regenerations and upkeep it requires, a practice involving the costs, handling, and disposition of strong and dangerous chemicals. The problem of neutralizing and disposing of the waste solutions is also avoided.

8.14. REVERSE OSMOSIS

8.14.1. RO Membranes

Reverse osmosis (RO) is the pivotal technology of most modern high-technology water purification systems, both in pharmaceutical and semi-conductor applications. However, it is operated differently in these industries. It is perhaps these differences that merit the attention of the pharmaceutical practitioners. Let us first consider the genesis of the RO action. The solid state of matter is composed of molecules arranged in a fixed spatial pattern relative to one another. Yet each molecule is separated from its neighbors by some finite space. The dimensions of this separation depend on the particularities of the molecular composition. Thus, the various molecules may have different interstitial spacings. When the substance is a polymer in film form, the interstitial spaces may be just wide enough to permit water molecules to be forced through them under pressure but may be too small to allow the permeation of ions made larger by their skirts of hydration. The charged ions bond to the oppositely charged dipoles of water molecules to form aggregates of larger size. The semipermeable nature of the RO membrane derives largely from its ability to discriminate on the basis of size between water and its dissolved or suspended accompaniments. (It is interesting to note that the smaller the crystallographic size of the ion, the larger its envelope of hydration, and the larger the actual size.) In the case of organics, solubility factors operate as well. Very few polymeric structures serve the RO function. The polyamides, and the polymers of the cellulose acetates, are the principal materials that are currently utilized for manufacturing RO membranes.

Present RO membrane usage centers largely on the asymmetric polyamide types. Their structure consists of a thin section of dense polymer overlying a thicker, far more open polymer section. The density of the thin film provides the discrimination against ion, organic, and particle passage. Its thinness minimizes its resistance to permeate flow. The more open layer is a support for the functional thin layer, and its openness is designed not to impede water flows. Depending on the method of manufacture, the support structure may be polyamide or polysulfone. Its composition is not critical to the RO operation.

Earlier RO membranes were made of cellulose acetate (CA) polymers. They have largely been replaced by their polyamide (PA) counterparts, which offer higher rejection, notably of silica and organics, and which operate at somewhat lower pressures, thereby reducing pumping costs. However, cellulose acetate–based RO has its adherents. It is less expensive than the PA. More importantly, it tolerates chlorine oxidation, whereas polyamide is ruinously susceptible to chlorine. It is permeable to chlorine to some extent, and this provides a modest protection to the permeate against microbial growth. To minimize its deterioration by hydrolysis, CA-based RO is operated at pH 5.5 to 6. At this pH, CO_2 is released from bicarbonates or carbonates, necessitating its removal by decarbonaters. The RO made of PA rejects the bicarbonate ion at pH 8 to 8.5.

8.14.2. Tangential Flow Filtration

The filter action of the RO membrane reflects the size of its "pores," the interstitial distance. These are on the order of 25 to 100 Ångstroms, small enough to avidly sieve-retain even very minute particles. Were the RO membranes to be used in the usual dead-end filtration mode, they would become blocked early on because of filter cake buildup. Therefore, tangential flow filtration is used for RO purposes. The feedwater as it emerges from the pretreatment stage with its remaining particulate load is flowed tangentially across the membrane. Some of the water permeates the RO, appropriately purified of its contaminants. This is the RO product water. The rest of the water, the reject stream, by its tangential sweep across the membrane surface, serves to cleanse it of accumulated deposits. Thus, the tangential flow mode, unlike dead-end filtration, limits the extent of filter blockage caused by filter area preemptions. This prolongs the useful life of the RO operation.

8.14.3. Reuse of Reject Water

The reject water stream, now more concentrated in impurities because its volume is decreased by permeate removal, is usually discarded. On occasion, however, some fraction of it is mixed with new feedwater and is again fed to the RO. Hopefully, it is recirculated at a rate swift enough to slow the accumulation of foulant on the surface of the RO membrane. Nevertheless, depending on its cumulative exposure to particulate contamination, the RO does eventually require being cleaned of its surface-blocking foulants. The reuse of the reject water inevitably shortens the time to the next RO cleaning. RO cleaning involves downtime, can detract from the membrane's longevity, and has its costs. It is a disruptive procedure that is best avoided. Although strongly guarded against in the electronics industry, it is accepted in pharmaceutical water purification practices to an astonishing degree.

There is a reluctance to discard the reject. It has already undergone pretreatment. It is presumably free of the hardness elements, including iron and manganese, and of chlorine, chloramines, and possibly of much TOC. It represents an operational cost. The chief drawback of the RO operation is the need to discard reject water. The conservation of water is a serious concern, and efforts to utilize it, as for instance in cooling towers, are unending. Where two-pass, product-staged RO operations are practiced, reuse of the reject water from the first pass is not made. The permeate is fed to the second RO. The reject stream resulting from the second-pass RO is considered clean enough to be mixed with fresh feedwater to feed the first RO. However successful such practices, the reuse of reject water is best avoided.

8.14.4. Concentration Polarization

An unwelcome phenomenon that is the inevitable result of the essential RO action also contributes to a diminishment of the RO discrimination—namely, concentration polarization. As the ionic, organic, and other matter is rejected by the RO, their concentration increases in the areas immediately adjacent to the membrane where their rejection occurs. The RO separation of water from its solutes and suspensions is more difficult in proportion to their concentration. For this reason, too, the crossflow sweep of the circulating liquid stream requires a vigorous flow, to disrupt the impurity concentrations of the boundary layer.

The point is that the less pure the feedstream encountering the RO membrane, the more attenuated the RO discrimination becomes. The aim, it would seem, is to furnish as clean a feedwater stream as possible to the RO. This calls for a pretreatment system of very high quality. It is the performance of the pretreatment stage in removing the oxidizing biocides, the particulate matter, the hardness elements, and organics from the feedwater that foreshadows the subsequent RO effectiveness.

8.14.5. The Permeate Stream

As stated, the feedwater stream is divided by the RO action. Part of it permeates the RO membrane purified of its ionic and organic contents, and of suspended entities, including organisms. The degree of purification attained is a function of the RO type and of the operational conditions. The RO greatly retains organic substances larger than about 300 in molecular weight; the weight being taken as a measure of size. The RO, being a filter with very fine pores, also retains organisms. So efficient are RO operations in their removal of ionic, organic, and organism contaminants, that UV and microporous membranes, the usual accompaniments of ion-exchange installations meant to kill and retain organisms, can be dispensed with; the UV units perhaps less so. However, it should be noted that the efficiency of removal of ionic impurities by ion-exchange, and especially by way of mixed beds, is superior to that of RO. As mentioned, the accumulation of retained deposits on the RO membrane necessitates periodic cleanings with their corresponding downtimes.

The RO membrane action is not absolute. Even the organism retention, presumably certain because of the minuteness of the RO "pores," the interstitial spaces within the polymeric membrane, cannot ensure sterile effluent. Organisms may on occasions be found on the membrane's downside, perhaps the result of flaws. Such are inevitable in any membrane manufacturing process, particularly where integrity testing of the sort common to microporous membranes does not exist.

8.14.6. Extent of Recovery

The RO operation results in the discard of a percentage of its feedwater. This is an expense that, although inherent in the process, is perhaps the chief drawback to the RO operation. It is understandable, therefore, that as much permeate as possible is sought. The problem is that the greater the recovery, the more impure the reject stream and the more its reuse necessitates the disruptive cleaning of the RO. Recoveries of 50%, 75%, 80%, and 90% result respectively in 2-, 4-, 5-, and 10-fold increases in contamination concentrations. Usually the recovery will run to 75% or even higher. Figure 7 illustrates the progressive increase in contamination in the reject water that accompanies larger recoveries, particularly above 75%. It is advised that a well-run RO unit should initially be operated at a 50% recovery. Its ability to produce higher recoveries should be exercised when supported by trouble free operations, the result of a proper feedwater quality, and by an adequate pretreat-

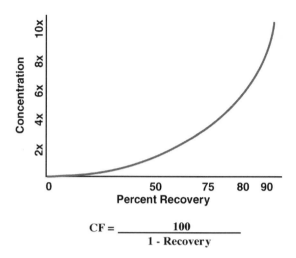

$$CF = \frac{100}{1 - Recovery}$$

FIGURE 7   Concentration factor. (Courtesy of Lee Comb, GE Osmonics.)

ment capacity. Larger RO units (above 15 gpm) must operate at higher recoveries, but careful system design and extensive pretreatment are needed to enable penalty-free recoveries of 60 to 70%.

Beginning at a 50% recovery, higher recoveries involving reuse of more highly contaminated reject waters should be made only after a lengthy trial period sufficient to demonstrate that stabilization of the RO unit and the pretreatment units has occurred. Not needing to clean the RO unit can be taken as the endpoint. At that point, a higher recovery becomes justified. A repeat of the stabilization period is made, and increases in the percentage of recovery are continued, as long as the RO unit and the pretreatments operations do not become insufficient to cope with the increasingly contaminated waste stream, as indicated by the need to clean the RO unit.

8.14.7. Discontinuous RO Operations

Anecdotal accounts have the RO composed of CA devoured by cellulose-digesting organisms when not maintained in constant flow. This, coupled to the still-current belief that nonflowing waters encourage organism growth, promote the practice of continuous RO operation whether permeate water is needed or not. These beliefs, whether correct or not, also have their detractors. The water produced in the continuous RO mode, when not needed, is stored in a break tank, as is also the reject water. However, when the water processed exceeds the available storage capacity, leading to its being discarded, the practice is unjustified. Such continuous RO operations waste water, and needlessly exhaust the carbon, softening, and other pretreatment units. It is also believed that an RO that has been shut down does not, in consequence, undergo deterioration. It requires only that the product water be flushed to drain for 20 to 60 seconds after being restarted to restore it to its suitable operating condition. What is argued is that there is no established reason to believe that continuous RO operations are necessary; no substantiating experimental data have been alleged. The practice, therefore, should be discontinued.

8.14.8. Size of RO Unit

The capacity of the RO unit should satisfy both the peak load and the daily demand, as called upon. There is a tendency to undersize RO units. A smaller RO unit is less expensive, but despite its lower capacity it can be operated continuously, even overnight, to furnish the needed amount of water. The permeate can be stored in a break tank from which the needed water can be withdrawn at the requisite rate. In the case of larger units, the operation of the RO would be intermittent. The length of the shutdown periods would depend on the demand for water and on its rate of production The continuous operation satisfies the belief, previously mentioned, that unless RO units are in continual operation problems ensue. In conformity with this belief, smaller RO units are championed as a way of securing a continuing RO operation. There are, however, disadvantages as well. Once the continual operation fills the storage capacity, further permeate production is directed to drain, an exaggerated waste of already expensively treated water. Exacerbating the situation is the continual, but unnecessary, consumption of the pretreatment facilities when continued beyond the storage capacity, as stated.

There may even be room for cynicism in the matter of selecting the RO size. The smaller units costs less and may be easier to sell; all in good conscience. But there is a subtle but significant risk that accompanies the choosing of smaller RO units: namely, one may conclude that proportionately smaller pretreatment facilities need be supplied. However, the continuous RO operation involving the reuse of the reject stream places an added and eventually an excessive load on the operation. The unwanted RO membrane fouling that results is progressive in its severity. It may not be apparent at first, but when evident can assert itself as a serious fouling problem.

Whatever choices are made regarding the size of the RO unit, concerning continuous or intermittent operations, with respect to recovery and water conservation, or in regard to the use of the reject stream, the quality and adequacy of the pretreatment components of the purification system should not be compromised. This will ensure that cleaning of the RO membrane will be kept to a minimum, as it is in the electronics industry. It can be expected that in consequence a higher quality RO product water with more reliable operation will eventuate.

8.15. STORAGE CONDITIONS

The above are examples of purification operations that can serve to minimize the presence of organisms during the preparation of pharmaceutical waters. The waters thus prepared must be stored and distributed in such manner as to avoid microbiological contamination. What should be of interest is that biofilm formation and its random shedding of organisms threaten the biological quality of the purified water regardless of its careful preparation. This is almost inevitable under most storage conditions and particularly so when ambient storage temperatures are involved. To avoid the contaminating consequences of biofilm shedding, the prepared water is best stored either under ozone or at temperatures above 65°C, preferably around 80°C.

8.16. CONCLUSION

Pharmaceutical purification system design can be undertaken from several different points of view. It is here advised that advantages over present approaches may be gained by applying certain of the techniques that are utilized in semiconductor rinsewater preparations. In particular, discontinuous RO operations are advocated. The key to this achievement is a moderation of the percent recovery that is sought in the reverse osmosis practice.

REFERENCES

1. Meltzer, T.H. *Pharmaceutical Water Systems*; Tall Oaks Publishing: Littleton, CO, 1997.
2. Collentro, W.V. *Pharmaceutical Water*; Interpharm Press: Buffalo Grove, IL, 1999.
3. Dey, A.; Thomas, G. *Electronics Grade Water Preparation*; Tall Oaks Publishing: Littleton, CO, 2003.
4. Chawla, G.L.; Varma, M.M.; Balram, A.; Murali, M.M. Trihalomethane removal and formation mechanism in water. Chemical Engineering Report, District of Columbia University, 1983.
5. Wiegler, N.; Anderson, C.C. Removing trihalomethanes from DI water: a consideration of the alternatives. Microcontamination 1990, *8* (10), 37–42, 108–112.
6. Wiegler, N.; Anderson, C.C. Removal of trihalomethanes by means of active carbon bed and UV light. Transcripts of Ninth Annual Semiconductor Pure Water Conference, Santa Clara, CA, Jan 17–18, 1990; 121–149.
7. Sweadner, K.J.; Forte, J.; Nelsen, L.L. Filtration removal of endotoxin

(pyrogen) in solution in different states of aggregation. Appl. Environ. Microbiol. 1997, *34*, 382–385.

8. Coleman, D.C.; Evans, R.W. Investigation of the corrosion of 316L Stainless Steel pharmaceutical water for injection systems. Pharm. Eng. 1991, *11* (4), 9–13.

9. Coleman, D.C.; Evans, R.W. Fundamentals of passivation and passivity in the pharmaceutical industry. Pharm. Eng. 1990, *10* (2), 43–49.

10. Husted, G.; Rutkowski, A. Control of microorganisms in mixed bed resin polishers by thermal sanitization. Watertech '91 Proceedings, San Jose, CA, Nov. 20–21, 1991; 451–453.

9

Airborne Contamination Control

Lothar Gail and Dirk Stanischewski
Siemens Axiva, Frankfurt/Main, Germany

9.1. INTRODUCTION

It is widely accepted that in open processing of sterile pharmaceuticals, product quality essentially depends on protection from airborne contamination. Such protection may be defined by specifying a certain tolerable level of airborne particulates/viable microorganisms and additional parameters such as airflow, temperature, and humidity.

Although the concentration of airborne viable microorganisms is understood to be the more critical contamination factor, the concentration of airborne particulates is the preferred specification for building, commissioning, and maintaining pharmaceutical cleanrooms. A correlation between the two factors is accepted, but there is no "scientific agreement on a relationship between the number of nonviable particulates and the concentration of viable microorganisms" (1).

The concentration of airborne particulates is used to demonstrate the performance and proper functioning of a facility because of the advantages of real-time measurement and the higher resolution possible in controlling particulates. The airborne concentration of viable microorganisms is employed as the main parameter in environmental monitoring programs.

To specify airborne particulate cleanliness some pharmaceutical standards (1) are referencing generic classification systems—for example, the former U.S. Federal Standard 209E, (2) now being replaced by ISO 14644-1 (3); see Table 1.

The basic requirements for contamination control have to be defined for each cleanroom or clean zone—for example, by assigning a standard cleanroom classification as well as additional specifications, such as for cleanroom segregation (differential pressure, airflow), temperature, and humidity.

Such classifications are described in ISO 14644, which includes specifications on testing the performance [see (1)] and continued compliance [ISO 14644-2, see (4)] of cleanroom installations.

In addition, sterile processing requires a qualification program to ensure:

A documented set of cleanroom user requirement specifications has been defined, together with the design of the sterile concept (design qualification/risk analysis).

The required equipment has been installed (installation qualification).

The operational requirements have been met (operational/performance qualification).

A program for safeguarding continuous compliance throughout the production process exists.

TABLE 1 Classifications for Airborne Particulate Cleanliness

	Particles equal to and larger than 0.5 μm		
		Nearest equivalent class of former U.S. Fed. Std. 209E (2)	
ISO class name	(per m^3)	(per ft^3)	Metric
1	—	—	—
2	—	—	—
3	35	1	M1.5
4	352	10	M2.5
5[a]	3,520	100	M3.5
6	35,200	1,000	M4.5
7[a]	352,000	10,000	M5.5
8[a]	3,520,000	100,000	M6.5
9	35,200,000	—	—

[a] Classes typically used for specifying pharmaceutical processing zones.

This qualification program requires the following:

Documented procedures for all stages of qualification
Validated test procedures
Calibrated test equipment
Documentation system linking qualification certificates and production reports

A crucial factor in planning sterile processing facilities is that almost every activity must be planned and every material used must be qualified according to a formal procedure. State-of-the-art sterile processing requires a fully understood and documented procedure. In sterile processing, virtually nothing can be carried out in the same way as in an R&D laboratory. For this reason, in planning a new facility or remodeling an existing one, it is necessary to consider all relevant production activities and situations that might influence the quality of sterile processing. Qualification and maintenance operations are therefore as relevant to the planning process as logistics, production, cleaning operations, break-down management, and hazard analysis.

The use of simulation tools is a suitable means of demonstrating the performance of a design concept at an early stage of development.

9.2. CLEANROOM DESIGN AND OPERATION

9.2.1. Processing Concept

The design of a sterile cleanroom facility starts with a processing concept and basic assumptions about the installation of production equipment.

The importance of this planning stage is emphasized by the fact that an airborne contamination risk is highly dependent on the following:

Whether processing is open or contained
The level of personnel activity
The level of airborne contamination

Continuous processing on an automated filling line with washing machine, tunnel sterilizer, and filling machine offers a high level of protection against airborne contamination. Batchwise processing—for example, sterilizing glass containers with dry heat in a chamber sterilizer and then transporting them to a filling machine—poses a higher contamination risk.

The risk of contamination from the activities of personnel (e.g., loading a freeze dryer) has to be countered by a very high air change rate or preferably by unidirectional airflow.

An especially high level of product protection is offered by isolator technology. The advantages of isolator technology depend, however, on a suitable level of automation and reliability. Frequent interventions may compromise the protection concept and jeopardize the advantages of this concept.

When a sterile processing concept has been devised, it must be completed by material and personnel flow, zoning, and building layout concepts.

Because any form of motion in a sterile processing facility poses a considerable risk of airborne contamination, a material and personnel flow concept is required, establishing a controlled procedure for the following:

Personnel entry and exit
Material entry, supply and exit
Consumables processing
Waste collection and disposal
Maintenance operations

Where required, this concept should provide a separate flow for entering (i.e., clean) and exiting (i.e., contaminated) personnel.

9.2.2. Zoning Concept

A zoning concept assigns the appropriate levels of tolerable airborne contamination to each processing step.

A zoning concept for sterile/aseptic processing that complies with most of the relevant regulations (1,3,5,6) is given in Table 2. Its particle numbers are taken from ISO 14644-1, the only international cleanroom classification standard. The airborne microbial contamination limits relate to USP 26 (1)

TABLE 2 Air Quality Specification for Aseptic Processing

Processing zones	Maximum permitted number of particles/m^3 equal to or above 0.5 μm[a] - in operation[b]	Microbial contamination air sample limit—cfu/m^3 -in operation
Critical zone	3 520[c]	USP 26 (M3.5): < 3
		EC GMP/Grade A: < 1
Sterile processing	352 000[c]	USP 26 (M5.5): < 20
		EC GMP/Grade B: 10
Support	3 520 000[c]	USP 26 (M6.5): < 100
		EC GMP/Grade C: 100

[a] EC GMP Annex 1 (7) also gives limit numbers for particles ≥5 μm.
[b] EC GMP Annex 1 also gives limit numbers for the "at rest" state.
[c] Classification limit numbers taken from ISO 14644-1—nearest equivalent classification to former U.S. Fed. Std.209E, see Table 1.

and EC GMP (7). It can be expected that the relatively small differences between the different classification systems will gradually disappear, while the ISO system will be overtaken by GMP and corporate guidelines. Because there are no fundamental discrepancies between the different GMP guidelines, international harmonization might proceed in this area.

9.2.3. Cleanroom Segregation Techniques for Sterile Processing

According to pharmaceutical GMP guidelines, a concept for controlling airborne contamination must include a protection concept based on different cleanroom segregation techniques. Detailed guidance on segregation measures is provided by FDA (6), EC GMP (7), and ISO (8).

According to FDA (6), "Rooms of higher air cleanliness should have a substantial positive pressure differential relative to adjacent rooms of lower air cleanliness. For example, a positive pressure differential of at least 12.5 Pascals (Pa) should be maintained . . . When doors are open, outward airflow should be sufficient to minimize ingress of contamination . . . Pressure differentials between cleanrooms should be monitored continuously throughout each shift and frequently recorded, and deviations from established limits should be investigated."

EC GMP (7) specifies that "A filtered air supply should maintain a positive pressure and an airflow relative to surrounding areas of a lower grade under all operational conditions and should flush the area effectively. Adjacent rooms of different grades should have a pressure differential of 10-15 Pascals (guidance values)."

Taking into account the whole spectrum of technological requirements, ISO 14644-4 (8) defines three different cleanroom segregation concepts:

The differential pressure concept, primarily used when ventilated rooms with different classifications are separated by doors or small openings
The displacement airflow concept, used for separating adjacent cleanroom zones with different classifications by a controlled and/or unidirectional airflow
The physical barrier concept, used for example, in isolator technology

The differential pressure concept may be chosen to protect a sterile processing room against environmental contamination from an adjacent, lower-classified support area. The displacement airflow concept may be advantageous in protecting a critical area of an automated filling line against ingress of airborne contamination from the ambient sterile room environment.

The advantage of differential pressure segregation is that even with large adjacent areas, a certain protective function can be provided simply by controlling (or even monitoring!) one parameter—the differential pressure. The advantage of the displacement concept is that airflow alone—without any physical barrier—provides safe separation of adjacent areas. The controlled airflow of a "laminar flow" unit may allow safe access/operation of personnel while the risk of air ingress from the lower to the higher classified area is efficiently under control.

A disadvantage of cleanroom segregation by pressure difference is that the chosen level of (for example) 15 Pascal (Pa) largely depends on the given control tolerances for room pressurization instead of on contamination control. For cleanroom segregation purposes, a pressure difference of more than 5 Pascal would be sufficient (8). Differential pressures above 20 Pascal between different graded cleanrooms may cause disturbances in continuous processing from the support to the adjacent sterile processing zone. Hence, it is an advantage to establish the lowest differential pressure that can be safely controlled between two adjacent cleanroom areas.

In this case, the differential pressures between the cleanrooms and airlock areas are too small and variable to be efficiently controlled. When the doors are closed, a pressure cascade from the higher to the lower classified area and, when the doors are open, an airflow in the same direction should be established to avoid ingress of air from the lower to the higher classified area.

For this reason, controlled overflow instead of pressure difference control has proved to be a more efficient and suitable technique for cleanroom/airlock segregation.

Figures 1 and 2 show how the different segregation concepts may be combined by providing the following:

A controlled 20 Pa differential pressure between the "sterile processing" cleanroom and the "support" cleanroom
A controlled overflow between the cleanrooms and the adjacent airlock zones as well as between the airlocks.

The advantages of combining the two segregation concepts are as follow:

Less variation in pressure differentials during entry/exit procedures
Improved protection against air ingress due to control and calibration effects
Fail-safe function
Continuous compliance with regulations.

The higher protection performance of cleanroom segregation using a physical barrier concept (e.g., isolator technology) makes it possible to reduce the air quality requirements for the cleanroom environment of such an installation. Sterile processing guidelines (1,6,7) recommend that at least "support" requirements (see Table 2) should be met in such an area.

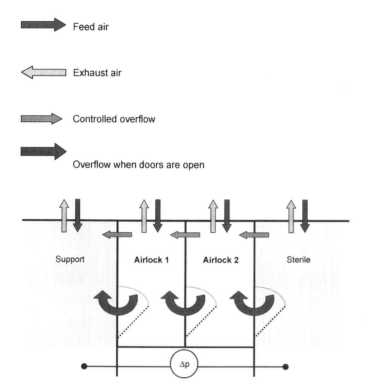

FIGURE 1   Cleanroom/airlock segregation.

Pressure Cascade					
Cleanroom/ Airlock	**Support**	**Airlock 1**	**Airlock 2**	**Sterile**	**Pressure**
Pressure cascade				+/-5 Pa	
					40 Pa
					30 Pa
	+/-5 Pa				20 Pa
					10 Pa

FIGURE 2   Pressure cascade.

Whenever airflow has to protect a processing area from airborne contamination, its design must be seen as an essential part of the process design. Airflow deficiencies should be identified as early as possible. To compensate for the risk of such deficiencies, it may be an advantage to simulate the airflow of a sterile processing facility prior to proceeding to project realization. Recently, the role of airflow control has been emphasized by the recommendation that weekly velocity monitoring should be carried out in aseptic processing clean zones.

Before progressing to the subsequent planning stages, it is advisable to discuss layout, zoning and segregation concepts with corporate experts, consultants, and the authorities to ensure that all relevant recommendations on sterile processing have been considered.

9.2.4. Airflow Simulation

During the design qualification of an installation, it is necessary—even though still not common practice—to verify or validate that the planned air handling system will provide the required quality and reliability in terms of flow patterns. A proven technology to achieve this in the design phase is to simulate the airflow—both unidirectional and non-unidirectional flows—with the help of a computational fluid dynamics (CFD) simulation. The only alternative is to study the installation experimentally (see sections on airflow and recovery) as soon as the installation or a model of it is available for testing.

Current CFD software is already highly sophisticated and so a very true picture of later reality can be predicted. (See example in Fig. 3a,b,c. A work-

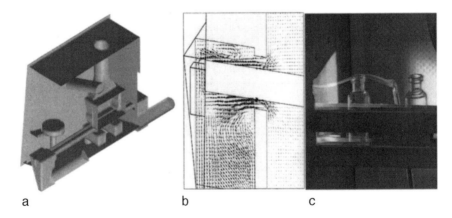

a   b   c

FIGURE 3   a. Workstation in a filling cleanroom. b, c. Simulated and experimental airflow. (Courtesy of Dr. W. F. Schierholz, Siemens Axiva.)

station in a cleanroom for filling processes has been simulated by CFD. The study shows good correlation between predicted flow patterns and measurements with the actual equipment.) This predictive capability is available at a very early stage in the planning process. A properly designed CFD study is able to point out all potential deficiencies in airflow patterns that might subsequently be detected (e.g., as microbiological deviations due to insufficient local air exchange).

For subsequent optimization, it is easy to create different variants of a concept in order to find a better layout (e.g., for air inlet locations). The intrinsically high resolution of CFD simulations is unparalleled in real-world measurement. Unlike with experimental measuring techniques, it is possible to study the conditions at virtually every point in a simulated volume. Hence, special expertise is needed to filter out the relevant locations that should be investigated, so that the flow rate or air velocity at critical points can really be defined.

Even a well conducted CFD study cannot guarantee that the final installation will meet all specifications. This is due to the fact that certain assumptions and simplifications have to be made for any simulation. The qualification process for a facility will reveal if some of these basic assumptions have been violated (e.g., in a CFD simulation walls are impermeable, whereas in the real world walls have joints, which may permit ingress of contamination). Nevertheless, the results of a simulation will always give valuable guidance in cases where a "common sense" judgment by design engineers is difficult or impossible.

It must be borne in mind that the additional financial outlay for conducting a CFD study leads to greater certainty in the planning or design process. These costs have to be balanced against the potential costs of remedying a deficient concept when a facility has already been built and then has to be modified to meet the specifications.

9.2.5. Qualification and Operation

An airborne contamination qualification program comprises a set of measurement activities covering initial qualification (commissioning), ongoing qualification, and monitoring.

Initial qualification activities are planned and performed prior to the start-up of a new facility to demonstrate compliance with a set of target data laid down in a supplier's contract. The supplier or another service provider may be responsible for demonstrating compliance, giving instructions on carrying out the measurement procedures, and delivering relevant target data.

Performing ongoing qualification activities is an essential responsibility of pharmaceutical manufacturers. The manufacturer has to initiate and evaluate such activities whereas the performance may be delegated to a service provider. Ongoing qualification activities are typically performed at a fixed time interval under "at rest" conditions.

Initial and ongoing qualification programs are typically concerned with the following:

Cleanroom classification
Installed filter leakage
Containment leakage
Airflow (including airflow visualization/smoke studies)
Air pressure difference
Recovery

Environmental protection parameters in monitoring programs typically include the following:

Particle concentration
Airflow velocity
Air pressure difference
Airborne viable contamination

9.3. METROLOGY AND TEST METHODS

9.3.1. General

The basic idea of separating a cleanroom or clean zone from its environment by means of flow barriers must be validated in order to verify the efficiency of the barrier. The relevant parameters that have to be checked are related to air quality and airflow.

In the following, a description is given of special points that need to be considered in conducting the required tests. Details of individual tests and test methods are given in ISO/DIS 14644-3, for example, and will not be dealt with here.

9.3.2. Cleanroom Classification

Measuring procedures for verifying airborne particulate cleanliness classification (using a discrete-particle-counting, light-scattering instrument) are defined in ISO 14644-1 and −2 (3,4). These definitions include requirements for measuring equipment, calibration, sampling procedures, recording, and statistical treatment of particle concentration data. Annex F of ISO 14644-1 deals with a sequential sampling procedure for environments in which the air being sampled is significantly more or less contaminated than the specified class concentration limit for the specified particle size.

The number of sampling point locations is derived from the square root of the cleanroom/clean zone area in square meters.

When identifying sample locations in a unidirectional airflow environment, it should be remembered that only the contamination level in the immediate vicinity of that location is being detected. In a turbulent (i.e., non-unidirectional) airflow environment, a considerably larger room area is contributing to the measurement.

When particle measurement is being carried out for initial or ongoing qualification purposes, it is convenient to use the classification limits and procedures of generic standards, such as ISO 14644-1 (see Table 9.2). For monitoring purposes, however, (i.e., under operational conditions), it may be more suitable to define individual limits, taking into account the specific needs of individual processing locations (e.g., sterile powder processing, aerosol and/or vapor release).

9.3.3. Installed Filter Leakage Testing

To maintain the specified level of air quality in a sterile processing facility, HEPA filters are used as a barrier between the nonsterile sections of an HVAC installation and the sterile processing cleanroom environment. Thus the air quality actually depends on two different parameters: the filter penetration rate and the rate of local leakages in a filter or filter outlet.

The filter penetration rate does not necessarily have to be tested at the filter installation site. To minimize the rate of defects at the installation site, some manufacturers prefer to install pretested filters. The FDA guideline "Sterile Drug Products Produced by Aseptic Processing" (6) requires sufficient attention to be paid to installed filter leakage testing, which includes the identification of any kind of local containment defect around a filter unit:

> "The purpose of performing regularly scheduled leak tests ... is to detect leaks from the filter media, filter frame, or seal."

A detailed procedure for installed HEPA filter leak testing is described by ISO/DIS 14644-3 (9), specifying the traditional aerosol photometer (AP) method and the more recent discrete particle counter (DPC) method. Although the equivalence of the methods has been demonstrated (10) for a given range, in certain applications preference should be given to the DPC method owing to its better resolution (11,12). The equivalence of using different test aerosol substances has been investigated (13,14).

HEPA filter leak detection normally requires the use of an aerosol challenge. Suitable aerosol substances and requirements for measuring equipment (particle counter, aerosol generator, dilution system) are specified by ISO 14644-3.

Figure 4 shows the results of a comparability study using the AP and DPC methods with different aerosol test substances (DOP, DEHS, PAO). In this study, the AP method produces consistently lower penetration rates than the DPC method (10). Figure 5 shows the correlation between the size of the "controlled leak" and the measured penetration rate. A penetration rate of 0.01%, as required (6) is caused by a leak of approximately 0.3 mm diameter (14). According to this definition, approximately the same penetration number [which characterizes the 'integral' efficiency of a HEPA filter (15)] has also been taken to represent the size of an "acceptable" local leak in the filter media, filter frame, or seal.

9.3.4. Airflow

A number of important requirements for a sterile processing installation are based on the proper design of airflow rates. It is therefore essential to verify—as early as possible within the qualification process—that flow rates or velocities meet the specifications.

As has been shown above, there are two basic segregation concepts: a displacement flow concept and a pressure difference concept. Both determine the proper airflow direction and flow distribution.

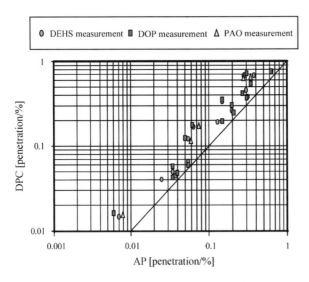

FIGURE 4 Comparison of penetration measured by the AP and DPC methods. (From Ref. 10.)

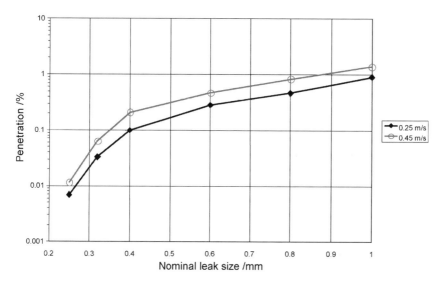

FIGURE 5 Correlation between penetration and leak size for different flow velocities. (From Ref. 14.)

When a displacement flow concept is used (e.g., under a laminar flow unit), the flow direction and flow distribution must be sufficiently homogeneous for the protected area or cross section of the workplace.

When a pressure difference concept is used (e.g., in a personnel airlock with turbulent ventilation), it is important to demonstrate that the airflow at any interface of the cleanroom with adjacent rooms or areas corresponds with the predefined direction and that the total flow rate into the room results in sufficiently high air exchange rates.

In both cases, the visualization of airflow patterns with the help of artificial fog or mist (produced by fog generators using organic liquid or DI water) makes it possible to determine the spatial and temporal consistency of the flow direction and to estimate the degree of turbulence (see Figs. 6 and 7 for examples). Flow direction alone can also be tested by observing, e.g., a silk thread.

The additional measurement of flow rates or local velocities is also necessary to demonstrate the compliance of an installation with design specifications.

During the initial qualification process, the number of sampling points for airflow measurement is high in order to demonstrate the required homogeneity and proper distribution, with typical measuring grid sizes of 300×300 mm. During this process characteristic locations must be chosen

FIGURE 6   Visualizing airflow direction at a door.

FIGURE 7   Visualizing airflow under a laminar flow unit.

(typically much fewer in number based on typical grid sizes of, for example, 600 × 600 mm), which are used for subsequent ongoing (re-)qualification. The initial qualification must define the actual target values and allowable ranges at these locations. This results in a "fingerprint" of an installation. As long as measured values during ongoing qualification at these locations stay within their defined ranges, the installation is (and has been) compliant with its specifications.

The time interval between ongoing measurements will depend on the probability of changes. Processes with high dust release, for example, will cause accelerated plugging of exhaust filters so that more frequent monitoring of flow will be necessary.

A common problem in determining airflow is that measurement variations between different locations are relatively high, whereas values taken at a well-defined location can be measured with high repeatability. Thus, a measured flow distribution (e.g., when testing the homogeneity of velocities under a laminar flow unit) will show greater variance. On the other hand, as soon as the velocity of a well-defined location (e.g., a reference point under a specified filter) is determined, it is possible to detect changes with high sensitivity. That is the reason why monitoring reference points is entirely adequate for detecting deviations from compliance status.

9.3.5. Pressure Difference

The designed pressure difference between a room and a reference point or adjacent room must be checked in order to demonstrate the ability of an installation to maintain the pressure within defined ranges. Pressure differences can be maintained only if the supply and exhaust airflows are appropriate and under control. This has to be verified prior to measuring pressure differences, otherwise the results may be misleading. Any disturbance in the flow will propagate pressure difference variations or deviations.

Pressure difference determines the flow direction between adjacent rooms and is therefore a critical quantity for avoiding contamination of cleanrooms or zones through ingress of contaminated air.

Because this measurement is easy to obtain, it is highly suitable not only for periodic but also for continuous measurement (monitoring) where it is chosen as an online indicator of the cleanroom status.

9.3.6. Recovery

One of the most important characteristics of an installation is its ability to return to the designed status after a disturbance within a finite time interval. Whereas under laminar flow conditions this is—intrinsically—a matter of a few seconds, in a room with turbulent ventilation the recovery is a function of room size, flow rates, air supply locations, installed equipment, exhaust, and so on.

As long as an installation remains constant in terms of equipment, flow rates, pressures, and so forth, its recovery characteristics are not likely to change. Therefore, this test is typically conducted during the initial qualification process.

The time taken to reduce an initial, artificially added aerosol concentration by a factor of 100 (the 100:1 recovery time) is the main result of this method.

The quality of room design has a major influence on this number: in the ideal case, a room with turbulent ventilation is completely mixed so that any contamination will be diluted and hence removed. Dead zones or poor airflow patterns will act as sources of contamination that prolong the recovery time. Such an influence can normally only be seen at concentrations close to the steady-state level.

In the following example, an airlock with conventional lockers has been tested for its recovery time. Although the decay of the aerosol concentration is

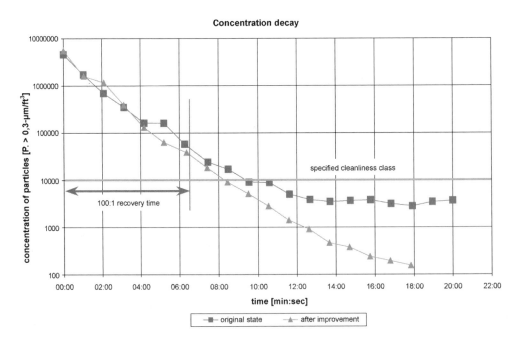

FIGURE 8   Measurement of recovery time.

fast and reaches the defined cleanliness class within a few minutes (see square symbols in Fig. 8), microbiological measurement shows severe deviation from the specification. An analysis of the situation reveals that the particles (and hence viable organisms) released from the lockers are causing the problem. When the particle sources from the lockers are removed (either by making them airtight or by providing local exhaust ventilation), the recovery not only is much faster but reaches lower levels of background concentration without any microbiological deviation (triangle symbols in Fig. 8).

9.3.7. Documentation

Specific requirements for documentation are laid down in ISO 14644 parts 3 and 4. Test documents should contain all the information necessary to fully understand the results of a measurement. Hence, it is essential to include all information about the measuring equipment used and its calibration status, conditions of measurement (e.g., occupancy state of a room, machines running or not, and so on) and test results. This allows an evaluation of the results without the help of additional documents.

REFERENCES

1. The United States Pharmacopeia, USP 26. United States Pharmacopeial Convention, Rockville MD, Jan 1, 2003.
2. Airborne Particulate Cleanliness Classes in Cleanrooms and Clean Zones. U.S. Federal Standard 209E. General Services Administration, Washington, D.C., September 11, 1992 (withdrawn Nov. 2001).
3. Cleanrooms and associated controlled environments—Part 1: Classification of air cleanliness, ISO 14644-1; ISO, Geneva, 1999.
4. Cleanrooms and associated controlled environments—Part 2: Specifications for testing and monitoring to prove continued compliance with ISO 14644-1, ISO 14644-2; ISO, Geneva, 2000.
5. Aseptic processing of health care products—Part 1: General requirements, ISO 13408-1: 1998-08, ISO, Geneva.
6. Guideline on Sterile Drug Products Produced by Aseptic Processing. Center for Drugs and Biologics and Office of Regulatory Affairs; Food and Drug Administration, Rockville, Maryland, Draft August 2003.
7. EC GMP Guide to good manufacturing practice. Revised annex 1: Manufacture of sterile medicinal products. European Commission: Brussels, 2003.
8. Cleanrooms and associated controlled environments—Part 4: Design, construction and start up, ISO 14644-4; ISO, Geneva, 2000.
9. Cleanrooms and associated controlled environments—Part 3: Metrology and test methods, ISO/DIS 14644-3; ISO, Geneva, 2002.
10. Gail, L.; Ripplinger, F. Correlation of alternative aerosols and test methods for

HEPA filter leak testing. Proc. Int. Symp. on Contamination Control; Institute of Environmental Sciences and Technology: Phoenix, Arizona, 1998.

11.   HEPA and ULPA Filters; IES-RP-CC001.3; Institute of Environmental Sciences and Technology, Mount Prospect, Illinois, 1993.

12.   Testing Cleanrooms, Clause 6.2: HEPA and ULPA filter installation leak tests, IES-RP-CC006.2; IEST Institute of Environmental Sciences and Technology, Mount Prospect, Illinois, 1997.

13.   Moore, D.R., Jr.; Marshall, J.G.; Kennedy, M.A. Comparative testing of challenge aerosols in HEPA filters with controlled defects. Pharm. Eng. 1994, 14 (2), 54–60.

14.   Gail, L.; Stanischewski, D. Installed HEPA Filter Leak Testing by Using Discrete Particle Counting—Investigation into Practicability, Proc. Int. Symp. on Contamination Control; Institute of Environmental Sciences and Technology, Anaheim, 2002.

15.   High efficiency particulate air filters (HEPA and ULPA)—Part 1: Classification, performance testing, marking, EN 1822-1; Beuth Verlag, Berlin, 1998.

10

Disinfection Practices in Parenteral Manufacturing

Vivian Denny
Peak to Peak Pharmaceutical Associates, Boulder, Colorado, U.S.A.

Frederic Marsik*
Microbiology Consultant, New Freedom, Pennsylvania, U.S.A.

10.1. INTRODUCTION AND DEFINITIONS

The objective of this chapter is to provide practical guidance in the selection, evaluation, control, and validation of chemical disinfectants for use in the pharmaceutical manufacturing environment. Suggestion for the design, implementation, and training necessary for a successful disinfection program are also presented.

The following definitions of terms used in the field of disinfection are presented to assist the reader in interpreting information from scientific and product literature. Additional definitions can be found in Ref. (1).

---

* Dr. Marsik has written this material in his own private capacity as a microbiology professional. No official authorization by the FDA is intended or implied.

Antimicrobial. A chemical agent used to inhibit the growth of or kill
  microorganisms. Sporicidal activity is not implied.

Biocide. An agent that kills all living microorganisms, pathogenic and
  nonpathogenic. A biocide kills spores as well as vegetative cells and is
  therefore considered a sterilizing agent.

Cleaning. The removal, usually with detergent and water, of adherent
  visible soils, blood, protein substances, and other debris from surfaces and equipment, by a manual or mechanical process that pre-
  pares the items for safe handling and/or further decontamination.

Disinfectant. A chemical agent that eliminates a defined scope of microorganisms, including in some cases microbial spores.

Disinfection. A process that eliminates a defined scope of microorganisms including in some cases microbial spores.

Fungicide. An agent that kills fungi. Sporicidal activity is implied but
  not inclusive.

High level. An agent that destroys all vegetative bacteria, including
  tubercle bacilli, lipid and nonlipid viruses, fungal spores, and bacterial spores. When spores are present in large numbers, not all spores
  may be killed.

Intermediate level. An agent that destroys all vegetative bacteria
  including tubercle bacilli, lipid and some nonlipid viruses, and some
  fungus spores, but not bacterial spores.

Low level. An agent that destroys all vegetative bacteria except tubercle
  bacilli, lipid viruses, some non-lipid viruses, and some fungus spores,
  but not bacterial spores.

Sanitization. A process in which microbial contamination is reduced to
  a safe level on inanimate surfaces as defined by public health criteria.

Sanitizer. An agent that reduces the number of specific bacterial
  contaminants on inanimate surfaces to safe levels as defined by public
  health criteria.

Sporicide. An agent that kills bacterial and fungal spores.

Virucide. An agent that kills viruses.

10.2. CURRENT REGULATORY ENVIRONMENT

A properly designed and maintained disinfection program that provides the
pharmaceutical manufacturing areas with an environment relatively free from
microorganisms on a consistent basis is an expectation of most regulatory
agencies worldwide. The Food and Drug Administration (FDA) has
acknowledged that their expectation is for all cleaning processes to be
"defensible and therefore grounded in sound scientific rationale, supported

by data, and considerate of the unique material, process, facility, and personnel characteristics" (2).

Stemming from an excess of "lack of sterility assurance" recalls in the late 1990s into early 2000, the FDA has become concerned that pharmaceutical facilities are not doing enough to prevent microbial contamination of products. In 2001, the FDA issued warning letters citing numerous compliance problems with cleaning practices. Equipment and cleaning problems were noted in more than a third of drug GMP warning letters. The citations described such things as deficient cleaning practices, inadequate validation of practices, and failure to sanitize sufficiently. The FDA has clearly stated that their expectation is to see cleary written and defined cleaning SOPs, documented evidence of cleaning validation studies, a well-maintained operator training and qualification program, and validated methods (2).

There are approximately 1200 disinfectant products on the market in the United States, manufactured by approximately 300 different companies. Thus, there is a wide array of products from which to choose. The choice, however, can be difficult because of the many different product formulations and claims made for these products. Therefore, the choice has to be made after a very thorough evaluation of the situation in which the disinfectant will be used. Under the United States Federal Insecticide, Fungicide, and Rodenticide Act (FIFRA), all disinfectants must be registered with the United States Environmental Protection Agency (EPA) before they can be marketed. The EPA has defined disinfectants (antimicrobials) as pesticides. With the passage of the Food Quality Protection Act (FQPA) in 1996, agents used to sterilize critical devices such as surgical instruments that penetrate the blood barrier and semicritical devices that contact (but do not penetrate) mucous membranes must be registered with the FDA. Neither, the EPA or the FDA, however, evaluates these products for efficacy or safety. Efficacy and safety claims are based on data provided by the manufacturer of the product to these regulatory agencies. Therefore, it is prudent to purchase these products from manufacturers who can provide the documentation of efficacy and safety of their product and the documentation that their product claims have been reviewed and cleared by the EPA and/or FDA.

In today's world many pharmaceutical companies have international marketing and manufacturing components. Therefore, it is important that one take into consideration not only United States guidances and/or regulations that deal with disinfectants and disinfection practices but also the guidances and/or regulations that are pertinent to the countries in which a product is made and/or sold. Also it is important to understand the methods used in other countries to evaluate the efficacy of the disinfectant if the disinfectant is not registered in the United States (3). Any disinfectant used in the pharmaceutical-manufacturing environment in the United States must be registered with the EPA regardless of where the pharmaceutical will be sold.

Information on a product or its active ingredient can often be found in the literature. However, it is difficult to find articles that address the efficacy of disinfectants in the pharmaceutical-manufacturing environment. Articles can be more easily found on the use of disinfectants in the hospital setting. These articles, although not directly applicable to the pharmaceutical-manufacturing environment, can provide insight into the efficacy of disinfectants. No matter what information is obtained from the literature or from commercial brochures about a disinfectant, the product should be appropriately evaluated before being used.

10.3. CHEMICAL DISINFECTANTS—ACTIVITY, MECHANISM OF ACTION, MECHANISM(S) OF RESISTANCE

10.3.1. Alcohols

| Ethanol | CH_3—$CHOH$ |
| Isopropanol | CH_3 CH_3 $CHOH$ |

Ethyl and isopropyl alcohols diluted to a concentration of 60 to 85% with water are commonly used as disinfectants. The most effective concentration is 60 to 70%. Because methyl alcohol has a lower bactericidal activity than either ethyl or isopropyl alcohols, its use is not recommended (4). Ethyl and isopropyl alcohols have been shown to be bactericidal, fungicidal, and tuberculocidal but they are not sporicidal (5). Ethyl alcohol has a broader spectrum of virucidal activity than isopropyl alcohol, being more effective against both lipophilic (Adenovirus, Herpesvirus, Influenza virus) and hydrophilic (Poliovirus 1, Echovirus 6, Coxsackie B1 virus) viruses (6,7). Both ethyl and isopropyl alcohols have been shown to have activity against human immunodeficiency viruses (5,8). Alcohols do not have activity against prions (9).

Alcohols are generally cidal at concentrations of 60 to 70%. The optimal contact time will vary with the particular physical situation and with the type of organism(s) that might be present. *Hepatitis* B virus takes longer to kill than the vegetative forms of most bacteria. It is recommended that the contact time be no less than 10 min. Alcohols work primarily by denaturing the protein found in living cells. The presence of organic matter diminishes the activity of alcohols. Alcohols are primarily used to disinfect clean inanimate surfaces and gloved hands. Because of their high volatility and flammability, particularly in the undiluted state, alcohols must be used in well-ventilated areas away from open flames and sparks. They are not the ideal candidates for disinfecting the inside of biological safety hoods because of their volatility. Alcohols are relatively noncorrosive to many surfaces and leave no residue.

There have been no reports of microorganisms becoming resistant to alcohols through constant exposure. Generally what occurs is that the concentration of the alcohol being used is incorrect or there is an excessive amount of organic material present causing inactivation of the alcohol. Because alcohols are not sporicidal, they may contain spores. Therefore, it is necessary that the use dilution of the alcohol be filter sterilized prior to use. Such filtration should be done through a filter matrix that is compatible with alcohol. The use of a noncompatible filter matrix may allow spores to go through the filter membrane. Compatibility of filter matrixes with chemicals can be obtained from filter manufacturers.

Alcohols have good compatibility with quaternary ammonium compounds, phenolics, and iodine. However, it must be remembered that if a proprietary product containing any of the above compounds or other compounds are diluted with alcohol and this is not indicated in the label of the product, this constitutes an off-label use of the product. In such a case the combination needs to be fully validated as to its efficacy and safety.

10.3.2. Aldehydes

Two aldehydes constitute the primary aldehydes used for disinfection purposes. These are formaldehyde and glutaraldehyde. A newer aldehyde o-phthalaldehyde is also available, but because of its newness only limited information about it is available.

10.3.3. Formaldehyde (Methanal)

Formaldehyde   H—C—HO

Formaldehyde is used as a disinfectant in the form of a liquid or vapor. A solution of 37% formaldehyde, referred to as formalin, can be used as a sterilant while concentrations of 3–8% can be used as a disinfectant. Its use as a vapor is generally limited to disinfection of sealed rooms. Formaldehyde is lethal to bacteria, spores, fungi, and many viruses. Its activity is diminished by the presence of organic material. Formaldehyde kills by reacting with the amino groups of nucleic acid bases and protein molecules by attaching itself to the primary amide and amino groups. Because of its irritating vapor and carcinogenicity (10), its use as a general disinfectant is not recommended.

Plasmid-mediated resistance to formaldehyde has been described (11). The resistance to the formaldehyde occurred due to degradation of the formaldehyde by the microorganism.

10.3.4. Glutaraldehyde (Pentanedial)

Glutaraldehyde $OH-CCH_2CH_2CH_2C-HO$

Glutaraldehyde, a saturated dialdehyde, is a relatively powerful disinfectant and can be used as a sterilizing agent. It is generally obtained as a 2%, 25%, or 50% solution at an acidic pH that must be activated by making the solution alkaline. Glutaraldehyde is more active at alkaline pH than acidic pH. Glutaraldehyde solutions at a pH of 8 generally lose their activity within 4 weeks. For disinfection purposes, glutaraldehyde is used at a 2% concentration. Alkaline (pH 7.4 to 8.5) 2% solutions have bactericidal, sporicidal, fungicidal, and virucidal activity (7,12–14). Although glutaraldehyde is sporicidal, this requires an increased exposure time to the glutaraldehyde. The literature indicates that it requires up to 10 hr exposure to a 2% glutaraldehyde solution at 20°C to kill dried spores of *Bacillus subtilis* (15). Concentrations of less than 2% glutaraldehyde may not be sporicidal. Glutaraldehyde is considered to be active against mycobacteria (16). However, such claims for a product containing glutaraldehyde must be evaluated carefully. Glutaraldehyde is mycobactericidal provided that there is adequate exposure time (17).

Glutaraldehyde cannot be used as a general surface disinfectant because of its irritating vapors and required long contact time. It is ideal for use on contaminated equipment that can be submerged for periods of time. Containers of this material should be kept closed to avoid release of irritating vapors. The exposure time needed to disinfect an item will depend on the organism(s) suspected of contaminating the equipment and the presence of organic material. Although glutaraldehyde is not inactivated by organic material, it does not penetrate organic material easily. Reuse of glutaraldehyde solutions is possible. However, decrease in the activity of glutaraldehyde solutions may occur due to the accumulation of organic material, dilution, and change in the pH.

Glutaraldehyde is biocidal as a result of its alkylation of sulfhydryl, hydroxyl, carboxyl, and amino groups of essential chemical constituents of microorganisms. Instances of microorganisms becoming resistant to glutaraldehyde have not been reported in the literature. Generally, what is mistakenly attributed to resistance is explained by the use of an inappropriate concentration of glutaraldehyde, incorrect pH of the solution, use of an outdated glutaraldehyde solution, or improper glutaraldehyde exposure time.

Glutaraldehyde does leave a residue, making proper rinsing of items exposed to glutaraldehyde very important.

10.3.5. Ortho-Phthaldehyde (OPA)

Ortho-phthalaldehyde was cleared by the FDA in 1999 as a high-level disinfectant at 20°C at an immersion time of 12 min. It is not meant for general surface disinfection but can be used to disinfect items that can be immersed. This material has several advantages over glutaraldehyde: it requires no activation, it has minimal odor, it is stable over a wide range of pH (3–9), it is not irritating to the eyes or nose, and it has excellent material compatibility. A disadvantage is that it stains proteins gray (including exposed skin). Solutions of OPA have been shown to be effective for up to 14 days (18). In a hospital study (18), it was shown to be effective in eradicating vegetative bacteria, fungi, and parasites on a variety of medical instruments. Studies have found that is tuberculocidal (19) even against glutaraldehyde-resistant mycobacteria (18), and it is sporicidal (18). There is limited literature on its effectiveness against viruses.

10.3.6. Halogens

The halogen group consists of bromine, chlorine, fluorine, and iodine. Chlorine and iodine, however, are the most commonly used halogens for general disinfection. The means by which chlorine and iodine kill micro-organisms have not been clearly elucidated. Because chlorine is a strong oxidizing agent, it is thought to kill by a process of oxidation (20). Iodine is thought to kill by reacting with N-H and S-H groups of amino acids as well as with the phenolic group of the amino acid tyrosine and the carbon-carbon double bonds of unsaturated fatty acids (21).

As pure substances, both chlorine and iodine are unstable, corrosive, and toxic. In an attempt to ameliorate these undesirable characteristics, these halogens have been combined with other chemicals. For example, chlorine has been combined with p-toluene-sulfonamide, with this combination being known as chloramine T, and iodine has been combined with carriers such as polyvinylpyrolidone, with these products being referred to as iodophors.

10.3.7. Chlorine and Chlorine-Containing Compounds

Chlorine is most commonly available as a solution of sodium hypochlorite ($NAOCl$) in a concentration range of 1% (equivalent to 10,000 ppm chlorine) to 5% (equivalent to 50,000 ppm chlorine). Such solutions are commonly referred to as household bleach. There are commercial bleach solutions available that contain as much as 35% sodium hypochlorite. It is important that the concentration be specifically stated when the solution is purchased. Sodium hypochlorite solutions are corrosive even to stainless steel. An aqueous solution of bleach leaves behind no residue.

Chlorine is bactericidal, fungicidal, sporicidal, mycobactericidal (22), and virucidal against both lipid and nonlipid viruses (23) when it is used in the appropriate concentration for the appropriate amount of time. Such cidal claims do not appear on the label of general household bottles of bleach. Therefore, it may be necessary to document such claims under local, state, and federal laws. The way in which these claims can be documented can vary between localities and states so that it is best to check with these entities. Under current Occupational Safety and Health Administration (OSHA) regulations, a disinfectant with a tuberculocidal claim may also be necessary when *Mycobacteria* may be present if chlorine is being used as the disinfectant (24).

Free chlorine at a concentration of 5 ppm kills vegetative bacteria. The amount of chlorine needed to kill bacterial spores is 10 to 1000 times greater than the concentration needed to kill vegetative bacteria (20). A concentration of 100 ppm has been reported to kill 99.9% of *Bacillus subtilis* spores in 5 min (25,26). The literature indicates that anywhere between 1000 ppm and 10,000 ppm (27,28) of chlorine is needed to kill mycobacteria. Viruses can be inactivated by 200 ppm in 10 min (29). It has been reported that fungal organisms are killed in 1 hr at a concentration of 100 ppm, and fungal spores are killed at a concentration of 500 ppm in 1 hr (20). A 1:10 dilution of a 5.25% solution of NAOCl that equates to approximately 5000 ppm available chlorine is recommended by the Centers for Disease Control and Prevention (CDC) for cleaning up blood spills (30). Because chlorine can be inactivated by organic material, it is best when disinfecting areas contaminated with organic material to assure that the area has been cleaned, the contact time is adequate, and the site disinfected numerous times. A contact time of no less than 5 min should be used. It is best to make up fresh solutions of bleach for cleaning up spills.

A solution of bleach made up with water at a pH 8.0 and kept in a closed opaque container is stable for up to a month. However, when the container is opened and closed repeatedly (e.g., once a day) over a period of a month, the concentration of bleach may decrease by as much as 50% (30). Thus a solution intended to have a concentration of 5000 ppm after 30 days should be diluted only 1:5 when made from a 5.25% stock solution of NAOCl.

There are a number of organic chlorine compounds (e.g., Chloramine-T, sodium dichloroisocyanurate). Such compounds generally contain the $=$N-Cl group. When added to water they tend to hydrolyze to produce an imino ($=$NH) group. These compounds generally have the same spectrum of activity as sodium hypochlorite (31–33).

There have been no published reports of organisms becoming resistant to chlorine-containing compounds. Cases of disease caused by organisms such as *Cryptosporidium* and *Acanthamoeba* were not due to these organisms having become resistant to chlorine. These organisms were able to survive in the chlorinated water because the level of chlorination was not sufficient to kill the organisms or the organisms were sequestered from the action of the chlorine.

10.3.8. Iodine (I_2)

Commercial preparations of iodine coupled with carriers (e.g., polyvinyl-pyrrolidone) are available as disinfectants. These preparations are referred to as iodophors. The most recognized of these combinations is povidone-iodine. Such combinations allow for the sustained release of iodine. However, it is critical to understand that in their undiluted state little active iodine (I_2) is present. It is the presence of free iodine that is responsible for activity against microorganisms (34). Therefore iodophors must be properly diluted in order for free iodine to be present in the solution. Several reports have documented microbial contamination of undiluted preparations of povidone-iodine and poloxamer-iodine (35,36). The advantage of an iodophor over iodine or tinctures of iodine (iodine diluted with alcohol) is that they are nonstaining and relatively free of toxicity and irritancy.

Iodine is bactericidal and fungicidal. Iodophors are generally used at a concentration of 500 mg active iodine/L. The contact time for iodine-containing preparations is at least 10 min for vegetative cells of most bacteria. Iodine is virucidal for certain viruses. It may not inactivate poliovirus (7) and rotavirus (37). Iodine is also considered sporicidal (38) and tuberculocidal against *Mycobacterium tuberculosis* but its cidal activity against other species of *Mycobacteria* is not well defined (39,40). Because bacterial spores and *Mycobacteria* are known to be more refractory to the action of disinfectants then most vegetative bacteria, these organisms generally require a longer contact time to iodine than vegetative cells. The exact mechanism of action of iodine against microorganisms is not known. It is believed that the iodine molecule combines with N-H and S-H as well as phenolic groups of amino acids and the carbon-carbon double bond of unsaturated fatty acids of various chemical constituents of microorganisms, making these compounds unusable for the normal metabolic processes of the microorganism. This leads to death of the microorganism.

The use of iodine-containing compounds as disinfectants is generally limited to those situations where items can be immersed in the iodine for prolonged periods of time. Iodine compounds are generally not used for surface decontamination. The efficacy of iodine is reduced in the presence of organic material such as dirt and serum. Iodophors do leave a residue. Very thorough rinsing of items exposed to iodophors is necessary.

10.3.9. Peroxygen Compounds

Included among the peroxygen compounds are hydrogen peroxide and peracetic (peroxyacetic) acid. Both of these compounds can be used for either disinfection or sterilization. In this section, their use as disinfectants will be addressed. Peracetic acid and hydrogen peroxide are oxidizing agents that attack the sensitive sulfhydryl and sulfur bonds in proteins of the cell wall, enzymes, and other metabolites. Catalase and peroxidases do not deactivate peracetic acid, unlike hydrogen peroxide, which is deactivated by these enzymes. Peracetic acid is more active at low temperatures than hydrogen peroxide, and the presence of organic material does not reduce the effectiveness of peracetic acid (41). Both hydrogen peroxide and peracetic acid are environmentally friendly in that upon decomposition both break down into oxygen and water. Neither compound leaves a residue.

Both compounds are considered thermodynamically unstable, with peracetic acid being less stable than hydrogen peroxide. A 1% solution of peracetic acid loses half its strength in 6 days (42). Resistance to hydrogen peroxide and peracetic acid has not been shown to develop in microorganisms. Resistance is generally mistaken for the use of incorrect concentrations of either compound or because the compounds have been inactivated by the presence of organic material or decomposition during storage.

10.3.10. Hydrogen Peroxide (H_2O_2)

Hydrogen peroxide has been shown to be bactericidal (43), virucidal (43,44), tuberculocidal (23), sporicidal (45), and fungicidal (45). Hydrogen peroxide kills vegetative forms of microorganisms much more quickly than the spore forms of microorganisms. Therefore, an increased contact time for spores is necessary (43). The activity of hydrogen peroxide is increased with temperature and concentration (43).

10.3.11. Peracetic Acid (CH_3COOOH)

Peracetic acid is available commercially as a 15% aqueous solution. It exists at this concentration in an equilibrium state between peracetic acid and its decomposition products of acetic acid and hydrogen peroxide. Peracetic acid has been shown to have activity against a variety of vegetative forms of microorganisms at concentrations of 50 to 500 ppm (.05 to .5%) when the microorganisms have been exposed to it for at least 5 min (43). Peracetic acid is virucidal (43,44), fungicidal (45), mycobactericidal (46), and sporicidal (47).

Both hydrogen peroxide and peracetic acid are generally used in their liquid or vapor forms for disinfection and sterilization in closed systems or disinfection of materials that can be immersed. They are not useful for general surface disinfection (walls, floors, equipment, surfaces) because of their rapid decomposition and toxic and irritating properties.

10.3.12. Phenolics

Phenol

Phenol itself is not used as a disinfectant because of its toxicity, carcinogenicity, and corrosive nature. Phenolic derivatives in which there is a functional group (e.g., chloro, bromo, alkyl, benzyl, phenyl, amyl) replacing one of the hydrogen atoms on the aromatic ring are commonly used as disinfectants. This replacement reduces the corrosive, toxic, and carcinogenic potential of the phenol. However skin irritation and absorption through the skin still occurs, so proper precautions need to be taken when working with these materials.

The three most common phenolic derivatives used as disinfectants are ortho-phenyl phenol, ortho-benzyl-parachlorophenol and para-tertiary amylphenol. The addition of detergents to the base formulation produces products that clean and disinfectant in one step.

Phenol compounds when used at a concentration of 2 to 5% for the appropriate amount of time are considered bactericidal, tuberculocidal, virucidal, and fungicidal (46,47). Phenol compounds are not sporicidal. Human immunodeficiency viruses (HIV) are inactivated by solutions containing as little as 0.5% of phenol (8,48,49). The concentration at which to use a phenolic will depend on the conditions under which it will be used. Because there are a wide variety of phenolic disinfectants available on the market with varying concentrations of active ingredient, it is prudent to very carefully evaluate label claims.

Phenolic compounds kill by inactivation of enzyme systems, precipitating proteins and disruption of the cell wall and membrane. There have been no reports of organisms becoming resistant to phenolic compounds. Generally, what is believed to be resistance is traced back to inappropriate use, concentration, or exposure time to the phenolic material.

10.3.13. Quaternary Ammonium Compounds (Quacs or Quats)

The general chemical formula for these compounds known as either "quacs" or "quats" [quaternary ammonium compounds (quacs)] is shown below:

$$R_1 - \overset{\overset{\displaystyle R_2}{\displaystyle |}}{\underset{\underset{\displaystyle R_4}{\displaystyle |}}{N^+}} - R_3 \quad x^-$$

The "R" group can be alkyl or heterocyclic radicals that may be alike or different. It is the type of "R" group linked to the N^+ that gives a quat its antimicrobial activity. The x^- is usually a chloride or bromide to form the salt. Quats are cationic surface-active detergents, often referred to as cationic agents. Quats are odorless, nonstaining, noncorrosive, inexpensive, and relatively nontoxic. Quats kill organisms by disrupting the cell membrane, inactivating enzymes, and denaturing protein (50). Disruption of the cell membrane appears to be the primary cause of cell death (51). The antimicrobial activity of quats is reduced by the presence of organic material. Anionic detergents (soaps) and items such as gauze and cotton pads can decrease the activity of quats. Quats are relatively nontoxic and noncorrosive to many materials.

Quats are bacteriostatic against a variety of bacteria. However, certain gram-negative bacteria are known to be intrinsically resistant to quats. Notoriously resistant organisms are *Pseudomonas aeruginosa*, *Burkholderia* (*Pseudomonas*) *cepacia*, and *Providencia stuartii* (52,53). Reports of *Pseudomonas* spp. growing in an ammonium acetate–containing quat (54) and disease outbreaks associated with gram-negative contaminated quats have been reported (55). Quats are sporostatic in that they inhibit the outgrowth of spores (the development of a vegetative cell from a germinating spore) but not the actual germination process (development from dormancy to a metabolically active state) (38). Because they are strongly surface active, they have good activity against lipophilic viruses (including human immunodeficiency virus and *Hepatitis* B virus) (32,56). The quats have poor activity against hydrophobic viruses (enterovirus—e.g., polio, coxsackie, and Echo) (56). Quats can be bactericidal at medium to high concentrations but they are not sporicidal or tuberculocidal at high concentrations (57). Quats are fungistatic (58) and mycobacteriostatic (59).

As noted above, some organisms such as *P. aeruginosa* are intrinsically resistant to quats. Reports in the recent literature have indicated that there is a possibility that organisms can acquire resistance to quats by acquisition of plasmids. The strongest evidence at this time for this is the acquired resistance to quats in *S. aureus* (60) and *Staphylococcus haemolyticus* (61). In *S. aureus* there are two gene families referred to as qacAB and qacCD. These genes are present on plasmids and appear to encode resistance to quats and to other substances (62,63). Evidence in the literature suggests that the plasmids containing the qac genes are transferable at least among the various species in the genus *Staphylococcus* (59,62,63). Evidence for the presence of plasmid-mediated resistance in gram-negative bacteria to quats is not as clear as in gram-positive bacteria (64). Gram-negative resistance to quats is most likely intrinsic or the result of mutation (64). Because quats are easily inactivated, they can become contaminated with microorganisms and, compared to other disinfectants, have a low level of activity against gram-negative bacteria. The use of quats should be limited to ordinary environmental decontamination of noncritical surfaces.

10.4. RESISTANCE TO DISINFECTANTS

In recent years more attention has been paid to the potential that microorganisms may develop or acquire resistance to disinfectants. It is important to understand that microorganisms may be naturally resistant to disinfectants ("intrinsic resistance"), may develop resistance by mutation, or may acquire resistance through transfer of genetic material. The resistance of spores to alcohol best exemplifies intrinsic resistance. Intrinsic resistance is easy to detect in that generally the entire genus or species of an organism is resistant to a particular class of a disinfectant. Development of resistance by mutation is more difficult to detect in that the incidence of such resistance is generally very low (e.g., one cell in 10^9 cells). Also when resistance develops through mutation, the resistance generally is only to a particular class of antimicrobial and the resistance involves a single genus or species of a microorganism. The acquisition of resistance by transfer of genetic material carrying genes (plasmids) that mediate the resistance generally occurs at a higher incidence (e.g., one cell in 10^4 cells). It is not unusual for genes mediating resistance to a variety of antimicrobials to be transferred at the same time and for the plasmid to be transferred among various genera and species of microorganisms.

It is important that if contamination of an area is detected after disinfection that resistance to the disinfectant not be immediately assumed. In the majority of cases, improper use of the disinfectant and not development of resistance to the disinfectant is the cause for existence of the contamination. A complete investigation should be done to determine why the contamination exists. The investigation should include, but not be limited to, the following items:

> A determination of whether the organism recovered was previously recognized as part of the environmental flora or a new entity
> If cleaning procedures were done correctly with approved product(s) and equipment
> A thorough check of all equipment being used for disinfection procedures
> If the proper disinfectant was used at the proper dilution for the correct contact time
> If the undiluted product contained the proper concentration of the active material
> If the disinfectant used was within its expiration date
> If personnel properly followed the disinfection protocol
> If ancillary equipment (mop heads, applicators) used in the disinfection procedure had been used properly
> Identification of the organism(s) to species and if possible to subtype
> The organism's susceptibility to the use dilution of the disinfectant
> Notification of the manufacturer of the disinfectant so that they have the opportunity to offer assistance

10.5. SPORES AND DISINFECTANTS

Bacterial spores are some of the most recalcitrant organisms to the action of physical and chemical means of destruction, being rivaled only by prions. Bacterial spores are more resistant than fungal spores, than yeasts, and considerably more resistant than vegetative bacteria to the actions of antiseptics and disinfectants (65). Although the members of the genera *Bacillus* and *Clostridium* are the most important bacterial sporeformers, other bacterial sporeformers belong to the genera *Sporosarcinae, Desulfomaculum, Sporolactobacillus,* and *Thermoactinomyces* (66,67).

Bacterial spores consist of an outer spore coat, an inner spore coat, a cortex, and a core. The outer and inner spore coats are made up mainly of proteins. The outer spore coat contains alkali-resistant protein fractions that are characterized by the presence of disulfide-rich bonds. The alkali-soluble inner spore coat consists primarily of acidic polypeptides. The cortex is made up primarily of peptidoglycan; the core contains DNA, RNA, dipicolinic acid (DPA), and divalent cations (68,69).

Spores are formed during the multiphase process of sporulation (38). During this multiphase process, there are a number of phases where antibacterial agents, such as disinfectants, can act to kill the microorganism as well as where resistance to the antibacterial agent can develop (38). The process by which spores become vegetative organisms is called germination. The initiation of germination is termed "activation." It is thought that this activation step can be induced by metabolic as well as nonmetabolic means and is reversible. Activation is followed rapidly by a number of degradative changes in the cell that lead rapidly to outgrowth (38). A variety of chemicals are known to inhibit germination (38). The concentrations of the chemicals shown to inhibit germination are similar to the concentrations of these chemicals needed to inhibit the growth of the vegetative cells of these spores (38).

A classification scheme based on the activity of various chemicals against spores exists (38). This classification scheme breaks down these chemical agents into two groups, A and B (see Table 1) (38). Group A contains those agents that are sporostatic and group B contains those agents that are sporicidal. The caveat to this grouping scheme is that the agents in group A at high concentrations at ambient temperatures are not sporicidal but may be sporicidal if used at elevated temperatures, whereas the agents in group B are sporicidal only at high concentrations but are sporostatic at low concentrations (38). All of the agents in group B are bactericidal but none of the agents in group A are bactericidal. It is necessary to realize that exposure time is a critical factor as to whether or not an agent is sporicidal. This has been demonstrated in a study that showed that a 2% alkaline gluteraldehyde solution will sterilize an inoculum of approximately 1×10^8 CFU/mL of *S. aureus*, *Escherichia coli*, and *Bacillus subtilis* vegetative cells in 10 minutes at 22°C, whereas *B. subtilis* spores require several hours (70). The message from

TABLE 1 Agents with Sporostatic (Group A) and Sporicidal (Group B) Activity

Group A
Phenols, quaternary ammonium compounds, organomercurials, alcohols
Group B
Gluteraldehyde, formaldehyde, iodine compounds, chlorine compounds, hydrogen peroxide, peracetic acid, ethylene oxide, β-propiolactone

Source: Ref. 38.

these data is that the parameters to achieve killing of spores are quite different from the parameters needed to kill vegetative forms of microorganisms. This fact must be recognized in all disinfection protocols.

10.6. PRIONS AND DISINFECTANTS

The word prion was coined in 1982 by the neurologist Stanley Prusiner for Pr(otein) + I(nfectious) + on agent. A prion particle is a protein particle similar to a virus but lacking nucleic acid. Prions are thought to be the infectious agent for a variety of degenerative diseases of the nervous system. Today the diseases in animals that appear related to prions are scrapie (known from at least the 18th century), bovine spongiform encephalopathy (BSE), chronic wasting disease (CWD) in deer and elk, and transmissible mink encephalopathy (TME). All of these diseases fall under the collective name of transmissible spongiform encephalopathy (TSE). There are at least four clinically distinct TSEs of humans known today. These are kuru and Creutzfeldt-Jakob disease (CJD) and its variant vCJD, fatal familial insomnia, and Gerstmann-Straussler-Scheinker (GSS) disease. All these diseases are eventually fatal. Prions have defied the basic tenets of our scientific knowledge. The prevailing view is that prions contain no informational nucleic acid and are capable of crossing the species barrier as witnessed by the jump from cattle to humans of BSE.

The roles played in the transmission to humans by accidental exposure to TSE agents through contaminated medicinal products and therapeutic devices (71) and through food products (72) are now reasonably clear. Some of the materials used in the manufacture of pharmaceutical materials that may contain prions are serum-derived materials, bovine pericardium in heart valves, injectible collagen in plastic surgery, gelatin derivatives, and fatty acids. The unique biology of prions makes them difficult to detect, and although some methods do exist, they are at this time experimental. Therefore, it is best to avoid the use of any materials in the manufacturing of a pharmaceutical product that has the possibility of containing prions.

No one single decontamination method is 100% effective against TSE agents (73–75). Dry heat (up to 360°C for 60 min) and various steam-sterilization procedures have been shown to reduce the infectivity of certain TSE agents but not to eliminate some of them (76,77). The infectivity of the scrapie agent has been shown to be stable over a broad pH range (pH 2–10) (78). The use of different concentrations of NaOH has been reported to inactivate prions, but residual infectivity was found for some strains. Alcohols and alkylating agents, phenolics, and other chemicals are relatively ineffective for inactivating prions (9,79). Boiling in 3% sodium dodecyl sulfate has been reported to be partially effective (80). Recommended decontamination methods include 1N NaOH pretreatment and prolonged steam sterilization at 134°C for CJD (81). The problem with this method is that delicate instruments and many materials cannot withstand this treatment. At this time there is no known disinfectant or disinfectant procedure that can be assumed to be effective against prions. The best methods for the elimination of prions appear to be a combination of chemical and steam sterilization methods but even this has not been conclusively proven.

10.7. REASONS FOR USING A DISINFECTANT

Destroy or remove microorganisms that are present
Prevent entry of microorganisms into a manufacturing facility
Prevent dissemination of microorganisms throughout a manufacturing facility
Eliminate and prevent buildup of pyrogens

10.8. WHAT TO CONSIDER WHEN CHOOSING A DISINFECTANT

The importance of using the proper disinfectants cannot be overstated. Selection of the appropriate disinfectants requires careful evaluation. The properties of an ideal disinfectant are broad spectrum, fast acting, not affected by environmental factors, nontoxic, noncorrosive to most surfaces, odorless, easy to use, stable, good cleaning properties, soluble in water, economical, and minimal exposure time. The most popular formulations used today meet very few of these characteristics but can do the job if used wisely.

The selection process requires numerous considerations (82) such as the following:

Identifying the number and type of microorganisms that need to be controlled
Determining the specificity of microbial action of commercially available disinfectants
Evaluating surface and disinfectant compatibility issues
Determining what precautions and provisions must be made for the safety of personnel involved with using the disinfectants
Understanding what physical and environmental factors influence the stability and effectiveness of the disinfectant
Evaluating the compatibility of the disinfectants with each other and cleaning agents
Cost

10.9. IDENTIFYING THE NUMBER AND TYPE OF MICROORGANISMS THAT NEED TO BE CONTROLLED

Knowing the bioburden levels and types of microbial flora that exist in the areas to be cleaned and disinfected provides a good starting point. This information can be obtained from facility environmental monitoring programs. Cultures of these in-house microorganisms should be propagated and stocks of them maintained for use in disinfectant efficacy studies.

10.10. DETERMINING THE SPECIFICITY OF MICROBIAL ACTION OF COMMERCIALLY AVAILABLE DISINFECTANTS

The most popular types of commercially available disinfectants used today are alcohol, phenolics, quaternary ammonium compounds (quats or quacs), aldehydes, chlorine compounds, hydrogen peroxide, and peracetic acid. Disinfectants today offer a broad spectrum of activity. Many new commercial formulations provide enhanced microbial action, a broader microbial spectrum, and allow for increased compatibility with each other and with a variety of surfaces. The previous section of this chapter provides an overview of each of the disinfectant types and their specificity of action.

10.11. SURFACE AND DISINFECTANT COMPATIBILITY

The selection of the appropriate disinfectant depends on surface characteristics such as texture, porosity, and durability of the material. Substrate impact studies that have been performed by the manufacturers of disinfectants can be useful in selecting the appropriate disinfectant. The disinfectant manufacturer should be considered as a valuable resource in the matter of substrate compatibility.

Most current disinfectants are formulated to be safe to use on most clean room surfaces, but they can become fairly corrosive to many surface types when used regularly and without residual removal. Damage from disinfectants varies depending upon the concentration and frequency of use. For the most part, damage is inevitable to rubber, nylon, hard plastic, fabrics, asphalt tile, and metal. Alcohol, hypochlorite, peracetic acid, and some of the aldehyde formulations will promote pitting and rusting of steel as well as stainless steel surfaces over time. The phenol, peroxide, and chlorine formulations are readily absorbed by rubber, making it brittle over time.

10.12. SAFETY AND PRECAUTIONS

Manufacturers must supply material safety data sheets (MSDS) and other pertinent information such as handling and disposal requirements with their products. The physical and chemical characteristics noted on a MSDS are important when determining special storage, handling, and disposal requirements of the material as well as identifying potential hazards to personnel, and determining what application method is safest and best to perform. As with any chemical, all disinfectants and sterilants must be used with caution. It is the responsibility of the employer that workers are provided with this information and that they understand any potential dangers of working with the material. Workers must also be provided with continuing training in the safe handling of the material and provided with the proper equipment for working with the material. In some states, such as California, it is a requirement that all EPA registered antimicrobials are used according to California worker safety regulations. Therefore, it is prudent to check with state and local officials to see if there are worker safety regulations mandated by the state or local agencies that pertain to disinfectants (pesticides).

   All of the commonly used disinfectants today are skin and eye irritants and are highly toxic. The oxidizers and odiferous compounds such as phenols, chlorines, peroxides, peracetic acid, and aldehydes are highly irritating to the respiratory system. Spraying, wiping, fogging, mopping, and immersion application methods create conditions that can produce aerosols, odors, residues, and personnel contact situations that can be hazardous. Therefore, personnel protective equipment and garments may become a necessary precaution when using disinfectants. In addition, disinfectants should be stored in compatible containers. Highly corrosive disinfectants should be stored in stainless steel rather than plastic containers. Pumps and spigots should be used to decrease the likelihood of spills or skin contact. Spills of disinfectants must be cleaned up immediately in accordance with established procedures. Unused concentrated disinfectant should be considered hazardous waste, and should be disposed of according to county, state, and federal guidelines.

10.13. PHYSICAL AND ENVIRONMENTAL FACTORS THAT INFLUENCE DISINFECTANT EFFICACY

Environmental factors such as pH, temperature, biofilms, surface soil, and concentration may have an impact on the antimicrobial activity exhibited by a disinfectant (83). In addition, the type of equipment, the material from which equipment and laboratory fixtures are made, and equipment and facility layout and configuration can have an influence on the efficacy of disinfectants.

10.13.1. pH

Antimicrobial activity may be influenced by small changes in the pH of the chemical agent. These changes often occur within the pH range that is compatible with microbial growth. Antimicrobial activity is linked to the availability of undissociated molecules between weak acids and pH. For example, a study of antimicrobial activity of sodium hypochlorite buffered at pH 7.2, 9.0, and 10.6 indicated that undissociated hypochlorous acid at pH 7.2 exhibited the most sporicidal activity and the least activity at pH 10.6 (84).

10.13.2. Surface Soils

Chemical disinfectants often show reduced antimicrobial activity when exposed to common soils and impurities such as salts, serum, dirt, and other organic materials. It has been demonstrated that certain aldehyde formulations are affected by salts in that sodium bicarbonate or sodium chloride increase sporicidal activity while the presence of lysine residues from protein decrease the sporicidal activity (84). The antimicrobial activity of paracetic acid can be reduced in the presence of serum and other proteinaceous materials due to a rise in pH. The activity of sodium hypochlorite drops dramatically in the presence of 2% serum or other organics. Quaternary ammonium compounds are inactivated by the presence of common organic residues.

10.13.3. Biofilms

For the most part, routine disinfection practices carried out in a pharmaceutical-manufacturing environment will not deal with biofilms. However, in the case of water storage tanks, water pipes, and areas where a surface remains wet, biofilms may occur. The activity of chemical compounds against microorganisms existing in biofilms is difficult to assess. Because organisms within biofilms are much more resistant to antimicrobial agents than free living cells or cells on hard carrier surfaces (85–88) extrapolation of a disinfectant's efficacy as determined by AOAC methods (89) to a disinfectant's efficacy against organisms in a biofilms cannot be made. A recently published study (90) compared sodium hypochlorite and bezalkonium chloride (BAC) for their ability to kill bacterial cells in a biofilm and a cell suspension. The data showed that 50 times more sodium hypochlorite and 600 times more BAC were needed to achieve 4-$\log_{10}$ killing of cells in the biofilm than in the cell suspension. In another study (87), the penetration of alkaline hypochlorite and chlorosulfamates into a biofilm was compared. It was shown that chlorosulfamate penetrated into the biofilm better than hypochlorite. The study, however, showed a very small log reduction of viable bacteria in the biofilm as compared to the log reduction for the same bacteria in a suspension. This study supports the theory that bacteria within a biofilm are better protected than cells in a suspension. No standardized method for determining the efficacy of disinfectants against organisms in a biofilm exists at this time. There are, however, published papers that address the issue of standardizing tests to assess the efficacy of disinfectants against biofilms (90). Therefore, when faced with disinfection of sites where biofilms may exist, a critical evaluation of the situation needs to be done and the approach taken needs to be different than the approach taken for disinfection of non-biofilm surfaces.

10.13.4. Temperature

Temperature may play an active role in antimicrobial activity of a chemical disinfectant. Temperature can affect the rate of the reaction between the chemical agent and microorganisms; however, these rates of reaction are not easily predicted. The rate of reaction of the chemical agent may increase with higher temperatures, but the growth of the bacteria may increase also (84).

10.13.5. Concentration

The concentration of a disinfectant affects its antimicrobial activity. The concentration of the chemical agent at the desired site of action is influenced by the ability of the agent to reach the target site. There is a minimum concentration that must be achieved at the target site to elicit the desired microbial response (84). Thus, the appropriate concentration of active ingredients in the use-dilution of the disinfectant must be present to achieve the desired concentration at the target site.

   It becomes apparent when choosing a chemical disinfectant that it is not enough to apply an adequate concentration of the chemical to the surface; instead to achieve optimal antimicrobial performance, certain physical conditions must exist. The chemical disinfectant must be buffered to a pH that elevates the amount of undissociated acid molecules that have an impact on the concentration of antimicrobial agents in the solution. Temperature changes may or may not come into play. Another physical factor that must be considered is the cleanliness of the surface to which the disinfectant will be applied. Inactivation does occur with exposure to organic materials, so removal of these materials by mechanical and chemical means such as scrubbing with detergents allows for better penetration and activity of the disinfectant.

10.13.6. Compatibility of Disinfectants

Compatibility of chemical disinfectants with each other and with detergent cleaners is a necessity to assure optimal results. As a general rule, quaternary ammonium compounds are not compatible with detergents and phenolic formulations. Hydrogen peroxide and hypochlorite are not compatible with some detergents. Aldehydes and hypochlorites are compatible with quaternary ammonium compounds and phenolics. Typically, the new chemical disinfectants available today remove the concern for compatibility by the creation of complex formulations that are available in both acidic and alkaline forms. Many current disinfectants are formulated as one-step disinfectants. They contain surfactants, chelating agents, pH buffers, and other ingredients that improve wetting and cleaning. By improving the cleaning ability, the antimicrobial effectiveness of the disinfectant is enhanced.

10.13.7. Economic Considerations

Cost is certainly a factor when choosing a disinfectant. The specified or proper disinfectant/diluent ratio must be used for optimal performance, thus the cost must be based on the use dilution. Correct use and application will give the desired results without adversely affecting people, surfaces, or equipment. From a cost perspective, consider the following when deciding on a disinfectant. First, consider the cost of the disinfectant itself. Second, consider the time and effort involved in proving efficacy. Third, consider the expiration dates of the concentrate and use-dilution. Short expiration dates mean the concentrate must be ordered more frequently and the use-dilution must be made more frequently. Finally, consider the damage costs over time if a less expensive disinfectant that is not compatible with equipment and surfaces is chosen.

10.14. PREPARATION OF DISINFECTANT SOLUTIONS

10.14.1. Water

The label and inserts for a disinfectant normally specify the use of "water" for preparing the use-dilution. Most users opt to use tap water, but the disinfectant manufacturer often means "purified" water. Because this is not perfectly clear to the user, the incorrect type of water is used to dilute the concentrate. The use-dilution is less stable than the concentrate and more susceptible to microbial contamination; therefore, potable or tap water is not a good choice. Hard water containing low levels of minerals such as calcium, magnesium, and other organics should not be used as the water source either.

The organics in hard water will reduce the antimicrobial effectiveness of the disinfectant. Sanitizers, on the other hand, are more effective if formulated with small amounts of calcium carbonate in the distilled water. EPA recommends a standard hardness of 200–400 ppm $CaCO_3$, but normally this information is not found on the label. For most pharmaceutical applications, WFI (Water for Infusion) or purified water is the diluent of choice.

10.14.2. Concentration

During preparation of the use-dilution, amounts of the concentrate to be added to the diluent should be measured accurately. Using a higher than recommended concentration of disinfectant does not make the disinfectant solution more efficacious. Excess disinfectant may increase the likelihood of damage to surfaces and of safety hazards and may increase cost.

10.14.3. Temperature

The stability of most disinfectants is reduced if water above 60°C is used for preparation. Alkaline surface-active agents in disinfectants exhibit a decrease in antimicrobial activity and cleaning ability when formulated using water above 60°C (91). Acid formulations are recommended to be formulated using water at 20°C. There is no documented benefit in elevating the temperature of the water during preparation. In addition, the likelihood of safety hazards is increased when hot solutions must be handled (91). Alcohols, chlorine compounds, and aldehyde formulations are best prepared and used at room temperature. Phenolics, quats, peroxide compounds, and peracetic acid formulations are best prepared and used at temperatures around 20°C.

10.14.4. Sterile Filtration

To assure that disinfectants are not contaminated with low levels of micro-organisms, most disinfectant concentrates and use-dilution solutions are filter sterilized prior to use. This is especially the case for alcohols. Since alcohol is not sporicidal and therefore may contain spores, it is best to filter the use-dilution of this disinfectant through a compatible filter matrix prior to use. Other disinfectants that are not sporicidal should also be filter sterilized prior to use, using a filter matrix that is compatible with the disinfectant. This practice has become the norm and does not affect disinfectant activity when done correctly.

10.14.5. Storage

Disinfectants should be stored in compatible containers with the appropriate closure systems. Disinfectant registration requirements include the submission of data to show there are no leachables from the container closure system, and the closure system remains intact and inert in the presence of the chemical over the established shelf life (92). Containers should be kept closed when not in use and not subjected to heat, humidity, light, or other physiochemical factors that might influence degradation of the ingredients. Improper storage can lead to a decrease in the activity of the active substance. Factors that can decrease the activity of disinfectants are as follows:

> Temperature
> pH of diluent
> Storage container material
> Number of times and/or length of time material is exposed to air
> Cross-contamination with other chemicals
> Number of times the disinfectant is filtered
> Length of storage

Use-dilution batches should not be replenished or topped off with fresh solution. When empty, the container should be completely emptied and cleaned before refilling. In some cases, the container should be sterilized before reuse.

10.14.6. Expiry Date

Stability data to support the shelf life (expiry date) of the disinfectant unopened stock concentrate and use-dilution must be provided by the manufacturer at the time of registration with the EPA. This information is normally found on the label or insert and will address the real-time stability of the formulation during storage, during use, and under the conditions specified in the labeling (92).

10.15. QUALITY CONTROL OF DISINFECTANTS

Sound quality control practices should be applied to disinfectants selected for a disinfection program. Quality control should begin with the qualification of a supplier to assure they can provide a stable and consistent product. Many companies will qualify a supplier by analyzing at least three different lots of disinfectant against the supplier's QC specifications to confirm consistent results. The most common tests used to evaluate a disinfectant are assays for pH, percentage of active ingredients in the concentrate, purity, and other stability-indicating tests (93). Physical properties such as color, odor, and clarity may also be important (93). Internal specifications should be developed for disinfectants based on the data gathered during the supplier qualification testing and performance validation. Once the internal specifications have been established, written instructions such as a formulation batch record should be developed to document the dilution and preparation steps. The batch record should identify specifically what quality of water is to be used, the precise quantities to be measured, what type of container and closure is acceptable, storage requirements, and dispensing instructions. Any diluted disinfectant should be labeled with the disinfectant name, date of preparation, initials of the person who prepared the solution, and an expiry date.

10.16. METHODS FOR VALIDATION OF DISINFECTANT EFFICACY

10.16.1. Evaluation of Disinfectant Efficacy Claims

Commercial disinfectants registered with the United States EPA have been tested for their activity against bacteria and fungi by the methods of the Association of Analytical Communities (AOAC) (89) or methods shown to be equivalent to the AOAC methods. Other countries may have their own methods for determining the efficacy of disinfectants (3). While there are pros and cons of the AOAC methods for determining the activity of a disinfectant, the results of testing done by the AOAC methods for the most part have been found to accurately reflect the in-use efficacy of disinfectants against bacteria and fungi. Methods for the determination of the efficacy of disinfectants against viruses and parasites are not well standardized. Descriptions of methods to determine the efficacy of disinfectants against viruses (94,95) and parasites (96) can be found in the indicated references. The activity of chemical compounds against prions (97) is extremely difficult to assess (98). Prions cannot be cultured, thus the activity of chemical compounds against prions is measured indirectly. Based on the fact that strain differences in the thermostability of prions have been described (76), it cannot be assumed that because a chemical compound has activity against a particular prion that it has the same activity against other prions. The bottom line is that when it is stated that a disinfectant has activity against viruses, parasites, or prions it is critical to understand what viruses, parasites, and prions and the methods by which the activity was determined. In the case of studies with prions, the studies vary in exposure times, temperatures, and the type of tissues studied (brain, reticuloendothelial). Other variables include the prion strains used, brain preparation methods (dried or macerated), and the kind of test system used.

Antimicrobial effectiveness of disinfectants can be evaluated using three types of testing (99):

Preliminary screening tests are in vitro laboratory tests used to determine if the disinfectant has activity against the microorganisms of interest. The most common of these tests is the time-kill study in which a variety of organisms are challenged in suspension using several different concentrations of a disinfectant and contact times. An example of this type of test is the AOAC phenol coefficient test that is a simple qualitative suspension test comparing disinfectant activity with that of phenol concentrations.

The second type of tests—in vitro real-use simulation tests—are also performed in the laboratory, but performed in conditions that simulate real-use situations. These determinations of the activity of disinfectants against bacteria and fungi is best done by AOAC methods (89). A review of in vitro methods that can be used for the evaluation of the efficacy of antibacterial and antifungal agents is given in Ref. 100. When testing is performed, the testing should include not only the organisms suggested in the protocol but also organisms that have been isolated from the pharmaceutical-manufacturing environment. Inclusion of the organisms recommended by the protocol allow for determining if the test is performed correctly, and the environmental organisms will allow for a determination of the activity of the disinfectant against pertinent organisms.

In the case of viruses, parasites, and prions, the methods used to determine the efficacy of chemical compounds against these microorganisms are not standardized. Many viruses and parasites are extremely difficult to culture. Thus determination of the in vitro activity of chemical compounds against viruses (6,101) and parasites (102) is difficult. It is recommended that any one who does not have the expertise, the facility, or the equipment to perform such testing refer this testing to a laboratory that has this capability.

The most common of these tests is the surface challenge test. An example is the AOAC Hard Surface carrier test, in which organisms are fixed and dried on a vehicle and then recovered to determine log reduction. In these tests, the disinfection procedure as well as the disinfectant is evaluated. The test determines which conditions and use-dilution are effective and whether the disinfectant is effective for the chosen application. The test requirements for this type of test are normally 60 carriers per sample, representing three different lots of disinfectant, one of which is at least 60 days old. The performance requirements for a general broad-spectrum disinfectant are that 59 out of each set of 60 surface carriers must show total kill to provide effectiveness at the 95% confidence level. For fungicidal claims, normally 10 carriers per sample, representing two different lots of disinfectant, are employed.

The third type of test used to evaluate disinfectant effectiveness is an in situ field test that is performed in the field. This test is often referred to as an in-use test and evaluates disinfectant effectiveness under practical conditions and uses. This test is not used to validate a disinfectant for routine use as it takes continuous data collection to make an assessment. The evaluation of disinfectants at the site of application while possible will not provide a statistically valid study unless hundreds of samples are taken over an extended period of time. This is because the areas in a pharmaceutical-manufacturing facility are generally devoid of large numbers of microorganisms against which the disinfectant is being evaluated. Negative cultures do not translate into organism eradication by the disinfectant unless it is known that organisms were there prior to use of the disinfectant. Thus, many samples over an extended period of time would need to be taken to get positive samples from which to make any conclusions about the activity of the disinfectant against specific microorganisms.

Such studies also are not practical because they involve closing down production areas for long periods of time. Thus, the in vitro evaluation of a disinfectant provides the most practical way of determining the activity of a disinfectant against organisms of interest. The ability of the disinfectant to work in situ is determined generally by monitoring for the presence of organisms in the facility in which the disinfectant is being used.

The artificial introduction into a manufacturing area of any organism for the evaluation of a disinfectant or for any other reason SHOULD NEVER BE DONE.

10.16.2. Validation Test Development and Efficacy Parameters

The normal course of action for most pharmaceutical companies in validating their disinfectants for routine use is to validate the effectiveness by conducting both use-dilution time-kill suspension studies and surface challenge tests. Protocols should be developed that will yield information that can be interpreted for practical use, provide repeatable and reproducible results, and provide adequate control (99). In order to do this, it becomes necessary to standardize equipment, media, challenge organisms, test manipulation, test temperature controls, and incubation temperatures of cultures and recovery subcultures. Also, accuracy in timing and preparation of dilutions is important. Several groups throughout Europe and the United States are working toward the goal of standardization of a quantitative disinfectant carrier test. So far, there is great diversity in the test methodology among all of the groups. There is also a lack of agreement on the standardization of the components of the testing method. A collaborative intra/inter-laboratory study of the AOAC use-dilution carrier test was performed in the United States and supported by the EPA found the test to be poorly reproducible and incapable of confirming the bactericidal label claims of some registered disinfectant products (103,104). However, many U.S. manufacturers of disinfectants have collectively modified the AOAC methods for assessing the effectiveness of their products against bacteria and fungi. Thus, the results from one disinfectant manufacturer to another are fairly reproducible and can be compared with some confidence.

To help with validation protocol development, the following should be considered:

10.16.2.1. Selection of Microorganisms, Challenge Levels, and Log-Reduction Efficacy Criteria

Challenge organism selection should include both the organisms recommended by the procedure and the microorganisms that are typical environmental isolates from the facility where the disinfectant is to be used. Maintenance of stock cultures of microorganisms and their propagation prior to use should be done according to the test protocol. The challenge levels for effectiveness testing are normally in the $6-7 \log_{10}$ range for suspension tests where a $3-5 \log_{10}$ reduction result is expected for bacteria and fungi, and $3-4 \log_{10}$ range for vegetative bacteria where a $3 \log_{10}$ reduction is expected, and $2-3 \log_{10}$ range for bacterial spores where a $2 \log_{10}$ reduction is expected in the surface challenge tests. These are the challenge levels used most often by testing laboratories.

10.16.2.2. Dried versus Suspension Inoculum

If a modified AOAC surface carrier test is designed, an important consideration is the drying step. During the drying step of the carrier test, some microorganisms such as the gram-negative bacteria may lose their viability. In addition, if the carrier surface is scratched, unpolished, or porous, the harvest of dried organisms may not be quantifiable. Studies have shown that a drying time exceeding one hour may allow the bacteria to become more firmly attached to a surface and create a biofilm that is considerably more resistant (105). Another consideration is that if a spray or aerosol application of the disinfectant is used on the dried inoculum, the volume of disinfectant available to the inoculum is often less and therefore adds a high degree of variability to the test results. For these reasons, one of the performance requirements of the "AOAC Hard Surface" test is that in order for the test to provide meaningful results, a concentration of at least 1×10^4 organisms should survive the carrier drying step (105). In suspension tests, the microorganisms are homogeneously dispersed, which makes the sampling and recovery process simpler and reproducible. Because suspension tests are acceptable only in the United States as screening tests to establish the "cidal" potential of a disinfectant, the considerations mentioned above for the surface carrier test should be taken into account when designing a test validation protocol.

10.16.2.3. Contact Time

Another consideration in the development of a hard surface carrier test is choosing or identifying an appropriate contact time. Consider the disinfectant manufacturer's recommended time and what is rationale for a real use situation. Most hard surface disinfectants such as the alcohols, phenolics, chlorines, quats, and peracetic acid formulations require 10 to 30 minutes of wet contact time, depending on the concentration used. Sterilants normally have longer contact times to allow for inactivation of bacterial spores. Gaseous disinfectant formulations normally have exposure times of 30 minutes to 6 hours, and also require a degassing period as a safety precaution.

10.16.2.4. Neutralization

Another important aspect of validation test design is neutralization. Disinfectant residues carried over into the subculture media can affect the results of disinfectant testing. To obtain a meaningful recovery of bacteria exposed to a chemical agent, it is recommended that a neutralizing agent that can prevent continued action of the disinfectant be included in the diluent or recovery media used. The most commonly used neutralizers are lecithin, polysorbate 80 (Tween), and thiosulfate. Lecithin is effective in neutralizing quats, polysorbate 80 is effective in neutralizing phenolic compounds and ethanol, and thiosulfate is effective in neutralizing chlorine.

Dilution and membrane filtration washing can also accomplish neutralization of a disinfectant. Dilution alone is acceptable for agents like alcohol, but because of the diversity of organisms and complex formulations, there is no real "universal neutralizer" available (106), so validation of this test step is difficult. Different combinations of neutralization techniques must be evaluated. Neutralizing agents must be shown not to inhibit the growth of the test organisms at the concentration that is used. It is recommended that the neutralization step be validated for each test rather than once for the neutralizer (106). Other considerations in test development are formulation, use, and test temperature as well as soil load and water quality for formulation. These have been covered in previous sections of this chapter.

Golden Rule: "Thou shall not change the disinfectant or the way a disinfectant is used without revalidation." All disinfectant materials must be used as indicated by the manufacturer or as originally validated by the user. Any change in the way a disinfectant is used requires a complete revalidation of that disinfectant. Failure to live by this golden rule is often the reason that companies are given deficiency letters.

10.17. APPLICATION OF DISINFECTANTS FOR CONTAMINATION CONTROL

10.17.1. The Three-Step Disinfection Process

There are three important steps in the sequence of cleaning and disinfection to maintain effective contamination control. The first step is the cleaning of the surface. Because some disinfectants can be inactivated by the presence of organic material or the presence of organic material can prevent the disinfectant from getting to organisms, it is essential that proper cleaning be done prior to disinfection. The cleaning material that is used must be compatible with the disinfectant or completely removed because the cleaning material may inactivate the disinfectant. For items that are heavily contaminated with organic material, the use of preparations containing enzymes to remove the organic material may be appropriate. Because cleaning is done prior to disinfection, proper precautions must be taken to avoid contamination of personnel and equipment with organisms.

When cleaning, a neutral detergent should be used that contains a surfactant to optimize the cleaning activity. In choosing a detergent, consider the compatibility to the surface to be cleaned, capability to remove the type of soil present, solubility of the detergent in water, and the rinse characteristics of the disinfectant to assure detergent residue can be easily removed. If using an alkaline detergent, rinse with soft water. If the water is hard, an acid rinse might need to be used to prevent scale.

Most cleaning detergents are more effective when used hot, but temperatures above 60°C will inhibit their ability to remove fat- or oil-containing films from surfaces (107). Alkaline detergents as well as acid cleaners are normally used at temperatures around 20°C. The cleaning and disinfection steps are often combined if using a specially formulated commercial product. For example, alkaline detergents are often combined with chlorine compounds or quats, along with surfactants, to provide increased cleaning ability and enhanced antimicrobial activity (107).

The second step of the disinfection process is the application of the disinfectant to the target surfaces. The goal of effective disinfection is to deliver the chemical agent safely to a surface for the determined contact time. The most common methods of application of a disinfectant are mopping, wiping, spraying, and fogging. Mopping and wiping provide a mechanical action that aids in the removal of particulates and attached microorganisms from surfaces. Mechanical action also aids in the removal of residues, thus minimizing chemical buildup. Mopping and wiping are normally employed for easy-to-reach, smooth surfaces. Spraying and fogging are normally employed for high ceilings, difficult piping configurations, and complex equipment. Spraying or fogging are most often used to apply sporicidal agents. Table 2 gives examples of where these delivery applications are normally utilized in the pharmaceutical environment (94,107).

Rinsing is the third important part of the disinfection process that is often overlooked. Rinsing can be problematic because there is a chance for recontamination from personnel and equipment from this activity, and if rinsing is performed too soon after applying the disinfectant, the contact time will be shortened and antimicrobial effectiveness will be reduced (91). Rinsing is important in the prevention of buildup of disinfectants and cleaning detergents. Detergent residues inactivate disinfectants such as quats, hydrogen peroxides, and hypochlorites. Sterile water and sterile alcohol are the most commonly used rinsing agents. If sterile water is used as a rinsing agent, surfaces should be dried with sterile lint-free wipes so as not to promote damage or microbial contamination. Efficiency in the removal of residues from surfaces may be checked by adding a fluorescent dye to the soiled surface before application of the detergent, disinfectant, and rinsing agent. After the cleaning and disinfection process, the surface is checked with UV light (107). Nonfluorescence of the cleaned surface indicates that there has been removal of the residue.

10.17.2. Application Techniques

The most popular mopping technique is the "triple bucket" technique. Three buckets are utilized; bucket 1 contains the disinfectant, bucket 2 contains the rinse solution, and bucket 3 has a wringer to squeeze the contaminated waste solution from the mop. The mop head is first dipped into bucket 1 and the

TABLE 2 Disinfectants

Chemical type	Application
Alcohol	Work surfaces, instruments, equipment, small scale treatment after maintenance during a production run
Phenolic	Floor maintenance, work surfaces, equipment, floor drains, building interior fittings
Quaternary ammonium compound	Work surfaces, floor maintenance, glassware, instruments, equipment, building interior fittings
Aldehyde	Air systems, enclosed rooms and cabinets
Hypochlorite	Work surfaces, glassware, equipment, floor drains
Peroxide	Work surfaces, floor maintenance, equipment
Paracetic acid	Work surfaces, floor maintenance, equipment

disinfectant is applied to the surface with short overlapping strokes or using an S-motion (108). The mop head is then wrung out into bucket 3 and then dipped into bucket 2 and again wrung out in bucket 3 (108). The mop head is dipped into bucket 1 again to obtain fresh disinfectant and applied to the surface (108). This sequence is repeated until all areas are covered. While performing this technique keep track of the wet contact time necessary for optimal antimicrobial performance of the disinfectant. It is recommend that the buckets and mop heads be sterilized prior to use. Sterile, single-use mop heads are convenient, but not necessarily the cheapest, to use. This technique has the advantage of mechanical action to remove particulates, films, and residues. The wiping technique to apply a disinfectant to the surface is less labor intensive and requires the use of sterile, low-lint wipes. Surfaces are wiped in one direction with short overlapping strokes using sterile wipes impregnated with disinfectant, or sterile water if rinsing. This technique also has the advantage of mechanical action to remove particulates, films, and residues.

Fogging applications disperse the disinfectant in a vapor phase on to the surface. This application requires less manual labor and utilizes equipment that provides the heat necessary to vaporize the disinfectant liquid. Contact times are somewhat longer and the concentration of disinfectant is normally much higher than for other application techniques. This is because sporicidal agents are most often applied using this technique. It is important to note that temperature and humidity must be monitored using this technique. For example, vaporized formaldehyde requires room temperature and 80–90 relative humidity to be effective (107).

Spraying applications disperse aerosolized disinfectant liquids. This application is also less labor intensive and time consuming. The aerosolized disinfectant is sprayed on to surfaces using trigger spray delivery devices. Whatever type of sprayer is used, it is important to wash and heat-sterilize the equipment prior to each use to prevent buildup of chemical residues that may clog the dispensing nozzle and prevent accurate delivery of the disinfectant to the surface.

For all techniques employed (excluding fogging) to apply disinfectants, the sequence of the application is very important. Cleaning and disinfection should always be performed starting from the cleanest to dirties areas moving from the ceiling (top) to bottom, and from the back of an area outward away from critical processes and equipment toward the area exit.

10.17.3. Using Sanitizers in Conjunction with Disinfectants

Sanitizers reduce the level of microorganisms on a surface but do not eliminate all of them. The most common sanitizer formulations contain quats, phenolics, hypochlorite, and alcohol. These formulations differ from disinfectants by the concentration of the active agent and label claims. They have a lower toxicity and are applicable for use on food-contact surfaces, as additives in laundry detergents, and as rinses on pharmaceutical surfaces. Sanitizers are often used as "maintenance" agents for contamination control. Sanitizers aid in the removal of disinfectant residue from surfaces and help maintain microorganisms at low levels on surfaces and equipment during periods of high activity. Alcohol is the most widely used sanitizer in the pharmaceutical environment. It is often used to sanitize skin, gloves, goggles, forceps, and cleanroom surfaces during shift operations.

10.17.4. Rotation of Disinfectants

Rotation of disinfectants is controversial. At this time it is not mandated by FDA regulations. Rotation is seen as a way of preventing the selection and development of disinfectant-resistant organisms. There have been no published papers of well-controlled studies that document that rotation of disinfectants prevents the development of disinfectant-resistant organisms. If organisms do appear after disinfection, the steps mentioned in this chapter on "Resistance to Disinfectants" should be carried out to determine if the organism(s) are truly resistant or whether there was a breakdown in the disinfection procedure.

If rotating two disinfectants is part of the disinfection program, it is prudent to consider using chemically compatible disinfectants to prevent the development and buildup of residues that can hamper the disinfection process by inactivating the disinfectants used, and require that extensive cleaning and rinsing be performed between disinfectant rotations. There is no set time frame for determining the frequency of rotation. Most facilities choose anything from weekly to monthly for their rotation schedule. Sporicidal agents are often used monthly or when the need is indicated from the facility environmental data.

10.17.5. Disinfection Program Design—Scheduling/ Frequency

Every facility offers a unique challenge to the design of an effective disinfection program with the variety of operations, area classifications, complex equipment, and diverse materials throughout its environment. Some of these areas housing isolators, work surfaces, and specialized equipment are critical because they come in contact with the product and its components. Others are less critical, such as floors, walls, ceilings, airlocks, pass-throughs, and gowning areas. These less critical areas cannot be ignored due to the heavy trafficking of people, equipment, and carts with supplies that have the potential to contaminate critical operations and areas. Therefore, a cleaning and disinfection schedule should be established for all critical and noncritical areas and surfaces. This includes general uncontrolled/unclassified areas and equipment such as halls, racks, and carts, and controlled/ classified areas such as 100 (Class A), 10,000 (Class B), and 100,000 (Class D). The cleaning schedule for areas such as floors, airlocks, gowning, and high traffic corridors should have a more aggressive cleaning regimen as opposed to the minimal-use areas. In general, isolators and laminar air flow (LAF) equipment should be disinfected prior to and upon completion of work activities. Floors in the controlled corridors and aseptic fill areas should cleaned once a day. Personnel contact surfaces such as benches, wall switches, and phones should be cleaned once a day. Aseptic areas, walls, and ceilings should be cleaned and disinfected between fill runs, batches, or shifts. The lesser-controlled areas in further proximity to the product should be cleaned and disinfected monthly or quarterly at a minimum.

The manufacturers of disinfectants often provide guidance and recommendations concerning cleaning schedules and frequencies commonly associated with their products that can aid in disinfection program design (108). Depending on the type of cleaning and disinfection schedule a facility develops, the frequencies and cleaning procedures should be consciously evaluated through the environmental monitoring program. This type of in situ evaluation will provide information as to whether the integrity of the controlled environments is being maintained or whether changes to the cleaning frequencies and/or disinfectants are needed.

10.18. DEVELOPING AND IMPLEMENTING A DISINFECTION PROGRAM

Although being knowledgeable about disinfectants and the way they should be used is important, this knowledge has no value without an adequate disinfection program. Such a program coupled with the properly trained personnel has a very high probability of producing satisfactory results. There are a variety of ways the development of a disinfection program can be approached. The critical thing to keep in mind when developing such a program is that it must be successful the first time it is used. This does not mean that the program has to be perfect. However, the program should have been developed in such a way that it will be acceptable to those using it so that problems can be quickly identified and corrected.

The keys to the development of a successful program are the early involvement of the personnel that will be directly involved with the program, their proper training, and continuous monitoring of the program to determine its effectiveness.

10.18.1. Steps That Can Be Taken to Develop a Program

10.18.1.1. Form a Select Team to Develop the Program

The "Select Team" may consist of two elements. One element of the "Select Team" may consist of the team leaders of production groups, managers of production areas, a regulatory individual, and a person involved with the purchase of materials. This element of the team would be responsible for the organization and development of the program. The second element of the "Select Team" may consist of those individuals who would be responsible for carrying out the directives of the program once developed. Members of all shifts should be involved with the writing of the program so that the program and its implementation will be seamless. The involvement of all these individuals will assist in the sharing of knowledge that will enhance the development of the program and its implementation. All members of the "Select Team" need to be held accountable for the development of the program.

10.18.1.2. Timeline

Prior to the beginning of writing and implementing a disinfection program, the "Select Team" members should develop a time line for completion of the project. The construction of a time line not only allows for the establishment of critical time points but also gives the project a sense of direction. Time lines need to be realistic for all members of the team. Agreeing on acceptable time points within the time line can be a challenge and will be a test of each member's commitment to completing the project in a timely manner. Good negotiation skills among the members of the Select Team are a must at this point.

10.18.1.3. Beginning the Development of the Program

The first step in the writing of a new program is to determine very explicitly what is currently being done. Reviewing the existing program and other documents relating to disinfection at the facility may not be enough. Reviewing copies of the existing program in production areas, speaking to those individuals doing the work, and actually witnessing how the current program is carried out is a necessity. To assume that everyone is on the same page from the beginning can lead to the failure of developing a program that will be successful. It is not uncommon for outdated programs to be in use. The review of actual written programs in specific locations can reveal numerous findings that can be helpful in understanding how to write and implement the new program. If the program copy in an area is clean and neat it may mean that it is not being used. If a review of the program copy reveals numerous notes in the margins, it is an indication that what is written is not adequate or there have been numerous changes made to the program. In both cases, one should be conscious of the fact that the program as it exists is probably not serving the purpose for which it was intended. Actually observing how the existing program is carried out may also indicate a number of situations that may need to be addressed in the new program.

10.18.1.4. Reviewing What Has Been Found

Once the list of the current disinfection practices is compiled, they should be reviewed to determine if they are appropriate for the situations in which they are being used. At this point, items that may be lacking in the current program can be identified and evaluated for their possible incorporation into the new program and items that are not appropriate can be deleted. It is also an ideal time to evaluate the consistency of the procedures. As few as possible disinfection programs and procedures should be used in a facility and between facilities. This has the advantage of having to monitor fewer programs, not having to train people in as many methods, reducing costs, and allowing for easier presentation to regulatory authorities.

10.18.1.5. Questions to Ask During the Review Process

How many disinfection programs exist?
Does the current program(s) produce appropriate results?
Is the current program safe and environmentally friendly?
Do the current practices meet regulatory requirements?
Is disinfection still needed or is just cleaning sufficient or is sterilization needed?
Can a more economical disinfectant be used that gives better or equivalent results?
Do we need to consider a backup disinfectant?
Would it be economical to validate a new disinfectant?
Can disposable items be used for cleaning and disinfection?
Would more automation of disinfection be appropriate?
Can we make the disinfection process more user friendly to perform and more understandable to those performing the disinfection?
Should the program be made available electronically?

Any inappropriate applications of disinfectants should be eliminated at this step. The selection of disinfectants for particular purposes requires careful consideration of various factors. The properties of disinfectant chemicals vary markedly and need to be taken into consideration when choosing a disinfectant for a job. The purchasing member of the "Select Team" should be asked to obtain specifications, use directions, safety data sheets, and cost of the disinfectants of interest.

10.18.1.6. Planning and Writing the Program

The planning and writing of the program should proceed in a timely fashion after the review of current methods is completed. The planning and writing of the disinfection program should be a top priority of the Select Team. This phase of the program needs to be completed in as short a time as possible.

The key to a successful disinfection program and its consistent implementation is a well-written program. In the program document, the process of procedure writing hinges on the identification of the critical steps in the procedure. These can be graphically outlined in a flow chart, and the flow chart can be used as the basis for writing the formal procedures and (if desired) a separate training program. Identifying the critical steps also aids in the development of learning objectives. At this point the team may find it beneficial to enlist the aid of an individual outside the "Select Team" who has a good command of English and writing procedures to participate on the team. This individual can help to organize the document and serve as the person responsible for incorporating changes in the document as they occur.

Main sections of the program document may include but not be limited to purpose, safety, disinfectants, equipment, instruments, specific procedures, procedures for corrective action, monitoring the effectiveness of the program, how to document actions, procedures for recommending changes to the program, and a section giving the names and contact information of individuals who can be contacted to answer questions about disinfection and the program. Subsections of the program document dealing with disinfection may include preparation of the disinfectant, use of the disinfectant, storage of the disinfectant, elimination of disinfectant residue, and testing for disinfectant residue. The document should have a table of contents. The document can also be made user friendly if it is indexed. If indexed, consideration should be given to cross-referencing within the index.

If the program is to be available to personnel electronically, some means of capturing the hits and identification of the person using the program should be incorporated into the electronic format. The number of hits on the program can serve as a means of determining whether the program is being reviewed on an appropriate basis. Capturing the identity of the individual accessing the program can be useful in determining the level of personnel utilizing the program. It can also serve to identify individuals who can best tell you about the utility of the program. Making the program available to personnel through electronic means can have many benefits. Some of these benefits are as follows: personnel can make comments directly to the program and these comments can be viewed immediately; responses to the comments can be made in a timely fashion; and the comments and responses can be a source of information when programs are reviewed. The electronic version of any program should not allow unauthorized individuals to make changes to the program. Any changes made to a document should be tracked and the portion of the program being changed saved.

10.18.1.7. Finalizing the Document

Once a draft of the document is completed, all members of the team should read it critically and make comments. In some situations it can be beneficial to have individuals of the team read sections for which they were not responsible and make comments. This can provide information on the clarity of the writing. Once a final draft has been agreed on by members of the team, the document should be given to other individuals who will be directly or indirectly involved for their comments. The individuals to whom the document is given should be carefully picked to assure that comments would be thoughtful and helpful. When the document is given to these individuals, they should be given a time frame in which to return their comments. The time line should be adhered to strictly. Any input from these individuals should be followed up on with the individual providing the input. Such action adds credibility to the document and the team. Although it is important to get feedback from a variety of individuals on the document prior to it being used, it is not generally beneficial to give a document to too many individuals since this can be time consuming. The true validation of the document will be when it is used in the real-time situation.

10.18.1.8. Implementing the Program

The implementation step is critical to acceptance of the document. The team can do the implementation of the program or the team may elect to enroll the help of individuals who are known for their expertise in training personnel and implementing programs. Members of the "Select Team" that were involved in writing the document should be recognized publicly by management so coworkers can appreciate the importance of the team's work. Such recognition will also show the individuals who will be involved with the new disinfection program the importance of the program to management and the company.

A good way to implement the program is to have individuals who will be involved with the program teach the program to their coworkers. This necessitates that a core group of individuals be identified who are willing and able to teach the new program to their coworkers. This core group of workers will initially need to be taught by members of the "Select Team." It is critical that each core team is taught in the same manner so that there is consistency to what is taught to each group. The need to have consistency between instructors needs to be stressed to those who will be teaching. Without this, there will be a lack of consistency in the way the program is carried out.

A specific person, preferably from the "Select Team," should be chosen as the contact point during the implementation of the program. Designating one person as the contact point to receive comments about how the implementation of the program is progressing allows for comments to be centralized and is easier for people to remember. Comments should be channeled to this person through a chain of communication that has been previously agreed upon. An example of a chain of communication is worker to supervisor to shift supervisor to the person designated to collect comments.

10.18.1.9. Critical Teaching Points

The purpose of the disinfection program should be clearly explained. Without a clear understanding of why the disinfection program was created and what is to be achieved with the program, those involved in performing the required duties of the program will not have a full sense of ownership of the program. This lack of ownership can hinder the implementation of the program. In addition, once the program is implemented this lack of ownership can impact on the how well the program functions.

Teaching should focus entirely on items that are in the disinfection document. There may be the need to discuss items relevant to the document but not in the document, but these items should be discussed in the context of the document. Policy directives in the document should be discussed as well as the rationale for doing specific procedures in a specific manner. Time for questions and discussion should be part of each teaching session. The teaching sessions should end with a summary of what has been taught to date and what is to be discussed in future sessions. Any assignments should be clearly explained and the completion date clearly stated.

To remove some of the anxiety individuals feel about doing new things, it should be clearly stated during training that it is recognized that mistakes will be made during the implementation of the program. It should be stressed that these mistakes need to be documented for regulatory and teaching purposes, not for disciplinary purposes. However, it should also be made very clear when the "training honeymoon" will end and people will be evaluated in a different manner related to their ability to perform their work correctly.

10.18.1.10. The Program Goes Real Time

Once the instruction phase is complete, the program should immediately be put into effect. Delays between completion of instruction and going real time with the program can be counterproductive. Generally, individuals are much more successful in implementing a program when there is a short time period between instruction and actually doing what has been taught.

10.18.1.11. Practice Makes Perfect

As with any new program, mistakes will be made. These mistakes should be documented by the individual and reviewed with the supervisor. During this review ways to avoid making the mistake again should be discussed. This is also an ideal time to determine if the mistake could have been avoided if the procedure were more clearly written.

10.18.1.12. Feedback on the Program

Informal feedback on how well the new program is functioning needs to be asked for in a reasonable amount of time after its implementation. The time interval should be such that the importance of the program is not forgotten; yet not too early that workers have not had a chance to evaluate the functionality of the program. Feedback should not be elicited at or near the time of an audit. This is because the particulars of the program will get lost in the excitement of the audit period and good feedback will not be obtained. All feedback should be discussed with those involved in a timely manner so that the issues do not become cloudy. In addition, timely feedback will instill in the mind of those giving the feedback the importance of the program.

10.18.1.13. Formal Internal Auditing

Internal auditing of the program at regular intervals is critical to assure that the program is being implemented as intended and that the program is achieving the desired end results. Again, items identified during an audit should be addressed within a timely fashion. If an audit reveals a deficiency, it should be addressed immediately. If major changes in the program are necessary to address the deficiency, it may be beneficial for the individuals who originally wrote the program to meet and discuss what modification(s) to the program may be necessary. If it is determined that the program needs to be modified, it should be done in a timely manner. The modifications to the program should be explained to all individuals involved with the program. The modifications should then be audited both separately and as part of the entire program to determine if the modifications have corrected the previous deficiency.

10.18.1.14. External Auditing

Audits performed by individuals from the outside can be very revealing about how a program is functioning. This is because the people doing the audit are not only unfamiliar with the program itself but they are also unfamiliar with the way things are normally done in the facility.

10.18.1.15. Routine Review of the Program Document

The program document should be reviewed on a routine basis. It is suggested that this review be conducted at frequent intervals after the initial implementation of the program. After the program has been in place for several years, it may be appropriate to cut back on the reviews. Whether this can be done is determined by how the program is functioning as determined by the individuals who are responsible for carrying out the program as well as formal internal and external audits. It is advisable to review all program documents on a yearly basis and as needed.

10.19. PERSONNEL TRAINING AND DEMONSTRATION OF COMPETENCY

10.19.1. Training and Regulations

Training is a regulatory requirement as noted in 21 CFR 211.25. Training is considered by the FDA to be directly related to the manufacturing of a quality pharmaceutical product, and thus a central part of cGMP. This training needs to cover:

> Assigned tasks
> cGMP regulations
> Written procedures that are maintained by the firm in compliance with regulations

It is important to recognize that each person engaged in the manufacturing, processing, packing, or holding of a drug product shall have:

> Education, training, and experience, or any combination thereof, to enable that person to perform the assigned functions.
> And the following should apply:

>> Training shall be in the particular operations that the employee performs and in cGMP.
>> Training in cGMP practice shall be conducted by qualified individuals on a common basis and with sufficient frequency to assure that the employees remain familiar with cGMP requirements applicable to them.

Each person responsible for supervising the manufacture, processing, packing, or holding of a drug product shall have the education, training, or experience, or any combination thereof, to perform assigned functions in such a manner as to provide assurance that the drug has safety, identity strength, quality, and purity that it purports or is represented to possess.

There shall be an adequate number of qualified personnel to perform and suprvise the manufacture, processing, packing, or holding of each drug product.

The FDA does not determine the level of compliance with CFR 211.25 merely by an analysis of the company's training program. The existence of cGMP violations may be taken as evidence of the lack of effective employee training. However, the mere fact that errors and improper procedures are not observed during a FDA inspection does not relieve a company of the obligation to establish or maintain a training program for employees, backed by documentary support, which is open to FDA inspection.

In summary, the FDA views a company's training program as a critical component of the company's compliance profile. The ability of a company to demonstrate the results of that training may be crucial in determining whether the company meets compliance criteria.

10.19.2. Developing a Training Program (109)

No federal regulation specifically describes what training needs to be done, how the training is to be done, or how employees will be judged as to their competency to do a job after training. Also, no specific literature can be referred to as to what type of training needs to be done at a facility since this will be specific to each manufacturing facility. In order to determine the type of training that is necessary, one must be very knowledgeable about the facility, personnel, and manufacturing process. Since it is impractical for one person to know all these aspects, particularly at large facilities, it is imperative to have individuals knowledgeable about specific operations and aspects of a facility involved in writing training documents as well as the actual training. There are a variety of documents available from a variety of sources that can be helpful in providing guidance for the development of personnel training programs and assessment of the competency of individuals in what they were taught (109–111).

The training program should be designed around the duties and responsibilities of individuals as described in the disinfection program. This will make the training program relevant and efficient. The education level of the trainees must be taken into consideration when designing the training. Generally the training sessions should be geared to a 6th grade reading and comprehension level. The content of training sessions should be clearly stated at the beginning of the training session. It is also important that all individuals at the beginning of the training session be made aware of what they will be required to learn from the training session. They should also be made aware at the beginning of the training session if they will be evaluated and how they will be evaluated at the end of the training session. It should also be stated at the beginning of the training session what will constitute a passing grade and what the consequences will be of not passing. Prior to the beginning of the training it is advisable to determine what remedial training will be given to individuals who have difficulty learning during training. What the consequences will be for an individual who may fail competency testing a number of times should also be determined and made known to trainees prior to beginning training sessions. All individuals receiving training in similar duties and responsibilities must be trained and evaluated in a similar manner.

There are a variety of measurement tools that can be used to determine the competency of an individual to perform a function. In some cases it may be desirable to evaluate an individual on the ability to perform a function prior to a training session. This can accomplish two goals. It can help to characterize the level at which the training sessions need to begin and it can establish a baseline against which to evaluate the effectiveness of training. This approach can cause anxiety among workers. It should be made clear to the individuals why they are being evaluated on the performance of a job prior to receiving training.

Some of the tools that are available for evaluating competency after training include the following:

Administration of a written test including mathematical calculations as they relate to making various concentrations of disinfectants
Observations of procedures and outcome
Assessment of responses to case studies, problems, or situations related to the procedure
Documentation of response to actual incidents that may have occurred during the performance of the procedure ("critical incidents")
Assessment of responses to oral queries related to procedures

10.19.3. Documentation-Documentation

While not explicitly mentioned in any regulation, documentation of all training activities and personnel performance is necessary. All training documents should be kept and evidence of review and updating of the documents should be easily identified. The format of the document used to record training should be clear. All training dates noted in an individual's file should be clearly linked to a specific training session and trainer. Training evaluations should be saved and evidence that the results were reviewed with the trainee clearly evident. Any remedial training and its outcome should be clearly documented. If an individual is removed from performing a specific job, the reason for the removal should be clearly stated in the person's file. If an individual is fired for not being able to perform his or her duties or responsibilties after attending training sessions, the reason for the firing should be clearly stated and all training records of that individual kept with the dismissal document (109–111).

10.20. SUMMARY

There are several key elements to developing and maintaining a successful disinfection program:

> Proper selection and sound quality control of the disinfectants
> Demonstration of the efficacy and consistency in the disinfection process via validation
> Development of an effective training program that provides continuous education and fosters a "team" approach to assessing disinfection process performance
> Adequate documentation of the disinfection process and program

As stated at the beginning of this chapter, a successful disinfection program that is regulatory compliant includes clearly written SOPs for cleaning/disinfection, documented evidence of cleaning/disinfection studies, a well-maintained operator training and qualification program, and validated methods. If the items discussed in this chapter are implemented, the disinfection program will be current, consistent, and successful and there is a high probability that regulatory compliance in this area will be achieved.

REFERENCES

1. Block, S.S. Definition of terms. In *Disinfection, Sterilization, and Preservation*; Block, S.S, Ed.; Lippincott, Williams, and Wilkins: Philadelphia, 2001; 24–25.
2. "The Gold Sheet", Pharmaceutical and Biotechnology Quality Control, Apr 2002; 36 (4).
3. Cremieux, A.; Fleurette, J. Methods of testing disinfectants. In *Disinfection Sterilization and Preservation*; Block, S.S., Ed.; Lea & Febiger: Philadelphia, 1991; 1009–1027.
4. Tilley, F.W.; Schaffer, J.M. Relation between the chemical constitution and germicidal activity of monohydridric alcohols and phenolics. J. Bacteriol 1926, *12*, 303–309.

5. Larson, E.L.; Morton, H.E. Alcohols. In *Disinfection, Sterilization and Preservation*; Block, S.S., Ed.; Lea & Febiger: Philadelphia, 1991; 191–203.

6. Sattar, A.S.; Sprinthorpe, S. Activity against human viruses. In *Disinfection, Preservation and Sterilization*, 3rd Ed.; Russell, A.D., Hugo, W.B., Ayliffe, G.A.J., Eds.; Blackwell Science: London, 1998; 168–186.

7. Klein, M.; DeForest, A. The inactivation of viruses by germicides. Chem. Specialists Manufacturing Assoc. 1963, *49*, 116–118.

8. Sattar, S.A.; Springthorpe, V.S. Survival and disinfectant inactivation of the human immunodeficiency virus: a critical review. Rev. Infect. Dis. 1991, *13*, 430–437.

9. Ernst, D.R.; Race, R.E. Comparative analysis of scrapie agent inactivation methods. J. Virol. Methods 1993, *41*, 193–202.

10. Formaldhyde: evidence of carcinogenicity. NIOSH Current Intelligence Bulletin 34. DHEW (NIOSH) Publication No. 81-111, April 15, 1981.

11. Heinzel, M. The phenomena of resistance to disinfectants and preservatives. In *Industrial Biocides Critical Reports on Applied Chemistry*; Payne, K.R., Ed.; John Wiley & Sons: Chichester, 1988; Vol. 22, 56–67.

12. Borick, P.M.; Donershine, F.H.; Chandler, V.L. Alkalinized glutaraldehyde, a new antimicrobial agent. J. Pharm. Sci. 1964, *531*, 1273–1275.

13. Gorman, S.P.; Scott, E.M. A quantitative evaluation of the antifungal properties of gluaraldehyde. J. Appl. Bacteriol. 1977, *47*, 463–468.

14. Power, E.G.M. Aldehydes as biocides. Progress in Medicinal Chemistry 1997, *34*, 149–201.

15. Rubbo, S.D.; Gardner, J.F.; Webb, R.I. Biocidal activities of gluteraldehyde and related compounds. J. Appl. Bacteriol. 1967, *30*, 78–87.

16. Russell, A.D. Activity of biocides against mycobacteria. J. Appl. Bacteriol. Symposium Supplement 1996, *81*, 87S–101S.

17. Scott, E.M.; Gorman, S.P. Gultaraldehyde. In *Disinfection, Sterilization and Preservation*; Block, S.S., Ed.; Lea & Febiger: Philadelphia, 1991; 377–384.

18. Walsh, S.E.; Maillard, J.Y.; Russell, A.D. Ortho-phthalaldehyde: a possible alternative to gluteraldehyde for high level disinfection. J. Appl. Microbiol. 1994, *86*, 1039–1046.

19. Gregory, A.W.; Schaalje, B.; Smart, J.D., et al. The mycobacterial efficacy of ortho-phthalaldehyde and the comparative resistances of *Mycobacterium bovis, Mycobacterium terrae*, and *Mycobacterium chelonae*. Infect. Control Hosp. Epidemiol. 1999, *20*, 324–330.

20. Dychdala, G.R. Chlorine and chlorine compounds. In *Disinfection, Sterilization and Preservation*; Block, S.S., Ed.; Lea & Febiger: Philadelphia, 1991; 131–151.

21. Gottardi, W. Iodine and iodine compounds. In *Disinfection, Sterilization and Preservation*; Block, S.S., Ed.; Lea & Febiger: Philadelphia, 1991; 152–166.

22. Trueman, J.R. The halogens. In *Inhibition and Destruction of the Microbial Cell*; Hugo, W.B., Ed.; Academic Press: London, 1971; 135–183.

23. Favero, M.S. Chemical germicides in the health care field: the perspective from the Centers for Disease Control and Prevention. In *Chemical Germicides in Health Care*; Rutala, W.A., Ed.; Polyscience: Morin Heights, 1995; 33–42.

24. Occupational Safety and Health Administration. Department of Labor. Occupational exposure to hazardous chemicals in the laboratory. U.S. Government Printing Office: Washington, DC, 1990, 29 CFR 1910.1450.

25. Bloomfield, S.F. Chlorine and iodine formulations. In *Handbook of Disinfectants and Antiseptics*; Ascenzi, J.M., Ed.; Marcel Dekker: New York, 1996; 133–158.

26. Williams, N.D.; Russell, A.D. The effects of some halogen-containing compounds on Bacillus subtilis endospores. J. Appl. Bacteriol. 1991, *70*, 427–436.

27. Rutala, W.A.; Cole, E.C.; Wannamaker, N.S., et al. Inactivation of *Mycobacterium tuberculosis* and *Mycobacterium bovis* by 14 hospital disinfectants. Am. J. Med. 1991, *91* (suppl 3B), 267S–271S.

28. Best, M.; Sattar, S.A.; Springthorpe, V.S., et al. Efficacies of selected disinfectants against *Mycobacterium tuberculosis*. J. Clin. Microbiol. 1990, *28*, 2234–2239.

29. Klein, M.; DeForest, A. The inactivation of viruses by germicides. Chem. Specialists Manufacturing Assoc. Proc. 1963, *49*, 116–118.

30. Centers for Disease Control. Guidelines for the prevention of transmission of human immunodeficiency virus and hepatitis B virus to health care workers and public-safety workers. MMWR 1989, *38* (S-6), 1–37.

31. Coates, D. Comparison of sodium hypochlorite and sodium dichloroisocyanurate disinfectants: neutralization by serum. J. Hosp. Infect. 1985, *11*, 60–67.

32. Bond, W.W. Activity of chemical germicides against certain pathogens: human immunodeficiency virus (HIV), hepatitis B virus (HBV) and *Mycobacterium tuberculosis* (MTB). In *Chemical Germicides in Health Care*; Rutala, W., Ed.; Polyscience: Morin Heights, 1995; 135–148.

33. Hoffman, P.N.; Death, J.E.; Coates, D. The stability of sodium hypochlorite solutions. In *Collins, C.H. Disinfectants: Their Use and Evaluation of Effectiveness*; Allwood, M.C., Blood, S.F., et al., Eds.; Academic Press: London, 1981; 77–83.

34. Allawala, N.A.; Riegleman, S. The properties of iodine in solution of surface active agents. J. Am. Pharm. Assoc. Sci. Ed. 1953, *42*, 396–401.

35. Craven, D.E.; Moody, B.; Connolly, M.G., et al. Pseudobacteremia caused by povidone-iodine solution contaminated with *Pseudomonas cepacia*. N. Engl. J. Med. 1981, *305*, 621–623.

36. Parrott, P.L.; Terry, P.M.; Whitworth, E.N., et al. *Pseudomonas aeruginosa* associated with contaminated poloxamer-iodine solution. Lancet 1982, *2*, 683–685.

37. Suttar, S.A.; Raphael, R.A.; Lochnan, H., et al. *Rotavirus* inactivation by chemical disinfection and antiseptics used in hospital. Can. J. Microbiol. 1983, *29*, 1464–1469.

38. Russell, A.D. Bacterial spores and chemical sporicidal agents. Clin. Microbiol. Rev. 1990, *3*, 99–119.

39. Berkelman, R.L.; Holland, B.W.; Anderson, R.L. Increased bactericidal activity of dilute preparations of povidone-iodine solutions. J. Clin. Microbiol. 1982, *15*, 635–639.

40. Russell, A.D. Mycobactericidal agents. In *Russell, A.D. Principles and Practice of Disinfection, Preservation and Sterilization*; Hugo, W.B., Ayliffe, G.A.J., Eds.; Blackwell Sciences: Malden, Mass, 1998; 321–332.

41. Greenspan, F.P.; Johnsen, M.A.; Trexler, P.C. Peracetic acid aerosols. Proc 42nd Ann Mtg Chem Special Manufacturers Assoc, 1955; 59–64.

42. Schaeffer, A.J.; Jones, J.M.; Amundsen, S.K. Bactericidal effect of hydrogen peroxide on urinary tract pathogens. Appl. Environ. Microbiol. 1980, *40*, 337–340.

43. Block, S.S. Peroxygen compounds. In *Disinfection, Sterilization and Preservation*; Block, S.S., Ed.; Lea & Febiger: Philadelphia, 1991; 167–181.

44. Mentel, R.; Schmidt, J. Investigation on rhinovirus inactivation by hydrogen peroxide. Acta. Virol. 1973, *17*, 351–354.

45. Terleckyi, B.; Axler, D.D. Quantitative neutralization assay of fungicidal activity of disinfectants. Antimicrob. Agents Chemother. 1987, *31*, 794–798.

46. Wardle, M.D.; Renninger, G.M. Bactericidal effect of hydrogen peroxide on spacecraft isolates. Appl. Microbiol. 1975, *30*, 710–711.

47. Alasri, A.; Valverde, M.; Roques, C., et al. Sporicidal properties of peracetic acid and hydrogen peroxide, alone and in combination, in comparison with chlorine and formaldehyde for ultrafiltration membrane disinfection. Can. J. Microbiol. 1993, *39*, 52–60.

48. O'Connor, D.O.; Rubino, J.R. Phenolic compounds. In *Disinfection, Sterilization and Preservation*; Block, S.S. Ed.; Lea & Febiger: Philadelphia, 1991; 204–224.

49. Martin, L.S.; McDougal, J.S.; Loskoski, S.L. Disinfection and inactivation of the human T lymphotropic virus type III/lymphadenopathy-associated virus. J. Infect. Dis. 1985, *152*, 400–403.

50. Hugo, W.B. Some aspects of the action of cationic surface active agents on microbial cells with special reference to their mode of action on enzymes. S.C.I. monograph no. 19: Surface Active Agents in Microbiology; London Society of the Chemical Industry: London, 1965; 69–82.

51. Franklin, T.J.; Snow, G.A. *Biochemistry of Antimicrobial Action*; Chapman & Hall: London, 1984.

52. Sakagami, Y.; Yokagama, H.; Nishimura, H., et al. Mechanism of resistance to benzalkonium chloride by *Pseudomonas aeruginosa*. Appl. Environ. Microbiol. 1989, *55*, 2036–2040.

53. Russell, A.D.; Gould, G.W. Resistance of *Enterobacteriaceae* to preservatives and disinfectants. J. Appl. Bacteriol. Symp. Suppl. 1988, *65*, 167S–195S.

54. Adair, F.W.; Geftie, S.G.; Gezler, J. Resistance of *pseudomonas* to quaternary ammonium compounds. Growth in benzalkonium chloride solution. Appl. Microbiol. 1969, *18*, 299–302.

55. Dixon, R.E.; Kaslow, R.A.; Mackel, D.C., et al. Aqueous quaternary ammonium antiseptics and disinfectants: Use and misuse. JAMA 1976, *236*, 2415–2417.

56. Grossgebauer, K. Virus disinfection. In *Disinfection*; Bernarde, M., Ed.; Marcel Dekker: New York, 1970; 103–148.

57. Merianos, J.J. Quaternary ammonium compounds. In *Disinfection, Sterilization and Preservation*; Block, S.S., Ed.; Lea & Febiger: Philadelphia, 1991; 225–255.

58. D'Arcy, P.F. Inhibition and destruction of molds. In *Inhibition and Destruction of the Microbial Cell*; Hugo, W.B., Ed.; Academic Press: London, 1971; 613–686.

59. Broadley, S.J.; Jenkins, P.A.; Furr, J.R., et al. Antimycobacterial activity of biocides. Lett. Appl. Microbiol. 1991, *13*, 118–122.

60. Sasatsu, M.; Shirai, Y.; Hase, M., et al. The origin of the antiseptic-resistance gene ebr in *Staphylococcus aureus*. Microbios 1995, *84*, 161–169.

61. Anthonsien, I.L.; Sunde, M.; Steinum, T.M., et al. Organization of the antiseptic resistance gene quacA and Tn552-related β-lactamase genes in multidrug-resistant *Staphylococcus haemolyticus* strains of animal and human origins. Antimicrob. Agents Chemother. 2002, *46*, 3606–3612.

62. Littlejohn, T.G.; DiBeradino, D.; Messerotti, L.J., et al. Structure and evolution of a family of genes encoding antiseptic and disinfectant resistance in *Staphylococcus aureus*. Gene 1990, *101*, 59–66.

63. Sidu, A.S.; Heir, E.; Leegaard, T., et al. Frequency of disinfectant resistance genes and genetic linkage with β-lactamase transposon Tn552 among clinical staphylococci. Antimicrob. Agents. Chemother. 2002, *46*, 2797–2803.

64. McDonnell, G.; Russell, A.D. Antiseptics and disinfectants: activity, action and resistance. Clin. Microbiol. Rev. 1999, *12*, 147–179.

65. Russell, A.D.; Furr, J.R. Biocides: mechanisms of antifungal action and fungal resistance. Sci. Prog. 1996, *79*, 27–48.

66. Doores, A.S. Bacterial spore resistance—species of emerging importance. Food Technol. 1983, *37*, 127–134.

67. Russell, A.D. The Destruction of Bacterial Spores; Academic Press: New York, 1982.

68. Ellar, D.J. Relations between structure and function in the prokaryotic cell. Sym. Soc. Gen. Microbiol. 1978, *28*, 295–325.

69. Warth, A.D. Molecular structure of the bacterial spore. Adv. Microb. Physiol. 1978, *17*, 1–45.

70. Power, E.G.M.; Russell, A.D. Gluteraldehyde: its uptake by sporing and non-sporing bacteria, rubber, plastic, and an endoscope. J. Appl. Bacteriol. 1989, *67*, 329–342.

71. Brown, P.; Pierce, M.A.; Wills, R.G. "Friendly fire" in medicine: hormones, homografts, and Creutzfekdt-Jakob disease. Lancet 1992, *340*, 24–27.

72. Will, R.G.; Ironside, J.W.; Zeidler, M., et al. A new variant of Creutzfeldt-Jakob disease in the U.K. Lancet 1996, *347*, 921–925.

73. Antloga, K. Prion diseases and medical devices. ASAIO J. 2000, *46*, S69–S72.

74. Steelman, V.M. Prion diseases: An evidence based protocol for infection control. AORN J. 1999, *69*, 946–967.

75. Taylor, D.M. Inactivation of unconventional agents of the transmissible degenerative encephalopathies. In *Principles and Practice of Disinfection, Preservation and Sterilization*; Blackwell Scientific: Oxford, 1999; 222–236.

76. Taylor, D.M. Transmissible subacute spongiform encephalopaties: practical

aspects of agent inactivation. In *Transmissible Encephalopathies, Prion Diseases,* Proceedings of the Third International Symposium on Transmissible Subacute Spongiform Encephalopathies: Prion disorders (Protection Against Transmission Risks); Paris, March 18–20, 1996; 479–482.

77. Brown, P.; Liberski, P.P.; Wolff, A., et al. Resistance of scrapie agent to steam autoclaving after formaldehyde fixation and limited survival after ashing at 360°C: Practical and theoretical implications. J. Infect. Dis. 1990, *161*, 467–472.

78. Mould, D.L.; Dawson, A.M.; Smith, W. Scrapie in mice: The stability of the agent to various suspending media, pH, and solvent extraction. Res. Vet. Sci. 1965, *6*, 151–154.

79. Dickinson, A.G.; Taylor, D.M. Resistance of scrapie agent to decontamination. N. Engl. J. Med. 1978, *229*, 1413–1414.

80. Walker, A.S.; Inderlied, C.B.; Kingsbury, D.T. Conditions for the chemical and physical inactivation of the K.Fu strain of the agent of Creutzfeldt-Jakob disease. Am. J. Public Health 1983, *73*, 661–665.

81. World Health Organization. Infection Control Guidelines for Transmissible Spongiform Encephalopathies. Report of a WHO Consultation, Geneva, March 23–24, 1999 WHO/CDS/CSR/APH/2000.3.

82. Denny, V.F.; Marsik, F.J. Current practices in the use of disinfectants within the pharmaceutical industry. PDA J. Parent. Sci. Tech. 1997, *51*, 227–228.

83. Rutala, W.A.; Weber, D.J. Draft Guideline for Disinfection and Sterilization in Healthcare Facilities. *CDC* and *DHHS, HIPAC 2b*; February 20, 2002; 9 pp.

84. Russell, A.D. Factors influencing the efficacy of antimicrobial agents. In *Principles and Practice of Disinfection, Preservation, and Sterilization*; Russell, A.D., Hugo, W.B., Ayliffe, G.A.J., Eds.; Blackwell Science: Oxford, 1999; 95–123.

85. Costerton, J.W.; Lewandowski, Z.; Caldwell, D.E., et al. Microbial biofilms. Annu. Rev. Microbiol. 1995, *45*, 711–745.

86. Whitely, M.; Bangera, M.G.; Bumgarner, R.E., et al. Gene expression in *Pseudomonas aeruginosa* biofilms. Nature 2001, *413*, 860–864.

87. Stewart, P.S.; Rayner, J.; Roe, F., et al. Biofilm penetration and disinfection efficacy of alkaline hypochlorite and chlorosulfamates. J. Appl. Microbiol. 2001, *91*, 525–532.

88. Donlan, R.M. Biofilms: microbial life on surfaces. Emerg. Infect. Dis. 2002, *8*, 881–890.

89. Disinfectants. In *Official Methods of Analysis of AOAC International*, 17th Ed., AOAC International: Washington, DC, 1995.

90. Luppens, S.B.; Reij, M.W.; van der Heijden, R.W.L., et al. Development of a standard test to assess the resistance of *Staphylococcus aureus* biofilm cells to disinfectants. Appl. Environ. Microbiol.; Godalming, Surrey, UK, 2002, *60*, 4194–4200.

91. Kopis, E. Disinfectants program. In *Microbiology in Pharmaceutical Manufacturing*; Prince, R., Ed.; Davis-Horwood: Godalming, Surrey, UK, 2001, 419–433.

92. US Food and Drug Administration. Guidance for Content and Format of Premarket Notification [510(k)] Submissions for Liquid Chemical Sterilants/ High Level Disinfectants. FDA, CDRH; January 3, 2000; 14–17.

93. Denny, V.; Kopis, E.; Marsik, F. Elements for a successful disinfection program in the pharmaceutical environment. PDA J. Parent. Sci. Tech. 1999, 53 (3), 118 pp.
94. Chen, J.H.S. Methods of testing virucides. In *Disinfection, Sterilization and Preservation*; Block, S.S., Ed.; Lea & Febiger: Philadelphia, 1991; 1076–1093.
95. Steinmann, J. Some principles of virucidal testing. J. Hosp. Infect. 2001, 48, S15–S17.
96. Leland, S.E. Jr. Methods of testing protozoacides and antihelminthics. In *Disinfection, Sterilization and Preservation*; Block, S.S., Ed.; Lea & Febiger: Philadelphia, 1991; 1094–1096.
97. Sy, M.S.; Gambetti, P.; Wong, B.S. Human prion diseases. Med. Clin. North. Am. 2002, 86, 551–571.
98. Taylor, D.M. Transmissible degenerative encephalopathies. In *Disinfection, Preservation and Sterilization*, 3rd Ed.; Russell, A.D., Hugo, W.B., Ayliffe, G.A.J., Eds.; Blackwell Science: London, 1998; 222–236.
99. Croshaw, B. Disinfectant testing—with particular reference to the Rideal-Walker and Kelsey Sykes Tests. In *Disinfectants: Their Use and Evaluation of Effectiveness*; Hoffman, P.N., Ed.; Academic Press: London, 1981; 1–3.
100. Reybrouck, G. Evaluation of antibacterial and antifungal efficacy. In *Disinfection, Preservation and Sterilization*; Russell, A.D., Hugo., W.B., Ayliffe, G.A.J., Eds.; 3rd ed.;. Blackwell Science: London, 1998; 124–144.
101. Quinn, P.J.; Carter, M.E. Evaluation of viricidal activity. In *Disinfection, Preservation and Sterilization*, 3rd Ed.; Russell, A.D., Hugo, W.B., Ayliffe, G.A.J., Eds.; Blackwell Science: London, 1998; 197–206.
102. Jarrol, E.L. Intestinal protozoa. In *Disinfection, Preservation and Sterilization*, 3rd Ed.; Russell, A.D., Hugo, W.B., Ayliffe, G.A.J., Eds.; Blackwell Science: London, 1998; 251–257.
103. Cole, E.C.; Rutala, W.A. Disinfectant testing using a modified use-dilution method: collaborative study. J. Assoc. Off. Anal. Chem. 1998, 71 (6), 1187–1194.
104. Groschel, D.H. Disinfectant testing in the USA. J. Hosp. Infect. 1991, 18 (suppl A), 274–279.
105. van Klingeren, B. Disinfectant testing on surfaces. J. Hosp. Infect. 1995, 30 (suppl), 397–408.
106. Cremieux, A.; Fleurette, J. Methods of testing disinfectants. In *Disinfection, Sterilization, and Preservation*; Block, S.S., Ed.; Lippincott, Williams, and Wilkins: Philadelphia, 2001; 1305–1323.
107. Russell, A.D. Good manufacturing practices. In *Principles and Practice of Disinfection, Preservation, and Sterilization*; Russell, A.D., Hugo, W.B., Ayliffe, G.A.J., Eds.; Blackwell Science: Oxford, 1999; 376–393.
108. STERIS Corporation, Technical Tip # 4014, Disinfectant Application Guidelines for Cleanrooms and Controlled Environments, February, 1999.
109. Nevalain, D.E.; Berte, L.M. Training verification and assessment: Keys to quality management. Laboratory Management Association. Paoli, PA.
110. Tetzlaff, R. Systematic approach to GMP training. Pharm. Tech. 1982, 6, 42–51.
111. Vesper, J. *Training for the Healthcare Manufacturing Industries*. Interpharm Press: Englewood, CO, 1993.

11

Sterile Filtration

Maik W. Jornitz
Sartorius Corporation, Edgewood, New York, U.S.A.

11.1. INTRODUCTION

Sterile filtration is widely used in the biopharmaceutical industry to remove contaminants, especially microorganisms from liquids, air, and gases (1). Microorganism removal is required either to achieve a sterile filtrate or, if the drug product is thermally sterilized, to reduce the bioburden level to avoid elevated levels of endotoxins, the remains of gram-negative organisms. Such sterilization or bioburden removal filter systems require thorough validation and qualification to confirm that the particular filter is working for its intended purpose.

Filter configurations are manifold depending on the particular use and requirement. Sheet or modular depth filter types are utilized for prefiltration to remove larger quantities of contaminants or to protect sterilizing grade filters. Flat filter membranes are used mainly for microbial detection and specifications. In process filtration most commonly used are filter cartridges containing either depth filter fleeces or membrane filters. Both membrane and prefilters are available in a large variety of polymers and configurations for different applications.

Sterilizing grade membrane filters are defined by the FDA Guideline on Sterile Drug Products Produced by Aseptic Processing (2) by being able to retain 10^7 *Brevundimonas diminuta* (formerly *Pseudomonas diminuta*) organisms per square centimeter of filtration area at a differential pressure of 29 psi (2 bar) (3). Such retention efficiency has to be validated using the actual drug product and the process parameters, due to the possibility of an effect on the filters compatibility and stability and/or the microorganism size and survival rate (4,5). Performing these so-called product bacteria challenge tests became a regulatory demand (6) and, therefore, they are associated with standard filter validation. Before these challenge tests can be performed, the appropriate challenge methodology has to be evaluated via viability tests. These tests determine the mortality rate of the challenge organisms due to product or process parameters. If the mortality rate is high, greater than 1 log during exposure time, parameters require change to perform the challenge test. PDA Technical Report No. 26 (7) describes the individual parameters, the possible effects, and mechanisms to be used to perform challenge tests. Additionally, the report discusses filtration modes, sterilization, and integrity testing.

11.2. TYPES OF FILTRATION

11.2.1. Prefiltration

Prefilters are most commonly depth filter types and are most often constructed of nonwoven or melt-blown fiber materials such as polypropylene, polyamide, cellulosic, glass fiber, metals, and (before the interdiction of its use on account of its carcinogenicity) asbestos. These fiber materials are constructed into mats by the random deposition of either individual or continuous fibers whose permanence of positioning is sought through pressing, heating, gluing, entanglements, or other forms of fixing. The pores of such filter constructions are the interstices among the fibers. As shown in Figure 1, the random deposition of the fibers during construction of the filter mat results in a broad pore-size distribution. Such pore-size distribution can be influenced by the thickness of the individual fiber or the compactness of the matrix. Therefore, there are a large variety of prefilter types that can be selected for any kind of application. The resulting advantage is that processes can be optimized by using the large range of prefilters.

Because these filters are composed of fibers and/or of other discrete particles, they are properly regarded as being potentially fiber-releasing. This property is not necessarily eliminated by liquid flushing. Therefore, at least in the case of injectables, their use must be followed by a final membrane filter among whose purposes is the capture of fibers generated by the migration of the medium from the depth-type filter.

A major advance in depth filter design technology was made of melt-spun depth filter types and the introduction of heat stabilization of fiber

FIGURE 1   Random depth filter matrix. (Courtesy of Sartorius Group.)

fleeces (8). These treatments avoided the release of particulate matter and were utilized to stabilize the final filter fleece. Additionally, these technologies allowed producing fleece construction of different fiber sizes within a filter matrix. This allowed improving the total throughput and protection performance of these filters due to fractionate retention of a large spectrum of particle sizes. Furthermore, the longer melt-spun fibers coupled with some thermal fusion that occurs in the process reduces concerns about fibers coming loose and passing into the filtered effluent, whereas fiber migration can occur with the staple fiber yarn-wound filter design.

The melt-spun filter design offers several other advantages over traditional textile winding technology. First, the process produces a filter free of lubricants or finishing agents, which avoids the criticality of potential leachables (extractables). These agents are additives that are processing aids necessary to process the fiber and yarn used to make yarn-wound filters. Second, the extrusion process produces a distribution of fiber diameter sizes. Although its distribution is relatively uncontrolled in this process, the mean fiber size can be smaller than the traditional staple fiber diameters. The smaller mean fiber diameter coupled with the graded density method can produce filters down to the 0.5 μm nominal range, commonly claimed to be 99% retentive.

A further advance in depth filter technology occurred with the advent of the first melt-blown type of cartridge that incorporated various fiber diameters as the filter was manufactured to achieve a graded pore design by means other than varying the fiber packing density. This design is based on variation of standard melt-blowing equipment. In this process, the polymer is extruded through a multihole die and the polymer stream is stretched and attenuated by a high-velocity heated air stream. The mean fiber diameter is changed as the filter is being made by adjusting the air velocity or one of the other variables that contribute to the formation of the fiber sizes (e.g., temperature or polymer pumping rate). This technology is becoming more and more advanced, such that some manufacturers are naming the fibrous fleece constructions nano-fiber fleeces.

Using a graded or changing pore size to enhance filtration performance is a desirable concept. This technique involves incorporating a series of prefilters into a single stage to maximize the use of the entire filter and extend filter life (dirt-holding capacity). The factor of fractionate retention is especially important for applications with a wide particulate spectrum, such as water pretreatment. Prefilters can also contain membranes, porous or fibrous, commonly from cellulose, mixed esters, or borosilicate. These prefilter types are utilized to remove a very fine band of particulate or contaminants from the fluid to specifically protect sterilizing-grade membrane filters.

11.2.2. Membrane Filtration

Membrane filters commonly contain a defined pore structure and porosity band (Fig. 2). The narrower the porosity band, the more defined the retention rate of the membrane. The filtration obtained by the use of such membrane filters (more properly, microporous or controlled-pore membrane filters) is often referred to as microfiltration, or MF. Microporous membrane filters offer a much finer degree of porosity than is available from the conventional depth filters commercially available.

Depth filters are produced under controlled conditions; nevertheless, the randomness of the fibrous material does not allow it to produce a defined porous structure as in membrane filtration. Membranes are produced by an evaporation (air casting), quenching (immersion), stretching, or track-etched process (9). In the evaporation process the casting solution is applied onto a belt. Due to defined temperature, belt speed, and air conditions (air flow and humidity), the solvent from the casting solution starts evaporating and leads eventually to phase inversion and formation of the wet-gel form of the microporous membrane. Changes in the described conditions and the casting solution mix lead to different pore structures, porosities, and membrane

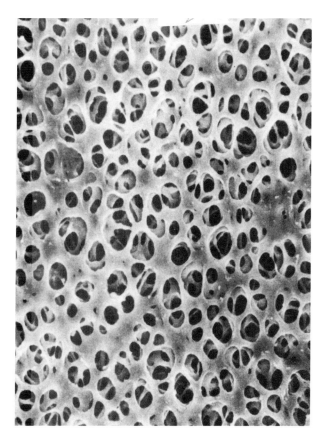

FIGURE 2 SEM of the porous structure of a celluloseactetate membrane. (Courtesy of Sartorius Group.)

structures. In the quenching process the polymer/solvent mix is applied onto a drum or belt, which immerses into a solvent or extraction bath. The polymer starts precipitating and forms a porous membrane. This membrane will be dried in further steps. A stretching production process to form a membrane is mainly used for polytetrafluorethylene polymers. Melt-extruded films are stretched under very defined process conditions to create a thin (commonly 60 to 100 μm) membrane. Such membranes have distinctive nodes, which are connected by filaments (Fig. 3). The thinnest (10 to 20 μm) membrane films are created by track-etched manufacturing process. Commonly polycarbonate is subjected to a bombardment of high-energy particles. The membrane polymer is damaged at the bombardment track and after the submittal to an etching bath, pores are formed along the damage.

FIGURE 3 SEM of the porous structure of a PTFE membrane. (Courtesy of Sartorius Group.)

The pore structure of track edge membranes is very defined. Nevertheless, the pore volume is far less than that of conventionally produced membrane structures due to the limit bombardment density. If the density of the high-energy particles is too high, a double or multiple hit could result in an enlarged pore.

Membrane filters can be formed in a variety of structures for specific application purposes. For example, the formation of asymmetric membrane structures (the pore structure on the upstream side of the membrane filter is larger than the downstream side) can enhance the dirt load capacity of such filter. Some applications require very distinct pore shapes to avoid premature blockage; or, in the case of the use of a membrane as microbiological test filter, the pore structure has to be very even to achieve appropriate nutrient distribution. Membrane filters are the most common filtration devices used in aseptic processing to remove organisms from liquids or gases. Due to the defined structure, these filters are highly reliable in respect to the retention requirements and, furthermore, can be integrity tested.

11.2.3. Prefilter and Membrane Filter Comparison

Depth-type filters cannot dependably be used to produce sterile filtrates; membranes can. This dissimilarity is due to the difference in the pore-size distributions and the stability of the pore structure within both filter types. By whatever manufacturing technique filters are prepared, not all of the pores produced within a filter are of the same size. Given the relatively homogeneous sizes of a suspension of particles (organisms) whose filtrative removal is being sought, the broader the pore size distribution, the more likely the encounter of a particle penetrating the filter.

Depth filters are manufactured by technologies involving the incorporation of discrete particles or fibers into some matrix or fixed form. These constitute the structured depth filters. The fabrication almost always requires the use of insoluble particles or fibers and a rather viscous dispersing medium. Uniform dispersal is a problem; the viscosity of the matrix, the preferred orientation of the fibers, or the agglomeration of the primary particles, insolubility of the fibers, insolubility of the heterogeneous phase, the usual mechanics of the mixing or lay-up, all work against it. The tendency to diffusional equilibration that is the response to concentration gradients in the porous membrane-casting solutions is absent here. In principle, individual fibers, for example, are deposited on a surface until the complete fiber mat becomes constructed. Each fiber falls largely in accordance with the laws of chance. The fiber mat irregularities reflect this random deposition. The spaces among the fibers constitute the filter pores. As indicated in Fig. 1, a modeled representation of the randomness of fiber deposition, the interstices vary greatly in size, reflecting localized low- or high-fiber population densities. Because the fiber, or other particle, deposition follows the random pattern, the consequent pore-size distribution is broad. The melt-spun and melt-blown processes randomly position the constituting fibers as well.

The breadth of the pore-size distribution of a depth-type filter will depend on the thickness of the fiber (particle) mat. Thicker mats can be considered as consisting of repetitive layers of a thin "unit mat." Each successive layer or increase in mat thickness will serve to diminish the pore-size distribution of the composite. The larger pores of one layer will come randomly to be coupled with the smaller pores of succeeding layers. The overall effect will be a progressive narrowing of the pore size. Eventually, some constant value of pore-size distribution will be approached, perhaps asymptotically, but it will never reach the stability and specification of a membrane structure.

Additionally, depth filter structures can be subject to process conditions. It is essential that the process conditions, especially pressure differential or pressure pulses, fit the prefilter used. Such pressure conditions can either damage or loosen the filter structure and therefore have to be monitored accordingly. There have been examples of membrane filters being subjected to up to 72 psi (5 bar) of differential pressure and pulses. These membrane filters still passed the microbial retention and integrity test. A depth filter's fibrous structure could be damaged by such pressure conditions.

Depth filters, as is self-explanatory, remove any contaminants within the depth of the filter matrix, whereby membrane filters function mainly as surface retentive filters. This certainly depends on the contamination to be removed. The depth retention of prefilters make these the "workhorse" of filtration processes due to the high dirt load capacity of such filters. Surface-retentive filters' total throughput can only be enhanced by the porous structure (asymmetry), enlargement of the effective filtration area, or the use of depth filters as protection in front of the membrane filter. The goal of filter tests is to find the optimal filter combination of prefilters and final filters to achieve the desired retentivity, but also throughput need.

Membrane filters can be integrity tested, which is not possible with depth filters. To validate the membrane filter's performance and reach filtrative assurance, integrity testing of these filters is a must. Depth filters, though, commonly have the purpose to clarify and polish, but not to sterilize. For this reason an integrity test is unnecessary.

11.2.4. Cross-Flow Filtration

Cross-flow filtration differentiates itself from conventional "dead-end" filtration in that the fluid to be filtered flows parallel to the filtration surface rather than perpendicular to the filtration surface, the function shown in Fig. 4 (10). The cross-flow generates shear that limits the buildup of a filter cake or gel layer. In conventional dead-end filtration, the filter cake thickness increases with time, resulting in the eventual cessation of flow. In cross-flow filtration, the feed stream flows parallel to the surface of the membrane—that is, the feed flows tangential to the permeate or filtrate stream. A small fraction of the feed stream permeates the membrane (filtrate or permeate); the remaining fraction is retained by the membrane and exits as retentate or concentrate stream. The retentate or concentrate is recirculated over the membrane layers till the specified requirements are met.

In the biopharmaceutical industry, cross-flow filtration is used for both microfiltration (0.45, 0.2, and 0.1 μm) and ultrafiltration (1.000 to 300.000 MWCO (molecular weight cutoff)). The microfiltration devices are mainly used for cell harvesting or cell debris removal, downstream of a fermentation process. In some instances, cross-flow microfiltration devices are also used as

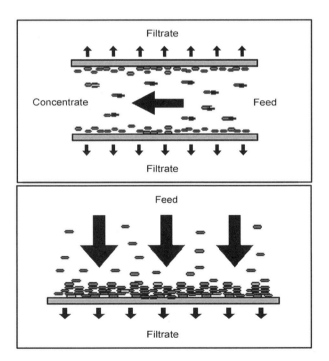

FIGURE 4 Schematic of "dead-end" and cross-flow filtration. (Courtesy of Brose and Dosmar, 1998.)

a prefiltration step before conventional membrane filtration. Ultrafiltration systems are mainly used for fractionation, concentration, and diafiltration steps of proteins, peptides, or viral vectors. This technology enables the removal of undesired contaminants, buffer exchange, and concentration of a target protein without compromising or stressing (via shear forces) the target.

Cross-flow filters have a variety of designs, which range from plate and frame cassette systems to spiral-wound and hollow-fiber modules. The individual designs have to be properly evaluated when cross-flow is chosen, due to performance differences in dead-volume, shear forces, cleanability, pressure chosen, due to performance differences in dead-volume, shear forces, cleanability, pressure conditions, energy inputs, and flow patterns. Cassettes or modules are placed into specific holding devices, which can be either manually driven or fully automatic systems. Plate and frame modules consist of flat-sheet membranes mounted into a framework, commonly silicone or polyurethane. In the assembly of these systems, each flow path is made up of two membranes that are facing each other. The upstream flow path must be sealed from the downstream permeate side of the membrane. Stacks of pairs of membranes are layered one on top of the other, and the permeate side of each membrane is supported by a rigid and porous spacer plate. The spacer plate may be smooth or have surface features that give the membrane an uneven surface for turbulence promotion. Flow paths are usually open and may be parallel and or in series. Spiral-wound modules utilize pairs of flat sheet membranes bound on the up and downstream sides by screens similar to those in cassette systems. The membrane sandwich is sealed at three edges so that the feed is isolated from the permeate. The fourth side of the membrane sandwich is attached to a perforated permeate collection tube. The membrane pairs are then rolled around the perforated collection tube, thereby creating the spiral. Feed flow enters at one end of the spiral, flows tangentially along the axis of the cartridge, and discharges at the other end. Permeate flows at a right angle to the feed flow towards the center of the spiral and is collected in the core of the spiral. Hollow fiber, as the name describes, is a tubular, porous design, which is commonly bundled into a module. Liquid permeates the fiber wall, as with flat-sheet membrane, and permeate is collected on the opposite side of the fiber. Depending on the manufacturer, hollow-fiber systems are fed from the outside or from the inside (most commonly inside flow). In the case where the rejecting layer is on the inside (lumen) of the fiber, the feed solution enters the lumen of the fiber at one end, flows down the length of the fiber, and retentate exits at the other end. Permeate is collected on the outside (shell-side) of the fiber.

11.3. MODUS OF FILTRATION

11.3.1. Sieve Retention

Sieve retention of particle capture is the one most evident in common filtration experiences. It occurs whenever a particle is too large to pass through a filter pore. It is a geometric or spacial restraint. This type of particle arrest is considered "absolute" (but only for the defined size of the particle) in that it is independent of the filtration conditions. The applied differential pressure does not influence it, unless the level is so high as to deform either the particle or the filter pore, an occurrence not alleged in pharmaceutical filtrations. Sieve retention is also free of the influences of the particle challenge level. Regardless of the number of particles confronting the filter, if each is too large to pass the filter pores then none will be able to do so, and all the particles, regardless of number, will be retained. Additionally, the particle retention will be independent of the suspending liquid vehicle as defined by its ionic strength, pH, surface tension, temperature, viscosity, and presence or absence of surfactant, and so forth.

11.3.2. Adsorptive Sequestration

As far back as 1909, Zsigmondy pointed out that the filter surface has a certain adsorbing capacity whose affinity must first be satisfied before unhindered passage of the dispersed phase through the filter may occur. Numerous investigators have since noted specific adsorptions of many entities. Elford (1933) (11) reported that dyes could adsorptively be removed from true solutions by collodion membranes (cellulose nitrate, one of the most adsorptive materials). The strong adsorption tendencies of the cellulose nitrate polymer had also been noted by Elford (1931) (12) in the case of viruses. The use of membrane filters adsorptively to collect and isolate nucleic acids, enzymes, single-strand DNA, ribosomes, and proteinaceous materials in scintillation counting operations is well established. Moreover, such adsorptive retentivity is currently utilized introducing chromatography and membrane adsorber steps into the downstream purification process. Bovine serum albumin, antigen/antibody, and antibody complex (13), and specific binding and receptor protein adsorption to cellulose nitrate has been shown to occur. Berg et al. (1965) (14) investigated the adsorption of both inorganic and organic compounds on polymers such as cellulosic filter papers, nylon, polyethylene, and cellulose diacetate dialysis membranes. That water-soluble organics could adsorptively be removed from aqueous solutions by filters was observed by Chiou and Smith (1970) (15). These investigators were thus led into a rather thorough study of such adsorptions by filters. Undani (1978) (16) and Brose et al. (1994) (17) studied the adsorptive sequestration of preservatives such as benzalkonium chloride, chlorocresol, and chlorhexidine acetate from their solutions by membrane filters. The adsorptive removal of flu vaccine impurities and antibodies onto membrane filters has been reported (18). Inorganic particulate matter can be removed filtratively through the adsorption mechanism (19). It is thus well documented that molecules and materials can be adsorbed onto filters, to become filtratively removed thereby.

There are several references in the literature pertaining to the retention of organisms by contact with filter surfaces. Pertsovskaya and Zvyagintsev (1971) (20) report that films of such polymer compounds as polyamide, polyacrylate, polyethylene, and cellulose triacetate adsorb different groups of different bacteria. Zierdt (1978) (21) demonstrated that bacterial adsorption could take place on the surfaces of membrane filters whose pores are many times larger than the organisms. During the laboratory development of a lyses-fractionation blood culture technique, Zierdt and his associates (1977) (22) at the National Institutes of Health noted that both gram-negative and gram-positive organisms were attracted to the membrane materials during filtrations. The filters were composed of polycarbonate and cellulose mixed esters. Furthermore, the arrested organisms resisted removal by the mechanical or adsorptive action of backwashing with buffer. These investigators were therefore able to use filter membranes with porosities much larger than would normally be expected to arrest the bacteria whose retention they wished. The organisms involved were *Escherichia coli* and *Staphylococcus aureus*. Sterility was neither sought nor obtained. Beyond a doubt, however, bacterial capture by membrane filters involves adsorptive arrest.

Zierdt et al. found that a higher percentage of bacterial retention occurs at low organism concentrations, about 500 to 100 CFU/mL. At higher levels of 10^8 to 10^9 CFU/mL, increasing percentages of *E. coli* pass through the membranes, although a larger total number is retained. These phenomena accord with adsorption. Retention was investigated as a function of the filter pore-size ratings. As expected, the larger the pore-size ratings of the filters, the greater the amount of bacterial passage. At low bacterial numbers, 6.2×10^2 CFU for *E. coli* and 7.3×10^2 CFU for *S. aureus*, apparently no *E. coli* pass a 3.0 μm filter, nor *S. aureus* a 5.0 μm filter. All of the above reflect the influence that organism concentration exerts on filter capture efficiency during adsorptive sequestration.

It also has been shown that *Brevundimonas diminuta* (formerly *Pseudomonas diminuta*) can be retained by adsorptive glass fiber filters (Fig. 5) (23). It is evident that many of the organisms are retained by contact capture rather than by sieve arrest; the filter pores, the spaces among the fibers, obviously are often too large to serve as retaining orifices.

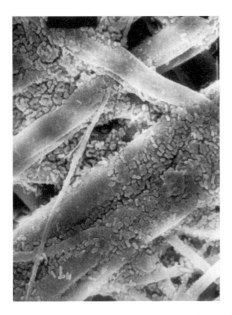

FIGURE 5 Microorganisms captured on a glass fiber depth filter matrix. (Courtesy of Sartorius Group.)

Surface phenomena, such as adsorptions, can be related to forces between molecules, especially to an asymmetry or imbalance of forces at an interface. The hydrogen bond is an example of an asymmetric force caused by the presence of unshared electrons within the water molecule. The forces are electrical in nature and are both attractive and repulsive. Between ions they are mostly electrostatic. The attractive forces are also short range and electrostatic and are usually characterized as van der Waals forces, such as govern the condensation of a vapor into a liquid. The double-layer distance is therefore very important in the adsorptive operation. Energy is required to effect the separation of a bacterium from a surface to which it is adsorbed. The energy level is an expression of the bonding strength, the adsorption, between the organism and the polymer surface. This, in turn, depends on the contributions made to the bond by the membrane surface and by the organism. It is not surprising, therefore, that different filter surfaces bond differently with a given organism, and that different organisms adsorb differently to a given filter surface. Additionally, product parameters (i.e., the filtrate properties) influence the adsorptive capture or attractiveness of and to a surface tremendously (24,25).

Ridgway (1987) (26) found that mycobacterial adhesions to polyamide-type reverse-osmosis membranes showed a five- to tenfold greater affinity than did their adsorptions to cellulose ester RO membranes. It may be speculated on the basis of this finding that strong bacterial adsorptions to polyamide (nylon) membranes account for the sterilizing effects of such 0.2 μm-rated membranes, even when they are more open than their counterparts not composed of this polymer. Ridgway also found that different organisms had different propensities to adsorb to surfaces, as gauged by biofilm formation. It is possible, however, that this adsorptive phenomenon reflects the particular morphological features of the different organisms rather than their molecular makeup.

TABLE 1 Comparison of Sieve Retention and Adsorptive Sequestration

Retention Mechanism	Advantages	Disadvantages
Sieve retention	Reliable at worst case product properties Reliable separation even at high flows and pressure conditions Blockage, (i.e., exhaustion) can be determined No unspecific adsorption, minimal loss of desired product, and little adsorptive fouling	Retentive only at the specific pore size rating
Adsorptive sequestration	It is possible to retain particles smaller than the filter's indicated pore size Separation of colloidal substances is possible In some cases, endotoxins can be removed	Highly influenced by product-specific properties Separated particles can be shed by varying process conditions Saturation of the active sites cannot be determined, no warning Unspecific adsorption will result in product losses and fouling Lower reliability in terms of absolute separation

In specific applications, adsorptive sequestration is sought, but it certainly requires in-depth validation. If adsorptive sequestration is a major function of the retentivity of a filter, such retentive effectiveness needs to be analyzed utilizing process conditions and the actual product to be filtered. Under no circumstances can filtrative efficiency be assumed if it not documented by bacteria challenge test results. This also is valid for claims of endotoxin removal by filtration. Such removal requires qualification over the filtration period at very defined process conditions. Any changes in the process conditions can alter the filtration result. For this reason, any sterilizing-grade filter needs to be validated using the product as the challenge test carrier and the actual process conditions. See Table 1 for a recap of the advantages and disadvantages of sieve retention and adsorptive sequestration.

11.4. MEMBRANE FILTER MATERIALS

11.4.1. Manufacturing Process

Most commercial ultrafiltration (UF) (Fig. 6), nanofiltration (NF), and many microfiltration (MF) membranes are manufactured by immersion casting, whereby a polymer is dissolved in an appropriate solvent along with appropriate pore-forming chemical agents. The polymer solution is cast into a film, either on a backing or free-standing, and then the film is immersed in a nonsolvent solution that causes precipitation of the polymer. The precipitate is dried afterwards. The casting machines in this case are commonly compact,

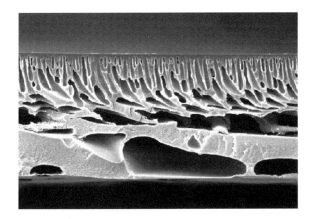

FIGURE 6 Ultrafiltration skin structure. (Courtesy of Sartorius Group.)

especially when the casting solution is applied to a rotating drum belt. Such membranes are polyamides such as nylon, polyethersulfon (PESU), or polyvinyldienefluoride (PVDF).

Cellulosic membranes, such as cellulose nitrate (CN), acetate (CA), or regencrated cellulose (RC), are casted as a cellulose/water/solvent mixture onto a belt and transported through heated tunnels. The resulting evaporation process produces the porous structure of the membrane. The casting lines require space, especially in length due to the slow evaporation process. Belt speed, temperature, airflow, and humidity conditions are the main parameters to adjust the pore size and structure.

In both casting procedures, the complexity of the individual parameters has to be well adjusted and monitored. Stringent process control is essential to achieve consistent membrane qualities and favored results. Any unaccounted change within one of the manifold parameters will result in a different membrane structure and performance.

Other techniques of membrane formation include stretching the polymeric film, commonly polytetrafluoroethylene (PTFE), while it is still in a flexible state and then annealing the membrane to "lock in" and strengthen the pores in the stretched membrane. The stretching process results in a distinctive membrane structure of PTFE nodes, which are interconnected by fibrils. PTFE membranes are highly hydrophobic and therefore are used as air filters. Air filters have to be highly hydrophobic to avoid water blockage due to moisture or condensate, especially after steam sterilization of these filters. Water blockage could be detrimental if the filters are, for example, used in a tank venting application to overcome condensation vacuum of a non–vacuum-resistant tank. If the filter does not allow a free flow of air into the tank, it may implode. Therefore, vent filters for this application have to be chosen and sized with care. PTFE membranes are also highly mechanical and thermal resistant, which is required because such filters are used over several months, withstanding multiple steam sterilization cycles. This is particularly true in large-scale fermentation, where these filters are used over several months, avoiding unwanted infections of the fermenter's or bioreactor's cell line.

Finally, track-etched MF membranes are made from polymers such as polycarbonate and polyester, wherein electrons are bombarded onto the polymeric surface. This bombardment results in "sensitized tracks" where chemical bonds in the polymeric backbone are broken. Subsequently, the irradiated film is placed in an etching bath (such as a basic solution), in which the damaged polymer in the tracks is preferentially etched from the film, thereby forming cylindrical pores. The residence time in the irradiator determines pore density and residence time in the etching bath determines pore size. Membranes made by this process generally have cylindrical pores

TABLE 2 Membrane Polymers Available and Their Advantages and Disadvantages

Membrane material	Advantage	Disadvantage
Cellulose acetate	Very low nonspecific adsorption (nonfouling)	Limited pH compatibility
	High flow rates and total throughputs	Not dry-autoclavable
	Low environmental impact after disposal	
Cellulose nitrate (nitrocellulose)	Good flow rate and total throughputs	High nonspecific adsorption
	Capture of smaller particles than the pore size	Limited pH compatibility
		Not dry-autoclavable
Regenerated cellulose	Very low nonspecific adsorption (nonfouling)	Limited pH compatibility
		Not dry-autoclavable
	Very high flow rates and total throughputs	
Modified regenerated cellulose	Very low nonspecific adsorption (nonfouling)	Ultrafilters not dry-autoclavable
	Moderate flow rates and total throughputs, especially with difficult to filter solutions	
	Broad pH compatibility	
	Easily cleanable (required in cross-flow applications)	
Polyamide	Good solvent compatibility	High nonspecific protein adsorption. Low hot-water resistance. Moderate flow rate and total throughput. Vacuole formation during casting can result in exaggerated pore sizes
	Good mechanical strength	
	Broad pH compatibility	
	Dry autoclavable	
Polycarbonate	Good chemical compatibility	Moderate flow rates
		Low total throughputs
		Difficult to produce
Polyethersulfon	High flow rates, total throughputs	Low to moderate unspecific adsorption depending on surface modifications
	Broad pH compatibility	
	Highest versatility	
	Mainly found as asymmetric membrane structure	Limited solvent compatibility

TABLE 2 Continued

Membrane material	Advantage	Disadvantage
Polypropylene	Excellent chemical resistance High mechanical resistance	Hydrophobic material High nonspecific adsorption due to hydrophobic interactions
Polysulfone	High flow rates and total throughputs Broad pH compatibility	Moderate to high nonspecific adsorption Limited solvent compatibility
Polytetra-fluorethylene	Excellent chemical resistance High mechanical resistance High hydrophobicity (used for air filtration)	Hydrophobic material High nonspecific adsorption due to hydrophobic interactions High-cost filter material
Polyvinylidene-difluoride	Low nonspecific adsorption Dry-autoclavable Good solvent compatibility	Moderate flow rate and total throughput. Hydrophobic base, made hydrophilic by chemical surface treatment; may lose hydrophilic modification due to chemical attack. High-cost filter material

with very narrow pore-size distribution, albeit with low overall porosity. Furthermore, there always is the risk of a double hit (i.e., the etched pore becomes wider and could result in particulate penetration). Such a filter membrane is often used in the electronic industry to filter high-purity water due to enhanced flushability.

11.4.2. Polymer Differences

There are distinct differences between the individual membrane and prefilter polymers. Table 2 lists the different membrane polymers available and the advantages and disadvantages, which depend on the properties of the polymer. The table shows that there is no such thing as a membrane polymer for every application. Therefore, filter membranes and the filter performance have to be tested before choosing the appropriate filter element.

11.5. FILTER CONSTRUCTIONS AND DESIGN

11.5.1. Disc Filters

Disc filters were the first filter configuration used in the biopharmaceutical industry, mainly as 293 mm discs within large stainless steel holding devices. These "process" filtration devices were replaced by pleated filter cartridge formats. Disc filters are stamped from the casted membrane sheet and are available in a large variety of sizes, either built into a disposable plastic housing or placed into a filter holder. Common diameter sizes to be placed in filter holders are 25, 47, 50, 90, 142, and 293 mm. Different sizes are used for different type of applications. The most common, 47 and 50 mm, are commonly utilized as microbial (analytical) assessment filters (Fig. 7).

11.5.2. Cartridge Filters

The primary motivation to develop pleated membrane cartridges was the need for an increase in the filter area sufficient to secure the engineering advantages of lower applied differential pressures and larger volume flows (particularly advantageous with more viscous liquids). Achieving this goal in the pleated filter cartridge form meant, moreover, that less plant space needed to be

FIGURE 7 Different disc filter types. (Courtesy of Sartorius Group.)

allocated for filter installations. As described above, 293 mm discs utilized before pleated filter cartridges required large floor space due to the low effective filtration of 0.5 ft^2 (0.05 m^2). To replace a common 10" filter cartridge and to achieve its same effective filtration area, fifteen 293 mm discs would be needed. Therefore, the footprint of such system is by far larger than the need for a 10" filter housing.

Now available are pleated filters composed variously of cellulose acetates, Teflon, polyvinylidene fluoride, polysulfone, polyethersulfon, nylon, and so forth. The pleating arrangement, the back-and-forth folding of the flat membrane filter upon itself, permits the presentation of a large filter surface area within a small volume. A pleated membrane cartridge of some 2.75" (70 mm) plus in diameter and 10" (254 mm) in length can contain from 5 to 8 ft^2 (0.5 to 0.8 m^2) of filter surface, depending on the membrane thickness, prefiltration layers, and construction detail. Pleated membrane cartridges are also offered in various lengths from 2" to 40" effective filtration areas from 0.015 m^2 to 1.8 m^2 (Fig. 8). This range of sizes and effective filtration areas are required for scale-up and down within the process and development steps. A pleated filter

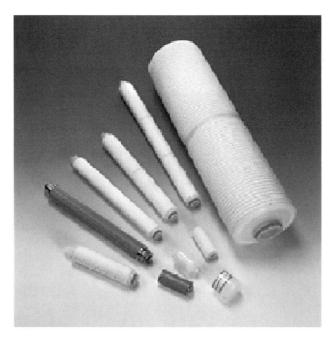

FIGURE 8 Different filter cartridge structures and types. (Courtesy of Sartorius Group.)

device should be able to scale-up linear from the preclinical volume size to process scale.

Typical construction components of the pleated filter cartridge are as follows:

End caps are the terminals for the cartridge and the pleat pack and are responsible for holding the cartridge contents together. The end caps are also responsible for providing the seal between the cartridge and the O-ring nipple adapter on the cartridge-housing outlet plate. Polypropylene end caps are frequently adhered to the membrane pleat pack by the use of a polypropylene melt, softened preferably by fusion welding. In some instances, polypropylene end capping can cause hydrophobic areas on the pleat pack—for example, with nylon membranes. Therefore, polyester end caps and melt are used, which is not completely problem-free, due to the lower chemical and thermal compatibility of the polyester. It has been reported that the polyester material became so brittle that one could rub it to dust. Therefore, such filter cartridges should be inspected on a regular basis if used in applications with multiple uses. In the past, polyurethane adhesives were also used in end cap materials. In conjunction with polyurethane sealant, the use of polypropylene end caps has sometimes resulted in the falling off of end caps; therefore, fusion welding is the most common bondage of end caps. Besides using similar components, the process gives a low extractable level. Polysulfone end caps are also used when required, as an inert polymeric material that can be adhered dependably to the pleat pack/outer support cage without creating hydrophobic spotting problems.

A stainless steel ring stabilizes the cartridge orifice against steam-induced dimensional changes and so preserves the integrity of the O-ring seal against bypass. Use of such dimension-stabilizing rings is made in the construction of pharmaceutical-grade cartridges intended for sterilization(s), especially when polypropylene end caps are involved. Nevertheless, it has been found that such stainless steel rings, with different expansion rates during temperature changes, can also cause problems in respect to hairline cracks and fissures within the adapter polymer or the welding sites. This could go so far that the adapter damage no longer allows proper O-ring sealing (Fig. 9). This effect often has been seen with an adapter that has not been molded from one piece. The welding starts cracking, liquid penetrates into the stainless steel ring cavity and expands during the next steaming. To avoid the differences in expansion of the support ring and the adapter polymer, most adapters are constructed with a polymer support ring.

The outer support cage is responsible for forming the outer cylinder of the cartridge and for holding the pleated internal contents together. The outer support cage also provides for a back-pressure guard in preventing loss of filter medium integrity as a result of fluid flowing in the opposite direction

FIGURE 9   Filter cartridge code 7 adapter damage. (Courtesy of Sartorius Group.)

under excessive back-pressure. Additionally, it eases the handling of the filter cartridge during installation. The user does not come in direct contact with the pleats, and damage can be avoided.

The outer filter pleated support layer serves as a multipurpose constituent. Pleating, and the assembly of the membrane into cartridge form, requires its inclusion in the cartridge. The supportive outer pleated layer aids in protecting the filter medium throughout the cartridge pleating and assembly operation. The material also serves as a prefilter to extend the useful service life of the final membrane that lies beneath it. Lastly, the support maintains the structure throughout fluid processing. Without this layer, the pleats under pressure might be compressed, limiting the filter area available to the fluid processing.

The drainage or downstream screen, similar to the outer filter pleat support, stabilizes the pleating of the pleat pack. Additionally, it keeps the filter medium pleats separated during fluid processing to assure that maximum filtration area is open for optimum flow rates and drainage of remaining filtrate—that is, reducing the dead volume or otherwise trapped fluids. The filter arrangement of the microporous membrane sandwiched between the support and drainage layers, all simultaneously pleated, is often called the filter pack or the pleat pack.

As the sealing between the pleat pack, drainage fleeces, inner core, and outer cage and the end caps, low-melting polypropylene sealants are widely used. Early work on nylon cartridges attributed to this material the generation (possibly through wicking) of hydrophobic spots that frustrated attempts to bubble-point the sealed cartridge. At least one company therefore adopted polyester sealants. A less general heating of a more restricted area seems to avoid the (wicking) problem. Use of a low-melting sealant may involve some one half inch of the pleat pack at each end of the filter assembly. A newer sealing technique utilizing polyolefin end caps relies on fusion welding of the cap to approximately one eight inch of each of the pleat pack. Valuable effective filtration area is retained thereby. The tendency in cartridge sealing is to utilize as few different materials as possible. Polytetrafluoroethylene (PTFE) or polyvinylidene fluoride (PVDF) microporous membranes are applied for their hydrophobicity (vent and air filters), or for their resistance to aggressive reagents such as certain solvents and oxidizers, or hot acids (semiconductor etchants). Thermoplastic fluorinated polymers, preferably as fluorinated as possible, are used for the cartridge components and in its sealed construction. The melts supported are then usually made of a porous Teflon material or of PVDF, as is also the remainder of the cartridge hardware from the like polymer in its solid, impervious form.

The filter cartridge inner core serves as the inner hollow tube on which the pleated pack is supported. It confers strength upon the cartridge assembly. This component also determines the final assembly length of the cartridge. Finally, the core is the outlet port of the cartridge. Through its perforations, the filtered fluid passes to be guided to the outlet plate of the filter housing. The cartridge core should not be flow limiting but can be in high-flow applications i.e., air filtration or water filtration with prefilter cartridges. It can be seen that the flow rate will not drastically increase by using a 30" filter size instead of 20" filter (Fig. 10). The only benefit here is a higher service life, but not an increase in flow. For this reason, air filtration systems are commonly sized with 20" filter cartridges.

The filter membrane is the heart of the filter cartridge, responsible for removal of the contaminants. Solutions permeate into and through the filter medium and into the cartridge core, then proceed through the outlet assembly and effluent piping. Once the filter medium has become fully wetted, processing can be continued until one of several flow decay indicators signals the need for cartridge replacement, as customer preference dictates.

Cartridge designs can be manifold and fit for the application (27). Not only size differences are applicable, but also cartridge adapters (i.e., plug-ins), which fit into filter housing sockets and recesses. A single cartridge with an

Airflow Rate Comparison
10", 20" and 30"

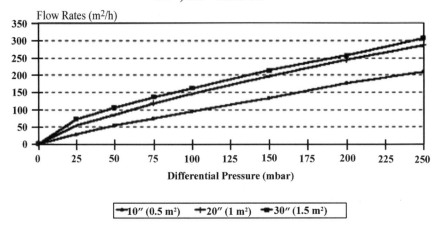

Atmospheric pressure conditions .02μm double bajonett adapter

FIGURE 10 Flow rate curves of 10", 20", and 30" filter cartridges sizes. (Courtesy of Sartorius Group.)

end plug is used as a 10" filter. Otherwise it can be joined by adapters to as many 10" double open-end cartridges as are necessary to form the ultimate unit length desired. The filter user need stock only three items, namely, the double open-end cartridges, the adapters, and end plugs. Single open-ended filter cartridges with bayonet locking are mainly used for sterilizing-grade filter cartridges due to the reliability of the fit into the housing (Fig. 11). Bypass situations have to be avoided, which can only be accomplished if the sealing between the filter cartridge and its holder is snug. In the case of string-wound cartridges, no end caps are used, because the avoidance of product bypass is not critical; only the double open-end cartridges and the adapters need be stocked.

In microporous membrane applications, frequent use is made of the single open-end 10" cartridge, usually in T-type housings. Therefore, such a unit is manufactured with an integral end cap. Such cartridges are also constructed in 20" and 30" lengths. Attempts have been made to offer pharmaceutical manufacturers the versatility of 10" single and double open-end units to be assembled via adapters with O-rings. Because such an arrangement increases the critical sealing area, its acceptance has been

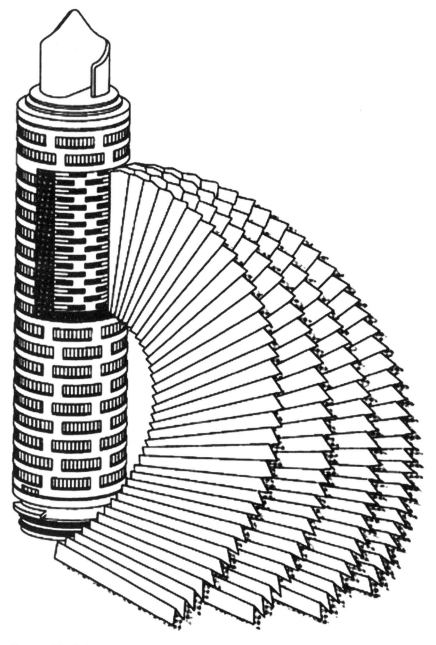

FIGURE 11   Schematic of filter cartridge. (Courtesy of Sartorius Group.)

limited. The more widespread use in critical pharmaceutical manufacture is that of single open-end 10", 20", and 30" cartridges.

The dimensions of the membrane cartridges are derived from those of the string-wound filters, roughly 10 × 2.5". Increasing the diameters of these cartridges serves to increase their effective filtration area (per unit number of pleats). Most manufacturers supply cartridges with a 2.75" (70 mm) diameter. Diameters as well as adapter types are commonly standardized or similar, which creates the opportunity for the filter user to choose. Additional capital investments into different filter housings are not necessary due to the common adapter types utilized. The resulting increase in the effective filtration area reflects two factors in addition to the cartridge diameter. The first consideration is the diameter of the center core of the cartridge. Each pleat consists of a membrane layer or of multiple membrane layers, sandwiched between two protective layers whose presence is necessary to avoid damage to the membrane in the pleating process, and which serve usefully in the finished cartridge as pleat separation and drainage layers. As a consequence of this sandwich construction, each pleat, naturally, has a certain thickness. Fewer of these thicknesses can be arranged around a center core of narrower diameter. Therefore, increasing the diameter of the center core increases the extent of its perimeter and the number of pleats that can surround it. This governs the number of pleats possible in the pleat pack that can comprise the membrane cartridge, thus increasing its effective filtration area.

One other consideration favors the use of center cores with larger diameters. Particularly in longer cartridges used under elevated applied differential pressures, the liquid flow through the microporous membrane may be so great as to encounter restrictions to its passage through long center cores of narrower diameters. Thus, in pleated cartridge constructions intended for the high water flows of the nuclear power industry, the outer cartridge diameter may be 12" to accommodate a maximum number of high pleats or greater arranged around a center core dimensioned at a 10" diameter. The concern, exclusive of pleat heights, is to increase the service life—the throughput of the filter—by increasing its effective filtration area. In this application, high flow rates are accommodated within the 10" core diameter. Such restrictions to flow within cartridge center cores are generally not the concern in critical pharmaceutical filtrations, where the applied pressure differentials are restrained in the interest of filter efficiency and longevity to yield.

To define a cartridge, therefore, designations must be made of such considerations as its pore-size designation, its diameter, its length, the type of outlet (e.g., the O-ring[s] sizes), the configuration of the outer end (e.g., open or closed, with or without fin), the type of O-ring or gasket seal (e.g., silicone rubber, EPDM rubber), and any nonstandard features. Manufacturer product numbers serve as shorthand substitutes for the detailed specifications.

The second factor governing the effective filtration area of a cartridge, in addition to its overall diameter and center core diameter, is the pleat height. Obviously, for any given pleat, the greater its height, the longer its surface area. Present pleating machines cannot fashion pleat heights beyond one inch or so. The designing of a cartridge usually begins with a defining of its overall outside diameter. Given a maximum pleat height of 1", the maximum size of the center core becomes determined. But if the pleat height is diminished in order for the center core diameter to be increased, the greater overall number of pleats that can be arranged around the wider core may more than compensate in effective filtration area for that lost through pleat height diminution.

The optimum number of pleats to be arranged about a center core of a filter cartridge may reflect the filtrative function for which it is intended. In the handling of rather clean, prefiltered liquids, as in most pharmaceutical final filtrations, relatively few particles require removal. A crowding of as large a number of pleats as possible in order to enhance the filter area may be acceptable because the pleat separation layers will operate to make even the crowded surfaces individually available to the liquid being filtered. Where there are high solid loadings in the liquid, or a viscous fluid, a different situation may result. The particles being removed may be large enough to bridge across a pleat, to block the interval between two adjacent pleat peaks. Or, being small, they may, after their individual deposition on the filter, secrete and grow large enough to cause bridging. Whatever the mechanism, the bridging serves to deny the liquid being processed access to useful flow channels bordered by membrane. In air filtration it is important that the pleat density is not too high: the pleating is not too tight together to avoid capillary actions and therefore water logging. The pleat density has to be optimal or a looser pleat pack, one containing fewer pleats, would more closely optimize the pleating arrangement. Decreasing the diameter of the center core will serve to lessen the number of pleats. In practice, pleated cartridges are built for general usage in what is still an artful construction. Nevertheless, it is said there is available an empirically developed formula that relates the outer cartridge diameter to the maximum core diameter, and to the number of pleats of given height that should be used.

Care must be taken to protect the surface of the membrane during the pleating operation and to avoid damage to the filter structure. Both these objectives are furthered by sandwiching the membrane between two support layers and feeding the combination to the pleater. The outlying support layers protect the membrane surfaces. Nevertheless, the fleeces have to be chosen properly; for example, a fleece too coarse could press too much on the membrane, at the pleating curvature, and start pressing into the membrane. In Fig. 12, one can see the result of coarse fleece compression on a PTFE membrane. On the other hand, a fleece that is too soft will not support the membrane sufficiently. Usually, soft fleeces have a high fiber density and a small fiber diameter, which means liquid would be bound within the fiber structure. Such a phenomenon needs to be avoided, for example, in air filtration, because it could cause water logging.

Additionally, the sandwich in its thickness minimizes opportunities for the membrane to be too strongly compressed at the pleat. What is required is a pleat having some radius of curvature rather than a sharp, acute angle of fold. This prevents the membrane from being subjected, at the pleat line, to forces in excess of its mechanical properties as expressed in the magnitude of its tensile and elongation values. Different polymeric materials will, of course, have different tensile and elongation qualities, as various materials differ in their brittleness.

11.5.3. Capsule Filters

The disc and cartridge filters available commercially are usually disposables. It is their housings and holders, usually of metal, that are permanent. However, filters encapsulated into plastic housings have been devised wherein the entire unit is disposable (Fig. 13).

There are advantages to these devices. Among them is that many are available in presterilized conditions, by gamma radiation, steam, or ethylene oxide. Another advantage, therefore, is their ready availability. They are in a

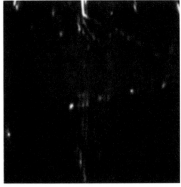

FIGURE 12 Prefilter fleece impression on a PTFE membrane. (Courtesy of Sartorius Group.)

FIGURE 13 Different types and styles of disposable capsule filters. (Courtesy of Sartorius Group.)

standby condition on the shelf, available when needed. That they are disposables does not necessarily argue against the economics of their usage. Calculations show that where labor costs are calculated, the installation of a single 293 mm filter disk in its housing is more costly than the equivalent filtration area in the form of a disposable filter device. The use of the disposables entails very little setup time, and no cleanup time. There is no need to sterilize the already presterilized units. Disposal after the single usage eliminates risks of cross-contamination.

One small volume parenteral (SVP) manufacturer adopted the use of disposable filter devices embodying flat disc filter design of essentially the same effective filtration area as a 293 mm disc to replace the latter. The cost savings, reckoned largely as labor, was considered significant. In making the substitution, there were such factors as flow rate versus differential pressure, throughput, rinse volume and time effect wetting and extractable removal, ability to be heat sterilized, confirmation by vendor of product nontoxicity, and freedom from pyrogenic substances. Another SVP manufacturer opted for the same type of replacements, selecting, however, the required effective filtration area in pleated filter capsule form. In both cases, the disposable device was equipped with sanitary connections, enabling a straightforward substitution. Pleated disposable devices commonly show better performance due to the prefilter fleeces and sometimes prefilter membrane in front of the final filter membrane. Therefore, 293 mm disc filters could potentially also be replaced by 150 or 300 cm^2 disposable devices, even when such have a smaller effective filtration area.

In one application involving the filtration of serum through a 0.1 μm–rated membrane, a pleated filter capsule replaced a 293 mm disk because a steam-autoclaved disk holder assembly required a much longer period to cool down to use-temperature than did the plastic-housed disposable filter. The savings in time was judged substantial enough to merit being addressed.

The venting of disposable filter devices has been the recipient of good design considerations. One disposable-capsule manufacturer has taken care to so position the vents that they are on the highest point of the containing shell, exactly where they are most effective. Another design utilizes a self-venting device in the form of a hydrophobic membrane. This permits the self-venting of air while safeguarding against the passage of liquid or contaminants (in either direction). This is particularly useful in water installations, where intermittent use serves repeatedly to introduce air to the system. The self-venting feature reduces maintenance and increases the system efficiency.

There are often ancillary advantages to the use of disposable filter devices. Some manufacturers construct their shells of transparent polymers so that the filtration process is observable. The instruments are compact and relatively lightweight, hence easy to handle. Nor does their construction lack the sophistication of their metal housing-contained counterparts. Thus, many of the disposable units are equipped with vent plugs and drain plugs. The identifying description they bear on their outer casings make their traceability, in accordance with FDA record requirements, rather certain. Product and batch numbers become part of the permanent operational record. Above all, the use of these disposables obviates the need to expense or amortize stainless steel filter holders. No capital expenditures are involved.

Furthermore, the use of disposable filters can reduce costs in terms of cleaning, which would occur with stainless steel filter housings after every use. Cleaning validation, which needs to be performed with fixed equipment like filter housings, will be greatly reduced. The disposable filters do not go through such a cleaning regimen and therefore the validation of cleaning exercises is avoided. For this reason, and the convenience of the use of disposable filters, the biopharmaceutical industry switches more and more to capsule filters instead of filter housings. The fact that the use of disposable equipment is becoming more common can also be seen in the fact that bags replace glass or stainless steel holding and storage vessels. Commonly, a disposable capsule filter is connected to such a bag; both are available in different sizes for the individual purposes. Once the capsule filter is connected, the bag and filter are gamma irradiated to sterilize the entire setup. Certainly the filter material and polymers need to be gamma stable, otherwise particle shedding or an excess of extractable can occur.

Another advantage is the fact that the user will not encounter the product filtered. This certainly could be the case when using cartridge filters within a housing. The cartridge has to be removed from the housing at the end of the filtration run—the user probably comes in contact with the filtered product remaining on the filter cartridge and housing, which may need to be avoided due to health hazards or biological activity. Disposable filters create the opportunity to replace a filter without being in contact with the product.

The disposable filter devices are available in a large variety of constructions, whether disc, multidisc, pleated cylinders of various lengths and of different effective filtration areas. The expanse of filter surface runs from 4-mm discs suitable for affixing to hypodermic needles to 30" capsules of about 180 ft^2 (1.8 m^2) (Fig. 14). The filters are made of a variety of polymeric filter materials, both hydrophilic and hydrophobic—namely, cellulose esters, polyvinylidene fluoride, polysulfone polyethersulfone, nylon, polyethylene, Teflon and so forth. Their shells are composed variously of polycarbonate, polyethylene, and—most often—polypropylene.

The versatility of these disposable filter instruments is increased by constructions involving integral prefilters, as in one capsule unit having approximately the effective filtration area of a 293 mm disc. This is appropriate, as single disc filtrations most often involve applications that require the use of a prefilter. Repetitive final filter constructions are also available in disposable unit form. These are used, for instance, in tissue culture medium filtrations where repetitive final filter arrangements are common.

The increase in the tailoring of disposable filter device constructions to specific application needs helps explain the mounting popularity of their use

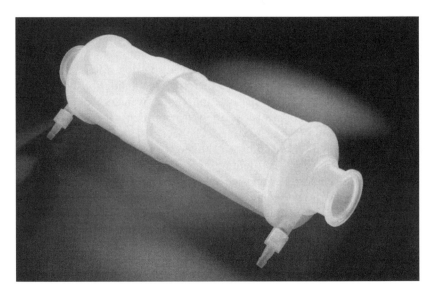

FIGURE 14 Large scale disposable capsule filters. (Courtesy of Sartorius Group.)

and heightens predictions of their continuing replacement of at least part of the more conventional filter/holder market.

The use of most cartridge filters facilitates compliance with FDA emphasis on record keeping. Despite all the care with which filter manufacturers pack flat disc filters, the membranes themselves are unlabeled. Cartridge filters are, however, available with identifying data. Most are identified with some code: if not on the cartridge itself, then on its container. Some manufacturers stamp the cartridge end cap with the part number, its pore size identity, and its lot number as well. Indeed, some manufacturers even number each cartridge consecutively within each lot. Should the need ever arise to trace the components and history of these filters, and of their components, the ability to do so exists. Batch records in concert with the appropriate manufacturing QC records make this possible.

Because of the fragility of most membrane filters, appropriate and even extreme care is to be used in their handling. In the case of cartridge filters, this practice continues. However, the actual membrane surface of these instruments is out of reach during ordinary handling. There is, therefore, far less possibility of damage to the filters. Overall, cartridges are used mostly for the more rapid flow rates and/or the large-volume filtration productions they enable, a consequence of their aggrandized effective filtration areas. Cartridges are increasingly constructed so that their in situ sterilization can be effected by the convenient use of the steam-in-place technique.

11.6. FILTER VALIDATION

11.6.1. Guidelines and Documents

Probably the most useful document is PDA Technical Report No. 26 (7). It thoroughly describes filter structures, usage, purpose, and integrity testing. Most important, though, is the description of the filter validation needs within the actual filtration process. The document defines the needs for viability; product bacteria challenge; extractable, particulate, and adsorption testing. It is not meant as guidance, certainly not as a compliance document, but one can rest assured that regulatory authorities also utilize the report.

Before the PDA Technical Report was written, the FDA Guideline on Sterile Drug Products Produced by Aseptic Processing (2) had been the guidance document of choice. This document, from 1987, is outdated and there is a need for a new guideline. In September 2002, the FDA published a concept paper, probably the first draft for a new aseptic guideline. This concept paper addresses new requirements as listed in Technical Report 26. Being a draft or concept paper, it probably will still take a considerable amount of review and one can only speculate when the final version will become effective.

Similarly, ISO 13408 is in a draft format. This guideline leans very much toward Technical Report 26 and describes appropriate filter validation very much in the fashion of the mentioned report. Again, it has yet to become effective, but utilizing the PDA report will avoid any filter validation surprises.

The USP (United States Pharmacopeia) 25 as well as any other pharmacopeia are closely monitored, due to the descriptions of required limits for particulate, endotoxins, and biocompatibility testing. Within the filter manufacturers filter qualification tests, pharmacopeial limits are analyzed and must be met by the filter products distributed. These tests commonly cover toxicological, endotoxins, extractable, and particulate tests, which are well defined with the pharmacopeias and any filter utilized within the biopharmaceutical industry must comply. These tests are the basic requirements to be fulfilled and should not be misinterpreted as appropriate filter validation studies. Filter validation must be performed with the actual drug product to be filtered under process conditions. Most of the pharmacopeial tests are performed with water or other pure solvents.

A guideline of considerable importance, especially in regard to revalidation or second filter vendor implementation, is the FDA Guidance for the Industry—Changes to an Approved NDA or ANDA, section VII, Manufacturing Process (28). This guideline describes distinctively the different needs of prior approvals, if changes have been made to the actual processes. It defines what is a minor, moderate, or major change in respect to filtration devices and changes to sterilizing grade filters and what are the consequences.

Minor change (Annual Report): filtration not mentioned.

Moderate change (Supplement—Changes Being Effected): CBE 30:

> Changes to filtration parameters for aseptic processing (including flow rate, pressure, time, or volume, but not filter materials or pore size rating) that require additional validation studies for the new parameters
>
> Filtration process changes that provide for a change from a single to a dual product sterilizing filters in series, or for repeated filtration of bulk
>
> For sterile drug products, elimination of in-process filtration performed as part of the manufacture of a terminally sterilized product

Major change (Prior Approval Supplement):

> Changes in the sterilization method (e.g., gas, dry heat, irradiation). These include changes from sterile filtered or aseptic processing to terminal sterilization, or vise versa.
>
> Addition, deletion, or substitution of sterilization steps or procedures for handling sterile materials in an aseptic processing operation.
>
> Changes in materials or pore size rating of filters used in aseptic processing.
>
> Changes in the virus or adventitious agent removal or inactivation methods. This is applicable to any material where such procedures are necessary, including drug substance, drug product, reagents, and excipients.
>
> Filtration to centrifugation or vice versa.

A guideline that causes confusion and insecurities in respect to redundant 0.2 μm filtration is the EMEA CPMP/QWP/486/95 Guideline (29). This guidance document defines a maximum allowable bioburden level of 10 cfu/100 mL in front of a 0.2 μm sterilizing-grade filter. If this level is exceeded, a bioburden-reducing filter has to be used in front of the sterilizing-grade filter. Although the guidance leaves room for interpretation with respect to what type of filter this could be, it also states that the use of a second 0.2 μm in front of the final 0.2 μm filter does not require additional validation. It is now debatable whether the bioburden limit defined is reasonable, as well as the excessive reliance on pore size.

11.6.2. Bacteria Challenge Tests

Before performing a product bacteria challenge test, it has to be assured that the liquid product does not have any detrimental, bactericidal, or bacterio-

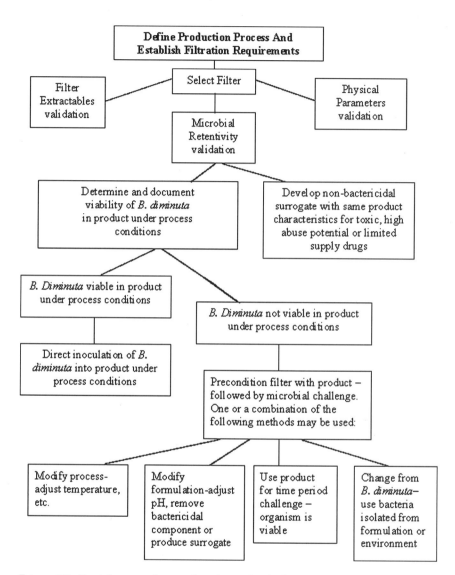

FIGURE 15 Decision tree for product bacteria challenge testing. (Reprinted, by permission, from Technical Report No. 26, Sterilizing Filtration of Liquids, © 1998 by PDA.)

static effects on the challenge organisms, commonly *Brevundimonas diminuta*. This is done utilizing viability tests. The organism is inoculated into the product to be filtered at a certain bioburden level. At specified times, defined by the actual filtration process, the log value of this bioburden is tested. If the bioburden is reduced due to the fluid properties different bacteria challenge test modes become applicable (Fig. 15). There are three bacteria challenge methodologies described within the PDA Technical Report No. 26 (7); high organisms challenge, placebo (modified product) challenge, and product recirculation with a challenge after recirculation. If the mortality rate is low, the challenge test will be performed with a higher bioburden, bearing in mind that the challenge level has to reach 10^7 per square centimeter at the end of the processing time. If the mortality rate is too high (common definition, >1 log during processing time), the toxic substance is either removed or product properties—for example pH, temperature—are modified. This challenge fluid is called a placebo. The third methodology would be to circulate the fluid product through the filter at the specific process parameters for as long as the actual processing time would be. Afterward the filter is flushed extensively with water and the challenge test, as described in ASTM F838-38 (3), performed. Nevertheless, such challenge test procedure would be more or less a filter compatibility test.

Sterilizing-grade filters are determined by the bacteria challenge tests. This test is performed under strict parameters and a defined solution [ASTM F 838-83 (3)]. In any case, the FDA requires evidence that the sterilizing-grade filter will create a sterile filtration, no matter the process parameters, fluid properties, or bioburden found (6,7). This means that bacteria challenge tests have to be performed with the actual drug product, bioburden, if different or known to be smaller than *Brevundimonas diminuta* and the process parameters. The reason for the requirement of a product bacteria challenge test is threefold. First of all, the influence of the product and process parameters to the microorganism has to be tested. There may be cases of either shrinkage of organisms due to a higher osmolarity of the product or prolonged processing times or starvation due to the extreme low organic properties of the fluid. Second, the filter compatibility with the product and the process parameters has to be tested. The filter should not show any sign of degradation due to the product filtered. Additionally, assurance is required that the filter used will withstand the process parameters—for example, pressure pulses should not influence the filter performance. Third, there are two separation mechanisms involved in liquid filtration; sieve retention and retention by adsorptive sequestration (1,4,5,7,11–26). In sieve retention the smallest particle or organism size is retained by the biggest pore within the membrane structure. The contaminant will be retained, no matter the process parameters. This is the ideal. Retention by adsorptive sequestration depends on the filtration conditions. Contaminants smaller than the actual pore size penetrate such and may be captured by adsorptive attachment to the pore wall. This effect is enhanced using highly adsorptive filter materials—for example, glass fiber as a prefilter or polyamide as a membrane. Nevertheless certain liquid properties can minimize the adsorptive effect, which could mean penetration of organisms. Whether the fluid has such properties, will lower the effect of adsorptive sequestration, and may eventually cause penetration has to be evaluated in specific product bacteria challenge tests.

11.6.3. Extractable Test

Besides the product bacteria challenge test, tests of extractable or leachable substances have to be performed (7,30,31). Previous reliance on nonvolatile residue (NVR) testing as a method of investigating extractable levels was dismissed by the regulators in 1994 (32). Since then, extractable/leachable analysis from filters and other components are routinely done by appropriate separation and detection methodologies. Extractable measurements and the resulting data are available from filter manufacturers for their individual filters (Figs. 16, 17).

These tests are performed with a specific solvent, in this case ethanol and water at "worst case" conditions. Such conditions do not represent true process realities. Therefore, depending on the process conditions and the

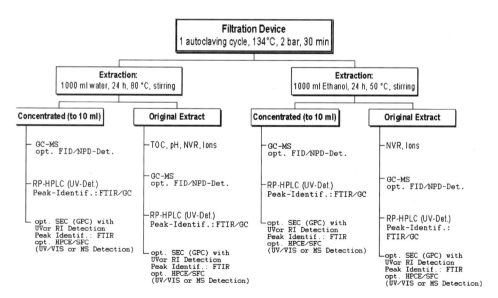

FIGURE 16 Extractable test schematic for ethanol and water. (Courtesy of Reif, 1996.)

Cartridge A PP-Fleece Prefilter	Cartridge B GF-Fleece Prefilter	Cartridge C Membrane Prefilter	Cartridge D* PP-Fleece Prefilter	Cartridge E* GF-Fleece Prefilter	Cartridge F* PP-Vleece Prefilter
Dibutylphthalate	Succinic acid	Methyl-2,4-pentadiol	Caprolactame	Caprolactame	Propionic acid
2,6-Di-tert-butylphenol	8 oligo. aliphates	Glycerol	Propionic acid	Dodecanol	Butyl-1-methoxybenzene
Hydroxybenzoic acid	2,4-Bis(1,1-dimethylethyl)phenol	2,4-Bis(1,1-dimethylethyl)phenol	4-(1-Methyl-1-phenylethyl)phenol	High MWN-cont. Compound	Naphthalenic compound
2,6-Di-tert-butylcresol	Ethoxyethylbenzoate	Butylphenoxyacetic acid	2,6-Di-tert-butylcresol	2,6-Di-tert-butylcresol	Polyether
Stearic acid	Diethylterephthalate	Ethoxybenzoic acid	Benzoic acid	High MW aromate	2,6-Di-tert-butylcresol
8 oligo. aliphates	Myristic acid	4 oligo. siloxanes	Dibutylphthalate	Cyclododecane	Dibutylphthalate
7 oligo. siloxanes	Palmitic acid	6 oligo. aliphates	2,2-Dimethoxy-1,2-diphenylethanone	Butyl-4-Methoxyphenole	High MWN-cont. Compound
Polyalkylic ether	Octadecene	Diethylphthalate	Butyl-methoxybenzene	3,3-Thiobispropionic acid	Hexadecene
	Stearic acid	Polyalkylic ether	5 oligo. siloxanes	7 oligo. aliphates	
	Polyalkylic ether		7 oligo. aliphates	6 oligo. siloxanes	
	2 oligo. siloxanes		Caprolactame derivate	Polyether	
	Glassfibers		Triphenylphosphino xid	Glassfibers	

*Identification of the RP-HPLC peaks by FTIR is still in progress – extractables list of marked cartridges may be incomplete

FIGURE 17 Extractable listing of different sterilizing grade filters. (Courtesy of Reif, 1996.)

solvents used, explicit extractable tests have to be performed. Formerly, these tests were done only with the solvent used with the drug product, but not with the drug ingredients themselves, because the drug product usually covers any extractable during measurement. Nevertheless, recent findings have been presented that report the need to evaluate extractable utilizing the actual drug product as the extraction medium. Such tests are conducted by the validation services of the filter manufacturers using sophisticated separation and detection methodologies, such as GC-MS, FTIR, RP-HPLC, UV-VIS, GPC-RI, HPCE, and SFC. These methodologies are required due to the fact that the individual components possibly released from the filter have to be identified and quantified. Elaborate studies on sterilizing-grade filters, performed by filter manufacturers, showed that there is neither a release of high quantities of extractable (the range is ppb to max. ppm per 10" element) nor have toxic substances been found (30).

Authorities and organizations have changed their focus to other equipment used within the industry—for example, disposable media bags, plastic vials, tubing, or stoppers (32). Prefilters have also become a target. There are already extractable studies performed on a variety of pleated prefilter types of polypropylene and glass-fiber. Nevertheless, lenticular and string wound prefilters, widely used within the biopharmaceutical industry, still have to undergo such investigation.

11.6.4. Chemical Compatibility Test

The PDA Technical Report No. 26 describes very specifically "A simple chemical compatibility chart will often not provide enough information for predicting filter system compatibility, thereby requiring additional testing." Chemical compatibility has been underestimated in the past and reliance has been focused on chemical chart of pure solutions. The aim of chemical compatibility testing has to be to find subtle incompatibilities, which may happen due to a mix of chemical components and entities or specific process conditions. Elevated temperatures or prolonged filtration times may result in a filter incompatibility, which has to be investigated (Fig. 18).

If the filter membrane is compromised in respect to its retentivity, it can add to any extractable/leachables problem. Therefore, appropriate compatibility tests have to be performed with the actual drug product at the process conditions. Commonly, integrity tests before and after the submersion of the filter in the product to be filtered will show whether or not an incompatibility exists. Sole reliance, though, should not be on integrity testing. NVR testing parallel to integrity testing may be the procedure of choice, in case the filter is integral but shows elevated extractable levels (Fig. 19). In such cases, scanning electron microscopy should be utilized to see any chemical attacks

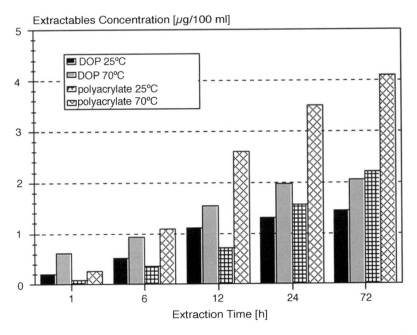

FIGURE 18 Example of leachable increases due to temperature increases. (Courtesy of Sartorius Group.)

on the membrane surface. The above-mentioned bacteria challenge tests and extractable analysis also contribute valuable information in respect to filter compatibility.

11.6.5. Other Tests

Particulates are critical in sterile filtration, specifically of injectables. The USP 25 (United States Pharmacopeia) and BP (British Pharmacopeia) quote specific limits of particulate level contaminations for defined particle sizes. These limits have to be kept and therefore the particulate release of sterilizing-

	Bubble Point [bar]	Burst Pressure [bar]	NVR [mg/l]	BC-Test Brev. Dim.
After extraction with RO-Water	3.6	0.42	11	sterile
After extraction with 0.1 M HCl	3.5	0.14	156	non sterile

FIGURE 19 Example of subtle incompatibilities of a filter membrane. (Courtesy of Reif, 1996.)

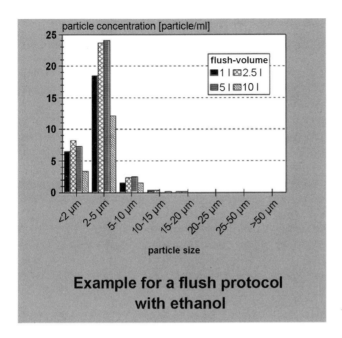

Example for a flush protocol
with ethanol

FIGURE 20   Example of a flush protocol of a filter cartridge. (Courtesy of Sartorius Group.)

grade filters has to meet these requirements. Filters are routinely tested, the filtrate evaluated with laser particle counters. Such tests are also performed with the actual product under process conditions to prove that the product, but especially process conditions, do not result in an increased level of particulates within the filtrate. Specific flushing protocol, if necessary, can be established for the filters used (Fig. 20). These tests are also useful for any prefilter as it reduces the possibility of a particulate contamination within the process.

Additionally, with certain products loss of yield or product ingredients due to adsorption shall be determined (34). Specific filter membranes can adsorb (for example) preservatives such as benzalkoniumchloride or chlorhexadine. Such membranes need to be saturated by the preservative to avoid preservative loss within the actual product. This preservative loss (e.g., in contact lens solutions) can be detrimental due to long-term use of such solutions. Similarly problematic would be the adsorption of required proteins within a biological solution. To optimize the yield of such proteins within an application, adsorption trials have to be performed to find the optimal membrane material and filter construction, but also flow conditions and prerinsing procedures. Any yield losses by unspecific adsorption can cost millions in respect to lost product and its market value. Adsorption studies are helpful to optimize downstream process in regard to any yield loss, and yield losses also correspond to capacity problems, which can be seen within the biotech industry. Yield losses have a detrimental influence, and most commonly such losses can also be targeted to nonspecific adsorption on the wrong choice of membrane polymers.

To summarize, most of the described validation effort have to be performed and are part of the validation master file of a particular process and drug product. Validation receives emphasis and attention, but one should not neglect training. Without appropriate personnel training, any validation effort is done in vain. Filter users should also test their staff's ability to handle filtration, the sterilization, and integrity test of such, installation, and sanitization. Training has to be the focus of all operations to really achieve a reliable and sustainable process.

11.7. INTEGRITY TESTING

11.7.1. Guidelines and Documents

Sterilizing grade filters require testing to assure the filters are integral and fulfill their purpose. Such filter tests are called integrity tests and are performed before and after the filtration process. Sterilizing-grade filtration would not be admitted to a process if the filter is not integrity tested in the course of the process. This fact is also established in several guidelines recommending the use of integrity testing, pre- and post-filtration. This is valid not only for liquid, but also air filters.

Examples of such guidelines are:

FDA "Guideline on Sterile Drug Products Produced by Aseptic Processing" (1987)

"Guide to Inspections of High Purity Water Systems," "Guide to Inspections of Lyophilization of Parenterals" and also in the CGMP document 212.721 Filters

ISO 13408 Draft, "Aseptic processing of health care products," 2002 (35)

USP (United States Pharmacopeia) 25 (2001)

Guide to Good Pharmaceutical Manufacturing Practice (Orange Guide, U.K., 1983)

PDA (Parenteral Drug Association), Technical Report No. 26, "Sterilizing Filtration of Liquids" (March 1998) (7)

Integrity tests, including the diffusive flow, pressure hold, bubble point, or water intrusion test, are nondestructive tests, which are correlated to the destructive bacteria challenge test with 10^7 per square centimeter *Brevundi-*

monas diminuta (2,7). Derived from these challenge tests specific integrity test limits are established, which are described and documented within the filter manufacturers' literature. The limits are water based: the integrity test correlations are performed using water as a wetting medium. If a different wetting fluid, filter, or membrane configuration is used, the integrity test limits may vary. Integrity test measurements depend on the surface area of the filter, the polymer of the membrane, the wetting fluid, the pore size of the membrane, and the gas used to perform the test. Wetting fluids may have different surface tensions, which can depress or elevate the Bubble Point pressure. The use of different test gases may elevate the diffusive gas flow. Therefore, appropriate filter validation has to be established to determine the appropriate integrity test limits for the individual process.

11.7.2. Bubble Point Test

Microporous membrane pores, when wetted out properly, fill the pores with wetting fluids by imbibing that fluid in accordance with the laws of capillary rise. The retained fluid can be forced from the filter pores by air pressure applied from the upstream side to the degree that the capillary action of that particular pore is overcome. During the bubble point test, the pressure is increased gradually, in increments. At a certain pressure level, liquid will be forced first from the set of largest pores, in keeping with the inverse relationship of the applied air pressure P and the diameter of the pore, d, described in the bubble point equation:

$$P = \frac{4\gamma \cos \theta}{d}$$

where γ is the surface tension of the fluid and θ is the wetting angle; P is the upstream pressure at which the largest pore will be freed of liquid; d is the diameter of the largest pore.

When the wetting fluid is expelled from the largest pore, a bulk gas flow will be evaluated on the downstream side of the filter system (Fig. 21). The bubble point measurement determines (to a certain degree) the pore size of the filter membrane (i.e., the larger the pore, the lower the bubble point pressure). Therefore, filter manufacturers specify the bubble point limits as the minimum allowable bubble point and correlate the bubble point test procedure to the bacteria challenge test (36). During an integrity test, the bubble point test has to exceed the set minimum bubble point.

The key for a successful bubble point test is the qualified wetting fluid and its surface tension. The bubble point will be highly influenced by surface tension changes within the wetting fluid. Table 3 shows different possible wetting fluids and the bubble point changes of such, utilizing the same membrane.

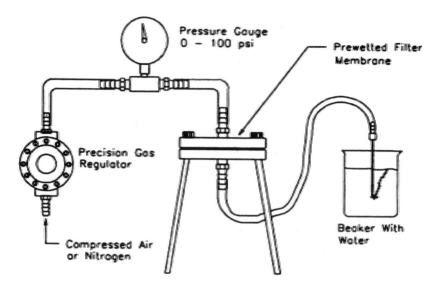

FIGURE 21 Manual bubble point test setup. (Reprinted, by permission, from Technical Report No. 26, Sterilizing Filtration of Liquids, © 1998 by PDA.)

However, the surface tension of the wetting liquid, as also its viscosity, diminishes with mounting temperature, while the angle of wetting increases, and its cosine decreases with the hydrophobicity of the filter polymer. In other words, the less hydrophilic the polymer, the less perfectly does it wet, particularly with aqueous liquids. Therefore, the bubble point is a specific product of each particular filter/liquid couple. It varies from one polymer to

TABLE 3 Bubble Point Values for Different Wetting Agents using Cellulose Acetate 0.2 μm

Product	Bubble point value
Water	3.20 bar
Mineral oil	1.24 bar
White petrolatum	1.45 bar
Vitamin B complex in oil	2.48 bar
Procainamide HCl	2.76 bar
Oxytetracyline in PEG base	1.72 bar
Vitamin in aqueous vehicle	2.07 bar
Vitamin in aqueous vehicle	2.69 bar

Source: Courtesy of Sartorius AG.

the other and therefore bubble point values given and obtained are not equal, even for the same pore size rating. That the bubble point of a filter differs for different wetting liquids is commonly known. That it differs also for polymeric materials is less appreciated.

The bubble point test can only be used to a certain filter size. The larger the filter suface, the larger the influence of the diffusive flow through the membrane. The diffusive flow would cover the actual bubble point due to the extensive air flow. Therefore the bubble point finds its ideal use with very small system to medium size systems (some mentioned the critical borderline to use the bubble point is a $3 \times 20''$ filter housing, depending on the pore size).

11.7.3. Diffusive Flow Test

A completely wetted filter membrane provides a liquid layer across which, when a differential pressure is applied, the diffusive airflow occurs in accordance with Fick's law of diffusion (Fig. 22). This pressure is called test pressure and commonly specified at 80% of the bubble point pressure. In an

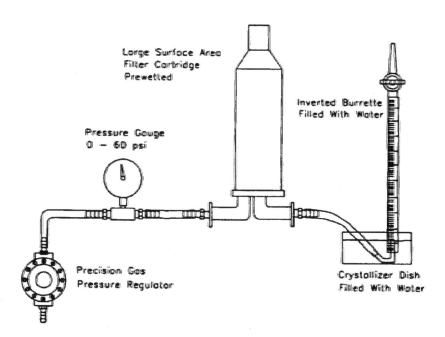

FIGURE 22 Manual diffusive flow test setup. (Reprinted, by permission, from Technical Report No. 26, Sterilizing Filtration of Liquids, © 1998 by PDA.)

experimental elucidation of the factors involved in the process, Reti (37) simplified the integrated form of Fick's law to read:

$$N = \frac{DH(p1 - p2)}{L} \cdot \rho$$

where N is the permeation rate (moles of gas per unit time), D is the diffusivity of the gas in the liquid, H is the solubility coefficient of the gas, L is the thickness of liquid in the membrane (equal to the membrane thickness if the membrane pores are completely filled with liquid), P (p1−p2) is the differential pressure, and ρ is the void volume of the membrane, its membrane porosity, commonly around 80%.

The size of pores only enters indirectly into the equation; in their combination they comprise L, the thickness of the liquid layer, the membrane being some 80% porous. The critical measurement of a flaw is the thickness of the liquid layer (38). Therefore, a flaw or an oversized pore would be measured by the thinning of the liquid layer due to the elevated test pressure on the upstream side. The pore or defect may not be large enough that the bubble point comes into effect, but the test pressure thins the liquid layer enough to result into an elevated gas flow. Therefore, filter manufacturers specify the diffusive flow integrity test limits as maximum allowable diffusion value. The larger the flaw or a combination of flaw, the higher the diffusive flow.

The diffusive flow cannot be used for small filter surface, due to the low diffusive flow with such surfaces. The test time would be far too extensive and the measured test value too unreliable to be utilized. Nevertheless, the diffusive flow, as well as the pressure drop test, are best used for larger filtration surfaces, where the bubble point test finds its limitations.

11.7.4. Pressure Hold Test

The pressure hold test is a variant of the diffusive airflow test (39). The test setup is arranged as in the diffusion test except that when the stipulated applied pressure is reached, the pressure source is valved off (Fig. 23). The decay of pressure within the holder is then observed as a function of time, by using a precision pressure gauge or pressure transducer.

The decrease in pressure can come from two sources: (a) the diffusive loss across the wetted filter (since the upstream side pressure in the holder is constant, it decreases progressively all the while diffusion takes place through the wetted membrane) and (b) source of pressure decay could be a leak of the filter system setup.

An important influence on the measurement of the pressure hold test is the upstream air volume within the filter system (Fig. 24). This volume has to be determined first to specify the maximum allowable pressure drop value.

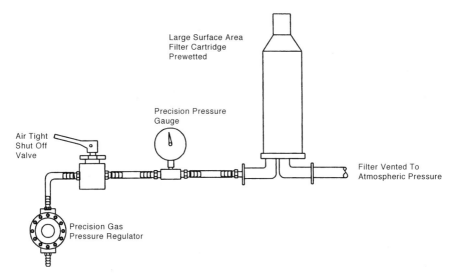

FIGURE 23 Manual pressure-hold test setup. (Reprinted, by permission, from Technical Report No. 26, Sterilizing Filtration of Liquids, © 1998 by PDA.)

The larger the upstream volume, the lower will the pressure drop be. The smaller the upstream volume, the larger the pressure drop. This means also an increase in sensitivity of the test, but also an increase of temperature influences, if changes occur. Filter manufacturers specify maximum allowable pressure drop values, utilizing their maximum allowable and correlated diffusive flow value and convert this diffusive flow maximum with the upstream volume into a maximum allowable pressure drop.

Another major influence, as mentioned, is the temperature. Any temperature change during the test will distort the true result, as an increase in the temperature will lower the pressure drop and a decrease will artificially elevate the pressure drop. Therefore, the temperature conditions during the test should only vary slightly. This also means that the wetting agents used should have a similar temperature as the environmental temperature surrounding the test set-up. Temperature differences between the wetting solution and the test gas and the temperature of the environment will influence the true test result.

The pressure hold test is an upstream test, even when performed manually. Both tests, bubble point and diffusive flow, require downstream manipulation and therefore cannot be used after steam sterilization of the filter system. The pressure hold, as it measures the pressure drop on the upstream side, can be used without downstream evaluation.

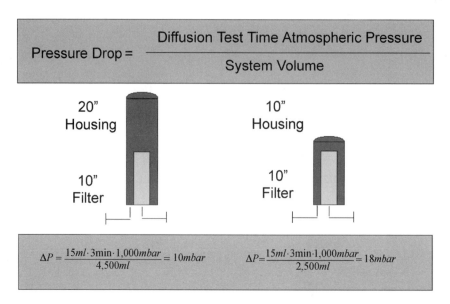

$$\text{Pressure Drop} = \frac{\text{Diffusion Test Time Atmospheric Pressure}}{\text{System Volume}}$$

20" Housing

10" Housing

10" Filter

10" Filter

$$\Delta P = \frac{15ml \cdot 3\min \cdot 1,000mbar}{4,500ml} = 10mbar \qquad \Delta P = \frac{15ml \cdot 3\min \cdot 1,000mbar}{2,500ml} = 18mbar$$

FIGURE 24 Pressure-hold test volume influence. (Courtesy of Spanier, 2000.)

11.7.5. Water Intrusion Test

The water intrusion (also called water pressure hold) test is used for hydrophobic vent and air membrane filters only (40–43). The upstream side of the hydrophobic filter cartridge housing is flooded with water. The water will not flow through the hydrophobic membrane. Air or nitrogen gas pressure is then applied to the upstream side of the filter housing above the water level to a defined test pressure. This is done by way of an automatic integrity tester. A period of pressure stabilization (during which the cartridge pleats adjust their positions under imposed pressures as recommended by the filter manufacturer) is necessary. After the pressure drop stabilizes, the test time starts and any further pressure drop in the upstream pressurized gas volume, as measured by the automatic tester, signifies a beginning of water intrusion into the largest (hydrophobic) pores, water being incompressible (Fig. 25). The automated integrity tester is sensitive enough to detect the pressure drop. This measured pressure drop is converted into a measured intrusion value, which is compared to a set intrusion limit, which has been correlated to the bacteria challenge test. As with the diffusive flow test, filter manufacturers specify a maximum allowable water intrusion value. Above this value a hydrophobic membrane filter is classified as nonintegral.

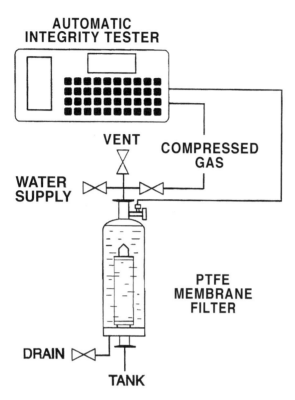

AUTOMATIC
INTEGRITY TESTER

VENT

COMPRESSED
GAS

WATER
SUPPLY

PTFE
MEMBRANE
FILTER

DRAIN

TANK

FIGURE 25 Water intrusion test schematic of different pressure conditions. (Courtesy of Tarry, 1993.)

11.7.6. Multipoint Diffusion Test

In single-point diffusive flow testing, the test is performed at a defined test pressure, which is commonly around 80% of the bubble point value. Therefore the area between the diffusive flow test pressure and the bubble point value is not tested and stays undefined (44). In comparison, the multipoint diffusive airflow test is performed at a multitude of test pressures. Usually this test is performed with an automated test machine, which allows defining the individual test pressure points. In any case, the multipoint diffusion test should be performed right to the bubble point. Therefore, the entire graph with its linear and exponential section is plotted (Fig. 26).

The additional benefit of an automated test machine is the accuracy of its measurement. Moreover, once the pressure points are defined the machine performs the test without the need for supervision. Therefore, valuable time

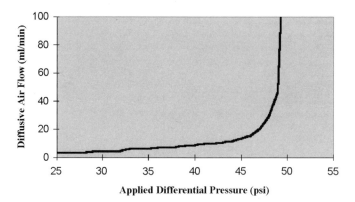

FIGURE 26 Multipoint diffusive flow graph at different test pressures. (Courtesy of Sartorius Group.)

and resources are not expended. To the benefit of data storage, the test machines also print an exact graph of the test performed; therefore, any irregularities will be detected.

Multipoint diffusive testing has advantages over single-point diffusive testing because it can more rapidly detect a pending product failure due to gradual filter degradation. A multipoint integrity test could indicate a trend of increasing diffusion over time that might be overlooked with single-point diffusion testing and even through bubble point testing (Fig. 27). Take as an

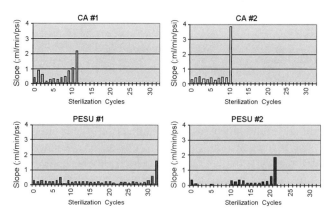

FIGURE 27 Multipoint diffusion test slope at multiple steaming cycles. (Courtesy of Jornitz et al., 1998.)

example the case of a hydrophobic vent filter cartridge on a water-for-injection tank. If the system is in-line steam-sterilized daily, potentially stressing the filter membranes with each cycle, the filter may eventually lose its integrity and fail both a single-point diffusive airflow test and a bubble point test. The bubble point value in this example may also never quite decrease to the point at which the filter actually loses its integrity. The same may be true for the case of single-point diffusion testing. However, a trend may be elucidated if a reduction in membrane integrity is demonstrated as a function of time and not as a single stressful incident. Better estimates of the service life of these vent filters may be made available through such validation of the filters over their operating service life. Such tests could be performed within the Performance Qualification (PQ) stage, where the vent filter would be subjected to multiple steam sterilization cycles to evaluate the resistance of such filters to the individual steaming cycle used in the process. The life span of the vent filter could be evaluated during such test series, using the multipoint diffusion test.

These tests were performed at a steaming cycle temperature of 134 °C. The results of these tests showed that initially these filters fell within the acceptable air diffusion range suggested in the literature. CA filter #1 showed an initial increase after the first sterilization then remained lower until the seventh cycle. At this point the air diffusion rate slowly increased to the tenth cycle, at which point the rate exceeded 15 ml/min at 36.8 psi. In this case eventual filter failure was forewarned by the increase in slope. The second CA filter (CA #2) was within acceptable limits until the tenth cycle, in which failure was abrupt and not preceded by an increase in slope. PES filter #1 had an initial decrease in air diffusion after sterilization, then began to show an increase in slope between the 17th and 20th cycle. This slope increase indicated a pending filter failure. PES filter #2 did not show a marked change in air diffusion after sterilization but eventually did show an increase in slope prior to filter failure. It shows that the multipoint diffusion test creates a possibility to predict filter failure at certain steaming conditions. The steaming cycle performance given by the filter manufacturer can only be an implication. Due to the individual steaming procedures within the users facility, one should perform a filter steaming qualification. When multiple steaming cycles are used, the multipoint diffusion test can be a useful tool to support such qualification efforts.

Additionally, multipoint diffusive testing is invaluable in the characterization of a filter's diffusive flow when wetted with a drug product (44,45). Instead of using a single-point determination, which can cause inaccuracies, one measures the diffusive flow graph for water and for the product to be used. The measurement especially evaluates the slope of the linear section of the diffusive flow measurement and the shift of the Bubble Point. The slope will arise from the differences in diffusivity and solubility of the test gas in the different wetting media (Fig. 28). The linear section of the diffusive flow will follow the described equation:

$$N = \frac{\mathcal{D} \cdot H \cdot \rho}{L} \cdot P_1 + \frac{-\mathcal{D} \cdot H \cdot \rho \cdot P_2}{L}$$

$$N = slope \cdot P_1 + (y - intercept)$$

The slope of the line is $(DH\rho/L)$ and the line's y-intercept is $(-DH\rho P_2/L)$. The values for the filter porosity and thickness are identical for any of the wetting agents (water and product). Therefore, the differences in slope will arise from differences in diffusivity and solubility of air in the wetting liquid; and these differences should be constant over a pressure range if D and H are constant over this pressure range. Indeed, if D or H changes with pressure, then we would not observe a line at low pressure, but a curve. Therefore, to predict a value for N (diffusion rate) with a product as the wetting agent, one would use this equation:

$$\frac{N_{product}}{N_{water}} = \frac{(\mathcal{D} \cdot H)_{product} \cdot P_1 + (-\mathcal{D} \cdot H)_{product} \cdot P_2}{(\mathcal{D} \cdot H)_{water} \cdot P_1 + (-\mathcal{D} \cdot H)_{water} \cdot P_2}$$

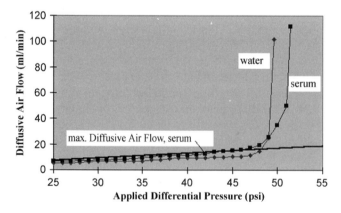

FIGURE 28 Example of different diffusive flow slope using different wetting media. (Courtesy of Jornitz et al., 1998.)

This equation reduces itself to the ratio of the slopes, which is required to evaluate the correction factor for the maximum allowable product wet diffusion:

$$\frac{N_{product}}{N_{water}} = \frac{(\mathcal{D} \cdot H)_{product}}{(\mathcal{D} \cdot H)_{water}}$$

For example:

Slope value for Serum with CA =	0.264 ml/min/psi
Slope value for Water with CA =	0.343 ml/min/psi
Serum/Water Diffusion Ratio of Slope = 0.264 + 0.343 = 0.769	

This ratio is then multiplied by the maximum allowable diffusion limit set by the filter manufacturers at a certain test pressure, which is correlated to the bacteria challenge test. Once the proper diffusion curve limit is defined by multipoint diffusive testing, done during the performance qualification (PQ) phase, the reliability of the single-point diffusive airflow test becomes established.

For example, the maximum allowable air diffusion through water at 36.8 psi described in the validation guide of the filter vendor and correlated to the BC Test, to determine the maximum acceptable air diffusion through serum at 36.8 psi:

Maximum Allowable Air Diffusion through Serum in

CA at 36.8 psi = 0.769 × 15 mL/min =   11.5 mL/min

The value of 11.5 mL/min would be the maximum allowable product diffusion value used in production for a single point diffusion test at a test pressure of 36.8 psi (2.5 bar). The same can be done with any other filter material, wetting agent, and test pressure. Nevertheless the foundation for this maximum product diffusion value is the bacteria challenge test correlated maximum allowable water diffusion value, which can be obtained from the individual filter manufacturers. In any case the determination of the maximum allowable diffusion value using the multipoint diffusion test instead of a single point determination has by far a higher accuracy, due to the multitude of test points. The ratio of slopes is measured at several test pressure points, within a fixed frame, set by the user and the linearity of the graph. These data create a statistically firm basis, contrary to the product wet single point test.

Furthermore, the multipoint diffusion test seems to have the ability to test multiround housings reliably. As described in the bubble point and diffusive flow test section, both tests have their limitations in integrity testing multiround filter housings. A single-point diffusive flow test may not be able to find a flawed filter within the multitude of filters. The bubble point may be covered by an excessive diffusive flow.

In any case the multipoint diffusive flow test seems to be able to find a flawed filter due to the change of the slope of the linear section of the diffusive flow. As seen in Figure 29, a single flawed filter cartridge can be detected within a three-round filter housing, where a single-point test would not have determined the defect. Such test will take longer but will add to the overall accuracy of integrity testing multiround housings. Certainly, like the other tests, the multipoint diffusion test has its limits, with increasing size of the filter system. At one point the automatic integrity test machine will not be able to test the size of housing or the amount of filters used. According to the claims of one filter manufacturer, such testing can be performed using the five-round 30-inch size filters.

In some instances, the multipoint diffusion test finds its usefulness in the analysis of failed filter integrity tests. For instance, when a filter fails the single point diffusive flow test or bubble point test, one should aim for testing the filter with a multipoint diffusion test to see the entire graphic. This result could be compared to the graphs established during the performance qualification

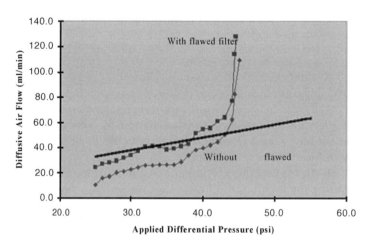

FIGURE 29 Multipoint diffusion test with multiround housing. (Courtesy of Jornitz et al., 1998.)

phase. Commonly there are distinct test graphics, which show whether the filter has a flaw or not and if so what the cause of the flaw could be. Often enough, failed filter integrity tests are caused by wetting problems or product residues within the filter membrane matrix or contaminants in the steam. Such problems can be evaluated by using the multipoint diffusion test and by running the graphic of the failed filter in comparison to a passed filter. The user has the opportunity to discover the reason for failure or is able to send such graphs to the filter manufacturer for evaluation and answers. Single point diffusion testing and bubble point testing are not able to show the reason for a failure in the same scale as the multipoint diffusion test.

REFERENCES

1. Meltzer, T.H., Jornitz, M.W., Eds.; Filtration in the Biopharmaceutical Industry; Marcel Dekker: New York, 1998.
2. FDA, Center for Drugs and Biologics and Office of Regulatory Affairs, *Guideline on Sterile Drug Products Produced by Aseptic Processing*, 1987.
3. American Society for Testing and Materials (ASTM), Standard F838-83, Standard Test Method for Determining Bacterial Retention of Membrane Filters Utilized for Liquid Filtration, 1983, Revised 1988.
4. Meltzer, T.H.; Jornitz, M.W.; Mittelman, M.W. Surrogate solution attributes and use conditions: effects on bacterial cell size and surface charges relevant to filter validation studies. PDA J. Pharm. Sci. Technol. 1998, *52* (1), 37–42.
5. Levy, R.V. The effect of pH, viscosity, and additives on the bacterial retention of membrane filters challenged with *Pseudomonas diminuta*. In *Fluid Filtration: Liquid*; ASTM: Washington, DC, 1987; Vol. 2.
6. PDA Special Scientific Forum, Bethesda, MD: Validation of Microbial Retention of Sterilizing Filters, July 12–13, 1995.
7. Technical Report No. 26, Sterilizing Filtration of Liquids. PDA J. Pharm. Sci. Technol. 1998, 52 (S1).
8. Shucosky, A.C. Prefilter constructions. In *Filtrations in Biopharmaceutical Industry*; Meltzer, T.H., Jornitz, M.W., Eds.; Marcel Dekker: New York, 1998.
9. Zeman, L.J.; Zydney, A.L. Membrane formation technologies. In *Micofiltration and Ultrafiltration Principles and Applications*; Marcel Dekker: New York, 1996.
10. Dosmar, M.; Brose, D.J. Crossflow ultrafiltration. In *Filtrations in Biopharmaceutical Industry*; Meltzer, T.H., Jornitz, M.W., Eds.; Marcel Dekker: New York, 1998.
11. Elford, W.J. The principles of ultrafiltration as applied in biological studies. Proceedings of the Royal Society, London, 1933; 112B, 384–406.
12. Elford, W.J. A new series of graded colloidion membranes suitable for general microbiological use especially in filterable virus studies. J. Path. Bact. 1931, *34*, 505–521.
13. Tanny, G.B.; Strong, D.K.; Presswood, W.G.; Meltzer, T.H. The adsorptive

retention of *pseudomonas diminuta* by membrane filters. J. Parent. Drug Assoc. 1979, *33* (1), 40–51.

14. Berg, H.F.; Guess, W.L.; Autian, J. Interaction of a group of low molecular weight organic acids with insoluble polyamides. I. Sorption and diffusion of formic, acetic, propionic, and butyric acids into Nylon 66. J. Pharm. Sci. 1965, *54* (1), 79–84.

15. Chiou, L.; Smith, D.L. Adsorption of organic compounds by commercial filter papers and its implication quantitative-qualitative chemical analysis. J. Pharm. Sci. 1970, *59* (6), 843–847.

16. Udani, G.G. Adsorption of preservatives by membrane filters during filter sterilizations, Thesis for B.Sc. Honours (Pharmacy), School of Pharmacy, Brighton Polytechnic, Brighton, U.K., 1978.

17. Brose, D.J.; Hendricksen, G. A quantitative analysis of preservative adsorption on microfiltration membranes. Pharm. Tech. 1994, *18* (3), 64–73.

18. Tanny, G.B.; Meltzer, T.H. The dominance of adsorptive effects in the filtrative sterilization of a flu vaccine. J. Parent. Drug Assoc. 1978, *32* (6), 258–267.

19. Johnston, P.R. Submicron filtration. Filtration and Separation 1975, *12* (4), 352–353.

20. Pertsovskaya, A.F.; Zvyagintsev, D.G. Adsorption of bacteria on glass, modified glass surfaces, and polymer films. Biol. Nauk 1971, *14* (3), 100–105.

21. Zierdt, C.H. Unexpected adherence of bacteria, blood cells and other particles to large porosity membrane filters [abstr]. *American Society of Microbiologists*, 78th Annual Convention, Las Vegas, May 1978; Article Q-93, 10.

22. Zeirdt, C.H.; Kagan, R.I.; MacLawry, J.D. Development of a lysis-filtration blood culture technique. J. Clin. Microbiol. 1977, *5* (1), 46–50.

23. Leahy, T.J.; Sullivan, M.J. Validation of bacterial retention capabilities of membrane filters. Pharm. Tech. 1978, *2* (11), 64–75.

24. Mittleman, M.W.; Jornitz, M.W.; Meltzer, T.H. Bacterial cell size and surface charge characteristics relevant to filter validation studies. PDA J. Pharm. Sci. Technol. 1998, *52* (1), 37–42.

25. Mittlemann, M.W.; Kawamura, K.; Jornitz, M.W.; Meltzer, T.H. Filter validation: bacterial hydrophobicity, adsorptive sequestration and cell size alteration. PDA J. Pharm. Sci. Technol. Nov/Dec 2001, *55*, 422–429.

26. Ridgway, H.F. Microbiological fouling of reverse osmosis membranes: genesis and control. In *Biological Fouling Of Industrial Water Systems, A Problem Solving Approach*; Mittelman, M.W., Geesey, G.G., Eds.; Water Micro Associates: San Diego, 1987.

27. Soelkner, P.G.; Rupp, J. Cartridge filters. In *Filtrations in Biopharmaceutical Industry*; Meltzer, T.H., Jornitz, M.W., Eds.; Marcel Dekker: New York, 1998.

28. FDA, Center of Drug Evaluation and Research (CDER), Guidance for the Industry—Changes to an Approved NDA or ANDA, November 1999.

29. EMEA, CPMP/QWP/486/95, Note for Guidance on Manufacture of the Finished Dosage Form, London, April 1996.

30. Reif, O.W.; Sölkner, P.; Rupp, J. Analysis and evaluation of filter cartridge extractables for validation in pharmaceutical downstream processing. PDA J. Pharm. Sci. Technol. 1996, *50* , 399–410.

31. Jornitz, M.W.; Meltzer, T.H. *Sterile Filtration—A Practical Approach*; Marcel Dekker: New York, 2000.

32. Human Drug cGMP Notes, Motisse, FDA, Ref. 21 CFR 211.65, Equipment Construction, 1994.

33. PDA Special Scientific Forum, Rockville, MD, The Extractable Puzzle: Putting the Pieces Together—Resolving Analytical, Material, Regulatory, and Toxicological Issues to Find Solutions, November 2001.

34. Hawker, J.; Hawker, L.M. Protein losses during sterilization by filtration. Lab. Practises 1975, *24*, 805–814.

35. ISO/DIS 13408-2 Draft, Aseptic processing of health care products—Part 2: Filtration, 2002.

36. Jornitz, M.W.; Agalloco, J.P.; Akers, J.E.; Madsen, R.E.; Meltzer, T.H. Filter integrity testing in liquid applications, revisited; Parts I & 2. Pharm. Tech. 2001, Part I, *25* (10), 34–50; Part II, *25* (11): 24–35.

37. Reti, A.R. An assessment of test criteria in evaluating the performance and integrity of sterilizing filters. Bull. Parenter. Drug Assoc. 1977, *31* (4), 187–194.

38. Meltzer, T.H.; Madsen, R.E.; Jornitz, M.W. Considerations for diffusive airflow integrity testing. PDA J. Pharm. Sci. Tech. 1999, *53* (2), L 56–59.

39. Madsen, R.E. Jr.; Meltzer, T.H. An interpretation of the pharmaceutical industry survey of current sterile filtration practices. PDA J. Pharm. Sci. Tech. 1998, *52* (6), 337–339.

40. Jornitz, M.W.; Waibel, P.J.; Meltzer, T.H. The filter integrity correlations. Ultrapure Water Oct. 1994, 59–63.

41. Meltzer, T.H.; Jornitz, M.W.; Waibel, P.J. The hydrophobic air filter and the water intrusion test. Pharm. Tech. 1994, *18* (9), 76–87.

42. Tarry, S.W.; Henricksen, J.; Prashad, M.; Troeger, H. Integrity testing of ePTFE membrane filter vents. Ultrapure Water 1993, *10* (8), 23–30.

43. Tingley, S.; Emory, S.; Walker, S.; Yamada, S. Water-flow integrity testing: a viable and validatable alternative to alcohol testing. Pharm. Tech. 1995, *19* (10), 138–146.

44. Waibel, P.J.; Jornitz, M.W.; Meltzer, T.H. Diffusive airflow integrity testing. PDA J. Pharm. Sci. Technol. 1996, *50* (5), 311–316.

45. Jornitz, M.W.; Brose, D.J.; Meltzer, T.H. Experimental evaluations of diffusive airflow integrity testing. PDA J. Pharm. Sci. Technol. 1998, *52*, 46–49.

12

Process Development of Alternative Sterilization Methods

Volker Sigwarth
Skan AG, Basel, Switzerland

12.1. INTRODUCTION

The use of standard sterilization technologies in modern pharmaceutical manufacturing processes is often limited. In this chapter, the use of alternative sterilization methods will be considered and a method of proving their suitability will be demonstrated. Whereas correlation of physical parameters with bacterial reduction is well established in standard sterilization technologies, such a correlation is currently neither generally recognized nor possible for alternative sterilization processes. Furthermore, there is a lack of information regarding basic data including process mechanisms, physical influences and boundaries, suitable test organisms, and qualification strategies. Consequently, the efficiency and reproducibility of alternative sterilization methods is difficult to validate in comparison to standard sterilization processes. Because of this, alternative sterilization methods are often confusedly discussed and are poorly accepted in pharmaceutical industries and by authorities.

In this chapter, a complete and systematic method for process development of alternative sterilization technologies is introduced. This method of development is based on statistical procedures such as the Design of Experiment (DoE) for process evaluation and the Fraction Negative Method (FNM) of determining D-values of a microbiological system (1). By the consequent combination of the physical process with the response of a microbiological system, this tool set leads to scientifically based and statistically proven data, giving a complete description of the influences of process parameters on sterilization kill. The chronology of the single steps systematically seeks to form a deeper process comprehension by analyzing the physical interrelationships of the process and by assessing the suitability of a microbiological system. Finally, both systems are used for process development, excluding any secondary effects on the results and thus ensuring that the experimental data can be properly interpreted. This information makes it possible to define the important parameters, to include their boundaries, to design the validation strategies, and to control the whole process in routine application. Furthermore, a suitable microbiological system is established reflecting the process performance in concordance with the regulatory requirements to be used during process validation and revalidation. On this basis, additional questions, such as studies on surface effects of materials and resistances of different microbes, can be carried out easily to further improve the database of the process. As an example, surface decontamination with gaseous hydrogen peroxide is completely characterized in this chapter by using each single step of the introduced method.

12.2. PROCESS DEVELOPMENT—GENERAL

The purpose and aim of an inactivation method is to reduce an existing microbiological contamination load by a defined level. Based on the defined reduction level, the methods are split into disinfection, sanitization, decontamination, and sterilization processes. Whereas for a wide field of applications it is enough to know the assumed process efficiency, in the pharmaceutical industry the use of such methods has to be more deeply known, validated, and controlled. This necessitates an understanding of the link between the applied process and the defined inactivation performance. Only by knowing the correlation between physical or chemical process parameters to the inactivation effect is it possible to choose the right process and to control the relevant parameters. These tools should be applied not only to new applications of alternative inactivation methods, but also for applications used for many years in the field of pharmaceutical manufacturing, where there is often a lack of information regarding such a correlation.

It is difficult to validate and apply alternative sterilization processes to be in compliance with today's pharmaceutical requirements. Furthermore, most alternative processes are used for specific applications inside their own tight boundaries and it is often neither useful nor possible to describe the applied sterilization method completely.

The following methodology focuses on the need to establish the necessary correlation of process parameter to microbiological inactivation in the application of alternative sterilization methods. Step by step, this methodology leads from the microbiological inactivation requirements and the physical possibilities of a chosen process to the definition of a suitable microbiological system for process development and quantification. The knowledge and description of these two systems—(a) the physical process and (b) the microbiological system—make it possible to design experimental runs focusing on the required correlation with a minimum of test effort. Based on the results generated, the process and its boundaries can be defined and a strategy for validation can be established. In order to get the two systems correlated, different statistical tools are implemented into the methodology. First, the Fractional Negative method of D-value determination of biological indicators is used to quantify a microbiological inactivation effect. Second, the Design of Experiment is used as a method to define the required minimum experimental runs needed and to enable a statistically based interpretation of the results.

These two tools in combination with some practical aspects of the application are the key factors in process development of applications that require the use of alternative sterilization methods. For a better understanding of the introduced method of process development, the surface decontamination of an isolator system with gaseous hydrogen peroxide (H_2O_2) is described from the bottom up using each single step of the development tool set.

12.2.1. Steps of Process Development

The structure of the introduced methodology consists of the following steps:

12.2.1.1. Process and Application

First, the applied process has to be defined in consideration of its performance, requirements, and expectations relative to the sterilization effect. Second, the process has to be characterized and its parameters have to be identified in a useful range. Third, the equipment of the application has to be evaluated considering the possibility of controlling the identified process parameters for routine use and the experimental runs.

12.2.1.2. Microbiological System

A method to quantify the sterilization effect is developed based on biological indicators. The suitability of the chosen method and of the biological indicator is proved by experimental trials in the light of the defined process performance. The development of the target value for quantitative process justification is an important part of the whole methodology. Great attention is directed toward a stable and reproducible microbiological system in order to evaluate the statistically significant result.

12.2.1.3. Main Experiment

For planning, execution, and evaluation of the experimental runs, the Design of Experiment (DoE) method is used. The DoE method requires a detailed consideration of the expected target prior to the execution of the experimental trials. Parameters that may influence the process are defined within a useful range for the application. The influences of the chosen parameters on the sterilization effect are estimated and their interrelationship to the process has to be understood. This defines the basis to choose an experimental design focusing exactly on the defined parameters and the expected result. This experimental design further develops the empirical model for the statistical analysis of the results. With such an optimization, the DoE method makes it possible to reduce the number of experiments to an minimum without losing any required information.

12.2.1.4. Process Description

Based on the analyzed data of the experimental runs, the whole process is characterized, including a detailed description of the single process steps and their meaning relative to the process performance. Furthermore, the important process parameters including their boundaries are defined for each process step. In conclusion, the whole process is completely described and an understanding of the correlation between process parameters and resulting performance is reached. Based on this knowledge, it is possible to adjust and improve the applied process if necessary.

12.2.1.5. Validation Strategy

Based on the understanding achieved of the process and the suitable microbiological system, a complete validation strategy for the application can be designed. The validation strategy includes the control and monitoring of the known process parameters within their boundaries and a method to develop and quantify sterilization cycles. The validation strategy uses the fundamental data and the practical tools established during process development.

12.2.1.6. Additional Studies

Further information required for particular applications can be gained by carrying out additional studies based on the established understanding of the process.

12.3. PHYSICAL FACTORS AND VALIDATION

The following steps will establish the physical factors needed to understand and proceed with the validation and use of an alternative sterilization method:

> Identify and describe process factors and range of process factors used.
> Use the requirements of the pharmacopoeias as a basis.
> Characterize the process, separated into different steps and phases.
> Estimate their influence on the sterilization effect.
> Establish the relationships in the process.
> Split out the factors fixed by the application and the factors that can be adjusted.
> Construct a suitable experimental system to represent the application.

At the beginning of process development, the chosen alternative sterilization method has to be described in detail. First, it is important to define the requirements and expectations of the process regarding its performance. This consideration leads to a justification to determine if the process is suitable for the application or not. Often alternative sterilization methods are applied without any rationale between process expectations and performance. Because of this, results are observed during validation that either are not reproducible or may not even be possible. Without having such a justification of process suitability, it is useful to look for other sterilization technologies. If suitability of the chosen method is stated, it is important to characterize the different steps and phases of the sterilization process. Consequently, each single process phase has to be described in consideration of its target effect and its relevant process factors. Furthermore, the influence of all factors on the sterilization effect and their interaction in the process has to be estimated. During this estimation process, it is important to pay attention to the current standards of pharmaceutical manufacturing. Finally, the above-mentioned thoughts lead to a transparent characterization of the process and the application. This makes it possible to differentiate between factors fixed by the application and such factors that it may be useful to consider adjusting in process development. After identifying the relevant factors, the application equipment has to be evaluated considering the possibility of controlling the identified factors for routine use and experimental runs. This first step of process development will be described in detail next, using the surface decontamination with gaseous H_2O_2 of an isolator system as an example. Bear in mind that each mentioned point of decision and argument is suitable for transfer to other alternative sterilization methods.

12.3.1. Surface Decontamination with Gaseous Hydrogen Peroxide

Surface decontamination with gaseous hydrogen peroxide (H_2O_2) has gained widespread acceptance, especially in the field of isolator technology, but also for the decontamination of transfer devices, disposables, cleanrooms, and different surfaces of equipment for pharmaceutical manufacturing. It is applied where surface decontamination with steam or heat is not possible because of the heat resistance of the enclosure or the product. In isolator technology, H_2O_2 decontamination is often used to reduce the microbiological contamination on the inner surface of the isolator system, where later aseptic manufacturing or testing takes place. The term decontamination is used to separate this process clearly from the term sterilization with its absolute definitions and requirements. Whereas the process target is clearly quantified for sterilization, the required target for a decontamination process is not absolutely defined. Also, during a decontamination process, a quantified inactivation is reached but the level of inactivation can be selected based on the requirements of the application; if necessary, even the definition of sterilization can be defined as a process target.

12.3.2. Isolator System

The isolator system to be discussed in this example represents state-of-the-art isolator technology and is comparable to modern systems available on the market. The construction is mainly stainless steel and glass; the isolator volume is 1.4 m^3. It features unidirectional air supply via terminal HEPA filters; see Fig. 1 (1).

For H_2O_2 decontamination the isolator system can be closed by flaps, and the H_2O_2 is directly vaporized by an integrated decontamination system into the recirculated air flow. The whole system including the H_2O_2 decontamination cycle works in a fully automated manner, with predefined values determined by a PLC controller. The chamber is controlled for temperature [°C], humidity [% rH], air velocity [m/s], and pressure [Pa]. For the H_2O_2 gas concentration [ppm] measurement during decontamination, a electrochemical sensor is applied.

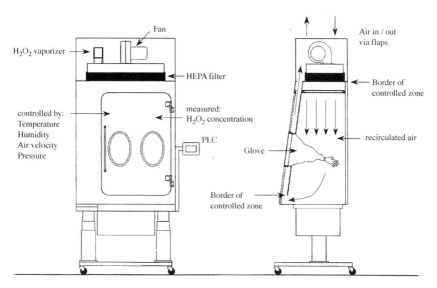

FIGURE 1 Isolator system with integrated H_2O_2 decontamination. (From Ref. 1.)

12.3.3. H_2O_2 Decontamination Process (From Ref. 2)

In H_2O_2 surface decontamination, the overall bacterial reduction is obtained from the release of gaseous H_2O_2 and the effect of the lethal dose over time. H_2O_2 decontamination is subdivided into four cycle phases, as follows:

Phase 1: Preconditioning. In the preconditioning phase, the initial conditions required for decontamination are created in the chamber air.

Phase 2: Conditioning. In this phase, the dose of gaseous H_2O_2 (necessary to reach the desired decontamination effect) is generated in the chamber. For this purpose an initial quantity of H_2O_2 [g] is vaporized from an aqueous solution at a certain dosage rate [g/min].

Phase 3: Decontamination. In this phase, the obtained effective dose is kept stable for the period of time necessary to achieve the desired decontamination result. In addition, the quantities of gaseous H_2O_2 that are no longer available in the chamber air (due to adsorption on surfaces used in bacterial reduction and decomposition) are continuously made up. Therefore the dosage rate can be adjusted [g/min].

Phase 4: Purging. In this phase, the required maximum residual concentration of H_2O_2 in the chamber is achieved through purging with fresh air over a certain time [min] (Fig. 2).

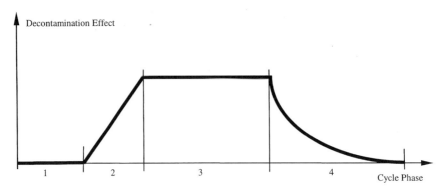

FIGURE 2 Decontamination effect as a function of cycle phase. (From Ref. 2.)

By subdividing the decontamination cycle into four functional steps, a transparent basis is established as a first step toward further process development. From a technical point of view, such an isolator system with an integrated H_2O_2 decontamination system seems to be complete and well done. The system is controlled by defined factors, and for the decontamination cycle the first assumptions are made linking physical factors to decontamination effect. Under the microbiological aspect of decontamination process there is no proven correlation between physical values and an inactivation effect defined. At this point in time, it is not possible to select the required parameters of factors to control the process and to validate the application (1).

12.3.4. Process Performance, Requirements, and Expectations

In this example isolator system, the decontamination process should reduce the microbial contamination on the inner surfaces of the isolator chamber. The borders of the controlled zone of the chamber are defined by the terminal HEPA filter at the top of the chamber and the return air duct near the bottom (see Fig. 1). In this area the decontamination process should finally be validated. Surface decontamination means that the decontamination effect takes place on each surface of the defined area. It does not mean that the effect penetrates through any surface or into a closed volume or to a covered surface. This is the first important restriction for this application, with an impact on the following steps of process development. For the requirements of the overall microbial reduction on the surfaces of an isolator system, various values can be found in the literature (3,4). Reduction levels of 3 logs to more than 12 logs are used by the industry. The USP (3) mentions a proven total kill of a 5-log population as a requirement for an isolator used for sterility testing. For production isolators there has been no regulatory requirement committed until now. All reduction levels are defined for a highly resistant microorganism or biological indicator (5–8). The requirement for the overall microbial reduction of the isolator system of the example is set to a total kill of a 6-log population of a highly resistant biological indicator on each position of the defined area. The requirement for the residual concentration of H_2O_2 after purging is based on the TLV level and set to 1 ppm or lower.

Most problems during validation of such systems are obtained by the proof of the microbial reduction required for the application. At this final step of validation, the whole physical process and the microbiological system comes into question by a failing microbiological result. While it is not possible to attribute influence and effect completely, a traceable reaction aimed at improving the inactivation effect can be established based on procedural transparency and an understanding of the factors influencing the inactivation effect. In the following process development, it will be shown that it is possible to reach a level of procedural understanding that enables a proper interpretation of microbiological results and to focus them to influence factors of the process. Moreover, the process can be individually adjusted to specific overall microbial reduction requirements, even for each single application.

12.3.5. Process Factors, Influences, Range, and Relationships (From Ref. 1)

In general, an influence on the sterilization effect is expected from physical conditions, specific factors of the process equipment, and the microbiological system. The selection of factors for further consideration should be focused on the relevant regulatory requirements to ensure the control and validation of the process. Furthermore, a basis should be formed for interpretation of the inactivation method, to evaluate process-specific assumptions, and to improve the application. The following conditions, factors, and systems will be discussed: (a) principal and physical factors, (b) influences of process-related factors on decontamination effect, (c) system-related factors, and (d) factors related to the microbiological system.

12.3.5.1. Principal and Physical Factors

The concentration of the sterilizing media, the temperature, the humidity, and the duration of the decontamination are mentioned in the USP (9) as critical factors for a sterilization method. All the mentioned factors, excpet the duration of the sterilization, are considered as general physical influences on a gaseous process like surface decontamination with vaporized H_2O_2. The main process factors for the H_2O_2 decontamination will be identified and described herein. The relationship of those factors and their influence on the target decontamination effect will be estimated. Last, the factors that are fixed by the application and that are adjustable will be differentiated.

a. Concentration of the Sterilizing Media, Dose of H_2O_2

For H_2O_2 decontamination, the vaporized quantity [g] is generally accepted as an indirect measure for the concentration of the decontaminating media (2). It represents the lethal dose for obtaining the decontamination effect and is therefore the main factor in the H_2O_2 decontamination process. According to this assumption, the decontamination effect is enhanced by increasing the vaporized quantity of H_2O_2. Lower quantities of H_2O_2 will result in a low decontamination effect. If the applied quantity of H_2O_2 is too small, no effect will be established at all. Furthermore, it can be assumed that the relationship between increasing the H_2O_2 quantity and the established decontamination effect is not linear. By vaporizing H_2O_2 into a closed chamber, the air becomes more and more saturated until the dew point is reached. At this point it is not possible to add more H_2O_2 to the air without condensation appearing on the surfaces. This shows that by vaporizing H_2O_2, the process changes from gaseous into saturated and finally into condensate conditions. This general change in process may influence the decontamination effect. With the introduced isolator system, it is possible to adjust the initial quantity (q1) of H_2O_2 in steps of 1 g.

b. Temperature

In the literature, a temperature range between 4 and 80°C is mentioned for the application of H_2O_2 decontamination (10,13). The influence of the temperature on the decontamination effect is assumed in relationship with the capacity of the gaseous phase for decontamination media. At high temperatures the air is able to contain more decontamination media before reaching saturation or condensing H_2O_2 onto surfaces. The influence from different temperatures on decontamination effect reached by a defined quantity of H_2O_2 is not described in the literature. The air temperature in the chamber of the isolator system can be adjusted and controlled in a range between 25 and 45°C.

c. Humidity

If condensation occurs in the progress of H_2O_2 decontamination, it is assumed that the decontamination effect is immediately aborted. The entire H_2O_2 in gaseous phase is absorbed into the condensation because of its huge affinity to water (11). Therefore, the initial humidity in the chamber seems to be critical because of the possibility of condensation. On the other hand, in different sterilization methods a higher humidity often increases or even ensures the sterilization effect. Another opinion points out that the effect of H_2O_2 decontamination takes place only by condensation of H_2O_2 to surfaces (12). With the isolator system, the chamber air can be adjusted by humidity in a range from 5 up to 80 %rH. But it seems to be ingenious to start with a lower initial humidity in the chamber to ensure a good capability of the air for H_2O_2 and further adjust the saturation level of the process by vaporizing H_2O_2. Temperature and relative humidity itself are linked by the water capacity of the air. This is important to know for further steps of process development.

d. H_2O_2 Gas Concentration

In the literature, the H_2O_2 gas concentration is often used as a direct value for the decontamination effect (10). In practice, measuring H_2O_2 gas concentration is not standardized, no useful calibration procedure is possible, and no correlation between gas concentration and microbiological kill has been stated until now. Furthermore, a relation to the capacity of the air is assumed that links the H_2O_2 gas concentration to the temperature, as it is assumed for relative humidity. The isolator system is equipped with a sensor for gaseous H_2O_2 but no process step is controlled by the value.

e. Pressure

In isolator technology the operating differential pressure between the surrounding and the chamber is generally set in a range between -100 and $+100$ Pa based on the application. Compared to the atmospheric pressure this differential pressure is too low ($\sim 1‰$) to assume an influence on the decontamination effect by the chamber pressure.

f. Duration of the Decontamination

The duration of the decontamination is contained in the target value for sterilization effect.

12.3.5.2. Influences of Process-Related Factors on Decontamination Effect

Other influences on the decontamination effect are process-related factors of the system, including:

> Air distribution and air velocity in the chamber
> Temperature of the H_2O_2 vaporizer
> Vaporizing velocity of the H_2O_2 aqueous solution
> Release of the H_2O_2 vapor into the chamber before or behind the HEPA filter

The variations of the process-related factors are limited by the given system. The unidirectional air flow of the isolator provides a good distribution inside the chamber. As the air is assumed to be the carrier for the H_2O_2 vapor, the decontamination media is distributed in the same way. Concerning the air velocity, a hypothesis can be formed in a way that with higher air velocities the decontamination effect on surfaces is decreasing by an assumed wiping effect. For production mode of the isolator, the air velocity is required at 0.45 m/s. For the decontamination mode it is able to switch it to a lower air velocity at 0.25 m/s. The temperature of the vaporizing plate is fixed and controlled in a range that supports enough energy for vaporization of the H_2O_2 quantities. The vaporizing velocity is set to the maximum value for the initial dose of H_2O_2; the dosing rate during decontamination phase can be adjusted. It seems logical that releasing the H_2O_2 directly into the chamber would increase the decontamination effect in comparison with pushing it first though a HEPA filter. But for isolator technology, it is recommended that any media enter the isolator through a suitable filter system (4), so the assumed improvement does not apply.

Two further process-related factors should be discussed in greater detail, because the knowledge of their influence relating to the decontamination effect can be important for the practical application and the adjustment of the process. Also, these factors support the process hypothesis formed by discussing the principal and physical factors.

a. Concentration of the Aqueous H_2O_2 Solution

H_2O_2 is commercially available in aqueous solutions up to 70%. By vaporization of the required quantity of pure H_2O_2, a fixed quantity of water is vaporized in relation to the concentration of the used aqueous solution. When the vaporized quantity of pure H_2O_2 is kept stable, the variation of the concentration of aqueous solution establishes different partial pressure ratios of H_2O_2 and water in the gaseous phase. Moreover, the concentration of the aqueous solution influences the saturation of the air. At constant doses of the decontamination media, the vaporized volume changes and thereby the established air saturation changes also. The influence of a variation of the pressure ratio between H_2O_2 and water at a constant dose of pure H_2O_2 on the decontamination effect is not easily estimated. In the literature a relationship between a higher H_2O_2 partial pressure and an increase in decontamination effect is assumed (10). The concentration of the aqueous H_2O_2 solution can be adjusted by dilution in a range from 30 to 70%.

b. Dosage During the Decontamination Phase

In comparison to all the other factors discussed until now, the dosage during decontamination phase influences the decontamination effect after it is finally established by the condition phase. Primarily, the dosage of H_2O_2 during decontamination phase is assumed to influence the stability of the decontamination effect by replacing quantities of H_2O_2 no longer available due to absorption, adsorption, or decomposition. On the other hand, it is also possible that an established decontamination effect doesn't change much in a closed system like an isolator, but there is no basis for those assumptions. An influence of the dosage during the decontamination phase is assumed in the same way as for the main dose. Higher dosages leads to an increase in decontamination effect. Based on observed influences of small doses after decontamination effect is established, it is possible to interpret the stability of the reached effect. On a stable decontamination effect this dosage will have a lower influence than on a more unstable effect. The dosage during decontamination phase (q2) is stated as a percentage of the dosage rate per hour [%/h]. The system allows for easy adjustment of the dosage (q2).

12.3.5.3. System-Related Factors

Influences of the isolator system hardware on the decontamination effect can also be assumed. Possible influences include the following:

Size of the isolator chamber
Materials of construction
Ratio between surface and volume of the system
Different air filter systems

To figure out the influence of these factors requires a huge expense in hardware changes of the actual system. But the influence caused by the ratio between surface and volume and different filter systems is interesting to investigate.

12.3.5.4. Factors Related to the Microbiological System

There are numerous factors relating to the microbiological system influencing an observable decontamination effect. The following list focuses only on commercially available biological indictors, but the mentioned factors should also be considered by specially prepared biological systems. Possible influences include the following:

Kind of test organisms
Selected carrier material
Preparation of the inoculation
Primer package used
Chosen culture media

In this chapter it becomes clear that the microbiological system itself is a complex system for which suitability of the process has to be known in advance in order to produce reliable results. To conduct a study of factors influencing the sterilization effect, the knowledge of suitable and reliable microbiological systems provides a big advantage. On the other hand, experiments to understand the influences of variation in the composition of biological indicators on a responding sterilization effect requires a well-known and reproducible sterilization system. Often, for alternative sterilization methods, neither is a reliable and reproducible sterilization system available nor is a suitable biological indicator commonly known.

To provide this information, the most effective way is to select and combine all available subject-matter knowledge. In this case, first comes the analysis of process performance and expectations. Second comes an understanding of a microbiological system, and finally comes the proper experimental techniques. The following steps of this chapter focus on the microbiological system and its use for the quantification of a sterilization effect.

12.4. TARGET VALUE FOR STERILIZATION EFFECT

As mentioned above, the target of a sterilization method is to reduce a microbiological contamination by a defined value. In pharmaceutical applications, this reduction has to be quantified and validated. For standard sterilization methods there are microbiological tools available to prove a sterilization effect of a process and quantify its reduction performance. In general, biological indictors are used for process validation (5,6,8,9). Even when a clear correlation of physical process and sterilization effect is stated, process validation is not possible without microbiological tests.

In this case, the inactivation performance of an alternative sterilization method should be determined and quantified. First a figure is needed to represent the sterilization effect. Moreover, a empirical model and practical tools have to be found, based on which determination and quantification of a sterilization effect is possible. In the following section, based on biological indicators and the empirical survival time model of microbial reduction, a measure, a method, and the tools needed to quantify sterilization effect are established and discussed.

12.4.1. Biological Indicators (BIs)

BIs consist of a carrier inoculated with a defined number of a specified test organism. This carrier is sealed into primary packaging that is permeable to the sterilization medium. A BI generally indicates a defined resistance to a specified inactivation method. Therefore, a BI is a ready-to-use device used to challenge sterilization processes (5,6,8,9). The test organism used and the carrier material depend on the applied process. The primary packaging should protect the carrier against destruction and contamination without affecting the penetration of the sterilization media to the test organism. The test organism should represent a highly resistant organism against the applied process. The population is chosen in accordance with the required microbial reduction of the applied process. For standard sterilization methods in general, a population of 1×10^6 CFU/carrier or greater is defined in accordance with the definition of sterility (9). An adequate composition of a BI including the identification of the test organism is well characterized for standard sterilization methods (6–8). Such BIs are commercially available from several suppliers.

To determine the sterilization efficiency of an application, a number of BIs are placed in the zone to be sterilized in a range of different positions. After the sterilization cycle, the BIs are evaluated using the growth test. Therefore, the carrier of the BI is transferred aseptically into a suitable growth media and incubated at the specific temperature. Based on the turbidity of the media as a result of the subsequent growth, the test can be interpreted as follows (6–8):

Turbidness observable $\Rightarrow$ growth $\Rightarrow$ positive $\Rightarrow$ no kill $\Rightarrow$ not sterile

No turbidness observable $\Rightarrow$ no growth $\Rightarrow$ negative $\Rightarrow$ kill $\Rightarrow$ sterile

An evaluated BI using the growth test leads generally to a qualitative result. It is not possible to quantify the sterilization efficiency based on this result because neither the numbers of surviving microbes causing the turbidness of the media is known nor can the exact time of the transition of a BI from positive to negative be observed. To establish a method for the quantification of a sterilization effect, it is necessary to understand how the resistance of a BI is described and defined. The principles of this can be found in the survival time model of microbial reduction.

12.4.2. Survival Time Model of Microbial Reduction
(From Ref. 2)

The goals of the survival time model of microbial reduction include the following:

Understanding the model of microbial reduction
Recognizing the relationship of D-value and model behavior
Measurement of sterilization effect (can be the D-value of biological indicators)

The resistance of BIs to a defined inactivation method is expressed as a decimal reduction per unit of time [min], the D-value. The D-value thus specifies the time it takes to reduce the population of the test organism by 90% (5,6,8,9).

If N_0 is defined as the initial number of test organisms at time $t = 0$ and $N_{(t)}$ is the number of surviving test organisms at time t, then the survival time model of the BI is defined as follows:

$$N(t) = N_0 10^{-t/D} \text{ (where } D = \text{D–value)} \tag{1}$$

When the population is expressed on a log scale, as is customary, this produces the following:

$$\log N(t) = \log N_0 - t/D \tag{2}$$

When displayed graphically in semi-logarithmic form, the survival curve appears as a straight line whose origin is $\{t = 0 \text{ [min]}, \log N_0\}$ and whose slope is $-1/$D-value; see Fig. 3 (14,15).

The D-value is used to define a time window (survival/kill window) inside which the transition from reliably positive to reliably negative results takes place (5–8).

The survival time is defined as follows:

$$< (\log N_0 - 2) \times \text{ specified D–value [min]} \tag{3}$$

The kill time is defined as follows:

$$> (\log N_0 + 4) \times \text{ specified D–value [min]} \tag{4}$$

This means that BIs that are exposed for less than the survival time to the specified inactivation then test reliably positive in the subsequent growth test. Exposure for longer than the kill time produces reliably negative results. Between the survival time and the kill time lies the "fractional field." This time window represents the later stages of microbial inactivation (i.e., only a small number of surviving microorganisms are left on the carrier), some of which test positive and some test negative in a growth medium; see Fig. 3 (2,5–8).

The definition of the survival/kill window and hence the transition from reliably positive results through the fractional field to reliably negative results is based on a probability distribution (14,15). If N_0 is defined as the initial number of microorganisms, the average number of surviving microorganisms $N_{(t)}$ after t minutes exposure is given by the following formula (from (1.1)): (15)

$$N(t) = 10^{(\log N0 - t/D)} \tag{5}$$

If $N_{(t)}$ represents the average number of surviving microorganisms, the probability $P(N_{(t)})$ that a very small $N_{(t)}$ will produce a negative result is given by the following formula: (15).

$$P_{(N(t))} = e^{-N(t)} \tag{6}$$

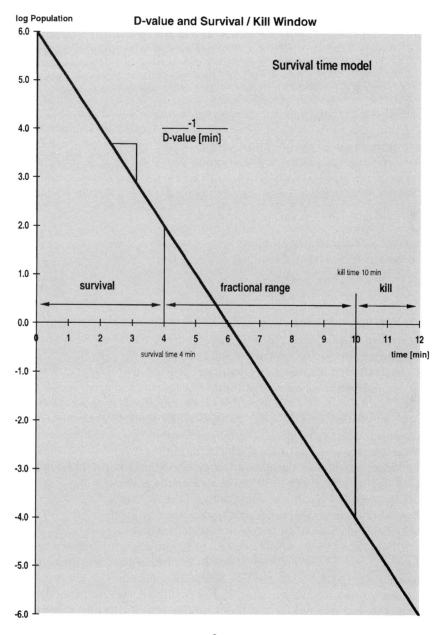

FIGURE 3 Survival time model, N_0 10^6, D-Value 1.0 min. (From Ref. 2.)

On this basis, the survival curve is now obtained and a probability distribution of positive and negative results expressed as a function of exposure time can be made.

12.4.2.1. Example 1

If we assume an initial number of microorganisms N_0 of 1×10^6 and a D-value of 1 min, the probability of observing a negative result is near 0% for exposures equal to the duration of the defined survival time, and near 100% for exposures equal to the duration of the kill time. It is interesting to consider an exposure duration of 6 in, which in this example corresponds to a bacterial reduction of 6 orders of magnitude. From this we obtain the following:

Exposure time:	6 min
$N_{(t)}$:	$10^0 = 1$
$P(N_{(t)})$:	0.367

BIs with an average of one surviving microorganism have a 37% probability of testing negative in the growth test and a 63% probability of testing positive. The evaluation of BIs using the growth test provides no basis for quantifying the residual number of microorganisms on the carrier and hence the bacterial reduction obtained. In order to be able to observe reliably negative results, the initial population of a BI must be reduced by more than the number of inoculated microorganisms. If the range of sizes of the initial population is reduced, positive and negative results occur in the ratio 2:1.

Independent from the inactivation time and the D-value, the boundaries of the fractional field can be defined based on the population (see Fig. 3). A population of 1×10^2 thus defines the transition from reliable positive to the observation of fractional results. At populations of less than 1.0×10^{-4} the fractional field ends and only negative results are observable.

Populations between 1×10^1 and 1×10^2 are defined to perform growth promotion tests in microbiological quality control of pharmaceutical manufacturing (16). This is one widely accepted application of the surviving boundary of the survival time model. For fractional negative methods (e.g., population recovery tests or D-value determination of BIs) the international standards (5,8) define a population of less than 5×10^0 to observe the required fractional result. Therefore, also the fractional field of the survival time model is thus applied in the routine of pharmaceutical production. Even the definition of sterility (9) as the probability of 10^{-6} for a positive result can be understood based on the survival time model.

For the quantification of a sterilization effect, it is important to understand that a observed total kill of a defined population indicates a reduction of 4 log steps higher than the count of its initial population. Moreover, the resistance of a BI is defined based on the relationship of two factors. First, the D-value as the figure for the inactivation velocity, and second, the transition from reliable survival over the fractional field to reliable kill as the model behavior or reactive pattern of a BI during inactivation.

12.4.2.2. Measure for Sterilization Effect (From Ref. 2)

The exponential survival time model (the survival curve as a semi-logarithmic straight line) is generally recognized as a means of describing the inactivation of microorganisms (8,14,15). The reduction rate that an initial population experiences during a sterilization cycle is a measure of the resistance of a BI to the selected sterilization process with the D-value as the slope of the survival curve being the defining parameter. If a strong sterilization effect is shown on the BI, the slope of the survival curve is steeper (i.e., the D-value drops and reliably negative results are observed earlier). Conversely, with a weak sterilization effect, negative results occur only very late, the survival curve is flat, and hence the D-value is high.

If a BI behaves as predicted by the survival time model during inactivation with a sterilization method, the D-value for this BI (and the change in its D-value depending on changes of the sterilization parameters) can be used to describe and quantify the sterilization capacity of the applied process. Therefore, the D-value is also a measure to quantify the inactivation effect of sterilization methods. In the next section, the different methods of D-value determination of BIs are introduced and discussed considering the suitability of the different results for the interpretation of a sterilization effect.

12.4.3. D-Value Determination

The following list serves to facilitate the discussion of methods and application of D-value determination:

Description of different methods for D-value determination
Selection of suitable method
Considering first the responding information versus process target
Extent and suitability for the application
Detailed description of the LSK method
The Minimized LSKM method
Iterative test procedure

12.4.3.1. Different Methods for D-Value Determination (From Ref. 1)

In general, two different standard methods for D-value determination of BIs are described (5,6,8,17). The first is the Survivor Curve Method (SCM) and the second is the Fractional Negative Method (FNM). For the FNM two different applications are described: the Limited Spearman Karber Method (LSKM) and the Stumbo Murphy Cochran Method (SMCM). All these methods are based on the survival time model, the assumption that the survivor curve forms a straight line over the whole course of inactivation and its origin $\{t = 0$ [min]; log $N_0\}$ used to calculate the D-value (see Fig. 3). Therefore, the initial population N_0 of the BI has to be known for all D-value determination methods. In practice, by all methods, several BIs are exposed to the inactivation and later removed at different time intervals or exposed to identical sterilization cycles with different durations. After the exposition the probes are evaluated using the defined method.

12.4.3.2. Survivor Curve Method (SCM)

With the SCM, the BIs are evaluated by counting the survived microorganisms of each exposed BI. Based on the relation of the exposure time and the recovered population after exposure, the survival curve is fitted and the D-value derived from this. The accuracy of this method, based on the numbers of BIs for each exposure time and the numbers of selected exposures, can be estimated using statistical standard methods. Valid for D-value determination with the SCM are results showing populations between less than 50% of the initial Population N_0 and A more than 5×10^1 microorganisms after the exposure (8,14,15). With this restriction the SCM for D-value determination covers only the survival window of the survival time model. Therefore, by using this method, it is possible to determine a D-value but not to give any evidence about the model behavior of the tested BI. Another disadvantage of this method is that the population recoveries require a huge amount of laboratory work and equipment.

12.4.3.3. Fractional Negative Methods (FNM)

With the FNM the final stages of microbial inactivation are captured. At this point of inactivation, the exposed populations are reduced to less than 5×10^0 organisms. Such low populations produce a ratio of positive and negative results as described in the model above. When the initial population and the exposure time of the test organism to the inactivation method are known, this fractional ratio is used to calculate the D-value. In contrast to single populations, the probability to observe (r) negative and (n-r) positive results

$$P_n(r) = nCr \times [P(m)]^r \times [1 - P(m)]^{n-r}$$

FIGURE 4 Probability distribution of fractional negative methods. (From Ref. 1.)

out of a inactivation run with (n) BIs, as it is used in FNM, can be expressed as shown in Fig. 4.

12.4.3.4. Stumbo Murphy Cochran Method (SMCM)

To perform SMCM, one group with a large number of BIs (e.g., 50) is exposed to bacterial reduction. The whole group is removed from the decontamination cycle at the final stages of microbial inactivation and evaluated using the growth test. The resulting ratio of BIs showing no growth to the number of exposed BIs is used to calculate the D-value and the 95% confidence interval for the D-value. The result is only valid for D-value determination by observing a defined fractional ratio (8,14,15). This makes it necessary to know a good estimation of the D-value in advance in order to achieve a valid result. Advantages of this method are, first, the evaluation of the BIs using the growth test as an uncomplicated tool, and second, the high statistical accuracy of the calculated D-value is caused by the large number of BIs used. But the SMCM covers the survival time model only up to the fractional field. Also, by using the SMCM, failures in the model behavior of BIs or in the sterilization effect of the applied process (which occurs only in the killing window—e.g., a late positive BI results) cannot be observed.

12.4.3.5. Limited Spearman Karber Method (LSKM)

For LSKM, several groups of BIs are exposed simultaneously to bacterial reduction. The groups are sequentially removed from the sterilization cycle at a constant time interval (d) and then evaluated using the growth test. The exposure times for the individual BI groups (and hence the intervals at which they are removed) are chosen so that the results of the LSKM provide a "map" or "reactive pattern" of the entire survival time model, from reliably positive results through the fractional field to reliably negative results. The accuracy of this method depends on the number of BIs chosen for each group and the time interval between each exposure. The LSKM enables one to calculate the D-value and the 95% confidence interval for the D-value. In the standards (5,6,8,17), different requirements are described for a result of a LSKM to be valid for D-value calculation. At least one group showing only positive BI results has to be observed in front of the fractional field and at least one or two all-negative groups have to be occur after the fractional field. Two to four groups have to cover the fractional field and thus show a result relevant for the calculation. The number of BIs per group is stated for up to 20 samples. In the USP (17), 10 BI per group are mentioned, but lower group numbers are explicitly permitted. Also, for the LSKM a known estimation of the D-value to be determined is a requirement in order to be able to observe a valid result. The big advantage of the LSKM is the visualization of the entire survival time model including the transition to all-negative results. Furthermore, the result is evaluated using the growth test without the need for great microbiological lab resources.

12.4.3.6. Selection of Method

All introduced methods for D-value determinations can be performed in modern systems utilizing an alternative sterilization method. It is only necessary to expose the BIs to the inactivation and remove them at selected time intervals. The main advantage of the FNM is the evaluation of the results using the growth test without great lab resources. The described methods are generally defined and used for specification and quality assurance of BIs for commercial supply. To use his method as a tool to determine a sterilization effect during process validation, it is required to be able to evaluate both resistance factors (first the D-value and second the corresponding model behavior). Moreover, the final target of validating a sterilization process is to observe a total kill of the exposed BIs. This target has to be included in the applied method. Only the LSKM provides this information and, therefore, enables a clear interpretation of the result.

12.4.3.7. Detailed Description of LSKM

In the following, the LSKM is described in more detail concerning the interpretation of results and the D-value calculation. Moreover, based on the LSKM, an experimental tool is established for the visualization of the reactive pattern of BIs and the quantification of the microbial reduction rate obtained.

12.4.3.8. D-Value Calculation Using the LSKM

With the LSKM, the "Mean Time to Sterility," U_{SK}, is calculated using the final stages of microbial inactivation from which the D-value can be derived in consideration of the initial population N_0 (5,6,8,17).

Equation legends:

U_i: = Exposures

U_1:   = Longest exposure with none of the BIs negative, all BIs showing growth

U_k: = First exposure with all of the BIs negative, all BIs showing no growth

U_{k-1}:   = Exposure prior to U_k

d: = Time interval between exposures
n: = Number of replicate BIs at each exposure
r_i: = Number of negative replicates at exposure U_i
N_0:   = Initial population of BIs
U_{SK}:   = Mean time for sterility
$V(U_{SK})$:   = Variance of U_{SK}

The Mean Time to Sterility U_{SK} [min] is calculated as follows:

$$U_{SK} = U_k - \frac{d}{2} - \frac{d}{n}\left(\sum_{i=1}^{k-1} r_i\right)$$

The D-value [min] is derived using the initial population N_0:

$$D-value = \frac{U_{sk}}{\log N_0 + 0.2507}$$

The variance of U_{SK} is expressed as follows:

$$V(U_{sk}) = s^2 U_{sk} = \frac{d^2}{n^2(n-1)}\sum_{i=1}^{k-1} r_i(n - r_i)$$

Finally a 95% confidence interval for the D-value [min] can be estimated:

$$95\%CI(D_{value}) = D_{value} \pm \frac{2 * \sqrt{V(U_{sk})}}{\log N_0 + 0.2507}$$

In the first of the two formulas above, the possibilities of increasing the accuracy of a D-value estimation with the LSKM are evident. First, increasing the number of BIs per group (n) decreases the variance of U_{SK} and therefore the accuracy of the estimation. Second, a lower time interval (d) between the exposures increases the resolution of the method and, therefore, also the accuracy of the estimation.

12.4.3.9. LSKM in Practice (From Ref. 2)

Figure 5 shows the results of a LSKM that complies with USP (17) requirements as to its execution and the trend of the individual group results. The experiment was carried out using a well-defined H_2O_2 decontamination cycle similar to those of the applied isolator example and second, commercially available BIs selected as described below.

Group 1 shows an overall positive result preceding the fractional field, which is formed by groups 2 through 6. Then come four all-negative groups. Based on these results, a D-value of 2.06 ± 0.2 min can be estimated with 95% confidence limits.

Group	1	2	3	4	5	6	7	8	9	10	Pos.	Neg.
Exposure time (min)	6.0	8.5	11.0	13.5	16.0	18.5	21.0	23.5	26.0	28.5		
Result 1	+	+	+	+	+	+	−	−	−	−	+	−
2	+	+	+	+	+	+	−	−	−	−	+	−
3	+	+	+	+	−	−	−	−	−	−		
4	+	+	+	+	−	−	−	−	−	−		
5	+	+	+	−	−	−	−	−	−	−		
6	+	+	+	−	−	−	−	−	−	−		
7	+	+	+	−	−	−	−	−	−	−		
8	+	+	+	−	−	−	−	−	−	−		
9	+	+	−	−	−	−	−	−	−	−		
10	+	−	−	−	−	−	−	−	−	−		

FIGURE 5 Results of LSKM. + = growth; − = no growth. (From Ref. 2.)

The results shown for the LSKM reflect the survival time model extremely well and permit accurate estimation of the D-value and of the 95% confidence limits for it. In order to be able to achieve such a result, the expected D-value must be known prior to performing the LSKM so that the time window for removal of the BIs can be precisely set. The narrow confidence limits for the calculated D-value results from the large number of groups and the short removal interval, which requires more than 100 BI samples (2).

12.4.3.10. Minimized LSKM

For the purposes of process development and as an estimating pretest prior to D-value determination, a precise D-value is less important than the mapping of the whole survival time model toward the bacterial reduction throughout the sterilization cycle. Figure 6 shows the results of a minimized LSKM. Three BIs were used per group and a removal interval of 3 min was selected. The BIs

Group	1	2	3	4	5	6	7	8	9	10	Pos.	Neg.
Exposure time (min)	6.0	9.0	12.0	15.0	18.0	21.0	24.0	27.0	30.0	33.0		
Result 1	+	+	+	+	−	−	−	−	−	−	+	−
2	+	+	+	−	−	−	−	−	−	−	+	−
3	+	+	−	−	−	−	−	−	−	−		

FIGURE 6 Result of minimized LSKM. + = growth; − = no growth. (From Ref. 2.)

and cycle parameters used were identical to those previously described for the LSKM in Fig. 5 (2). Groups 1 and 2 both test all-positive, groups 3 and 4 constitute the fractional field, and then come six groups with all-negative results. Using the LSKM formula, a D-value of 2.0 min can be estimated from these results.

A LSKM minimized in this way produces a good estimate of the D-value and maps the reactive pattern of the BI very clearly, even when compared with the complete LSKM. All in all, it provides a comprehensive description of the resistance behavior of the BI and a good estimation of the sterilization effect it was exposed to. Because of the lower number of groups and the larger intervals at which the BIs are removed, the resolution of the fractional field is not so detailed; however, the rapid transition and the large number of all-negative groups provide clear evidence of the effectiveness of the inactivation process. The sterilization effect obtained can be quantified from the estimated D-value. If necessary, a more precise D-value can be determined based on the result of the minimized LKM with a high confidence level to observe a valid pattern. Such an iterative test procedure enables an effective way to quantify a sterilization effect even when it is not possible to assume its range in advance (2).

The selected BI and its corresponding reactive pattern to the inactivation throughout a sterilization process plays a key role in process development. Only by knowing the reactive pattern of the BI used is it possible to interpret the results during the process development of an unknown sterilization system proper (2). The minimized LSKM is a suitable tool for visualizing the reactive pattern of BIs and for estimating and quantifying the sterilization effect obtained. Further on in this chapter, the minimized LSKM will be shown to be sensitive to changes in the parameters of the sterilization process and to the resistance behavior of BIs. In practice, the results obtained from a minimized LSKM with 30 BIs scores well in terms of cost versus benefit (2).

12.4.4. Selection of Biological Indicator

The following list summarizes the salient points in the selection of a suitable BI:

> Considering model behavior and resistance.
> Suitability for the applied process.
> BI versus process performance and requirements.
> Statement: The best BI reflects the process and reacts as the model described.
> Difficulties/problems.

At this stage of process development, a suitable microbiological system (i.e., BI) for the specific sterilization method has to be found in order to be able to quantify sterilization effect. This is one of the most important steps in process development and requires a deep understanding and correlation with the process performance and expectations.

A suitable microbiological system reflects the process performance and expectations defined and reacts during inactivation throughout the process as it is described by the survival time model. Each single component in the composition of a BI as microbiological system has to be properly chosen and justified to be useful to challenge the process. The requirements of the current standards [e.g., USP (9)] has to be considered as does the individual process target. In comparison to standard sterilization methods, a suitable microbiological system is neither defined in standard literature nor commercially available for most alternative sterilization methods. Consequently, during the development of an alternative sterilization method, a suitable microbiological system has to be established in parallel.

To handle this complex situation of describing an unknown sterilization process coupled with an unknown microbiological system, it is important to define first only one suitable microbiological system. With such an established microbiological system it is further possible to perform a screening experiment combining the inactivation process and the microbiological system (i.e., BI). Based on the results of this screening experiment, appropriate changes and adjustments of both systems can be defined to enable further process development steps.

In the following, a suitable microbiological system to be used for the H_2O_2 decontamination is first discussed and selected, then a screening experiment is conducted verifying the main assumptions of the process and the microbiological system. All single steps are described in detail, so it is possible to transfer the decision to different sterilization processes. In summary, a BI consists of a test organism with a defined population, a carrier and a primary package. All components have to be chosen in the light of the regulatory standards and the required process performace, even if the BI is commercially ordered and, therefore, already specified for the applied process.

12.4.4.1. Test Organism

In general, the test organism has to be highly resistant to the applied process (5–8). For standard sterilization methods, a suitable test organism is specified in the regulatory standards (5,7). The use of this specified test organism for process validation ensures first, a high process level regarding the sterilization effect, and second, a general acceptance by the authorities. If no test organism for an alternative sterilization method is stated in the standards or generally recognized, the first step is to define an adequate organism for the first trials. In general, spores of *Bacillus* species are known to be highly resistant to most sterilization methods. A further advantage of working with *Bacillus* spores is their high resistance to environmental influences, which makes it possible to handle spore preparations that are very stable and uncomplicated. By using spores of *Bacillus stearothermophilus* cross-contamination can be excluded by its high incubation temperature, which is unsuitable for most other organisms. If no test organism is predefined, starting with spores of *Bacillus* species seem to be a well-based approach. If such a high process level is not required or if the flora of microbes to be inactivated are exactly known, also these organisms (i.e., typical isolates during pharmaceutical production) can be used for process development.

For H_2O_2 decontamination according to international standards, spores of *Bacillus stearothermophilus* are specified for use in process validation (7). Two strains of this kind of spore are generally recognized. The strain ATCC 12980 is commonly used in the United States and the strain ATCC 7953 is often used in Europe. The difference in resistance of the two strains is shown later in this chapter. For the following example, *Bacillus stearothermophilus* ATCC 12980 spores are used as the test organism.

12.4.4.2. Initial Population

The definition of the initial population of the BI has to reflect first of all the defined target log reduction of the sterilization process. For most applications, the initial population is chosen at a level of 1×10^6 [CFU/carrier] or greater in consideration with the definition of sterility (9). For alternative sterilization methods the required process target is often different from those of standard sterilization processes and, therefore, BIs with lower initial populations are also required. However, especially for alternative sterilization methods, there are no standardized BIs commercially available reflecting the required process. Therefore, such special BIs often have to be manually prepared. Here a possible interaction between the initial population and the model behavior of a BI is important to understand. In the survival time model, the D-value of a BI is independent of its population. To reduce a 10^6 population by one log step requires the same time to reduce later in the inactivation process the same population from 10^{-3} to 10^{-4}. But often an interaction between the initial population and the D-value is observable when using manually prepared BIs. BIs with a lower initial population lead to lower D-values than BIs with a higher initial population of the same test organism.

This shift in D-value can be interpreted as a failure in the preparation of those BIs. Often such preparation failures lead to an enormous shift in D-value or to random late positive results even after a long exposure to the inactivation. To recognize such effects it is recommended to prepare BIs with different initial population for the first approach. For the H_2O_2 sterilization process development of the example, a initial population of 1×10^6 [CFU/carrier] or greater is chosen.

12.4.4.3. Carrier Material

The carrier material of the test organism has to be chosen in a way to be relevant for the sterilization process and the application. For a penetrating sterilization media like heat, the test organism can be added to a porous structure such as a thick paper strip or even to a liquid in a closed vial or ampule without protecting it from the sterilization media. This preparation will reflect the assumed process performance for the sterilization media, including the penetration into a structure or liquid. A surface decontamination process such as the H_2O_2 decontamination is applied to reduce a microbiological contamination on a exposed surface, not to penetrate into a surface or a closed volume. Consequently, the test organism has to be added to the surface of a carrier material, avoiding penetration of the test organism into the carrier material to reflect the process performance. Furthermore, the carrier material should be relevant for the surfaces to be sterilized. If the system to be sterilized consists of more than one surface, the authorities demand detailed information on the extent to which the carrier material used is associated with the sterilization effect on different surfaces in the system (18). In the example, the surfaces of the isolator system are mainly stainless steel and glass; therefore, as carrier material for the BI, stainless steel is chosen. Studies to estimate the effect of different carrier materials on the resistance of a test organism should be conducted later on.

12.4.4.4. Primary Package

Finally, the primary packaging of the BI has to be suitable for the applied sterilization process also. For this it is important that the primary packaging is permeable for the sterilization media. As mentioned above, the sterilization media heat penetrates even through glass; this is not possible for gaseous sterilization media. For most BIs specified for gaseous sterilization media, the semipermeable material Tyvek is used as primary packaging. Tyvek is permeable for gaseous media but not for liquids; therefore, it is called semipermeable. Furthermore, it is a sterile barrier during test handling and so avoids contamination of the carrier. As primary packaging for the BIs for the H_2O_2 example, Tyvek is selected. The influence of the Tyvek packaging on the sterilization effect could be evaluated later on, but the primary packaging can be assumed to be a worst-case situation compared to an unpacked carrier. If it is not possible to use primary packaging, unpacked carriers could also be used for process validation.

12.4.4.5. Commercially Available BIs

The advantage of commercially available BIs is that they are tested for their suitability for the specified sterilization process in advance. For standard sterilization methods, the test methods for BIs are standardized and so commercially available BIs for those processes show a high quality and a reproducible performance. For alternative sterilization methods, there are often neither specified BIs commercially available nor standard test methods described. Further on in this chapter it will be shown that this leads to incoherent results while testing such BIs and even a tested and specified commercial BI may not show suitable model behavior. Often the only solution to validate an alternative sterilization method is to prepare a suitable BI manually or to use a useful BI composition specified for another similar sterilization method. For the H_2O_2 decontamination, there are different compositions of BIs commercially available from different suppliers.

12.4.4.6. Selected BIs

A BI selection of this kind leads to the best assumable BI for the specific process by a consequent consideration of each BI component regarding its impact on the process. For the surface decontamination of the isolator system with gaseous H_2O_2, the following commercially available BI composition is selected:

Test organism:	Spores of *Bacillus stearothermophilus* ATCC 12980
Initial population:	$\geq 1 \times 10^6$ [CFU/carrier]
Carrier material:	Stainless steel
Primary packaging:	Tyvek
Specified inactivation:	Gaseous Hydrogen Peroxide

With the BI selection, the first step in process development, a screening experiment, is carried out to attain an idea of the assumed process performance and reaction of the selected BI. Furthermore, all other commercially available BIs for H_2O_2 decontamination are tested later on in process development, to compare their resistance and model behavior in consideration to the survival time model.

12.4.5. Screening Experiment (From Ref. 1)

The following activities will make up the screening experiment for the selected BI:

Identify suitability of the combined systems.
Provide sensitivity of target value.
Identify useful range for main experiment.

The screening experiment is the first step in process development, where the applied sterilization process and the selected BIs are tested in combination. The purpose of this experiment is to identify any sterilization effect by using the selected process and BI and also the suitability of the combined systems for further steps. A sensitivity range of the target D-value is provided by changing the main factor of the process. This makes it possible to identify the suitability of the D-value as target value for the sterilization effect and to choose a useful experimental range for the main experiment. On the other hand, the suitability of the selected BI for the sterilization process has to be proved.

By the screening experiment, both unknown systems are combined: The applied sterilization process and the selected BI. To be able to properly interpret the resulting sterilization effect and to identify which systems fail and cause a poor sterilization effect, the variables in the screening experiment have to be reduced to a minimum. A possible way to proceed is to use only the selected BI and to keep all process factors of the system stable in a useful range, then to conduct D-value determinations at different values of the assumed main factor of the applied sterilization process. The suitability of the process and the selected BI can then be interpreted using the resulting D-values and the corresponding reactive pattern of the BIs; the changes in D-value corresponding to a change in the main process factor can be interpreted as an influence of the main factor on the sterilization effect. On the other hand, the corresponding reactive pattern to the observed D-values can be interpreted in consideration of the suitability of the selected BI. If the D-values change but the reactive pattern of the BI doesn't reflect the survival time model, the selected BI is not suitable for the process and has to be adjusted or replaced. If the resulting D-value doesn't change with the main factor, the process is in a steady state, or the values for the main factor are not adequately chosen (if no sterilization effect at all is observable), the process development has to be stopped and all development steps leading to the screening experiment need to be repeated.

12.4.5.1. Example for Gaseous H_2O_2

In the following example, the screening experiment for the introduced surface decontamination of an isolator system with gaseous H_2O_2 is shown in its design, performance, and result. The result is discussed considering the suitability of the D-value as target value for the main experiment, the definition of a useful experimental range for the main experiment, and the suitability of the selected BI. All decisions and details included in this example are adaptable to other sterilization methods.

12.4.5.2. Experimental Design

The main factor in the H_2O_2 decontamination, the initial quantity of vaporized H_2O_2 per volume (q1), is assumed. Three values for the main factor (q1) are chosen to be used in the screening experiment: 4.3, 6.5, and 9.8 g/m^3. These values are chosen to cover the whole range of air saturation during the process, from a low saturation at 4.0 g/m^3 to an oversaturation at 10.0 g/m^3. All other parameters of the system are kept stable during the screening experiment, especially the initial conditions of the chamber such as temperature at 35°C and humidity at 15% rH. The air velocity is set to 0.25 m/s. A liquid H_2O_2 solution of 35% is chosen and the vaporizing rate during sterilization is set to 30%/h.

12.4.5.3. Experimental Performance

For each value of the main factor (q1) a D-value is determined using LSKM and iterative testing. To do so, the BIs are exposed under the HEPA filter where the sterilization media enters the chamber. At this position, all conditions are shown to be homogeneously distributed and a good sterilization effect can be expected. The influence of secondary effects on the experimental results is thus excluded. The performance of such experiments, especially of LSKM, in sterilization systems requires some special tools to make the experiments reproducible and, therefore, the results trustworthy. In an isolator system, the BI handling during decontamination can be easily done by using the glove ports. To expose all groups of BIs at a defined time, the groups are sealed in gas-tight foil prior to beginning decontamination. To remove the BI out of the chamber a special port is assembled on the isolator.

The test handling for all trials of the screening experiment and for all other experiments shown in this chapter can be described as follows:

The gas-tight sealed BIs are placed into the isolator chamber.
The required initial conditions in the chamber are established
The vaporization of H_2O_2 is started to the required quantity (q1)

After reaching the required value for (q1) the BIs are exposed by
opening the gas-tight foil and placing them at the defined position

During the decontamination phase, H_2O_2 is dosed at the required
vaporizing rate (q2)

The BIs are removed from the inactivation atmosphere during
decontamination phase at a defined time interval.

After minor training such test handling can be done routinely in each isolator
system. For systems different from isolators, a similar test procedure can be
assumed and realized. It is important to find a position in the system where the
conditions are well distributed and secondary effects can be excluded. This
position should represent a best place for sterilization effect. Furthermore,
tools for a defined exposition and removal of the BIs are needed.

12.4.5.4. Result of Screening Experiment and Interpretation

After the execution of the single trials, the result for the screening experiment
is determined as shown in Table 1.

All D-value determinations show a valid result in the light of the conventions for LSKM and also by a reactive pattern according to the survival
time model. Furthermore, the D-values are shown to be reproducible at the
same experimental conditions by using the iterative test method. Therefore,
the selected BI can be assumed to be suitable for the sterilization method
applied. Finding a suitable BI for an alternative sterilization method at the
first trials happens only by a consequent selection of each BI component in
coordination with the applied process as described above. The kind of
observations that can be done during the selection of a BI, especially with
unsuitable BIs, and the interpretation of the observations according to the
goal of describing a sterilization method is discussed below.

For further interpretation of the screening experiment, the D-value
versus the quantity (q1) is plotted on a graph (see Fig. 7). In this graph the
relationship between the vaporized quantity of H_2O_2 and the D-value is
visible. The D-value changes to nonlinear by increasing the quantity of
vaporized H_2O_2 in a range from 1.4 to 7.9 min. By vaporizing H_2O_2 into

TABLE 1 Results of Screening Experiment

(q1), Quantity of H_2O_2 [g/m^3]	D-value [min]
4.3	7.9
6.5	1.6
9.8	1.4

Source: Ref. 1.

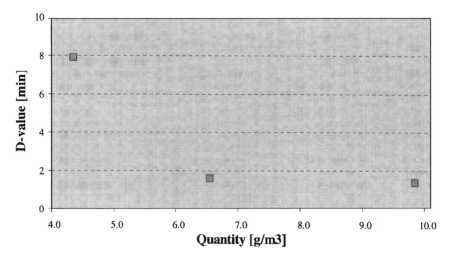

FIGURE 7 Plot of D-values versus quantity of H_2O_2, (q1). (From Ref. 1.)

the chamber, a decontamination effect is established up to a certain level. After reaching this level of inactivation, vaporizing more H_2O_2 does not change the established decontamination effect significantly. This shows a useful sensitivity of the D-value to changes in the decontamination effect and a wide range of D-values over the selected process range, thus providing a good resolution of the quality of the changes in decontamination effect.

All in all, based on the results of the screening experiment, it can be stated that a D-value determined using the LSKM provides a useful measure to quantify a decontamination effect and, therefore, it provides a suitable target value for experiments focused on the decontamination efficiency.

A suitable range of the following main experiment can be selected based on the results of the screening experiment. It can be assumed that secondary influencing factors on the decontamination effect are not observable in the process range of stable D-values, where even the assumed main effect does not show any influence any longer. This information is useful for the design of the following main experiment, especially for the selection of a useful range for the experimental factors.

By using the D-value as a measure for the quantification of a sterilization effect and LSKM as methodical tool for the determination of this target value, the following basic conventions have to be kept in mind:

The determined D-value represents the entire process, including the applied microbiological system and the sterilization method used, its parameter and equipment.

The assumptions of the empirical survival time model have to be in accordance with the real inactivation of the microbiological system used.

In the end, the interpretation of results and data based on D-values has to be done with the consideration that the D-value is not an absolute figure, but it is a value including deviations.

12.4.6. Suitability of BI (From Ref. 1)

12.4.6.1. Examples and Interpretations of Reactive Pattern Recognition (From Ref. 2)

During process development of alternative sterilization methods and finally during routine validation and revalidation of an applied process, the suitability of the used BI is an important milestone. In order to be able to interpret the entire results properly, if the BI is to be used as a sensor for specifying the sterilization process, its resistance and model behavior must be known in advance. Performing a reactive pattern recognition of the model behavior of the BI using the minimized LSKM allows its D-value to be estimated and hence assess its suitability for use in development of the process and during routine validation.

The results of reactive pattern recognition performed on different BIs are presented and interpreted below. All BIs used are commercially available and are specified for gaseous H_2O_2. The reactive pattern recognitions were all carried out in a test chamber with identical cycle parameters. Where necessary, the LSKM parameters (exposure time, removal interval) were adjusted to the specific resistance of the BIs. The examples shown focus on different strains of the selected test organism and variations between different lots of the same BI composition. In comparison to those results, a reactive pattern recognition of a BI not specified for gaseous H_2O_2 is also shown. The observed results and the interpretations of the representative examples for gaseous H_2O_2 are helpful to determine a suitable BI for other sterilization methods (2).

12.4.6.2. Example: *B. stearothermophilus* ATCC 7953

In the example below two different lots of the same BI composition using *B. stearothermophilus* ATCC 7953 are shown. The lots in the example represent the observed variation between different lots of the same BI over more than 3 years (1).

a.   Example: BI A, Lot 1 (1)

Test microorganism/ATCC no.	*B. stearothermophilus*/7953
Initial population [CFU]	3.5×10^6
Carrier material	CrNi steel
Primary packaging	Tyvek

Group	1	2	3	4	5	6	7	8	9	10	Pos.	Neg.
Exposure time (min)	6.0	9.0	12.0	15.0	18.0	21.0	24.0	27.0	30.0	33.0		
Result 1	+	+	+	+	+	−	−	−	−	−	+	−
2	+	+	+	+	−	−	−	−	−	−		
3	+	+	+	−	−	−	−	−	−	−		

+ = growth; − = no growth.

BI *A*'s reactive pattern is consistent with the model, and a D-value of 2.4 min can be estimated from the results of the minimized LSKM.

b.   Example, BI A. Lot 2 (1)

Test microorganism/ATCC no.	*B. stearothermophilus*/7593
Initial population [CFU]	3.2×10^6
Carrier material	CrNi steel
Primary packaging	Tyvek

Group	1	2	3	4	5	6	7	8	9	10	Pos.	Neg.
Exposure time (min)	6.0	9.0	12.0	15.0	18.0	21.0	24.0	27.0	30.0	33.0		
Result 1	+	+	−	−	−	−	−	−	−	−	+	−
2	+	−	−	−	−	−	−	−	−	−		
3	+	−	−	−	−	−	−	−	−	−		

+ = growth; − = no growth.

This lot of BI *A* shows still a consistent reactive pattern with the survival time model, and based on the result a D-value of 1.3 min can be estimated.

12.4.6.3. Example: *B. stearothermophilus* ATCC 12980

In the example below two different lots of the same BI composition using *B. stearothermophilus* ATCC 12980 are shown in comparison to the first example. The lots in this example also represent the observed variation between different lots of the same BI over more than 3 years.

a. Example: BI B, Lot 1 (1)

Test microorganism/ATCC no.	*B. stearothermophilus*/12980
Initial population [CFU]	2.7×10^6
Carrier material	CrNi steel
Primary packaging	Tyvek

Group	1	2	3	4	5	6	7	8	9	10	Pos.	Neg.
Exposure time (min)	6.0	9.0	12.0	15.0	18.0	21.0	24.0	27.0	30.0	33.0		
Result 1	+	+	+	+	−	−	−	−	−	−	+	−
2	+	+	+	+	−	−	−	−	−	−		
3	+	+	+	+	−	−	−	−	−	−		

+ = growth; − = no growth.

BI *B* shows an acceptable reactive pattern with a D-value of 2.5 min. The lack of fractional groups results from minimizing the LSKM. However, the suitability of the BI for cycle development can nevertheless be assessed.

b. Example BI B, Lot 2 (1)

Test microorganism/ATCC no.	*B. stearothermophilus*/12980
Initial population [CFU]	2.5×10^6
Carrier material	Paper strip
Primary packaging	Coated paper

Group	1	2	3	4	5	6	7	8	9	10	Pos.	Neg.
Exposure time (min)	6.0	9.0	12.0	15.0	18.0	21.0	24.0	27.0	30.0	33.0		
Result 1	+	−	+	−	−	−	−	−	−	−	+	−
2	+	−	−	−	−	−	−	−	−	−		
3	+	−	−	−	−	−	−	−	−	−		

+ = growth; − = no growth.

The reactive pattern of lot 2 of BI *B* is also consistent with the survival time model, and a D-value of 1.3 min can be estimated from the results of the minimized LSKM. The all-negative group at an exposure time of 9 min followed by a fractional group at a 12 min exposure represents once more the random results in between the fractional field. BIs *A* and *B* are both shown to be suitable for the applied sterilization process. The variation in D-value between the different lots of both BIs can be estimated by an a factor of 2, whereas a difference between the used strains of *B. stearothermophilus* cannot be stated. Further tests with other spores of *Bacillus* species and the same BI composition also showed no significant difference in D-value compared with each other. Here the variation between lots is larger than the variation between various test organisms. By using these suitable, commercially available BIs for process development and further validation work, the variation between lots must be recognized and implemented into the process.

c. Example BI C; From Ref. (2)

Test microorganism/ATCC no.	*B. stearothermophilus*/12980
Initial population [CFU]	4.5×10^5
Carrier material	Glass fiber pad
Primary packaging	Tyvek

Group	1	2	3	4	5	6	7	8	9	10	Pos.	Neg.
Exposure time (min)	5.0	10.0	15.0	20.0	25.0	30.0	35.0	40.0	45.0	50.0		
Result 1	+	+	+	+	+	−	+	−	−	−(*)	+	−
2	+	+	−	−	−	−	−	−	−	−		
3	+	−	−	−	−	−	−	−	−	−		

+ = growth; − = no growth.

BI *C* has a large fractional field, extending from group 2 to group 7, which, along with the estimated D-value of 2.6 min, departs from the survival time model. This casts doubts on the reliability of the all-negative groups 8 to 10. In an additional experiment using a larger time window, positive results were also obtained for this BI at exposure times of up to 50 min (*) (2).

d. Example BI D (2)

Test microorganism/ATCC no.	*B. stearothermophilus*/7953
Initial population [CFU]	1.0×10^6
Carrier material	Glass fiber pad
Primary packaging	Tyvek

Group	1	2	3	4	5	6	7	8	9	10	Pos.	Neg.
Exposure time (min)	6.0	9.0	12.0	15.0	18.0	21.0	24.0	27.0	30.0	33.0		
Result 1	+	−	+	+	+	+	+	−	+	+	+	−
2	+	−	+	+	+	+	+	−	−	+		
3	−	−	−	+	−	+	−	−	−	+		

+ = growth; − = no growth.

BI *D* shows the first negative result after exposures of 6 min in group 1; however, a clear transition to all-negative results is not observed even after exposure times of 33 min in group 10. In additional trials to determine the resistance of this BI, this stochastic pattern of positive and negative results was found at exposure times of up to 70 min.

It is not possible to specify or develop a sterilization process with BIs of type *C* and *D*. When using BIs with such reactive patterns, random late-positive results are found over the whole course of development, preventing unambiguous interpretation of the experimental data observed. Changes to the cycle variables do not produce any reliable effect on the bacterial reduction. On the other hand, BI *A* and *B* show a reactive pattern which is in line with the model specified in the referenced standards. Random, late-positive results that are attributable to the model behavior of the BI can be excluded with these BIs. Hence, the experimental results can be projected directly to the bacterial reduction obtained. Thus, it is possible to detect insufficient bacterial reduction and, if appropriate, inhomogeneities in the distribution of bacterial reduction (2).

e. Example BI E, Not Specified for Gaseous H_2O_2 (8)

Test microorganism/ATCC no.	*B. stearothermophilus*/12980
Initial population [CFU]	2.5×10^6
Carrier material	Paper strip
Primary packaging	Coated paper

Group	1	2	3	4	5	6	7	8	9	10	Pos.	Neg.
Exposure time (min)	10.0	20.0	30.0	40.0	50.0	60.0	70.0	80.0	90.0	100.0		
Result 1	+	±	+	+	+	+	+	±	+	+	+	--
2	+	±	+	+	+	+	+	±	±	--		
3	±	±	±	+	±	+	±	±	±	--		

+ = growth; − = no growth.

With BI *E*, the first negative results are observable after the maximal exposures time of 100 min in group 10; a clear transition to all-negative results cannot be observed in between the applied exposures times. The D-value has to be estimated as more than 14.8 min.

With BI C, D, and E, a reliable inactivation of the applied test organism is not observable, whereas the same organism is inactivated within the convention of the survival time model with BI A and B. Looking at the results of BI C, D, and E could lead to the assumption the applied sterilization method is not sufficient to reach the required sterilization target. Only by reflecting the previous set requirements of the sterilization application in consideration of the composition of the used BI, can the observed results be clearly interpreted. BI C and D both use a glass fiber pad as carrier material for the test organism. Glass fiber is a porous structure into which applied organisms may be absorbed; therefore, prevention of the accumulation of organisms during the preparation is unreliable. To inactivate such a BI, the sterilization media has to penetrate into the glass fiber pad or be pushed actively through the pad to reach the test organisms. The target of the applied H_2O_2 decontamination is to inactive organisms on surfaces, not to penetrate into surfaces. BI E consisting of a paper strip as carrier material leads to the same problem as the glass fiber pad. Furthermore, the coated paper used as primary packaging is not permeable to gaseous H_2O_2.

BI C, D and E do not represent the required process target for the surface sterilization with gaseous H_2O_2. Moreover, they all represent a higher challenge for the applied process than the decontamination with gaseous H_2O_2 is able to perform. Only a penetrating process like heat sterilization is able to inactivate this BI within the described model. Therefore, the results of BI C, D, and E show, first, that the BIs are not suitable for H_2O_2 sterilization; second, they show clearly the borders of the performance of the applied surface decontamination with gaseous H_2O_2.

A suitable BI for a sterilization method represents, on the one hand, the process requirements and expectations of the applied method; on the other hand, it reacts during its activation as described in the survival time model. Furthermore, it must be understood that too low a sterilization effect leads to an inadequate BI result even with a suitable BI.

With the identification of a suitable BI and the knowledge of the influence of the main factor on the sterilization effect, a level of understanding is reached; based on this, it is possible to conduct further experiments. These experiments should focus more deeply on the influence of single-process factors on the sterilization effect and their meaning for the course of the sterilization cycle (1).

12.5. PROCESS-INFLUENCING FACTORS (From Ref. 1)

For each pharmaceutical process, especially sterilization processes, the knowledge of factors influencing its performance is a basis for a proper validation and control of the process. Furthermore, for each established process factor, a valid range with defined boundaries has to be determined. With the development tools established up till now, including the D-value of a biological indicator as the measure for sterilization effect and its determination, the selection of a suitable BI, and the proven sensitivity of the D-value to the specific process, it is possible to conduct studies focusing on process factors influencing the sterilization effect. Such an experiment should establish a significant, quantitative database including all necessary process factors in order to be able to choose the optimal parameters and control them in a defined range. Further, this database should enable one to optimize the process between the individual application and to finally describe the process concerning the sterilization effect. Therefore, the experiment has to cover all previous identified process factors within their relevant range. The target value of such an experiment has to be the D-value of a suitable BI. To ensure that such a study leads to the required information and the result can be interpreted statistically significant with respect to the extent of time and costs, the Design of Experiment method (DoE) is applied. The factors influencing the decontamination effect of the H_2O_2 surface decontamination process will be established as an example using the DoE method and the introduced process and equipment. All necessary steps are described in a way adaptable to other, different sterilization processes (1).

12.5.1. H_2O_2 Surface Decontamination Process (From Ref. 1)

12.5.1.1. Process Factors

As described above, the following factors of the H_2O_2 surface decontamination are identified as process critical or useful for the interpretation of the process performance; see Table 2.

The selected factors represent main physical factors and process related factors of the applied sterilization method. Based on this selection it would be later possible to control and monitor the sterilization process in compliance with the standards for pharmaceutical manufacturing. Additional process factors, especially hardware-related factors, should be analyzed later on, if required. During the study, all other factors are kept stable based on predefined values as described above. All selected factors, including their defined range, have to be implemented into the study to determine their influence on the decontamination effect. The factors can all be adjusted continuously within the defined range. For factor A, the quantity of vaporized H_2O_2, a nonlinear relationship to the D-value is established in the screening experiment. A nonlinear relationship of the other factors and also interaction between the single factors cannot be excluded. As a target value for the study the D-value of a defined BI is used. The D-value is determined by using the

TABLE 2 Relevant Process Factors of the H_2O_2 Decontamination

Factor ID	Description	Range low	Range high	Unit
A	Quantity of vaporized pure H_2O_2 (q1)	4	10	g/m^3
B	Rate of vaporized H_2O_2 during sterilization (q2)	20	120	%/h
C	Temperature	25	40	°C
D	Humidity	5	25	%rH
E	Concentration of aqueous H_2O_2 solution	30	55	%

Source: Ref. 1.

LSKM method and the iterative test procedure as described previously. Furthermore, all experimental trials of this study are carried out using the procedure described above (1).

By looking at the five selected factors and the defined range for each factor, it becomes clear that the study has to be properly designed and planned in order for one to interpret the results usefully. All combinations between factors and values of factors has to be exactly selected so that finally the required information can be extracted without any doubts. Moreover, the number of experimental trials should be minimized to an extent balanced in cost and information. These requirements can only be fulfilled by using the DoE method.

12.5.2. Design of Experiment, DoE (From Ref. 1)

The Design of Experiment will accomplish the following:

Show the statistical significance of the parameter of each single factor
Be exactly adjustable for the needed information
Result in extremely effective information with an acceptable numbers of trials

For planning, execution, and evaluation of the experimental runs, the Design of Experiment (DoE) method is used. By using this method an empirical experimental model for the study is formed, based on the defined parameter and the expected result. This empirical model defines a theoretical environment in which the study takes place and the resulting parameters are valid. In this empirical model the experimental factors are arranged in a way that all parameters of the needed effects can be independently estimated. The number of required experiments is reduced to a minimum by combining all factors and factor levels without losing any information. The data determined by practical experimental runs are analyzed using statistical methods.

The advantages of applying the DoE method are (a) the statistically significant estimation of the effect of each single factor, including its direction and relationship to influences of all other implemented factors, (b) the DoE allows building an experimental model exactly for the needed information using acceptable numbers of trials, and (c) results in statistically based and extremely effective information.

In the next section, the application of the DoE method, especially the model analysis and justification, is not described in detail; however, each major step in proceeding with the DoE is mentioned and commended in order to be able to follow the process development of the decontamination method. For a deeper explanation of the DoE, see Ref. (19).

12.5.3. Main Experiment

The following forms an outline for the steps involved in the main experiment:

> Definition of the experimental range
> Definition of the kind and extent of the used information
> Definition of a suitable DoE model
> Choosing the required module structure leading to plan of experimental runs
> Runs and statistical evaluation of results
> Identification and characterization of main effects versus process

12.5.3.1. Definition of the Experimental Range and the Empirical Model (From Ref. 1)

For the five selected experimental factors (see Table 2), an empirical model has be to built. The following information is available for model building:

> All five factors are continuous in their range.
> The main factors are assumed to be nonlinear.
> Interaction between factors cannot be excluded.

The following effects are all of equal interest and have to be independently estimated by the selected model:

> All main effects of the factors
> All quadratic effects of the factors
> All effects of two factor interactions

Using the DoE method, an experimental design with the required properties can be chosen. For the stated example a Central Composite Design is selected as an empirical model. A Central Composite Design is a full quadratic model consisting of a Fractional Factorial Plan (FF), one Center Point (CP) and, for each dimension, two Star Points (SP). The structure of such a design in two dimensions is shown in Fig. 8 (1). The Fractional Factorial Plan is used for the estimation of the main effects, the Center Point is needed to estimate the quadratic effects, of the factors, and the Star Points are used to clearly separate the quadratic effect from each other (1,19).

The corresponding empirical model to the selected Central Composite Design is shown in Fig. 9. The model complies with a empirical model of the second order and enables the independent estimation of all main effects, quadratic effects and two factor interactions as required. Because of the spherical structure of the design, five levels for each factor are required. This is possible without a large effort with the selected factors (1).

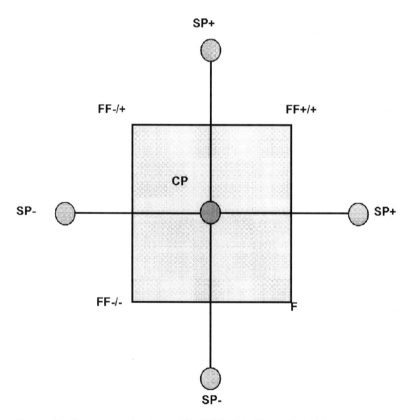

FIGURE 8 Structure of selected DoE Model. (From Ref. 1.)

$$Y = \beta_0 + \beta_1 x_1 + \beta_2 x_2 + \beta_{12} x_1 x_2 + \beta_{11} x_1^2 + \beta_{22} x_2^2 + \varepsilon$$

| Constant | Main Effects (linear) | Interactions | Quadratic Effects | Error |

FIGURE 9 Empirical model corresponding to the selected DoE. (From Ref. 1.)

The characteristic of such a design is to be orthogonal and able to rotate. The orthogonal structure enables the independent estimation of all effects, and the ability to rotate is linked to the error of the estimated effect and its direction (1,19).

12.5.3.2. Experimental Runs

The selected design structure defines the experimental runs required for analyzing the study. The five defined levels of factors corresponding to the design structure are shown in Table 3 (1).

For such an orthogonal design structure including five factors on five levels, a minimum of 31 experimental trials need to be executed: 16 trials using the levels of the Fractional FactorialPlan, 10 trials on Star Point levels, and 5 experiments using Center Point conditions. In practice, for an experiment with, say, Center Point Conditions, the following parameters for the factors are used (1):

A Quantity (q1) 6.5 g/m^3
B Redose (q2)   70%h
C Temperature   35°C
D Humidity   15%rH
E Aqueous H_2O_2 42.5%

TABLE 3 Factor Levels Corresponding to the Design Structure

Factor ID	Description	Unit	SP−	FF−	CP	FF+	SP+
A	Quantity (q1)	g/m³	4	5	6.5	8	9
B	Redose (q2)	%h	20	40	70	100	120
C	Temperature	°C	26	30	35	40	44
D	Humidity	%rH	6	10	15	20	24
E	Aqueous H_2O_2	%	30	35	42.5	50	55

Source: From Ref. 1.

For all 31 experiments the required factor levels are defined in a plan of experiments defined by the DoE method. The following conventions are defined for the execution of the experimental runs:

> To exclude external influences on the results, the experimental runs have to be randomized during the execution.
> At least one Center Point experiment has to be executed at the beginning, in the middle, and at the end of the study to exclude an influence over time.
> For each experiment, a D-value based on a valid result of the LSKM has to be determined.
> For the execution of the trials, the method described previously is applied.

After the execution of the experimental runs and the determination of the D-values, the data of the study are analyzed using statistical methods. For such a complex model, it is an advantage to use computer software to support the statistical work.

The model structure used in the example represents a more complex experimental design. Such designs are only useful if the sterilization process is described in real detail or the data should be used for a general description of a sterilization method. As an example, for the description of a BIER vessel (Biological Indicator Evaluator Resistometer) for standardized testing of BIs, such a complex design is adequate. For process development or improvement of a single application, experimental designs of lower complexity are often adequate to generate the required data. Also, the number of experimental runs corresponds with the complexity of the design. Therefore, the selection of the experimental factors, the required data, and the experimental design needs to be properly structured in order to find the best solution for the particular application. The single steps in proceeding with a study using the DoE method are independent from the finally chosen model. Next, the data of the main study using the introduced DoE are analyzed.

12.5.3.3. Statistical Analysis of Results

The following represent the steps involved in the statistical analysis of results obtained:

> Statistical work to do
> Choice of model for the process
> Plausible interpretation of process model

The data determined by practical experimental runs—here, the minimum 31 D-values of the different decontamination runs—are analyzed using statistical methods. The single steps of the statistical analysis are not described in detail but all steps necessary to follow the analysis are mentioned and commented on. As described above, the analysis focuses on the interpretation of the results in consideration of the decontamination process. After the experimental runs are analyzed using a Correlation Matrix and the range of the observed D-values justified by Summary Statistic, a complete model of the process, including all possible effects, is established. This complete model is then analyzed using the Analysis of Variance Method. Therefore, each single effect of the model is justified by its level of significance at which it describes the observed variations in D-value. By a stepwise elimination of insignificant factors the process model is reduced to only valuable factors needed to describe the observed variation in D-value significantly. The minimized process model is then justified by an analysis of the model residuals and a plausible interpretation of the model factors.

12.5.3.4. Final Statistical Process Model

After elimination of all insignificant effects, the complete model is reduced from 21 factors to 8 significant factors. This minimized process model is used for all further steps of process development. First, the model is justified using statistical tools. Then, the single effects of the process model are used for a plausible interpretation in consideration with the decontamination process. Figure 10 shows the standardized Pareto Chart for the process model finally chosen. For this model a Correlation Coefficient R^2 of greater than 90%

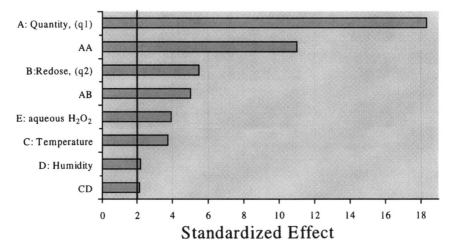

FIGURE 10 Final statistical process model. (From Ref. 1.)

between the selected model factors and the observed variations in D-value during the experimental runs can be estimated. This means that the selected process model describes closely the decontamination effect established in the system by the selected factor. Also, there are enough Degrees of Freedom left to estimate the error of the final process model properly. The Analysis of Residuals of the whole model and also of each single effect of the model shows no irregularity that would call into question the selected process model. Therefore, from the statistical point of view, the selected process model is highly significant for describing the observed variations in D-values during the main experiment. The next step is to analyze the selected process model in consideration of the decontamination process itself. Therefore, the previous process assumptions are used. At this point, the model analysis requires more subject matter knowledge than statistical tools to lead to the right decisions.

12.5.3.5. Complete Process Model

All selected factors of the main experiment are included in the final process model (see Fig. 10). This proves first the proper selection of the experimental factors prior to the study and, further, all assumptions made in consideration of the process. Factor A, Quantity (q1), is included in the model as a highly significant main effect and also as quadratic effect AA. This reflects the results of the screening experiment with its also nonlinear relationship. Additionally, factor A is included as part of the interaction AB. Together with the highly significant effect of factor B, Redose (q2), the quantity of pure H_2O_2 vaporized in the course of a decontamination cycle forms the main effect for the observed D-values. This corroborates once more the assumption made prior to the study.

A linear effect of factor E (aqueous H_2O_2) and factor C (temperature) is further significant for the process model. Factor D (humidity) and the interaction CD are part of the model too, their level of significance is low and near the border of significance of the process model. But especially factor D (humidity), although highly debated in H_2O_2 decontamination studies, should therefore remain in the model. The three factors E, C, D and also the interaction CD are all secondary factors in the decontamination process, compared to the main effects of factor A and B. All in all, the selected process model reflects the previous assumptions of all implemented factors and, therefore, the models seems to be suitable for describing the relationship between process factors and decontamination effect of the applied H_2O_2 decontamination.

12.5.3.6. Main Effects and Interactions

The next step in analyzing the process model is to interpret and justify each single effect of the model. Also, for the single effects, it is important to interpret the level and the direction of the effect in consideration of the process to come to a plausible justification for each single, significant model factor. During the interpretation of the single model factors, the impact of those effects and results on the decontamination process with gaseous H_2O_2 is directly interpreted too. Moreover, the results of the five Center Point experiments are interpreted in advance. So, during this analysis, the relationship between process factors and sterilization effect becomes very clear, step by step. Finally, one main interrelationship between all of the model factors should be found characterizing the decontamination process.

12.5.3.7. Results of Center Point Experiments (CP)

The following D-values and a 95% CI for D-value are determined as result of the five Center Point experiments:

CP 1 1.75 ± 0.12 [min]
CP 2 1.80 ± 0.10 [min]
CP 3 1.74 ± 0.11 [min]
CP 4 1.71 ± 0.14 [min]
CP 5 1.81 ± 0.12 [min]

The five observed D-values at Center Point conditions represent comparable results within a small range of deviation. The results indicate that the decontamination effect of the application is reproducible over the whole duration of the study. Therefore, as a characterization of the decontamination process, it can be stated, "Constant process parameters lead to a reproducible decontamination effect." Reproducibility is one of the main requirements of pharmaceutical processes. The H_2O_2 decontamination system used seems to fulfill this requirement. This is a major step in process development of the application, and once more the D-value is shown as a suitable measure for the decontamination effect.

12.5.3.8. Factor A, Quantity (q1)

Factor A, Quantity (q1), represents—with a change in D-value of 2.9 min between the factor levels—the highest effect of the model. The D-values decrease by increasing the vaporized quantity of H_2O_2 (see Fig. 11). This relationship makes sense, considering the process, and follows the previous assumptions. The nonlinearity observed in the screening experiment is confirmed with the quadratic character of the effect. Moreover, the relationship between quantities and D-value is comparable to those of the screening experiment.

By combining the results of the main experiment with those of the screening experiment, the character of the effect of factor A can be estimated in a range higher than the model (see Fig. 12). The stated quadratic fit of the

Main Effect Plot for D-value

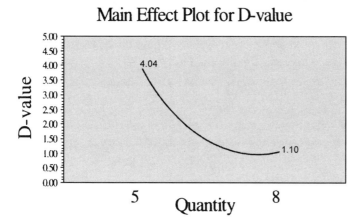

FIGURE 11 Main effect plot of factor A, quantity, (q1). (From Ref. 1.)

main experiment changes into a more asymptotic character. The sterilization effect is established by vaporizing H_2O_2 into the chamber. At a certain point of the process, the sterilization effect stays stable although the quantity of vaporized H_2O_2 is increased. This indicates that the sterilization effect in the system ends at a final maximum level but is not aborted by vaporizing more H_2O_2 than required.

Plot of D-value versus Quantity

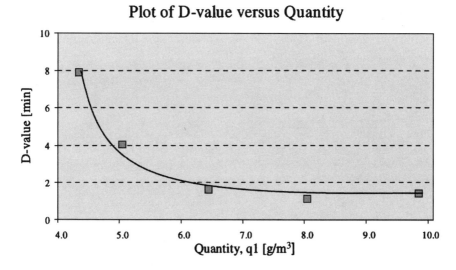

FIGURE 12 Main effect, factor A, including screening experiment. (From Ref. 1.)

For a characterization of the process, the effect of the vaporized quantity of H_2O_2 is summarized as follows (1): "By starting with constant initial conditions, increasing the initial vaporized quantity of H_2O_2 leads to a final maximal sterilization effect in the chamber."

12.5.3.9. Factor B, Redose (q2)

Factor B, Redose (q2), represents the rate of vaporizing H_2O_2 during the sterilization phase. It shows between the chosen levels an effect of -0.9 min on the D-value (see Fig. 13). In comparison to the effect of factor A, the direction and the observed value of factor B are plausible. Considering the process by looking on the effect of factor B, it would be possible to state that the higher the rate of Redose chosen, the better the sterilization effect in the chamber. But only the interpretation of the significant interaction of factor A and B shows the process-relevant relationship.

12.5.3.10. Interaction AB

The significant interaction AB, Quantity × Redose (see Fig. 14), represents the dependence of the single factors A and B. In accordance with this interaction, factor B lost its influence at the higher level of factor A, whereas factor B has the capability to change the D-value of around 2.0 min at low level of factor A. No influence at all is observable at the high level of factor A. This interaction, and also the described main effect of factor B, can be understood considering the main effect of factor A, Quantity (q1). If the decontamination effect reached the stable maximum by the initial quantity (q1), an additional vaporized quantity of H_2O_2 does not influence this established effect, whereas a lower initial decontamination effect can be improved by a later vaporization of H_2O_2. This interpretation of the interaction AB is plausible in the context of the other effects.

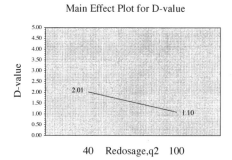

FIGURE 13 Main effect plot of factor B, redose, (q2). (From Ref. 1.)

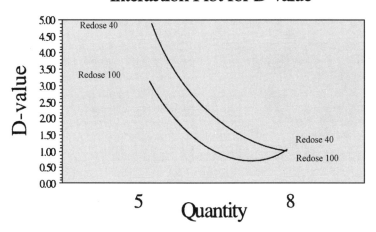

FIGURE 14 Interaction plot AB. (From Ref. 1.)

A much more important property of the process can be observed by looking at the interaction AB. At the lower level of Quantity A and also at the lower level of Redose B, the D-values are higher in comparison to the D-values of the independent factor A at the low level, but at the higher level of Redose B the D-value is lower than those observed at the independent factor A (see Fig. 11). This indicates that only high levels of factor B, Redose improves a low decontamination effect. Furthermore, with a low level of factor B, Redose, an achieved decontamination effect cannot be kept stable. The decontamination media has to enter the system at a certain velocity to keep a decontamination effect stable or even establish a decontamination effect. Therefore, as an important property of the described process, it can be stated, "The stability of a achieved decontamination effect depends on the rate of additional vaporized H_2O_2 during the decontamination phase" (1).

12.5.3.11. Factor C, Temperature

The effect of main factor C (temperature) is significantly lower than the effects of the main factors A and B. Between the factor levels a change of only +0.5 min in D-value is observable within a temperature range of 10 [°C] (see Fig. 15). Regarding the decontamination process it can be stated, "The lower the temperature the better the decontamination effect" (1).

Therefore, positions with higher temperature within the isolator system are "worst case" positions in consideration of the decontamination effect.

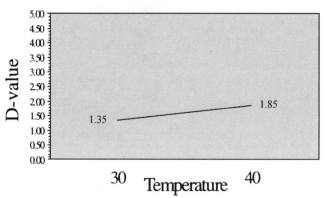

FIGURE 15   Main effect plot of factor C, temperature. (From Ref. 1.)

Because temperature and humidity are physically linked, the justification of the significance of both effects is discussed together after the presentation of the effect of factor D (humidity).

12.5.3.12.   Factor D, Humidity

The effect of the main factor D (humidity) is significantly lower than the observed effects of Quantity and Redose. Within a humidity range of 10 %rH a change in D-value of only −0.4 min is observable (see Fig. 16). In the Pareto Chart of the model (see Fig. 10), the effect of humidity is shown near the border of significance for describing the decontamination effect. This effect represents, therefore, a kind of resolution of the chosen model. But the interpretation of the influence on the process is nevertheless possible.

For the interpretation of factor D (humidity), it can be stated, "The higher the humidity the better the decontamination effect" (1). Therefore, positions with lower humidity within the isolator system are "worst case" positions in consideration of the decontamination effect.

12.5.3.13.   Interaction CD and Process Relationship

Both effects, temperature and humidity, influence the saturation level of the air during the process as discussed above. Both main factors show a better decontamination effect at higher saturation. Altogether this forms a plausible justification that the observed effects an the direction of those effects are

Main Effect Plot for D-valve

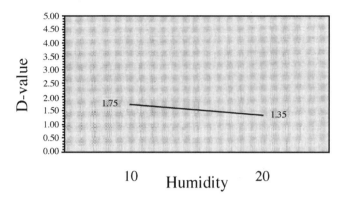

FIGURE 16 Main effect plot of factor D, humidity. (From Ref. 1.)

properties of the process itself. The significant interaction CD shows once more the relationship between temperature and humidity (Fig. 17.)

The interaction between a change in temperature within a 10°C range and a change in humidity within a 10 %rH on the saturation of the process is observable. At a temperature of 30°C the changes of the humidity level is equivalent to a water content of 2.5 g/kg in the air. At 40°C, the same change

Interaction Plot for D-value

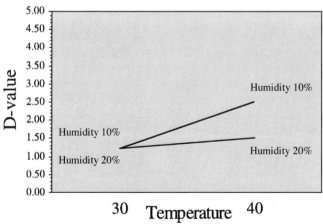

FIGURE 17 Interaction plot CD. (From Ref. 1.)

in humidity is equivalent to a 4.5 g/kg water content in the air. The decontamination effect follows those relationships of difference in saturation levels (1).

Two things are important to recognize. First, the discussed relationship of the process and the saturation level, which is clearly shown with the three effects of temperature, C, humidity, D, and of the interaction, CD. Second it is shown that the previous assumed physical interaction of temperature and humidity can be found later on in the analysis of the data. This seems logical, but also requires the proper selection of experimental data. To analyze humidity and temperature, real independent absolute humidity [g/kg] has to be chosen as a factor. But the discussed effects represent the applied process and its relationship clearly, nevertheless.

12.5.3.14. Factor E, Aqueous H_2O_2

After the justification of all other effects, the effect of factor E (aqueous H_2O_2) is interesting to discuss. Between its factor levels, a change of $+$ 0.6 min in D-value, is observable in a way, that lower concentrations of aqueous H_2O_2 solution leads to a better decontamination effect (see Fig. 18).

This effect can be discussed considering the air saturation during the process. Vaporizing the same quantity of pure H_2O_2 produces a bigger volume of vapor at lower concentrations of the aqueous H_2O_2 Solution, because more solution has to be vaporized than at higher concentration. This

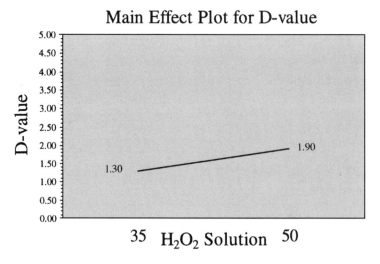

FIGURE 18   Main effect plot of factor, aqueous H_2O_2. (From Ref. 1.)

leads to a difference in the achieved air saturation during the process. The effect of factor E (aqueous H_2O_2) shows a better decontamination effect at higher saturation. Its direction fits, therefore, to the observations done at all other discussed factors. For the process interpretation of factor E, it is possible to say, "Vaporizing the same quantity of pure H_2O_2 leads to a better decontamination effect at lower concentrations of the aqueous H_2O_2 solution" (1).

12.5.3.15. Factor Interrelationship

After all factor effects are usefully analyzed by their own direction, effect value and process relevance, it should be possible to find a main interrelationship between all of the factors giving an explanation of the observed effects in consideration with the decontamination process. Such an interrelationship is, first, useful for understanding the main principles of the applied process; and second, it provides a further fact based on which the selected process model can be finally justified and accepted. For the H_2O_2 decontamination, the quantity of pure H_2O_2 vaporized into the isolator chamber is found to be the main process effect. But as the main interrelationship between all factors, the saturation of the air during the process can be established. By vaporizing a initial quantity of H_2O_2 into a chamber, the state with a maximal inactivation effect is reached at higher saturation of the chamber air. This achieved inactivation can be maintained stable over time by a useful dosage rate of H_2O_2 during the process in the same way as the air is kept saturated by this dosage rate. Starting the process at lower temperature levels leads to a higher saturation of the air than at higher temperature levels, because of the lower capability of the air at lower temperatures. Staring at higher humidity levels leads to higher saturation of the air, because the capability of the air is already limited. Using a lower concentration of the aqueous H_2O_2 solution leads also to a higher saturation of the air than using higher concentrations. Hence, more volume is vaporized at lower concentrations. With the saturation of the air, one main interrelationship is found that connects the all five main factors with the observed process performance. Considering the principle of the process, this interrelation shows, "The decontamination effect depends on the saturation of the gaseous phase" (1). Moreover, it seems to be important to establish and maintain a saturation gradient from the gaseous phase to the surfaces to be decontaminated.

12.5.3.16. Acceptance of Process Model

All observed effects of the main factors and the interactions of the chosen model can be clearly interpreted concerning the decontamination effect. Moreover, a main interrelationship between the factors could be found. This plausible process interpretation including the statistical proof of the final process model forms the basis on which the chosen model can be accepted as a description of the decontamination effect in dependence on the process parameters for the analyzed application. After the selection of a suitable biological indicator, this is the second major milestone in the process development of an alternative sterilization method. Such a model defines a quantitative relationship between the sterilization effect and the process parameters of the application. At this point the statistical work is finished. Further statistical steps, such as a regression quotation to calculate a D-value based on the used parameters prior to a decontamination run, do not seem to be useful for process validation. But, by analyzing the data of the main experiment, a basis of process comprehension is formed, based on which it is further possible to interpret the process in depth, select the right process parameters, and establish a validation strategy for the application.

12.5.3.17. Additional Data Analysis

After analyzing the main experiment, additional data collected during the study could be analyzed considering the control and validation of the process. Such indicative data could be collected during the study from different sensors (e.g., gas concentration sensors, dew point sensors, and all other sensors) collecting data of interest for monitoring the process. With the achieved knowledge of the needed parameters for a good decontamination effect, it is also possible to analyze this additional data considering the decontamination effect. Moreover, secondary influences first excluded in the main experiment could now be determined using a robust cycle and a min/max evaluation (e.g., air velocity during the decontamination cycle).

12.5.3.18. H_2O_2 Gas Concentration

In the following step, the measured H_2O_2 gas concentration during the main experiment is analyzed as an example for additional data analysis. The gas concentration is selected because it is mentioned by the USP as main process value for a gaseous decontamination process and because it is often assumed to be directly correlated with the decontamination effect. In addition to the one H_2O_2 gas concentration sensor integrated into the described isolator system, five different H_2O_2 gas concentration analyzers were implemented into the system during the study. Each sensor system is analyzed by its own values and in comparison to all other sensor systems. This analysis can be summarized in two main conclusions: One considering the correlation between H_2O_2 gas concentration and decontamination effect, the other focusing on the technical standard of H_2O_2 gas concentration measurement.

12.5.3.19. H_2O_2 Gas Concentration Versus Decontamination Effect

It is not possible to establish a correlation between the measured H_2O_2 gas concentration and the observed decontamination effect. Whereas the decontamination effect depends on the saturation of the gaseous phase, the measured gas concentration depends highly on the temperature of the air. The higher the capability of the air, the higher the measured H_2O_2 gas concentration. This relationship makes it possible to measure at a high air temperature, also a high H_2O_2 gas concentration, even when a low quantity of H_2O_2 is vaporized. But at low temperatures of the chamber air, a low H_2O_2 gas concentration is measured even when enough H_2O_2 is vaporized to achieve the required saturation for a good decontamination effect. Moreover, it is possible to observe, during a decontamination cycle, a low H_2O_2 gas concentration but determine during the same cycle an also low D-value representing a good decontamination effect. On the other hand, the parameters for the decontamination cycle can be selected in a way that during this cycle a high H_2O_2 as concentration is measured, but only a poor decontamination effect is achieved. This makes it impossible to use the measured H_2O_2 gas concentration as an independent measure for the decontamination effect.

12.5.3.20. Measurement Methods

By comparing the different H_2O_2 gas concentration analyzers and methods, it could be observed that all analyzers show the same shape of measurement curve but highly differ in values. Figure 19 shows the response of the applied analyzers in the same chamber at the same time while a defined quantity of H_2O_2 is vaporized into this chamber.

12.5.3.21. Interpretation of H_2O_2 Gas Concentration Measurement

The shape of the measurement curve is similar for all applied analyzers; the value of the measured H_2O_2 gas concentration in this trial differs from 200 ppm to approximately 3000 ppm depending on the selected analyzer. This observation could be finally traced on the different calibration methods of the single analyzers. At this time no standardized calibration method for gaseous H_2O_2 is established and, therefore, the different analyzers cannot be comparably adjusted. Although a correlation between the measured gas concentration and the decontamination effect is not possible, all sensors show reproducible measurement curves and values when the same cycle parameters are applied.

The H_2O_2 gas concentration measurement during the process is useful as an indicative measurement during a H_2O_2 decontamination cycle to ensure

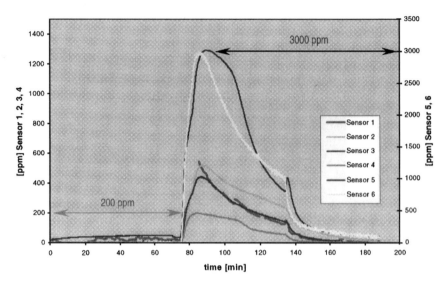

FIGURE 19 Measurement curves of different H_2O_2 gas analyzers. (From Ref. 1.)

a reproducible process in compliance with the USP. But it is not useful to describe an established decontamination effect in general. If the H_2O_2 gas concentration should be used in small range for process release, the control of the temperature becomes more important than the need to achieve a stable and reproducible decontamination effect.

Without the previous established process comprehension, such interpretations are neither possible nor traceably founded. With the introduced main experiment of process development, a level of process transparency is achieved based on which the whole process validation can be built. Such proven results can even be transferred to similar applications for a qualitative evaluation of the process performance and to solve problems during routine work. Moreover, this information enables one to conduct further studies of interest with a minimum effort. Finally, this process comprehension leads directly to the following step of process development. In this step, a new description of the decontamination process considering the important factors and their influences, the needed parameters and their boundaries of each cycle phase, is formed. Based on this achieved process description, the whole decontamination process can be properly validated. The methodology of process development changes at this point from a more scientific approach with experiments and statistical data analysis back to decontamination process and its application. By using all established tools and data, a method of quantification and development of alternative sterilization processes is finally achieved.

12.5.4. Interpretation of Process

The following represent the steps involved in the interpretation of the process:

Classify every single step of the process and its required parameters
Differentiate between process relevant or indicative parameters
Associate results of interpretation with the requirements of the pharmacopoeia

Based on the achieved process comprehension, a classification of every single process step and its required parameters has to be established. This classification defines all relevant or indicative process parameters required. Such a classification defines the requirements for the control, monitoring, and validation of the final course of the process. During this step of process development, all previous set assumptions, especially the requirements of the pharmacopoeia (9) have to be reflected and transferred into equivalent process steps or parameters. This classification also uses the previous established cycle phases to describe the corresponding parameters and effect.

12.5.4.1. Process Classification

As described above, the decontamination process using gaseous H_2O_2 is separated into four phases. For all phases, assumptions were made in advance in consideration of the process performance. Now, the determined influence of each single phase and the required parameters should finally be defined. Further process-relevant predefinitions and constants have to be made to complete the process definition. After the thorough analysis of the main experiment, process classification should be complete but also should be kept short and simple to form a suitable, not confusing, approach for process validation.

12.5.4.2. Predefinition, Constants, Indicative Parameters

First, useful process predefinitions and constants should be found in addition to the variable process parameters defined later on. All process-relevant influences that could be predefined and kept stable in a suitable range during routine application of the process simplifies the control and the validation of the process. Furthermore, all indicative measured parameters needed for the process control should be defined. For the introduced H_2O_2 decontamination of an isolator system, the following predefinitions could be made:

a. Aqueous H_2O_2 Solution

The aqueous H_2O_2 solution used during routine application of the process is selected at a concentration of 35[%]. This concentration level is commercially available without any difficulties and requires no special safety handling. To comply with the requirements of the standards (3, 4) the concentration of each lot of H_2O_2 solution could be controlled using a wet chemistry method, and an expiry date has to be established representing the application. The acceptance criteria of the aqueous H_2O_2 solution can be 35 ± 1 %. This concentration range is wide enough to be easily controlled and accurate enough to ensure a reproducible decontamination effect.

b. Air Velocity

The easiest way to handle the air velocity during the decontamination process is to use also the qualified air velocity for routine manufacturing or testing. This air velocity has to be already validated and therefore no additional effort is required. This air velocity is often selected to ensure a good air distribution in the chamber, which is also required for a good distribution of the decontamination effect. The air velocity in the described isolator system is set to 0.25 ± 0.05 m/s during decontamination.

c. Dosage Rate During Conditioning

The dosage rate of H_2O_2 during the conditioning phase is set to a maximum value to ensure the required fast vaporization of H_2O_2 at the beginning of the cycle. The system used is able to control dosage rates at a range of ±1 g/min.

d. H_2O_2 Gas Concentration

During the decontamination cycle, the H_2O_2 gas concentration is indicative of that measured.

These predefinitions are only examples focusing on the introduced isolator system. Other alternative sterilization methods may require different predefinitions and constants, but this gives an idea what kind of parameters of a alternative sterilization method are desired to be kept stable.

12.5.4.3. Phases Influence, Parameters, and Variables

Now, the single-process phases are classified considering their influences on the decontamination effect. The required parameters and variables to ensure the process performance are defined. A value for all these variables has to be finally defined during cycle development. Once more, the classification is kept easy and simple, focusing exactly on the influence of the phase and the corresponding parameter. As a reminder, the previous defined cycle phases, the decontamination effect as a function of cycle phases, is shown once again [Fig. 20; from Ref. (2)].

> Phase 1, Preconditioning. The preconditioning phase generates defined initial conditions in the chamber air to ensure a reproducible decon-

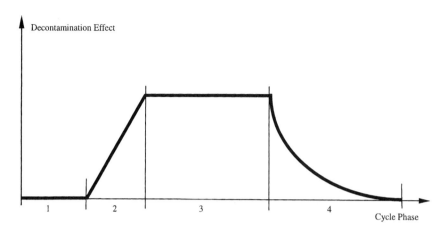

FIGURE 20 Decontamination effect as function of cycle phase. (From Ref. 2.)

tamination cycle. During this phase, the parameters humidity [%rH] and temperature [°C] of the chamber air must be monitored (2).

Phase 2, Conditioning. The conditioning phase is responsible for achieving a maximum bacterial reduction rate. The vaporized initial quantity (q1) of aqueous H_2O_2 [g] per chamber volume [m^3] is the critical process variable in this phase (2).

Phase 3, Decontamination. The decontamination phase has two important process parameters. First, the rate (q2) of continuously vaporized H_2O_2 [%/h] (q2). This rate ensures that the bacterial reduction rate previously achieved in the conditioning phase remains stable over the entire duration of the decontamination. The second is the duration of the decontamination [min]. It ensures the total bacterial reduction at a known reduction rate (2).

Phase 4, Purging. The purging phase [min] ensures that the maximum residual concentration of H_2O_2 [ppm] is reached in the chamber (2).

Accordingly, the parameters shown below in Table 4 must be ascertained and defined during cycle development.

The parameters, humidity and temperature, could be defined based on the result of the main experiment and physically validated and monitored. Also, the purge time is a physical parameter, which could be determined using a suitable measurement device. All the other parameters could not be defined using physical measurements. The initial quantity of H_2O_2 is required for the established decontamination effect, the rate of redose for its stability. Also, the duration of the decontamination phase depends on microbial reduction and, therefore, on the decontamination effect. To determine the needed value

TABLE 4   Parameters and Effects of Cycle Phases

Cycle phase	Parameters [units]	Effect
Preconditioning	Air humidity [%rH] Air temperature [°C]	Reproducibility of the decontamination cycle
Conditioning	(q1) [g/m^3] Initial quantity of aqueous H_2O_2 per volume unit	Bacterial reduction rate obtained
Decontamination	(q2) [%/h] Rate of redose	Maintain stability of the bacterial reduction rate obtained
Decontamination	Duration of the decontamination [min]	Overall microbial reduction obtained [log scale] during decontamination
Purging	Purge time [min]	Residual concentration of H_2O_2 [ppm] obtained in the chamber

Source: Ref. 2.

for those parameters, a suitable sensor for decontamination effect has to be used. As a sensor for the decontamination effect, the D-value of a defined BI was introduced and established during this process development. Using a suitable BI and the minimized LSKM as a methodical tool, it is possible to define the value for each of those parameters and finally quantify the overall microbial reduction obtained during decontamination.

By analyzing the main experiment, numerous possible interpretations and assumptions concerning the nature of the assessed process could be formed. Most of them would not be a major improvement of the process nor would they be useful for routine application. Often such interpretations and experiments lead to more questions and confusion than to answers or benefits regarding the use of the process in the highly controlled field of pharmaceutical manufacturing. Finally, what is missing is often a pragmatic, but also complete and transparent way to prove the performance of an alternative sterilization method. During this process development, the requirements, comprehension, and also the practical tools for a method of development and quantification alternative sterilization cycles are established. In the following, this cycle development method is introduced using the H_2O_2 decontamination as an example. This method is useful for all alternative sterilization methods, even when only a small database is available, as for the H_2O_2 decontamination of the example. The method focuses directly on the sterilization effect and therefore on the target of each sterilization method. Moreover, the method is based on the survival time model and on the LSKM to determine the D-value of a suitable BI and finally leads to a quantified sterilization effect (1).

12.6. METHOD OF CYCLE DEVELOPMENT AND PROCESS QUANTIFICATION

The following are the deliverables associated with the method of cycle development and process quantification:

> Define the value of each relevant process parameter.
> Recognize the suitability of the microbiological system.
> Show the achieved sterilization effect and evaluate required values for sterilization effect.
> Show stability of sterilization effect and evaluate the required values for stability of effect.
> Identify worst-case positions of the process.
> Quantify the sterilization time.
> Quantify the sterilization effect reached (e.g., the log-reduction).
> Show reproducibility.

Starting with the cycle development of an alternative sterilization method, it is important to remember that two main systems determine the sterilization cycle (2):

1. The microbiological system with BI and culture media
2. The system consisting of the chamber to be sterilized, the sterilization apparatus, and the procedure

Only if the behavior of one of the two systems is known is it possible to use this known system to describe and define the unknown system. With cycle development, the sterilization effect should be quantified and the right cycle parameters should be defined to ensure and prove the performance of the sterilization cycle. During this process development method, the BI is established as a suitable sensor for the sterilization effect. Further, if the BI is to be used as a sensor for specifying the sterilization process, then its resistance and model behavior must be known in advance (2).

12.6.1. Experiment 0: Reactive Pattern Recognition

As a first step of cycle development, the BI has to be tested in the light of its suitability as sensor for the sterilization effect. As described above, this can be be done by performing a reactive pattern recognition. With the reactive pattern recognition, the behavior of the microbiological system is described and evaluated in a known decontamination cycle. In this sense, the reactive pattern recognition serves to calibrate the BI. Only then can the micro-biological system be defined and used to describe a corresponding decon-tamination system and to develop a cycle. If a similar decontamination system is already defined and available, this system can be used as a reference for the reactive pattern recognition. If no system is defined or available, then pro-ceeding as previously described ensures a proper selection of a suitable BI (2).

12.6.2. Cycle Development

What follows is a completely developed H_2O_2 decontamination cycle. The influence of the individual cycle parameters will be explained and discussed. Based on the survival time model, the cycle parameters are set and the achieved decontamination capacity is quantified (2).

12.6.2.1. Preconditioning, Initial Conditions of Chamber Air

a. Air Humidity [%rH]

The humidity of the chamber air is lowered to a defined value before starting H_2O_2 vaporization in order to ensure that the chamber air is capable of absorbing the H_2O_2 vapor, which will subsequently be introduced. Starting from a defined moisture content, a defined partial pressure ratio of gaseous H_2O_2, the water vapor is then set and hence a defined effective dose in the chamber is established. Starting humidities of 10–20%rH at normal chamber temperatures are usually adequate (2).

b. Air Temperature [°C]

To avoid incurring extra power consumption and time while preconditioning the chamber, it is recommended that the initial temperature of the chamber air be set as either the working or operating temperature of the chamber. H_2O_2 decontamination is used over a broad range of temperatures, so the optimal process temperature is the one that incurs the lowest cost for the operator (2).

For the bacterial reduction rate achieved in the subsequent conditioning phase, the initial chamber air conditions constitute secondary effects that are small compared with the main effect, the quantity of H_2O_2 vaporized. If the quantity of vaporized H_2O_2 is held constant and the initial humidity is varied (10%rH, 20%rH), slightly better bacterial reduction rates are obtained at higher initial humidities than with lower ones. If the temperature is varied (30°C, 40°C), better bacterial reduction rates are obtained at lower temper-atures than at higher initial temperatures.

It is shown above that if a suitable initial quantity of H_2O_2 (q1) is chosen in the conditioning phase, the impacts of the side effects, in the range stated above, on the bacterial reduction rate achieved are so low that they can be ignored. For both temperature and humidity, a range of ±5 around the selected target value ensures a reproducible decontamination effect (2).

12.6.3. Experiment 1—Bacterial Reduction Rate Achieved, Quantity (q1)

The initial quantity of H_2O_2 per chamber volume (q1) $[g/m^3]$ vaporized during the conditioning phase establishes a killing effect per unit of time [min] in the chamber. This bacterial reduction rate can be visualized through a minimized LSKM and quantified through the estimated D-value. To perform the minimized LSKM, a position is chosen in the chamber at which a good killing effect can be expected. The influence of local gradients in the decontamination effect on the experimental results is thus excluded at the beginning of cycle development so that the bacterial reduction rate obtained can be assessed under optimum conditions. Based on the reduction rate thus observed, a relationship can be established between positions where decontamination is good and positions where bacterial reduction is poor. To establish the bacterial reduction rate with the selected quantity (q1), the BIs are exposed immediately following the end of the conditioning phase (2).

Next, two experimental results show the effect of the initial quantity (q1) on the bacterial reduction rate to be determined. In experiment 1.1 (q1) was 5 g/m^3, in experiment 1.2 it was 7.5 g/m^3. All other cycle parameters were held constant (2).

Experiment 1.1: Bacterial Reduction Rate achieved with Quantity (q1) = 5 g/m^3 (from ref. 2)

Group	1	2	3	4	5	6	7	8	9	10	Pos.	Neg.
Exposure time (min)	6.0	9.0	12.0	15.0	18.0	21.0	24.0	27.0	30.0	33.0		
Result 1	+	+	+	+	+	+	+	+	−	−	+	−
2	+	+	+	+	+	+	−	−	−	−		
3	+	+	+	+	−	−	−	−	−	−		

$+$ = growth; $-$ = no growth.

Experiment 1.2: Bacterial Reduction Rate achieved with Quantity (q1) = 7.5 g/m^3 (from ref. 2)

Group	1	2	3	4	5	6	7	8	9	10	Pos.	Neg.
Exposure time (min)	6.0	9.0	12.0	15.0	18.0	21.0	24.0	27.0	30.0	33.0		
Result 1	+	+	+	−	−	−	−	−	−	−	+	−
2	+	+	−	−	−	−	−	−	−	−		
3	+	−	−	−	−	−	−	−	−	−		

$+$ = growth; $-$ = no growth.

Both experiments show no irregularities in the reactive pattern of the BI and, therefore, may be used to estimate the D-value. Based on the results of experiment 1.1 a D-value of 3.5 min is estimated; the results of experiment 1.2 produce an estimated D-value of 1.6 min.

Taken together, the two experiments show the clear dependence of the bacterial reduction rate obtained on the initial quantity (q1). An increase in quantity (q1) as in experiment 1.2 more than doubles the bacterial reduction rate. These experiments visualize the BI model behavior and also show how simple and easy it is to interpret the results obtained from the minimized LSKM. The D-value estimation makes the results and parameter effects quantifiable (2).

The dependence of the D-values on quantity (q1) is characterized by a nonlinear relationship as shown in the main experiment. An explanation of the impact of this relationship on cycle development is shown once more in Fig. 21. As the initial quantity (q1) of H_2O_2 is increased, the D-value falls to a minimum. At this point, the curve shows a sharp bend and a further increase in quantity (q1) does not improve the bacterial reduction rate significantly; the observed D-values remains stable (2).

At bacterial reduction rates below the maximum bacterial reduction rate, small changes in quantity (q1) result in large changes of the D-value. In this region, secondary effects (chamber conditions) are observed to have an influence on the bacterial reduction rate so that decontamination cannot be considered to be stable. If the initial quantity (q1) is set so that the bacterial reduction rate comfortably reaches its maximum, changes in quantity (q1) have no further effect on the D-value and the decontamination effect is insensitive to secondary effects. H_2O_2 decontamination is robust and the reproducibility of the bacterial reduction rate is thus ensured (2).

12.6.4. Experiment 2: Stability of Decontamination Effect, Quantity (q2)

By making up the H_2O_2 quantity in the decontamination phase at the rate (q2) [%/h], the stability of the bacteria-reducing effect over the entire duration of the decontamination phase is ensured. Parameter assignment for quantity (q2) is likewise performed using the minimized LSKM. To record data on the stability of the bacterial reduction rate, two minimized LSKMs (LSKM 1 and LSKM 2) are carried out over the maximum duration of the decontamination. The BIs for LSKM 1 are exposed immediately following the end of the conditioning phase. The results should reproduce the bacterial reduction rate obtained from the previous experiment, using (q1). Exposure of LSKM 2 takes place toward the end of the decontamination phase. The duration of the decontamination phase is set to the maximum, and the positioning of the LSKMs in the chamber is performed in a fashion analogous to the determi-

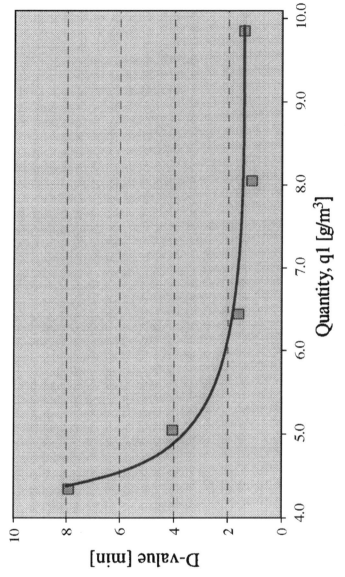

FIGURE 21 Plot of D-values versus quantity of H_2O_2, (q1). (From Ref. 2.)

nation of quantity (q1). The number of groups and removal intervals can be adjusted to the expected results (q1) (2).

Two sets of LSKMs are shown, with the number of groups reduced to five. In each case, LSKM 1 was exposed 5 min following the end of conditioning. In each case, LSKM 2 was exposed 30 min following the end of conditioning.

In experiment 2.1, the make-up quantity (q2) was set at 25% q1/h and in experiment 2.2 at 100%/h. All other parameters were held stable.

Experiment 2.1: Stability of Decontamination, Quantity (q2) = 25% q1/h
LSKM 1, exposure 5 min following the end of conditioning:

Group	1	2	3	4	5	Pos.	Neg.
Exposure time (min)	7.5	10.0	12.5	15.0	17.5		
Result 1	+	−	−	−	−	+	−
2	+	−	−	−	−		
3	−	−	−	−	−		

LSKM 2, exposure 30 min following the end of conditioning:
Experiment 2.1: from Ref. (2)

Group	1	2	3	4	5	Pos.	Neg.
Exposure time (min)	7.5	10.0	12.5	15.0	17.5		
Result 1	+	+	+	+	+	+	−
2	+	+	+	+	+		
3	+	+	+	+	+		

Experiment 2.2: Stability of Decontamination, Quantity (q2) = 100% q1/h
LSKM 1, exposure 5 min following the end of conditioning:

Group	1	2	3	4	5	Pos.	Neg.
Exposure time (min)	7.5	10.0	12.5	15.0	17.5		
Result 1	+	−	−	−	−	+	−
2	+	−	−	−	−		
3	+	−	−	−	−		

LSKM 2, exposure 30 min following the end of conditioning:
Experiment 2.2: from Ref. (2)

Group	1	2	3	4	5	Pos.	Neg.
Exposure time (min)	7.5	10.0	12.5	15.0	17.5		
Result 1	+	−	−	−	−	+	−
2	+	−	−	−	−		
3	−	−	−	−	−		

+ = growth; − = no growth.

A D-value of approximately 1.3 min was estimated for LSKM 1 in both sets of experiments. This reproduces well the good bacterial reduction rate obtained in the previous experiment with quantity (q1) in the conditioning phase. The results obtained from each of the LSKM 2s and their implications for the stability of the decontamination effect are evident. In experiment 2.1, LSKM 2 produced only all-positive groups and no killing effect was observed. In experiment 2.2, the LSKM 2 reproduced the results of the previous LSKM 1 very well, with an estimated D-value of approximately 1.3 min (2).

All in all, the results obtained from the minimized LSKM in the two sets of experiments reveal clearly the importance of the make-up quantity (q2) for the stability of the bacterial reduction rate. In experiment 2.1, the make-up quantity (q2) was not sufficient to sustain the bacterial reduction rate obtained during conditioning. The decontamination effect collapsed and no further bacterial reduction could be observed. The bacterial reduction rate obtained in experiment 2.2 and its stability over time provide the basis for a decontamination cycle (2).

The development of an H_2O_2 decontamination cycle with the objective of certifying a defined decontaminating effect is only possible if the bacterial reduction rate is known and stable over time. If conditioning does not produce a stable bacterial reduction rate, over the complete cycle, the reproducibility of the bacterial reduction and hence of the entire decontamination process cannot be assured. In practice, random results may then be observed in identical decontamination cycles when the BIs are evaluated (2).

It is essential to prove the stability of the bacterial reduction rate over time in order to design a decontamination cycle. If the rate is not stable and certified, it cannot be assumed that extending the decontamination phase will have the effect of increasing the overall achievable bacterial reduction. When assessing the decontamination capacity of the process on the basis of a single set of exposures of BIs, the bacterial reduction can only be assured by repeating the entire process and not by doubling the duration of decontamination. Changes in the reduction rate over time result directly in changes of the overall bacterial reduction. If these changes are not revealed, it is not possible to draw any firm conclusion as to the achievable bacterial reduction (2).

12.6.5. Estimation of D-value$_{best place}$

As the basis for the next steps in cycle development, the reduction rate achieved in experiment 1 was calculated and reproduced with the LSKM 1 from experiment 2.2. The stability of this reduction rate was confirmed in experiment 2.2 with LSKM 2. We therefore have three D-value estimations available, and from their mean we obtain the D-value$_{best place}$ (see Table 5) (2).

The D-value$_{best place}$ describes the bacterial reduction rate observed at a position in the chamber that can be well decontaminated.

Continuing the process of cycle development, the decontamination duration is calculated on the basis of the D-value$_{best place}$ and the survival time model. Positions in the chamber with poor decontamination effect (worst cases) are identified and tested. The results are then used to adjust the decontamination to the duration required to guarantee the target bacterial reduction (2).

12.6.6. Experiment 3: Worst-Case Study, Duration of Decontamination

In the worst-case study, the bacterial reduction is determined at positions in the chamber that are difficult to decontaminate. This is based on the calculated D-value$_{best place}$ and the BI survival time model.

The kill time is derived from the definitions of the survival time model. For a given D-value, it defines the exposure time in minutes after which the BIs used have to show reliably negative results in the growth test. For the D-value$_{best place}$ in the above example (1.4 min) and the initial population of microorganisms of 1×10^6 [CFU], the required kill time is calculated to be 14 min (corresponding to $10 \times$ D-values). This means that BIs that are exposed

TABLE 5 D-value$_{best place}$

Experiment	Estimated D-value (min)
Experiment 1.2	1.6
Experiment 2.2, LSKM 1	1.3
Experiment 2.2, LSKM 2	1.3
D-value$_{best place}$ (mean)	1.4

Source: Ref. 2.

to a bacterial reduction rate equal to the calculated D-value$_{best\ place}$ will show reliably negative results after a decontamination period of 14 min. If the isolated positions in the chamber show D-values greater than the calculated D-value$_{best\ place}$, then the decontamination period of 14 min will no longer assure sufficient reduction of the test microorganism population. In the subsequent growth test, these BIs do show fractional and/or all-positive results (2).

12.6.6.1. Definition of Worst-Case Positions

For the worst-case study, the critical positions to be considered in the chamber are first identified. Particular attention should be paid to places where large deviations from the average physical conditions in the chamber may be expected. To determine worst-case positions in the chamber, the following physical tests are adequate.

Smoke tests to evaluate the air distribution in the chamber
Temperature mapping during a decontamination cycle to evaluate the temperature distribution
Chemical indicator mapping to evaluate the distribution of the decontamination media

By selecting the worst-case positions based on physical tests, it is useful to consider the whole observed range. As an example, regarding the temperature distribution, the highest and the lowest observed temperature is selected as worst-case position. This ensures that during the worst-case study the whole temperature range representing the application is covered. Furthermore, it is important to consider the final manufacturing or testing process taking place in the system. Positions that are critical for the single-process steps or that have an obviously high risk to contaminate the process have to be selected as worst-case positions. Such a consideration covers the whole physical range of the system and reflects the needs of the process. Therefore, it builds a transparent basis for the following worst-case study (2).

12.6.6.2. Worst-Case Study Procedure

Three BIs are placed in each of the previously defined worst-case positions. This allows one to observe all-positive, fractional, and all-negative results at the individual position and, based on the result, to estimate the decontamination effect achieved at that specific position. In the worst-case study, the BIs are subjected to a complete decontamination cycle with preconditioning, conditioning, and decontamination. The duration of the decontamination phase, as explained above, is set to equal the kill time of the BI used based on the mean D-value$_{best\ place}$. For all other cycle parameters, the previously determined values apply (2).

12.6.6.3. Interpretation of the Results

On the basis of the survival time model, the results of the worst-case study are interpreted as follows:

> If all three BIs used in a specific position test negative, then this chamber position assures a bacterial reduction similar to that calculated for the best-place position; the D-value at that position therefore corresponds to the D-value$_{best\ place}$. Given the bacterial reduction obtained, this position does not constitute a worst case.
>
> The bacterial reduction at positions at which the BIs produce a fractional result (positive and negative results in the ratio 2:1 or 1:2) is estimated to be equal to the test microorganism population.
>
> Positions with all-positive results show no or only poor bacterial reduction, corresponding to the definition of the survival time, and therefore quantification is not possible (2).

On this basis the bacterial reduction rate achieved for positions with fractional results is estimated to be the D-value$_{worst\ case}$ as follows:

$$D-value_{worst\ case}\,[min] = \frac{Duration\ of\ decontamination\ [min]}{\log N_0}$$

In the above example, with a decontamination duration of 14 min, a test microorganism population of 1×10^6 CFU and a D-value$_{best\ place}$ of 1.4 min, the estimated bacterial reductions based on the results are as shown in Table 6.

To confirm the estimates made, a second worst-case study is carried out on the basis of the first D-value$_{worst\ case}$ calculated, but this time only positions previously found to correspond to the worst case are considered. With this iterative procedure, the duration of decontamination is adjusted to the worst-case positions observed and at the same time the maximum D-value$_{worst\ case}$ is determined (2).

TABLE 6 Estimation of Microbial Reduction

Biological indicator results		Bacterial reduction [log scale]	Estimated D-value [min]
All-negative	(3:0)	≥ 10	1.4 (D-value$_{best\ place}$)
Fractional	(2:1, 1:2)	$\geq 4, \leq 10$, (evaluated with 6)	2.3
All-positive	(0:3)	≤ 4	Cannot be estimated

Source: Ref. 2.

The final duration of the decontamination phase depends on the overall bacterial reduction the process should guarantee, and is derived from the maximum D-value$_{worst\ case}$ and the target bacterial reduction as follows:

$$Duration\ of\ decontamination\ [min] = D\text{-}value_{worst\ case}\ [min]$$
$$\times\ target\ bacterial\ reduction$$
$$[log\text{-}scale]$$

Once the worst case study has been completed and the parameter decontamination duration has been established, all parameters of the decontamination cycle affecting kill rate have now been described and quantified (2).

12.6.7. Experiment 4: Determination of Purge Time

It is appropriate to generate a purge curve to calculate the purge time. Suitable measuring methods are gas test tubes and H_2O_2 gas sensors with appropriate measuring range. The residual gas concentration required [ppm] of H_2O_2 in the chamber is determined with reference to the system application. If the chamber will be opened or entered by personnel after decontamination, the residual concentration must satisfy legal requirements for personal safety and limit values before the chamber is opened. Where the systems to be decontaminated are used in the manufacture and testing of products, the residual concentration achieved must not affect the quality of the product or the test to be carried out. The maximum permitted H_2O_2 residual concentration in these applications can only be established through appropriate tests (2).

12.6.8. Experiment 5: Determination of D-Value

As the final step of the cycle development, an LSKM is performed to determine the definite D-value using the final decontamination cycle parameters: The procedure adopted here is similar to the minimized LSKM in experiment 1. The D-value thus calculated with its 95% confidence limit supports the D-values estimated throughout the course of the cycle development and, as the characteristic figure for the process in the specific equipment, describes the bacterial reduction rate obtained by the complete system.

When the decontamination process is requalified, the reactive pattern recognition used to check the BI batches can be based on the D-value for the equipment/process. In this way, in the event of fluctuations in the D-value and irregularities in model behavior of the BIs, appropriate action can be taken prior to starting any qualification work (2).

12.6.9. Summary of Cycle Development Method

The method presented here to develop alternative sterilization cycles describes and quantifies the influence of every process parameter relevant to the decontamination effect using a defined microbiological system. The chronology of the series of experiments systematically excludes any secondary effects on the results and thus ensures that the experimental data can be properly interpreted. The presented cycle development method is adaptable to mostly all inactivation methods, even if the process is not defined in such detail. Single experiments of this cycle development method can also be used to debug or improve any inactivation system in a short time period and with minimal cost. The resulting D-value estimations during such experiments can then be used to quantify and justify any change or improvement of an inactivation process.

12.7. ADDITIONAL STUDIES

Often the application of alternative sterilization processes requires additional qualification studies to justify the suitability of the applied process. Even during the presented process development method, many possible influencing factors were excluded or kept stable to be able to interpret the results properly and traceably and, at least, to reduce the experimental effort. Now, based on the process comprehension achieved and the tools established to quantify the inactivation effect, such additional studies can be planned and carried out in a useful and effective way.

On one side, possible additional studies could consider specific factors of a single application influencing the sterilization effect. Such factors could include the following:

> The resistance of different microbes to the inactivation process, especially the resistance of microbial isolates found in the environment of an application, in comparison to the used BI. Such studies could be used to quantify the log reduction of the bioburden reached on the inner surfaces of an isolator system or on the product to be sterilized.
> The effect of different kinds of surfaces on the inactivation effect reached during the process. Such studies could be used to justify the suitability of the carrier material of the chosen BI. This is particularly useful in the field of isolator technology, but also for the sterilization of other different products and devices, to answer the question: is the inactivation effect observed with the chosen BI transferable to every kind of surface of the specific application?
> The transferability of results observed in one application may be used to support other applications using the same process approach. Such

studies are useful if an established process database can be used for more than one application. This is especially useful if a dedicated reference system is used for special trials, as the testing of BIs. Such studies are useful for establishing the relationship between the reference system and the production systems.

Further studies of the inactivation effect could be focused on the sterilization of special equipment and on a further improvement of the inactivation process or the design of the application.

Additional studies could focus on the effect of the applied sterilization process on further production steps or the performance and reliability of process devices and equipment, as follows:

- The influence of residuals of the sterilization agent in the environment on the quality of a pharmaceutical product, the results of a microbial environmental monitoring or the results of an sterility test of a product. Such additional studies are required to ensure the quality of the product and exclude any question of product quality.
- The influence of the sterilization process on the performance of growth media used for microbiological quality control. Here often special growth promotion studies or media compositions are required.
- The resistance of construction materials, equipment, and devices to the applied sterilization process. Often an effective sterilization process may affect the exposed materials and surfaces. Therefore, special attention on the compatibility of the material used during the design, production, and maintenance of the system is required.

The above-listed studies can only give an overview of additional questions that could arise during the development and use of an alternative sterilization process. The kind of additional studies actually required for a specific application will depend on the individual requirements and the properties of the process and pharmaceutical product.

12.8. CONCLUSION

The Process Development of Alternative Sterilization Methods presented, using the H_2O_2 decontamination as an example, describes the performance of the applied sterilization methods in a scientifically and statistically based manner. The methodical tools and chronology of the single steps introduced form a powerful and safe means to describe an alternative sterilization method by its target value, the sterilization effect. By the consequent consideration of the two systems of a sterilization method, the physical system and the microbiological system, this development method leads to a deep and complete process understanding on which the validation, the control and the routine application of any alternative sterilization methods in the field of pharmaceutical manufacturing may be based. On the basis of well-accepted microbiological and statistical methods, this process development of alternative sterilization methods becomes transparent and contributes to standard sterilization process validation.

ACKNOWLEDGMENT

Much of the information in this chapter is derived from Refs. (1) and (2).

REFERENCES

1. Sigwarth, V.; Skan, AG. Basel, Switzerland, Process Development of the Skan Integrated System, SIS 700, H_2O_2 Decontamination Method, Process Description and Qualification Guide Skan AG, 2002.
2. Sigwarth, V.; Moirandat, C. Development and quantification of H_2O_2 decontamination cycles. PDA J. Pharm. Sci. Tech. 2000, *54* (4), 286–304.
3. USP XXVI. Sterility Testing; Validation of Isolator Systems. General Information, Chapter ⟨1208⟩; 2431–2433.
4. Food and Drug Administration, FDA. Sterile Drug Products Produced by Aseptic Processing; Draft September 2002. 35–38.
5. International Standard, ISO 11138-1: 1994, Sterilization of health care products; Biological indicators, Part 1: General, Annex D.
6. European Standard, EN 866-1: 1997, Biological systems for testing sterilizers and sterilization processes Part 1: General requirements, Annex B.
7. USP XXVI. Biological indicators. General Information, Chapter ⟨1035⟩; 2244–2247.
8. International Standard, ISO 14161: 2000, Sterilization of Health Care Products; Biological Indicators; Guidance for the Selection; Use and Interpretation of Results.
9. USP XXVI. Sterilization and sterilization assurance of compendial articles. General Information, Chapter ⟨1211⟩; 2433–2439.
10. Amsco Scientific. *A Collection of Scientific Papers*, 2nd Ed.; May 1992.
11. Akers, J.E.; Agaloco, J.P.; Kennedy, C.M. Experience in the design and use of isolator systems for sterility testing. PDA J. Pharm. Sci. Tech. 1995, *49* (3), 140–144.
12. Watling, D.; Drinkwater, J.; Webb, B. The implication of physical properties of mixtures of hydrogen peroxide and water on the sterilization process. ISPE Symposium, Barrier Isolation Technology, Sept 23–24, 1998; Zürich, Switzerland.
13. Crozier, D.; Lang, G.; Ananth, S. On line analysis of vapor hydrogen peroxide for isolator technology. Pharm. Tech. 1996, 46–49.

14. Shintani, H.; Tahata, T.; Hatakeyama, K.; Takahashi, M.; Ishii, K.; Hayashi, H. Comparison of D_{10}-value by the Limited Spearman-Karber Procedure (LSKP), the Stumbo-Murphy-Cochran Procedure (SMCP) and the Survival-Curve Method (EN) Biomedical Instrumentation & Technology, March/April 1995, 113–124.

15. Holcomb, R.G.; Pflug, I.J. The Spearman-Karber Method of analyzing quantal assay microbial destruction data. Department of Food Science and Nutrition, University of Minnesota, MN 55455.

16. USP XXVI. Sterility Tests. Microbiological Tests, Chapter ⟨71⟩, 2013–2016.

17. USP XXVI. Biological Indicators—Resistance Performance Test. Microbiological Tests, Chapter ⟨55⟩, 2004–2006.

18. Pharmaceutical Inspection Convention; PIC/S; Draft Recommendation on Isolators used for Aseptic Processing and Sterility Testing April, 2002, 1–22.

19. Box, G.E.P.; Hunter, W.G.; Hunter, J.S. *Statistics for Experimenter*, 1st Ed.; John Wiley & Sons: New York, 1978.

13

Terminal Sterilization and Parametric Release

Klaus Haberer
Compliance, Advice and Services in Microbiology, Köln, Germany

13.1. ROLE AND SIGNIFICANCE OF THE STERILITY TEST IN THE RELEASE OF STERILE PHARMACEUTICALS

Terminally sterilized products are typically released for market distribution on the basis of a satisfactory review of the sterilization cycle records and compliance with sterility test results. This has been common and accepted practice in the pharmaceutical industry for decades. However, it has also been clear for many years that the test for sterility as described in the Pharmacopoeias is not a satisfactory test on which any strong reliance can be based for sterility assurance of the product.

13.1.1. Role of the Test

The test for sterility as described in the Pharmacopoeias is primarily intended as a reference test. This is obvious from qualifying statements in USP and European Pharmacopoeias. Ph. Eur. states, "A satisfactory result only indicates that no contaminating microorganism has been found in the sample examined in the conditions of the test. Guidance on the further requirements for demonstrating the sterility of the batch is given at the end of this text" (1). In

419

the (nonmandatory) annex it is further stated that "A manufacturer is neither obliged to carry out such tests nor precluded from using modifications...".

The difference between a reference test and a release test, however, seems not fully understood by most users of the test, nor are the texts completely free from passages that point toward the use as a release test. A reference test will typically be applied in case of serious doubt if a product is sterile. The aim of the test in this case is to look for insterility in case of problems encountered with the product, and this is exactly what the test is all about. It is a test to evaluate the product for gross insterility. There can be no sampling plan with respect to the manufacturing lot size, or any evaluation of manufacturing conditions, because the test is applied to retained samples of a product or to product withdrawn from the market. Should the result of a reference test indicate that a product may be nonsterile, it would be imperative to critically review the testing environment and eventually to test again, to verify that the product in question is indeed nonsterile. This is how the test was formulated years ago in the pharmacopoeias.

The aim of a release test, on the other hand is, to verify that a product complies with its release specifications. As the requirement is "sterile" and as sterility is defined as complete absence of microorganisms in the product with a sterility assurance level of a maximum of 1 nonsterile unit in a total of 1 million units, it is obviously beyond the significance of any sampling plan for a test for sterility to verify compliance with this requirement. The test cannot possibly show that a product is sterile in the sense of the definitions. Hence, release of a sterile product cannot solely rely on a sterility test. Release criteria must include the conditions under which a product was manufactured; they must include critical parameters of sterilization processes; they must include critical data about presterilization bioburden and environmental conditions of manufacturing; they must include assurance that there can be no mixing of sterilized and nonsterilized product.

Should the result of a release test indicate that a product may be nonsterile, it would be imperative to review the parameters of the sterilization process and all relevant manufacturing conditions in addition to the testing environment. Decisions based on retesting in case of a failed sterility test would be an attempt to mask a possible low-level contamination, because the significance of the test is so low that a relevant decision cannot be based on the result of a retest. For this reason, retesting in the case of a failed sterility release test is not deemed acceptable today except in a case where there is direct evidence for a testing error.

13.1.2. Statistical Significance of the Test

Sterility testing has been recognized for many years to be statistically not significant to detect a low level of nonsterility in a product to be released as sterile. Unless an enormous number of samples are used, the test is inadequate to detect anything but a complete failure of the sterilization procedure or of the precautions applied to maintain sterility. The statistical considerations applying to a sterility test were elaborated in detail by Spicher and Peters almost 30 years ago (2). The inadequacy of the test has also been acknowledged in the U.S. and European Pharmacopoeias for a long time by statements to the effect that release of such products may not be solely based on sterility testing but has to rely on validated procedures (3,4). Validated procedures are clearly seen to bear higher significance than the results of a sterility test.

The probability (p) of detecting a contaminated unit within an (unlimited) quantity of units can be expressed by the following equation.

$$p = 1 - (1 - c)^n \qquad (1)$$

where c is the fraction of contaminated units, and n is the total number of samples taken.

The considerations applying to the significance of the sterility test for sterility assurance have been discussed in PDA technical report No. 30 (5). Equation (1) can be used where c is taken as the sterility assurance level (SAL), and p is the confidence level achieved by the test. Plots of this equation are shown in Fig. 1. It is obvious that 20 samples, as required for the sterility tests of the Pharmacopoeias, are not even sufficient to maintain a SAL of 10^{-1} with a 95% confidence level. For aseptic manufacturing at least 3000 media units should be filled to demonstrate control of aseptic handling. The same number of samples would have to be tested for each lot to achieve a SAL of 10^{-3} by sterility testing. To verify a SAL of 10^{-6} as expected for terminal sterilization, 3×10^6 samples would theoretically be needed. Obviously, this number of samples cannot be tested, and hence the test for sterility is meaningless for product sterilized in the final container.

13.1.3. Performance and Correctness of the Test

The test for sterility is a classical microbiological cultivation test. Nutrient medium is added to a suitably prepared sample and after a specified time of incubation the assay is inspected for microbial growth as manifested by turbidity. Although such a test is highly sensitive and ideally has the potential to detect a single viable organism, it also has serious limitations. Microorganisms have highly diverse growth requirements. Some species need air, others grow anaerobically, some species grow at temperatures below 10°C, others need heat up to more than 100°C for optimal growth. Some species need complex nutrient media with serum or blood, others grow only at low nutrient concentrations. This means that there can be no universal growth

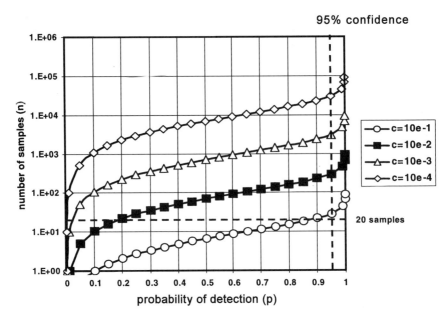

FIGURE 1 Probability of detection of insterility in a sterility test. Probability p to detect insterility occurring in a fraction $1/c$ units of the product as a function of the number n of samples tested. (From Ref. 5.)

medium or growth condition. There can always be microorganisms that are viable but cannot be cultivated with the methods applied.

The test for sterility as described in the European and U.S. Pharmacopoeia relies on two media and incubation conditions: casein-soybean digest Broth (CSB) incubated aerobically at 20 to 25°C and fluid thioglycollate medium (FTM) incubated at 30 to 35°C, which has oxygen-binding capacity and generates anaerobic conditions in the depth of a container of suitable geometry. Both media are a compromise and have their limitations.

Casein soybean digest broth is intended to grow yeast and mold as well as aerobic bacteria. It is not a specifically formulated medium for yeast and mold and although many fungi will grow in the medium, some species need more acidic conditions and/or carbohydrates as a carbon source. Many mesophilic bacteria will grow in CSB incubated at room temperature, but the medium is too rich for some environmental (e.g., water adapted) microorganisms and not rich enough for some other species (e.g., many human pathogens need media supplemented with serum or blood).

Fluid thioglycollate medium (FTM) is intended for aerobic bacteria in the upper layer and anaerobic bacteria in the depth of the container.

However, even if the medium is properly handled to prevent intrusion of air, the anaerobic conditions generated in FTM are usually not sufficient to sustain the growth of strict anaerobes (e.g., many Clostridia). Hence, while the compendial sterility test is certainly not a bad compromise and it will detect a broad range of mesophilic microorganisms, it will not detect each and every contaminant. There is simply no test available that has that capacity.

13.1.4. Performance of the Laboratory and the Rate of False Positives with Implications for the Rejection of Compliant Product

A second problem of the sterility test is its sensitivity to error during the testing procedure. Testing by membrane filtration is usually performed by filtration of the liquid or dissolved probes in closed filter canisters with subsequent addition of medium to the filters and incubation of the closed system. For testing where samples are handled and transferred to such closed systems in a Class A (Class 100) environment with a Class B (Class 10 000) background, contamination rates of 0.1% or 10^{-3} have been reported (6). A very similar positive testing rate has been found for negative controls tested for years in the microbiological quality control laboratory of a major pharmaceutical company, headed by the author of this chapter. This may not be considered a high failure rate in itself, but in order to use the test method to judge a product with a failure rate of 10^{-6} it is too high because this means that the majority of the tests that will become positive must be ascribed to testing error. If 20 units are tested from a theoretical batch with a theoretical contamination rate of 10^{-6}, one in 5×10^{-4} tests would theoretically be expected to become positive. With a false positive rate of 10^{-3}, the probability of a positive sterility test to be a testing error positive instead of a true positive would be 50 to 1. The wide majority of the batches rejected would be rejected solely for the reason of testing inadequacy. This is not an acceptable error rate for a release test. The assumption of a contamination rate of 10^{-6} is at the upper level of what is considered acceptable for a terminal sterilization procedure. Most sterilization procedures reach a much higher sterility assurance level (SAL). In this case, the relation of testing errors to true positives becomes even much higher.

The situation is improved by testing in isolators where false positive rates are much reduced, but not every manufacurer's laboratory or every laboratory of the surveying authorities operates such a piece of sophisticated equipment. With the limited statistical significance and performance of the test as discussed above, the question must be asked if it would make sense to require everybody who does a sterility test to acquire an isolator.

13.1.5. Contribution of the Sterility Test to Sterility Assurance for Products Terminally Sterilized in Their Sealed Container by a Validated Sterilization Process

The SAL reached by sterilization processes applied to product in the sealed terminal container is very high in most cases. Sterility assurance levels are typically calculated based on the inactivation of sterilization resistant bacterial spores. A process like the European standard process of 15 min at 121°C ($F_0 = 15$) theoretically inactivates heat-resistant spores with a D-value of 1 min by 15 orders of magnitude. Even if the product contained a bioburden of 10^3 CFU/container, and all of these microorganisms were heat-resistant spores with D_{121} equal to 1 min, the resulting theoretical expected probability of a unit containing a surviving spore (SAL) would be 10^{-12}.

Of course, more resistant bacterial endospores can be found (even if they are highly unlikely to occur in a presterilization bioburden). In the theoretical case that 10^3 bioburden microorganisms/container would consist of spores with $D_{121} = 2$ min, the resulting SAL would be only $10^{-4.5}$, not enough to meet regulatory expectations. (It should be emphasized here that except for bioindicator studies, it is not the objective of a sterilization process to inactivate preparations of resistant bacterial spores, but to inactivate a normal presterilization bioburden.) In pharmaceutical manufacturing the presterilization bioburden is usually much lower in numbers, the microorganisms present are to a high proportion vegetative forms with a D_{121} in the range of seconds, and in all experience spores characterized from the environment have D_{121} values of less than 0.5 min. But even if such a worst case would be assumed and the resulting SAL would be $10^{-4.5}$, the test for sterility would still be too limited in its statistical significance to be the right instrument to detect the problem: 45,000 samples would have to be tested to detect the contamination rate with 95% probability. Hence, it is clear that the sterility test is not a suitable test to detect a lack of sterility assurance caused by an unusually high presterilization bioburden.

13.1.6. Risk Consideration for Failure of Terminal Sterilization Processes and Significance of the Sterility Test for Failure Detection

In Table 1, possible failure modes of a terminal sterilization process are compiled, the significance of the sterility test to discover the failure is considered, and the best procedure to prevent such a failure is indicated. In cases of reduced SAL, the significance of the test to detect the process failure has been rated insignificant. This assumes a process that is basically functional as validated during process development and qualification of the routine process cycle. Of course, sterility testing would become significant if a failure was so

TABLE 1   Failure Modes of a Terminal Sterilization Process and Effectiveness of the Sterility Test to Detect the Failure

No.	Failure mode	Effect	Failure prevention	Significance of sterility test
1	High bioburden	Reduced SAL	Bioburden control	Insignificant
2	Insufficient sterilization effectivity	Reduced SAL	Sterilization cycle development	Insignificant
3a	Steam quality failure	None, if temperature profile is met	Parametric control (temperature)	Insignificant
3b	Inhomogeneous sterilization conditions	Reduced SAL	Sterilization cycle qualification	Insignificant
3c	Sterilization cycle failure	Reduced SAL	Cycle monitoring parametric control	Insignificant
4	Container/closure integrity failure during or post sterilization	Sporadic contamination, reduced SAL	Container/closure and sterilization cycle development	Insignificant
5	Mix-up of sterilized and nonsterilized product	Partial lack of sterility	General GMP measures	Depending on numbers
6	Total lack of sterilization	Total lack of sterility	General GMP measures	Significant

dramatic that it approached complete ineffectiveness of the process. Detection of such a dramatic failure should not be by a sterility test, but by cycle monitoring and control of the cycle parameters.

Steam quality has little influence on sterilization of products in their sealed final container, so long as the temperature profile during sterilization is not affected. Sterilization conditions inside the container will develop dependent on temperature input and product/container configuration depending on heat transfer to the closed container, which can be effectively measured by a thermometer. Hence, the only condition where sterility testing can contribute significantly to failure detection is mix-up or a complete lack of sterilization. (Mix-up would be reliably detected by a sterility test only if large numbers of nonsterile units were admixed to the sterilized product.) If large-scale mix-up or lack of sterilization was a realistic concern, the manufacturing company would be in severe violation of GMP. It would be irresponsible to rely on sterility testing to detect such a dramatic failure, and the reliability of the company to manufacture sterile product with or without a sterility test would have to be seriously questioned.

13.1.7. Sterility Testing and Release of Aseptically Manufactured Product

For aseptically manufactured products a sterility assurance level cannot be theroretically calculated as for terminally sterilized product. Aseptic processing is a highly complex process and there are many more possible failure modes than in terminal sterilization. The significance of the sterility test as a method to detect any of these failures is strongly dependent on the severity of failure.

While the list of failure modes in Table 2 is not intended to be complete, it is obvious that the significance of the sterility test to detect a failure can range from significant to insignificant depending on failure severity. When applied to products manufactured aseptically in modern cleanrooms, a contamination rate of 10^{-4} would be far too low to be detected reliably in a sterility test (see Fig. 1). Hence, for aseptic processing, the significance of a sterility test would also not be significant enough to solely rely on it for product release.

The potential of the sterility test to detect failure in an aseptic process is clearly higher than for a terminal sterilization process and hence, as long as the sterility test is the only test available to check for gross failure of the aseptic process, there is no possibility of abolishing it, insignificant as it

TABLE 2 Failure Modes of an Aseptic Process and Effectiveness of the Sterility Test to Detect the Failure

Failure mode	Effect	Failure prevention	Significance of sterility test
Nonsterile primary packaging materials	Partial lack of sterility	General GMP measures	Significant
Sterilization for primary packaging materials insufficient	Reduced SAL	Sterilization cycle development	Insignificant
High bioburden	Possible membrane filter failure	Bioburden control	Depending on failure severity
Filtration sterilization failure	Sporadic contamination	Sterilizing filter validation, integrity testing	Depending on failure Severity
Contaminated filling equipment	Sporadic contamination	Sterilization procedures, aseptic technique, barrier effectiveness	Depending on failure severity
Aseptic handling errors during filling	Sporadic contamination	Training of operators barrier effectiveness	Depending on failure severity
Failure of cleanroom barriers	Sporadic contamination	Barrier design, HEPA-filter integrity	Depending on failure severity

may be. For this reason, parametric release is considered only for product terminally sterilized in the final sealed container.

13.2. BASIC CONSIDERATIONS FOR PARAMETRIC RELEASE

13.2.1. Principle of Parametric Release

Parametric release has been defined in PDA Technical Report No. 30 as "the release of sterile pharmaceuticals without conducting a sterility test. Sterility of a lot produced by means of a validated process is ascertained solely by review of the parameters achieved during a sterilization cycle" (7). The document addresses specifically sterilization by moist heat in order to clearly discuss the principle in a single well-defined case. It was formulated out of the conviction and experience of the authors that sterility testing can contribute no additional relevant information for the release of a product that was sterilized in a correctly developed, validated, and controlled steam sterilization process, and that no additional information is needed to abolish the irrelevant sterility test in this case. To rely solely on a review of process parameters for release of a product to the market, all the failure modes of the process must be addressed with at least an equivalent diligence as in the sterility test. On the other side, it would be incorrect to introduce additional controls as a prerequisite for parametric release that have not been addressed by the test for sterility.

13.2.2. Elements Needed for Safe Release of Sterile Products

As sterility testing should (theoretically) indicate each failure in the manufacture of a sterile product, each possible failure mode should be addressed in a parametric release procedure. In Table 1, failure modes for a terminal sterilization process have been correlated to the significance of sterility testing in their detection. In the following paragraphs these failure modes are considered in greater detail with respect to their detection in parametric release.

Failure Mode 1: High Bioburden (*Excessive Input of Microorganisms, which Leads to a Failure of the Sterilization Process*) Vegetative forms of bacteria have very limited resistance toward most sterilization processes with D_{121} values in the range of less than 10 sec. For a population with D_{121} of 10 sec, 10^{90} microorganisms would be inactivated within 15 min at 121°C. Hence, the number of vegetative microorganisms in the bioburden is of very little consequence for sterility assurance. Endotoxins would be of much more concern with high bioburden numbers. However, although vegetative bacteria may release endotoxins during sterilization, these are not detected by a sterility test and hence, endotoxins should not be part of a consideration of parametric release. An increase in resistance against the sterilization process would be of much greater consequence than an increase of numbers.

Bioburden control measures will always depend on the safety margin of the sterilization process, which needs to be established during process development. A process that inactivates 15 orders of magnitude of resistant spores (D_{121} = 1 min) would not need a tight bioburden control if the maximum bioburden seen during process development and routine control measures is less than 100 organisms/container and the most resistant spore ever detected has a D_{121} value of 0.3 min. The considerations would be different if spores with a D_{121} value of 2 min. are routinely seen in the process.

In any case, if the safety margin of the sterilization process is sufficient and bioburden is adequately controlled, sterility testing of lots can add no relevant additional information for product release.

Failure Mode 2: Insufficient Sterilization Effectiveness in the Product (*Microorganisms Not Killed in the Product Due to Insufficient Access/Penetration of the Sterilant or Protective Action of the Product*)   It must be assured during validation of a sterilization cycle that homogeneous sterilizing conditions are reached in all parts of the product. For product in a closed container that is to be terminally sterilized by moist heat, sterilizing conditions will develop in each container that depend on the container/closure/product configuration together with the heat transfer during the process. The conditions in the containers are not the conditions of the sterilizer chamber, and thus, steam quality is not the primary consideration. The product itself can exert a protective or an additional destructive effect on microorganisms during the process. The geometry of the container/closure may have a protective effect (microorganisms hidden in a space where no sterilizing conditions [e.g., steam saturation] are reached). This may be relevant if microorganisms can be released after sterilization from their protective position into a sensitive product.

The effect of the chosen sterilization conditions on microorganisms in or on the product should be demonstrated in cycle development studies by use of resistant spores inoculated into relevant positions of the product. Sterility testing of lots must not be seen as a substitute for cycle development. For a properly developed sterilization cycle, sterility testing can add no relevant information for product release.

Failure Mode 3: Inhomogeneous Sterilization Conditions or Cycle Failure (*Sterilizing Conditions not Reached in Each Part of the Load or During Each Sterilization Run*)   Once process effectiveness has been demonstrated for the container/closure/product configuration, it must be shown during sterilizer qualification that sterilizing conditions are reached in each part of the load (worst-to-sterilize spot established in the load and the sterilizing cycle qualified so that product in the worst-to-sterilize spot is effectively sterilized) and for each run of the sterilizer (loading pattern established and sterilizer qualified to achieve reproducible sterilizing conditions in all parts of the load). Process parameters must be established that have to be met in order to verify that the process has homogeneously met the sterilizing conditions as specified in process development.

Temperature transfer to the container in the coldest position of the load should be measured in routine sterilization. If the temperature profile achieved in this position is correspondent to the profile achieved during cycle development, the necessary sterilizing conditions achieved in the closed containers can safely be assumed to have been met. Once a sterilizer and the cycle used have been qualified and the defined process parameters have been fully achieved in routine sterilization, sterility testing of lots can add no relevant information for product release.

Failure Mode 4: Container/Closure Integrity Failure During or Post Sterilization (Ingress of Microorganisms During or After the Sterilization Process) During the sterilization process, container/closure systems may be damaged and subsequently allow penetration of microorganisms. It is a question to be answered during product development if the container/closure system is fully compatible with the sterilization conditions by performance of meaningful container/closure integrity testing. Unless the container/closure system has been catastrophically misdeveloped, contamination via damaged container closure systems is expected to be a slow and sporadic process. The significance of sterility testing to detect sporadic leakers would be very low, and hence, it would be irresponsible to rely on the insensitive sterility test to detect container/closure failure.

Container/closure systems can also be deformed due to pressure differences between the inside and the outside of the container during a sterilization process. Contamination of the product could be introduced at the end of the sterilization process via contaminated cooling water. Again, resistance of the container/closure systems toward the rigors of the sterilization process should be verified during product and process development. If such deformation occurs frequently, the container/closure/process configuration should be reconsidered to eliminate the danger of product contamination. For a sporadically occurring event, the reliance on the insensitive sterility test would be irresponsible. Sterility testing must not be seen as a substitute for container/closure development. If the system is correctly developed, sterility testing of lots is too insignificant to add any relevant information concerning sporadic leakers.

Failure Mode 5: Mix-Up of Sterilized and Nonsterilized Product If sterilized product units cannot be clearly distinguished from nonsterilized units, there could be a danger that nonsterilized units are inadvertently mixed with sterilized units. Should this happen in relevant numbers, the nonsterilized units would eventually lead to failure of a sterility test if such units are contained in the sample taken for the sterility test. Depending on the number of admixed nonsterile units and the distribution of nonsterile units among the total number of sterile units, the probability of detecting the failure in a sterility test could be anything from high to very low. Sterility testing of lots must not be used as a substitute for GMP measures. If mix-up is not safely excluded by segregation measures and labeling with sterilization indicators, it would be irresponsible to rely on sterility testing, as the test does not have the significance to detect low numbers of admixed nonsterile units.

Failure Mode 6: Total Lack of Sterilization   Total lack of sterilization of an entire sterilizer load is the one fault that would clearly be detected in a sterility test. Any attempt to abolish the sterility test for routine release of terminally sterilized product must absolutely exclude the occurrence of this failure mode.

13.2.3. Elements Considered for Parametric Release

It was stated in the PDA report that "validation sterilization is the key to parametric release."

It was further stated that parametric release can be warranted if the points as stated in the following bullet points are met:

- Parametric release can be warranted if the sterilization process is properly designed and developed to achieve a SAL of $\geq 10^{-6}$ in the product;
- Process design and development are the key to parametric release. It must be assured that the process is suitable for the product to be sterilized and that any possible presterilization bioburden will be inactivated with a wide margin of safety. Thereby the failure modes 1 and 2 are excluded.
- It should be added that process development needs to also include suitability of the process to maintain integrity of the container/closure system. If this is correctly done, failure mode 4 is also excluded.
- Parametric release can be warranted if as a consequence of process development process parameters are defined and limits are set to allow proper control and documentation of the sterilization process.
- Definition of process control parameters is a prerequisite to verify and document that the process conditions as defined in the process are met. This is the necessary prerequisite to exclude failure mode 3.
- Parametric release can be warranted if the sterilization cycle is properly qualified to assure that each and every load of final product containers was subjected entirely and homogeneously to the sterilization process and that critical process parameters were met in every part of the load.

- Qualification of the cycle must ascertain that the process conditions as defined in process development are delivered to each unit of any load processed. If this is properly done, failure mode 3 is also excluded.
- Parametric release can be warranted if it is ensured that all relevant data are documented and reviewed during the process of releasing the product, and sufficient documentation is retained that contains all relevant data reviewed for the release decision.
- A formal review process must be in place to verify that the relevant process parameters have indeed been met. For later revision (e.g. in case of claims of insterility) a formal set of documents must be retained to give proof that the defined sterilization conditions were correctly met for each lot released parametrically.
- Parametric release can be warranted if procedures for maintenance, change control, and requalification ensure that the process remains under control.
- Measures must be in place to ascertain that the performance of the equipment as established remains unchanged and that attainment of critical parameters continues at all times to indicate that the process conditions as defined in process development were met. This is a further prerequisite to assure that failure mode 3 is permanently excluded.

The essential points given in the PDA technical report focus on control of the sterilization procedure. In addition, strong emphasis is needed on highly reliable mechanisms, to exclude mixing of sterilized and nonsterilized product (exclusion of failure mode 5) or complete omission of sterilization (exclusion of failure mode 6). The use of multiple chemical indicators distributed in the load is vital to allow immediate recognition of sterilized materials. Samples or evaluation records of indicators showing exposure to sterilizing conditions should be an integral part of the batch records. Double door autoclaves and clear physical separation of sterilized and nonsterilized product will effectively prevent mixup. In addition, clear operating procedures must be in place and strictly enforced to prevent the transport of any nonsterilized units to the area where sterilized product is stored. If all these elements have been considered and integrated into a manufacturing process for terminally sterilized sterile products, all the risks of releasing nonsterile units are minimized with much more diligence than by reliance—even partially—on a sterility test.

13.2.4. Parametric Release and Automated Manufacture

It is occasionally argued that automated sterilization procedures could minimize the risk of nonsterile product occurrence, and hence, automation could be an argument for parametric release. If the failure modes are considered, the role of automation can be clearly established for each process. Bioburden (failure mode 1) may be reduced by reduction of operator intervention; input from the environment may be reduced if automated equipment used to prepare and fill product is segregated from the environment (sometimes aseptic filling lines are used to fill product to be terminally sterilized). The input from starting materials including water and from primary packaging materials would be unchanged. While automation can certainly have an effect on bioburden reduction, it is the totality of measures taken to reduce presterilization bioburden that counts and not so much automation in itself.

Process effectiveness (failure mode 2) is clearly not a primary function of automation. Only where process effectiveness strongly relies on a low presterilization bioburden, would automation be seen as a factor (see failure mode 1). Sterilizer performance (failure mode 3) is also rather determined by the reliability of the sterilizing equipment and not by the degree of automation. Container/closure integrity (failure mode 4) is a function of process conditions and container/closure configuration and as long as it is guaranteed that the closures are placed and sealed correctly, automation does not add to further reduce the probability of the failure to occur. Mixup or lack of sterilization (failure modes 5 and 6) may be prevented if the automatic system includes effective segregation measures of nonsterilized from sterilized product. It is the principle how sterile and nonsterile units are kept separated that is important, not so much the degree of automation. While automation can be used to minimize human error to improve mix-up prevention, physical barriers and separate storage work as well in nonautomated facilities. Automation can be a factor to control bioburden that would be of special importance where parametric release is considered for a process with a limited safety margin with respect to effectiveness of the sterilization process.

13.2.5. Availability of New Technologies and Parametric Release

New technologies, especially in aseptic processing, are also expected to render parametric release more acceptable. New technologies have a vast potential to improve production processes for sterile products. It is certainly true that new segregation technologies, especially in combination with automation, can decrease the danger of contamination of a product by interference of personnel and thereby reduce presterilization bioburden. It is certainly true that application of isolator techniques can dramatically increase the assurance of sterility for an aseptically manufactured product. Novel technologies in microbiology can help to better understand the pathways of contamination and thereby improve cleanroom management.

However, the present discussion on parametric release is restricted to product terminally sterilized in the final container by means of overkill sterilization processes. This is the most easy to control situation, where presterilization bioburden has in all but exceptional cases only minor significance for sterility assurance. As long as a reasonable approach to parametric release is not possible for terminally sterilized products, parametric release of products manufactured by aseptic processing is out of question for the foreseeable future. Hence, new technologies can add very little to the application of parametric release of the sterility aspect of sterile products.

13.2.6. Aseptic Processing and Parametric Release

At present, parametric release is being discussed only for terminally sterilized products. The interest in parametric release would probably increase if it could also be applicable to aseptically manufactured products. Unfortunately, although terminal sterilization is a relatively simple and straightforward process and the delivery of the sterilizing principle to the product can be easily measured, aseptic processing is far more complex. Unlike for terminal sterilization, there is a lack of parameters that can be directly linked to the destruction of microorganisms. Removal of microorganisms by filtration cannot be measured; filter effectiveness can only be generally validated. Absence of contamination by microorganisms during aseptic processing cannot be correlated directly to physical measuring parameters. It can only be shown that under the conditions attained, such contamination will be improbable. In the absence of lot-specific parameters, the sterility test remains the only test to rely on.

Hence, in spite of the known shortcomings of the sterility test, parametric release is at present not an option to release the sterility assurance aspect of aseptically manufactured products.

13.3. HISTORY OF PARAMETRIC RELEASE AND PRESENT POSITIONS OF THE AUTHORITIES

13.3.1. Situation in the United States of America

13.3.1.1. Position of FDA

In 1987, conditions were formulated by FDA (8) under which parametric release can be accepted for terminally heat sterilized articles. These conditions include the following:

A sterilization process cycle validated to achieve a microbial bioburden reduction to 10^0 with a minimal safety factor of an additional 6-log reduction.

Integrity for each container/closure system has been validated to prevent in-process and post-process contamination over the product's intended shelf life.

Bioburden testing (covering total aerobic and total spore counts) is conducted on each batch of presterilized drug product.

Chemical or biological indicators are included in each truck, tray, or pallet of each sterilizer load.

If more than one critical parameter is not met, the batch is considered non-sterile despite BI sterility. These requirements may have seemed very strict at the time when the guideline was issued. But it contains nothing unusual for today's sterilization process validation, even if the details of the guideline are considered. The authority would certainly be expected to be very exacting in the verification that all the requirements are indeed met, but basically acceptance of parametric release seems attainable in the United States even today. Laboratory testing of sterile products is a legal requirement in the United States as stated in 21 CFR 211.167 under special testing requirements: "For each batch of drug product purporting to be sterile and/or pyrogen-free, there shall be appropriate laboratory testing to determine conformance to such requirements." While this seems to contradict FDA's policy guide, with the consent of the FDA evaluation of biological or chemical indicators in the laboratory may be accepted as sufficient to fulfill the testing requirement.

Only a few of the major pharmaceutical companies applied to the authority for authorization to release sterile products parametrically. The release procedures were accepted by the FDA, and these companies have meanwhile parametrically released hundreds of batches without any problems. Other companies did not follow. It was argued that the amount of validation work needed would justify parametric release only for major products, while the enormous validation work needed for a number of small products would make parametric release unattractive. The fear to be cited in a case of litigation for not having performed the sterility test (which is basically required in cGMP) was probably another important consideration. Whatever the reason, the vast majority of terminally sterilized products today are still released based on the results of sterility testing in the United States.

13.3.1.2. Position of United States Pharmacopoeia (USP)

Although the Test for Sterility as described in USP Chapter ⟨71⟩ (9) is a mandatory text, the compendium contains a long-standing recommendation in the advisory chapter ⟨1211⟩ on Sterility and Sterility Assurance (3). "If data derived from the manufacturing process sterility assurance validation studies and from in-process controls are judged to provide greater assurance that the lot meets the required low probability of containing a contaminated unit (compared to sterility testing results from finished units drawn from that lot), any sterility test procedures adopted may be minimal, or dispensed with on a routine basis." The statement first appeared in 1979 and has remained unchanged since. This recommendation was, however, not easily followed by the FDA or by the pharmaceutical industry.

In 1997 a USP advisory chapter specifically dealing with parametric release was proposed (10). The USP expert committee was clearly in favor of parametric release and felt that guidance to the industry was still needed 10 years after the FDA had issued their policy guide, to adopt the procedure. This chapter, however, has not been finalized until today.

13.3.1.3. PDA Technical Reports

PDA has issued two technical reports on parametric release. Technical Report No. 8 was published in 1987 (11). This document elaborated the technical basis for interpretation of the USP recommendation and set the stage for the FDA policy guide of that same year.

Technical Report No. 30 was written by a group of European and U.S. experts with the aim to propose parametric release procedures that would be acceptable and prepare the way towards International Harmonization (12). It was felt that with the globalization of the pharmaceutical industry there should be a global understanding of the procedures needed for parametric release, because it does not make sense for any company to release the same product parametrically for one market and to do sterility testing for another market. It was also intended to propose procedures from an industry point of view in order to assist the PIC/S working party, which was known to work on rules for parametric release in Europe and the PIC countries.

13.3.2. Situation in Europe and PIC Countries

13.3.2.1. Position of the Authorities

The situation concerning parametric release in Europe was very complex until the year 2001. Although there is a common European Pharmacopoeia together with a common basic rule for GMP regulations for medicinal products (13) in Europe, and there is the common EC-guide to GMP with its Annexes (14), interpretation of these rules is left to the individual European nations so long as there is no detailed guidance to the interpretation. Therefore, a company that applied to the individual "competent authorities" of the European nations to be granted permission to release their terminally sterilized products parametrically received answers that ranged from acceptance to conditional acceptance to rejection (15). This meant that it was practically impossible to obtain permission to proceed with parametric release in Europe except on a national basis.

It was the intention of the European authorities to mend this confusing situation with the finalization in 2001 of a CPMP Note (16) for guidance and an Annex to the European guide to GMP guide (17), which together regulate the procedures for parametric release in the EU. Authorization for parametric release will be given, refused, or withdrawn jointly by those parts of the authorities that are responsible for assessing products together with the GMP inspectors. The Note for Guidance sets the rules for application and revision of the applications, and the GMP annex gives the expectations that are to be met as verified during inspections conducted before parametric release is granted. So basically, parametric release is now acceptable in Europe under a common set of rules.

13.3.2.2. CPMP Note for Guidance on Parametric Release

Although the document deals with parametric release in a broader context and also includes other properties of products, the focus is on parametric release of sterile products to replace the test for sterility. It is stated that parametric release can only be applied to products terminally sterilized in their final containers by heat or radiation. The release of each batch would be dependent on the successful demonstration that predetermined, validated sterilization conditions have been achieved throughout the load. Parametric release is expected to be usually introduced as a variation of an existing market authorization when experience has been gathered with the product. This means that experience with sterility testing of the product is expected before changing to parametric release. Assessment of applications is stated to be in close collaboration between assessors and inspectors.

Any sterilization process must be as follows:

Performed in accordance with the requirements of European Pharmacopoeia

Well founded with regard to stability of the product and identification of critical parameters as defined during development studies

A heat sterilization process must be as follows:

Validated in accordance with GMP with heat distribution and heat penetration studies on established load patterns

Biologically validated in complementation to technical validation

Chosen in consideration of level and resistance of bioburden

Of demonstrated reproducibility

Emphasis placed on assurance of segregation of nonsterile from sterilized products

Documentation submitted for application should contain the following:

A description of the sterilization process (cycle type, loading pattern, specifications for cycle parameters and, if applicable, chemical indicators)

Specifications and methods/procedures for in-process testing (presterilization bioburden, cycle parameter monitoring, verification of load sterilization)

A process validation report (heat distribution, heat penetration, microbiological qualification, bioburden characteristics)

Package (container/closure) integrity data

13.3.2.3. EU Guide to GMP-Annex 17, Parametric Release

In the text GMP issues connected with parametric released are addressed that will be subject of a pre-approval inspection. The requirements stated in this Annex must be considered basic in the opinion of the European regulators. The positions of the CPMP note for guidance concerning the sterilization process are reiterated. Again it is stated:

"It is unlikely that a completely new product would be considered as suitable for Parametric Release because a period of satisfactory sterility test results will form part of the acceptance criteria. There may be cases when a new product is only a minor variation, from the sterility assurance point of view, and existing sterility test data from other products could be considered as relevant."

In comparison to the Note for Guidance, additional requirements are as follows:

Performance of a risk analysis of the sterility assurance system with respect to release of nonsterilized products

A history of good GMP compliance for the manufacturer

Availability of a qualified experienced sterility assurance engineer and a qualified microbiologist on the production site

Container/closure integrity validated to remain intact under all relevant conditions

Review of change by quality assurance under a change control system

Control procedure in place for presterilization bioburden

Mix-up prevention between sterilized and nonsterilized products (e.g., by physical barriers or validated electronic systems)

Sterilization record review by at least two independent systems (e.g., two people or a validated computer system plus a person)

Prior to release of each batch of product, additional verification is required:

completeness of planned maintenance and routine checks
approval of repairs and modifications by the sterility assurance engineer and microbiologist
instrument calibration
current state of validation for the product and load processed

13.3.2.4. Position of PIC/S

The Pharmaceutical Inspection Convention (PIC) is a multinational agreement or mutual recognition of GMP inspections. Formerly, PIC comprised many countries including the European Union member states plus other countries such as Switzerland and Australia. The rules by which PIC inspections are governed have been elaborated in working parties of the inspectors of the PIC member states. According to the rules of the EU, mutual recognition agreements cannot be maintained between individual EU member states and third countries, so the PIC is now strongly reduced in member states. However, European inspectors continued to collaborate in the former PIC working groups, which are now named Pharmaceutical Inspection Cooperation Scheme (PIC/S) in order to write recommendations to inspectors. These recommendations, even though not written in an official function, are highly influential and set the rules for inspectors in the EU and other PIC countries. As the same group of people is involved in writing European GMP documents, it is not surprising that PIC/S drafts often precede changes or amendments to the European GMP guidelines.

13.3.2.5. PIC/S Recommendation Guidance on Parametric Release

The purpose of the PIC/S guidance document (18) is stated as follows: "to provide guidance for GMP inspectors to use for training purposes and in preparation for inspections of company premises where Parametric Release has been approved or applied for". The document is intended to be in general accord with the CPMP Note for Guidance and EU GMP Guide Annex 17. It contains a summary and (in an annex) a detailed compilation of the aspects that the GMP inspectors should consider and an interpretation of the official guidelines. Although it is left to the assessors to evaluate product and process design and their initial validation, GMP aspects are to be checked by the inspectors on site. Interestingly enough, the specific requirements formulated in this document are meant to be general GMP requirements for sterile products: "Manufacturers of sterile products should comply with the principles expressed, whether or not they are successful in their application for

Parametric Release." Hence, the document is basic for the present thinking of the European authorities concerning sterile products. The document emphasizes a risk-based approach, but the risks are not correlated to the role a sterility test may have to eliminate the risk. Elements of the sterility assurance system are stated to be as follows.

Product and Process Design   Design and original validation of the manufacturing process are expected to ensure that product integrity can be maintained under all relevant conditions.

Bioburden Control   High emphasis is placed on bioburden monitoring: It is stated that a system to control bioburden in product streams and thus also presterilization bioburden would be needed in order to be eligible for parametric release. Control of the early stages of production is expected to extend to monitoring of chemical starting material, particularly for the presence of microorganisms that may be resistant to the sterilizing agent. This is also seen as an aspect of environmental control.

The filling systems are expected to be of sanitary type with regard to design, pipework, connecting joints, welds, internal structure of valves. Gases, solvents, lubricating fluids are expected to be controlled. Attention is also drawn to cleaning, sanitation, microbiological monitoring, planned preventive maintenance, breakdown repair, and operator error or noncompliance with procedures as reasons for loss of bioburden control. Reduction of bioburden testing would only be acceptable based on historical data of rigorous testing with a suitable method to determine also resistance of microorganisms to the sterilizing agent. While a lot of these considerations make sense for aseptic processing, the details of bioburden control expected seem quite out of proportion for product sterilized in a terminal overkill process.

Contamination Control During the Process   Contamination control is expected to comprise cleaning and sanitation, cleanroom control, time limitation, and filtration stages.

It is acknowledged that environmental monitoring can be considered a secondary measure; the primary focus of attention should be on presterilization bioburden. Nevertheless, some level of monitoring is expected. Setting of monitoring limits with a valid rationale and actions taken when limits are exceeded should be considered in conjunction with trend analysis, and the relevance should be evaluated by the inspectors.

Product Filtration   The text recommends control of filter grade, product/filter compatibility, microbiological pre-contamination of the filter, period of use, washing (prevention of pyrogen buildup), sterilization, and reuse, method of off-line or on-line integrity testing, filter storage post-integrity testing, process stage for filter integrity test, decisions taken in case of a failed integrity test, microbiological state of the test equipment—particularly product contact surfaces on the clean side, microbiological monitoring of product fluid after the filter. It remains unclear why product filtration is so strongly emphasized for a process which by definition of the scope of the text involves sterilization in a terminal overkill process.

The Sterilization Process   Parametric release is stated to be applicable only to terminal sterilization processes that incorporate large safety margins (pharmacopeial reference cycles or cycles delivering a minimum F_0 of 8 with a SAL of 10^{-6} or better). The sterilization process should be adequately validated and monitored with defined tolerances of acceptance derived from validation studies to show that minimum sterilizing conditions were delivered to each part of the load (revalidation at least annually). Validation is expected to follow appropriate guidelines. Validated loads should be precisely defined.

Failure modes of complex sterilization cycles should be known and evaluated, and preventive measures taken. Homogeneity and penetration of the sterilizing agent (e.g., steam) should be demonstrated (where there is no alternative by use of Biological Indicators). In addition, the cooling medium should be sterile (eventually verified by testing).

The monitoring instrumentation is expected to be calibrated and sufficient to confirm the delivery of the validated cycle, and be in the same position as for the validation.

Sterilizer Validation and Change Control   The sterilizer is expected to be in exactly the same mechanical, electrical, and software state as during the validation, and the state of the services should be similar as in the validation stage. Routine planned preventive maintenance programs should be documented. Sterilizer and services startup checks should be carried out successfully each day. Planned change and unplanned repairs should be approved or reviewed sufficiently by both the sterility assurance engineer and microbiologist.

Assurance of Container/Closure Integrity Under All Relevant Conditions   In process and finished product integrity testing methods should be capable of detecting and results should demonstrate that product into which microorganisms could penetrate will not be released for sale. It remains unclear how occasional leakers should be detected. While it is alleged that faults of integrity or other faults may be detected by the functional manipulation of the product during the sterility test and this might be lost by not carrying out the sterility test, this is not the experience of the author or other experts with practical experience in sterility testing, nor is any publication cited in the PIC/S document that would substantiate the claim.

Mix-Up Prevention Between Sterile and Nonsterile Product   Product that has not been exposed to the sterilization process is expected to be rigorously segregated from the flow of product after sterilization. It must be prevented that product can move directly to the stage of processing following sterilization without having passed through a valid sterilization cycle (e.g., by physical separation of process flows and/or by validated electronic systems). Mix-up prevention systems should be supported by comprehensive failure analysis, which also takes into account minor streams of product (e.g., samples) and contingency procedures to control breakdown situations. Deliberate attempts to bypass the system by an easier pathway should also be considered and avoided as far as possible.

Prevention of Complete Lack of Sterilization Comprehensive checks are expected to be carried out on completion of the cycle to assure that the validated sterilization process has been delivered. Should the cycle not be correct, measures need to be defined that assure that the failure is corrected and/or the product is quarantined without compromising mix-up prevention.

Product Release for Sterility Assurance Verification of numerous items is expected for each batch of sterilized product: Container/closure integrity, compliance with presterilization microbiological criteria including trends in bioburden, adherence to process time limits, filter integrity data, sterilizer maintenance records, unplanned repairs or modifications and coverage by change control, calibration of instruments, sterilizer qualification, reconciliation of units before and after sterilization, sterilization records' compliance with specification, record of activated process indicator. In case of an atypical cycle, approval of the release by the sterility assurance engineer and microbiologist assuring that cycle parameters are within tolerances that were accepted during the validation is necessary.

Sterility Assurance System The totality of the Sterility Assurance System (e.g., change control, training, written procedures, release checks, planned preventive maintenance, failure mode analysis, prevention of human error, calibration) must be verified. It is expected that a highly qualified sterility assurance engineer and a microbiologist with sufficient seniority and authority to enforce compliance for matters related to sterility assurance, should normally be present on site. The change control system should require their review of any change to ensure that small changes may not have an effect on the sterility assurance system that is not apparent to other reviewers.

General GMP Compliance As a general basis for authorization to apply parametric release, the documentation of the company's commitment to maintain a high level of compliance to GMP in general (not limited to the sterility assurance system) is strongly emphasized. The sterility assurance system should be fully capable and robust. It is expected to achieve the objective of assuring the sterility of the product without the additional challenge of the sterility test and, in addition, withstand variations that may reasonably be expected. It is advised to perform a risk analysis to judge how the system could break down and what are the means to reduce that risk. It is clear from the text that the PIC/S working group was aiming at a general increase in process controls for sterile products. What are described as requirements for controls of the environment and of the presterilization bioburden is not primarily relevant for terminally sterilized products or parametric release. What are described as requirements for development and control of the sterilization process is not specifically applicable to parametric release but must be seen as general expectations.

A lot of weight is placed on the special confidence in the GMP compliance of a company applying to employ parametric release. There is an interesting statement that if the assessor's or inspector's confidence in the elimination of sterility testing for a company's products is reduced, either group should have a mechanism to withdraw approval. Reduction in confidence may follow an inspection, or on receipt of other information. Is it really intended to say that a company that cannot be trusted to reliably adhere to GMP can continue to manufacture sterile product just because a sterility test is performed?

13.3.3. Situation in Japan

In the year 2002, parametrical release of the sterility assurance aspect became officially accepted in Japan. During the USP/PDA open conference on Sterile Product Manufacturing in Fort Myers, Florida in 2002 it was expressed by the Japanese delegate T. Sasaki that the test for sterility is considered to be of little significance for sterility assurance of terminally sterilized product by the Japanese authorities. Sterility assurance is expected to be founded on sound process validation and GMP procedures, and the Japanese authorities would expect the industry to apply for parametric release, which would be considered favorably. Permission would be granted after review of the validation documentation and an on-site inspection.

Although this may be considered not basically different from the European position, the detailed approach to assessment of the documentation and the detailed future approach the Japanese authorities will take towards assessment of the documentation and inspection of an applying company on site remains to be seen, to fully understand the Japanese situation. In a personal discussion with the author early in 2003, Japanese government officials expressed their disappointment that the Japanese industry was hesitant to apply after parametric release was officially accepted in 2002.

13.3.4. The Position of Official Reference Laboratories

The pharmacopoeial test for sterility is primarily meant as a reference test carried out by official reference laboratories. It is strongly maintained by representatives of such laboratories that the sterility test is needed as the only means to verify if a product on the market is sterile or not. Although this may be correct from a testing point of view, it should not be neglected that all the considerations on the limited significance and the inherent error probability of the sterility test apply in the same measure to tests conducted in reference laboratories. If there is any doubt about the sterility of a product, it is mandatory that the sterility assurance and GMP measures of the manufacturer be considered first and not the results of the sterility test, however carefully conducted.

13.3.5. Positions of Pharmaceutical Manufacturers

13.3.5.1. U.S. Manufacturers Who Apply Parametric Release

The few U.S. manufacturers (e.g., Abbott and Baxter) who have applied for and adopted parametric release have for almost 15 years continued to release sterile product without conducting a sterility test. No problems with sterility assurance have been reported. The sterilization processes used for these products have been meticulously validated and are controlled to the satisfaction of FDA. In personal discussion, representatives of the companies with responsibility for sterility assurance as well as FDA officials express their satisfaction with the agreement.

13.3.5.2. European Manufacturers Who Have Attempted Parametric Release

Before implementation of the new European guidelines, there were experience reports of companies who applied for parametric release in various European states. The attitude of the authorities was found to be very diverse. The answers ranged from a clear yes (e.g., Luxembourg), the request for specific additional information (e.g., Norway), a general request to submit all validation documentation (e.g., Germany), the request to dramatically increase basic validation (e.g., UK) to outright refusal (e.g., France). This was stated to be a main reason why pharmaceutical manufacturers felt it did not make any sense to further pursue parametric release. As long as sterility testing is required anyway for parts of the European market, there would have been no advantage to be granted permission for other countries. This situation was also seen to be unacceptable by the authorities especially of the Nordic countries, where the attitude toward parametric release was traditionally liberal. Thus, the initiative for new European guidelines was taken under the leadership of Sweden.

13.3.5.3. Reactions to the New European Guidelines

After release of the new guidelines, representatives of the European Authorities expressed their expectation to see many applications for parametric release. After 2 years, there is now the impression that not much has changed.

There continues to be a small number of applications, but no significant increase is apparently seen. As in Japan, the European industry failed to perceive the new guidelines as the wide opening of the gate towards parametric release that was intended by at least some of the authors from within the authorities.

13.4. THE FUTURE OF PARAMETRIC RELEASE

13.4.1. Undecided Manufacturers

What are the reasons for pharmaceutical manufacturers to wait and not to apply for parametric release? Sterility testing is recognized by industry as a problem due to its statistical insignificance and the error potential, which may lead to the rejection of perfectly compliant lots. However, the testing facilities are needed anyway by most companies as long as aseptically manufactured product must be tested for sterility. Many of the major manufacturers have invested in isolator testing facilities, where the problem of false positive tests is no longer so dramatic.

Validation requirements for terminally sterilized products are seen to be already at the upper limit, and each attempt of the authorities to further raise the bar is being opposed. Terminal sterilization is not seen as a highly error-prone manufacturing process, and this is clearly demonstrated to be correct by the data of routine sterility testing. Additional validation efforts are seen as a substantial additional burden that does outweigh the relief of not performing a sterility test, while they are not seen to significantly increase the safety of the products. Parametric release would be interesting if the effectiveness of the existing validated sterilization processes would be acknowledged based on internationally accepted harmonized principles. Any push toward increased validation efforts makes parametric release uninteresting.

13.4.2. Criticism of the Positions of the Authorities

The ambivalent position of the authorities has been very clearly highlighted by the PIC/S document on parametric release. On one side, a rational approach by use of a risk analysis is advocated. On the other side, alleged risks like presterilization bioburden are not properly regarded in their severity with relation to the outcome of the production process to be evaluated. The sterility test is ascribed the role of an "additional challenge" to the production process but it is not appreciated in its extremely limited capacity to recognize failure in the process. The general GMP attitude of a company is emphasized as an element to achieve confidence and trust on the side of the authorities. It is not appreciated that the test for sterility can never be a substitute for proper adherence to the rules of GMP. If a manufacturer cannot be trusted to properly control a process that involves terminal sterilization, the company should not produce sterile pharmaceuticals regardless whether a sterility test is conducted or not.

There seems to be emotional fear on the part of the authorities to possibly sanction a process where a nonrecognized defect of a product might cause a lethal infection to a patient. This fear leads to the use of killer arguments against deregulation in the field of production of sterile pharmaceuticals. Every step back in regulation might unleash an unrecognized threat to product safety. Hence, introduction of parametric release is not treated as a step toward deregulation, but as a step to further increase regulation. It is important that such an attitude be overcome by a meaningful and rationally applied risk analysis.

13.4.3. Need for a Risk-Based Approach

In Failure Mode Effect Analysis (FMEA), a method that evaluates the severity of the risk of individual process steps, the effect of failure is evaluated by three parameters: probability of occurrence of the failure p, severity of the effect of the failure s, and likelihood of detection of the failure l. It is important to use the tool prudently and not to overrate the risks. If FMEA is applied to any manufacturing process for sterile medicines, and the parameter s is indiscriminately set at maximum level because the effect of a failure could be death of a patient and l is also set at maximum because there is no secure method for failure detection available, the system is led ad absurdum.

It is certainly true that a single microorganism surviving in a unit of the product may, in given circumstances, proliferate and cause a life-threatening septicemia. But this is not the effect of a failure to meet a bioburden limit. This is not the effect of a failure to achieve saturated steam conditions in a sterilization process for closed containers. This is not even the effect of a failure to reach homogeneous conditions in an autoclave load by a deviation of 1 or 2°C. In all these cases, the effect of the failure would be a slightly reduced SAL, which considering that the sterilization process is an overkill process, somewhat reduces the ample safety margin. Certainly the effect of such failures must be considered quantitatively, but there should be no emotional overrating of risks.

It is certainly true that a single microorganism surviving in a unit of the product cannot be detected. But deviation of more than 1°C in the worst position of the load can be readily detected. Deviation from steam saturation can be detected with chemical indicators or specific measuring devices, but for closed containers this is not even a real risk to be considered. Where a comprehensive and rational risk analysis is conducted, it can easily be shown that correctly developed terminal overkill sterilization processes for product in their closed containers and conducted in qualified sterilizers are very safe processes. Irrational fears are unfounded, and the occasional occurrence of negative examples, where poorly designed processes in nonqualified sterilizers have led to disaster, are not a valid argument against this conclusion. In order to come to an atmosphere where useless and costly overregulation can be abolished, a well-defined rational risk analysis should be performed and sterilization processes should be shown to be validated to minimize the risks recognized.

If parametric release can be achieved and becomes acceptable to industry and regulators for terminal sterilization processes of closed containers, parametric release could still become a model for quality assurance of other processes.

REFERENCES

1. European Directorate for the Quality of Medicine. Chapter 2.6.1, Sterility. In *European Pharmacopoeia*, 4th Ed.; Strasbourg, 2003.
2. Spicher, G.; Peters, J. Mathematische Grundlagen der Sterilitätsprüfung. Zbl. Bakt. Hyg. I. Abt. Orig. A 1975, *230*, 112–138.
3. The United States Pharmacopoeial Convention Inc. Chapter ⟨1211⟩ Steriliza-Sterilization and Sterility Assurance of Compendial Articles. In *The United States Pharmacopoeia*, 26/NF 21, Rockville, MD, 2002.
4. PDA. Technical Report No. 30: Parametric release of pharmaceuticals terminally sterilized by moist heat. PDA J. Pharm. Sci. Technol. 1999, *53*, 217–222.
5. Van Doorne, H.; Van Kampen, B.J.; Van Der Lee, R.W.; Rummenie, L.; Van Der Veen, A.J.; De Vries, W. Industrial manufacture of parenteral products in the Netherlands: a survey of eight years of media fills and sterility testing. PDA J. Pharm. Sci. Technol. 1998, *52*, 159–164.
6. FDA. Parametric Release—Terminally Heat Sterilized Drug Products. FDA Compliance Policy Guides. Guide 7132a.13, 1987.
7. The United States Pharmacopoeial Convention Inc. Chapter ⟨71⟩ Sterility tests. In *The United States Pharmacopoeia*, 26/NF 21, Rockville, MD, 2002.
8. USP. Proposed chapter ⟨1222⟩ Terminally Sterilized Pharmaceutical Products—Parametric Release. Pharmacopoeial Forum 1997, *23*, 5074–5081.
9. PDA. Technical Report No. 8: Parametric release of parenteral solutions sterilized by moist Heat Sterilization. PDA J. Parenteral Sci. Technol. 1987, *41* (Suppl): S6–S10.
10. PDA. Technical Report No. 30: Parametric release of pharmaceuticals terminally sterilized by moist Heat. PDA J. Pharm. Sci. Technol. 1999, *53* (Suppl).
11. EEC Commission. Commission directive of 13 June 1991 laying down the principles and guidelines of GMP for medicinal products for human use (91/356/EEC).
12. EEC Commission. EEC Guide to GMP for Medicinal Products. (III/2244/87, 1989, Revised).

13. Schiebler, W. Erfahrungen mit der Parametrischen Freigabe. Presentation at Concept Symposium Trends in der Pharmazeutischen Mikrobiologie, Heidelberg 26/27.01.2000.
14. Committee for Proprietary Medicinal Products. CPMP/QWP/3015/99 Note for Guidance on Parametric Release. European Agency for the Evaluation of Medicinal Products, 2001.
15. Working Party on Control of Medicines and Inspections. Annex 17 to the EU Guide to Good Manufacturing Practice (ENTR/6270/00). EEC Commission, 2001.
16. Pharmaceutical Inspection Convention/Pharmaceutical Inspection Co-operation Scheme. Recommendation on the Guidance on Parametric Release. PIC/S Secretariat, Geneva (www.picscheme.org). 2001.

14

Raw Material Contamination Control

Lisa Gonzales
Amersham Biosciences, Sunnyvale, California, U.S.A.

14.1. INTRODUCTION

Usually, raw materials do not receive the attention that they should. Their identification, quality attributes, and characteristics should be examined and documented early in the development phase and carried out through to commercialization. Why are they so important in the manufacture of a drug product? It is extremely difficult, if not impossible, to manufacture a safe, pure, potent, effective, quality drug product if one start's out using inferior (or substandard) "building blocks" as materials. One must consider the saying, "garbage in, garbage out." Controlling the quality of the incoming raw materials will contribute to a higher probability of manufacturing a drug product that meets its quality attributes. Raw materials have been identified as the most common source of endotoxin contamination (1).

Why do you want to minimize bioburden for starting materials? Controlling bioburden prevents the growth of microorganisms that produce endotoxins. Low endotoxin, low bioburden is what you are striving to achieve for incoming raw material requirements. You are also looking to minimize or remove filth and contamination. These issues are good indicators of confidence that your suppliers are providing quality raw materials. It is important to have low endotoxin/bioburden levels in your raw materials so that when the manufacturing process has to be validated, a low level at the beginning of the process will allow removal and validation to be an easier task. Even though a process or process step has the ability to remove endotoxin, an unexpected increased level that is introduced could challenge the validated process and place undue stress on product quality.

It is easier to control the contamination of incoming raw materials and in-process steps than to remove contaminants during purification and the final stages of production. The manufacturing facility, equipment and process, raw materials, quality systems, and trained personnel are some of the key elements of cGMP (2). The introduction of endotoxin and other contaminants can be controlled by using Good Manufacturing Practices.

Quality Control (QC) and Quality Assurance (QA) programs should exert control over the manufacturing facilities, the manufacturing process, the validation efforts, and all testing of the raw materials, in-process material, bulk product, and final formulated product (2). The quality of raw materials used in the production of a pharmaceutical product can affect the safety, potency, and purity of the product. Therefore, qualification of raw materials is necessary to ensure the consistency and quality of all pharmaceutical products (2).

14.2. RAW MATERIAL REQUIREMENTS FOR GMP

An incoming raw materials inspection program is a GMP requirement. The minimum testing required is for identity. Depending on the component or container/closure, other testing may include safety-related issues like endotoxins and BSE/TSE. Raw materials with risk of endotoxin contamination should be tested with the receipt of every shipment or, depending on the material, they should be depyrogenated by a validated process. Alternatively, the supplier can provide materials that have been depyrogenated by a validated process and as such can be acceptable for receipt and approval once the supplier has been qualified.

There should be written procedures describing all actions of the raw material inspection program (21 CFR Part 211.80[a]) covering, at minimum, the parameters listed in Table 1.

There must be written procedures, test plans, and a minimum of at least one identity test is to be conducted. All equipment and instruments should be qualified. Analytical methods should be validated.

Each raw material should have a corresponding written specification that was developed to ensure the appropriate quality of material is used in the manufacturing process. In order to be released, the material must meet those specifications. (21 CFR Part 211.84[e]). Once raw materials are approved for use, the materials management department is responsible for using the oldest

TABLE 1 Raw Material Inspection Program

---

1. Describe how materials are received.
2. Describe how they are identified.
    a. Each lot of each shipment must be uniquely identified with traceability to the supplier manufacturing lot number. (21 CFR Part 211.80[d])
    b. Each lot is to be identified with its status: quarantine, approved, rejected. (21 CFR Part 211.80[d])
3. Describe how they are stored and what are the various storage conditions.
    a. Quarantine until tested or examined and dispositioned as approved for manufacturing use by the quality unit. (21 CFR Part 211.82[b] & Part 211.84[a])
    b. Storage should prevent contamination. (21 CFR Part 211.80[b])
4. Describe how raw materials are handled, sampled, and tested.
    a. Representative samples are to be taken: the number of containers sampled and the sample amount is to be statistically appropriate. (21 CFR Part 211.84[b])
    b. Sampling technique should prevent contamination. (21 CFR Part 211.84[c][2])
    c. Samples are to be appropriately identified. (21 CFR Part 211.84[c][5])
    d. A minimum of at least one identity test is to be conducted. (21 CFR Part 211.84[d][1])
5. Describe what the approval and rejection process is for raw materials.

---

material first. This is known as FIFO = first in, first out. There should be a procedure in place that describes how the Quality Assurance unit will handle raw material rejection.

14.3. INCOMING INSPECTION OF RAW MATERIALS PROGRAM

When raw materials are received, they should go directly into quarantine until they have been tested and approved for manufacturing use. Raw materials are brought in by the receiving department. This group should check for obvious damage to the shipping containers and match up the type and quantity of the material to the purchase order. If this information is correct, the material is moved to the designated quarantine area. Depending on the system in place, this may mean a locked-up area, or its status may be denoted by a bar code. The storage conditions (room temperature, 4°C, −20°C, −80°C, desiccated, hazardous chemical cabinet) should be appropriate for the type of material and any other type of handling that would prevent contamination. Bagged or boxed components of drug product containers, or closures shall be stored off the floor and suitably spaced to permit cleaning and inspection (21 CFR Part 211.80(c)). At this time, Quality Assurance or the incoming raw material inspectors are notified of receipt and its quarantine status. The material will remain in quarantine until it has been approved for use in manufacturing.

The QA inspector assigns a control (or lot) number to the raw material shipment and records this in the incoming inspection logbook or database. Shipments of more than one manufacturer's lot number must be assigned a separate control number for each lot within that shipment (21 CFR Part 211.80(d)). The control number log should capture date of receipt, date of control number assignment, incoming raw material part number, identity, manufacturer's lot number, quantity, purchase number and expiration date (if applicable). There should also be a column or field for the disposition (i.e., quarantined, approved, or rejected) of the material.

The inspector uses the raw material specification sheet document to collect all pertinent information about this specific lot of material and approve its use for manufacturing. The information the ICH Q7A guideline expects recorded is listed in Table 2 (13). The material should be visually inspected for labeling, identity, damage not detected during receipt, broken seals, evidence of tampering or contamination. When it has been determined that the material is suitable for identification testing and any other testing requirements, the lot is appropriately identified with a quarantine sticker and

TABLE 2 Items Needed to Identify Materials

Name of the manufacturer
Identity and quantity of each shipment of each batch of raw materials, intermediates, or labeling and packaging materials for API's
Name of supplier
Supplier's control number(s), if known, or other identification number
The number allocated on receipt
Date of receipt
The results of any test or examination performed and the conclusions derived from this
Records tracing the use of materials
Documentation of the examination and review of API labeling and packaging materials for conformity with established specifications
The final decision regarding rejected raw materials, intermediates, or API labeling and packaging materials

Adapted from Guidance for Industry Q7A Good Manufacturing Practice: Guidance for Active Pharmaceutical Ingredients.

sampled. Every container on a pallet need not be identified, the containers could be grouped for identification. Any individual units separated from the grouping, however, must be identified with the appropriate information (14).

The raw material is sampled according to the sampling plan indicated on the material specification. There should be designated areas suitable for collecting and preparing the sample(s) for each type of material (laminar flow hood, clean separated area, etc.) so as not to introduce contaminants into the sample. All sampled containers should indicate the date sampling was done, how much was or how many were removed, and who performed the sampling. The sample itself should be placed into a properly identified container that is nonreactive, nonadsorptive, or nonadditive—one that could not affect the result of any test performed. If testing for endotoxin, this would include not using stainless steel or certain types of plastic.

The sample is stored at the appropriate condition until it is delivered to the in-house testing lab or the outside reference lab. When the lab receives the sample, its storage conditions must also be maintained and it is logged as being received with the appropriate identification and information.

Testing each shipment of each lot is necessary even when a previous shipment of the same lot has been received, tested, and approved. The Commissioner feels that examination of each lot of each shipment received is necessary even though a portion of the same lot has previously been received, tested, and approved. Subsequent shipments may have been subjected to different conditions that may have caused changes in materials: although one shipment of a particular lot has met specifications, another may not. (preamble 228)

14.3.1. August 2001 ICH

The testing laboratory should have a system in place for tracking samples for receipt, status and result reporting. Analytical equipment should be qualified and test methods should be validated. There should be a procedure in place for handling out of specification results. In the event of an inspection, the FDA inspector will verify that all raw materials have been tested by quality control (7).

After all the acceptance criteria have been met for a raw material, it is dispositioned as approved for manufacturing (or quality control) use. The entire lot is physically moved from the quarantine area to the "approved" area. For bar code systems, the status in the computerized system is changed electronically and may not need a physical change in location. If appropriate, retention samples should be taken.

If any of the acceptance criteria have not been met for a raw material, it is dispositioned as MRB (Material Review Board) quarantined. The delegates representing various departments evaluate the results of the raw material testing and decide the disposition. Any disposition is appropriately labeled and segregated into the location corresponding to its status. All rejections should be thoroughly investigated and documented. A system and procedure should be in place for corrective and preventive action.

14.4. QUALIFICATION OF SUPPLIERS

It is the responsibility of the manufacturer of the product to ensure that all raw materials used in manufacturing are appropriately qualified (2). Manufacturers who purchase components from outside sources are required to establish adequate quality requirements and specifications for such components. The licensed manufacturer is ultimately responsible for ensuring that components conform to specifications and are acceptable for use. This may be done through inspections, sampling and testing, and/or through Certificate of Analysis from the supplier. Validity of the certificates should be established by the manufacturer through experience, historical data, testing, and/or audits of the supplier (11).

For components received from outside sources, either purchased or otherwise received, the firm should:

Have written, approved, specified requirements for the component(s)

Evaluate and select suppliers based on their ability to meet specified requirements

Define the type and extent of control needed over the component, which is based on the evaluation of the supplier (11)

14.4.1. Identification and Selection

Attention must be paid to issues such as suitability, toxicity, availability, consistency, contamination, and traceability. Raw materials that could be difficult to qualify may have to be investigated and identified in the early stages of product development (2).

In order to produce a safe and effective product, it is critical that the pharmaceutical manufacturer qualify and know the source of all raw materials. Every material employed in the manufacturing process should be accounted for. The source and intended use for each material should be established, and the necessary quantity or concentration of each material used should be determined. Primary sources, and when possible secondary sources, for each material should be identified. In all cases, suppliers should provide information regarding the traceability of each material, especially for human- and animal-derived raw materials (2).

The quality of incoming raw materials must be determined and documented to meet GMPs because it can impact the safety, purity, and potency of the drug product.

The following are the responsibilities of the drug manufacturer:

The minimum acceptance requirement for raw materials is to perform an identity test. (11)

Qualification of raw materials is necessary to ensure the consistency and quality of the drug product.

Test raw materials for pyrogen contamination. Documentation is maintained on all raw materials, in process and final product testing.

REFERENCES

1. Parenteral Drug Association Quality Assurance/Quality Control Task Group. Current practices in endotoxin and pyrogen testing in biotechnology. J. Parenteral Sci. Technol. 1990, *44*, 39–45.
2. USP 26-NF21 Supplement 1 < 1046 > Cell and Gene Therapy Products—Manufacturing Overview.
3. Kupp, G.D. Challenges, considerations, and benefits of raw materials testing. Pharmaceutical Technology—Analytical Chemistry & Testing 2003; 22–27.
4. Ibid.
5. Dept. of Health, Education, and Welfare Public Health Service Food and Drug Administration *ORA/ORO/DEIO/IB* Date: 3/20/85 Number: 40, http://www.fda.gov/ora/inspect_ref/itg/itg40.html.
6. Dept. of Health, Education, and Welfare Public Health Service Food and Drug Administration *ORA/ORO/DEIO/IB* Date: 1/12/79 Number: 32 http://www.fda.gov/ora/inspect_ref/itg/itg32.html.
7. Biotechnology Inspection Guide Reference Materials and Training Aids November 1991, http://www.fda.gov/ora/inspect_ref/igs/biotech.html.
8. Draft Guidance for Industry: Sterile Drug Products Produced by Aseptic Processing—Current Good Manufacturing Practice 9/3/2003, http://www.fda.gov/cder/gdlns/steraseptic.pdf.
9. Compliance Program Guidance Manual for FDA Staff: Drug Manufacturing Inspections Program 7356.002 http://www.fda.gov/cder/dmpq/compliance_guide.htm.
10. USP26-NF21 Supplement 1 < 1046 > Cell and Gene Therapy Products—Stability.
11. Compliance Program Guidance 7341.001, Inspection of Licensed Therapeutic Drug Products, 9/30/2000, (no longer on the FDA website).
12. Preamble to 21 CFR Part 211: Federal Register, Vol.43-No. 190, 9-29-78, pages 45013–45077.
13. Guidance for Industry Q7A Good Manufacturing Practice Guidance for Active Pharmaceutical Ingredients, U.S. Department of Health and Human Services Food and Drug Administration, Center for Drug Evaluation and Research

(CDER), Center for Biologics Evaluation and Research (CBER), August 2001, ICH section VI, Documentation and Records, C. Records of Raw Materials, Intermediates, API Labeling and Packaging Materials (6.3). http://www.fda.gov/cber/gdlns/ichactive.pdf.

14. PharmaNet- of cGMPs: Interpretations and Applications—John Lee rev 9/98 p. 41.

15. Journal of GXP Compliance. Introduction to the Good Manufacturing Practice (GMP) Tool box. Oct 2002, 7 (1), 74–77.

16. GMPs 21 CFR 211.94(c).

Company Logo

Your Company - RAW MATERIAL SPECIFICATION

Title: Monoclonal Antibody, Purified

Your Company Lot No.:

SOP No.: XXX-YYYY-ZZ	Issue Date: 11 August 2004	Version No.: 1.0
	Revision Frequency: 2 Years	Page 1 of 3

Storage Temperature: -73° C to -67° C (-70° C ± 3° C)

1. ACCEPTANCE CRITERIA - Quality Control:

Test	Specifications	Procedure	Results	Pass	Fail
a. Endotoxin Testing	≤ 2 EU/mg	LTM 960086	____	[ ]	[ ]
b. Protein Concentration					
(1) Manufacturing	For Information Only	LTM 960027	____		
(2) QC	For Information Only		____		
(3) Mfg ÷ QC	0.90 - 1.10		____	[ ]	[ ]
c. SDS-PAGE Gel - Coomassie					
(1) Reduced Sample	≥ 95% pure	LTM 960002 LTM 960232	____	[ ]	[ ]
(2) Non-reduced Sample	For Information Only	LTM 960002 LTM 960232		[ ]	[ ]
d. SDS-PAGE Gel - Silver Stain					
(1) Reduced Sample	Banding pattern equivalent to reference standard	LTM 960002 LTM 960233		[ ]	[ ]

CONFIDENTIAL - Do Not Photocopy Without Written Permission From Your Company

Company Logo

Your Company - RAW MATERIAL SPECIFICATION

Title : Monoclonal Antibody, Purified		
Your Company Lot No.:		
SOP No. : XXX-YYYY-ZZ	Issue Date : 11 August 2004	Version No. : 1.0
	Revision Frequency : 2 Years	Page 2 of 3

Test		Specifications	Procedure	Results	Pass	Fail
	(2) Non-reduced Sample	Banding pattern equivalent to reference standard	LTM 960002		[ ]	[ ]
			LTM 960233			
e.	IEF	Major banding pattern equivalent to standard	LTM 960003		[ ]	[ ]
f.	Microbial Detection	<1 CFU in 1 mg/mL	LTM 960083		[ ]	[ ]
g.	Antigen Binding Potency and Identity					
	(1) Reference Antibody	For Information Only	LTM 960197		[ ]	[ ]
	(2) Test Antibody	≥ 0.5 and ≤ 2.0 units/mg	LTM 960197		[ ]	[ ]
h.	Pristane Analysis	<0.1 µg/mL at 1 mg/mL Antibody	LTM 960151		[ ]	[ ]
i.	HPLC Test	For Information Only	LTM 960193		[ ]	[ ]
j.	Isotyping Test	Predominantly IgG_1	LTM 960194		[ ]	[ ]
k.	Murine DNA	For Information Only	LTM 960239		[ ]	[ ]

Recorded By:	Date:
Checked By:	Date:

Company Logo

Your Company - **RAW MATERIAL SPECIFICATION**

Title : Monoclonal Antibody, Purified

Your Company Lot No.: _____

SOP No. : XXX-YYYY-ZZ	Issue Date : 11 August 2004	Version No. : 1.0
	Revision Frequency : 2 Years	Page 3 of 3

2. SAMPLING PLAN

Parameter		Specification	Procedure	Result	Pass	Fail
a.	Identify number of containers to sample	$\sqrt{n} + 1$	SOP 970001			
b.	Identify quantity to remove	10% of container or 2 mL, whichever is greater	SOP 970001			

3. EXPIRATION DATE

a. 5 years from date of manufacturing

  i) Date of Manufacturing _____

  ii) Assigned Expiration Date _____

Recorded By: _____ Date: _____

Checked By: _____ Date: _____

Company Logo

Your Company - RAW MATERIAL SPECIFICATION

Title : Monoclonal Antibody, Purified

Your Company Lot No.:

SOP No. : XXX-YYYY-ZZ	Issue Date : 11 August 2004	Version No. : 1.0
	Revision Frequency : 2 Years	Page 4 of 3

4. RETENTION PLAN

Parameter		Specification	Procedure	Result	Performed	Yes	No
a.	Identify quantity to retain	2X the minimum amount required for testing- 2 mL	SOP 970002				
b.	Record in Retention log	Recorded	SOP 970002	N/A			

5. QA DISPOSITION:

Approved [ ]

Rejected [ ]

NCMR No.: _____   Disposition: _____

Date: _____

By: _____   Date: _____

15

Endotoxin:
Worst-Case Parenteral Pyrogen

Kevin L. Williams
Eli Lilly & Co., Indianapolis, Indiana, U.S.A.

15.1. INTRODUCTION

If ever a material seemed ill suited for use in analytical assays, it is endotoxin. As a standard it has been domesticated, but not entirely tamed; captured from the wild, grown up in captivity on rich media; chemically groomed (by solvent extraction), and trained to behave in a somewhat civilized manner in modern assays. But still it prances like a caged lion, back and forth, unable to escape its dual amphiphilic nature; unable to decide on the direction it should go in aqueous solution. The hydrophobic end would much rather aggregate with ends of its own kind or stick to the plastic or glass of a test tube or container in which it resides (or parenteral closure to which it has been applied for depyrogenation validation) rather than mingle with water. Furthermore,

Much of this chapter is derived from *Endotoxins, Pyrogens, LAL Testing, and Depyrogenation,* 2nd ed., by Kevin L. Williams, Marcel Dekker, 2001 (www.dekker.com).

the biological activity of endotoxin derived from different bacteria run the gamut from apyrogenic to highly pyrogenic (the extremes in variability holds true for endotoxicity also). Indeed, laboratories select different endotoxins for different purposes (i.e., product testing standards versus depyrogenation validation applications) given varying empirical recovery experiences. This chapter seeks to provide an overview for endotoxin as both a parenteral contaminant and as a standard used in modern assays.

15.2. ENDOTOXIN NOMENCLATURE AND CLASSIFICATION AS A PYROGEN

Although the terms have been used interchangeably, Hitchcock and others have proposed reserving the term "lipopolysaccharide" for "purified bacterial extracts which are reasonably free of detectable contaminants, particularly protein" (1) and the term "endotoxin" for "products of extraction procedures which result in macromolecular complexes of LPS, protein and phospholipid." Any study of Endotoxin requires definition as to its relative position as one of many pyrogens. Pyrogens include any substance capable of eliciting a febrile (or fever) response upon injection or infection (as in endotoxin released in vivo by infecting gram-negative bacteria (GNB)). Endotoxin is a subset of pyrogens that are strictly of GNB origin; they occur (virtually) nowhere else in nature. The definition of endotoxin as "lipopolysaccharide-protein complexes contained in cell walls of GNB, including non-infectious gram negatives" has also been used to denote its heterogeneous nature (2).

Exogenous pyrogens include any substance foreign to the body capable of inducing a febrile response upon injection or infection and would, of course, include microbial pyrogen, the most potent and predominate of which is endotoxin. Nonmicrobial exogenous pyrogen includes certain pharmacological agents or, for a sensitized host, antigens such as human serum albumin. The exactness of the term "pyrogen" has been eroded by (a) the replacement of the pyrogen assay wth the LAL test, (b) the characterization of a number of analogous microbial host-active by-products, (c) the identification of deleterious host responses that do not include fever, (d) the discovery of LAL reactive materials, some of which may be host reactive but nonpyrogenic, and (e) perhaps most significantly, the modern focus on cellular and molecular mechanisms that are not particularly concerned with fever as a measure of biological response. Fever is now known to be only one of a host of physiologically significant aspects of proinflammatory events occuring in response to infection, trauma, and disease progression. Many forms of infection and inflammation progress without the occurrence of fever.

TABLE 1 Bacterial Factors Capable of Stimulating Cytokine Synthesis

Components of Gram-positive species	Components of Gram-negative species
Lipoarabinomannan	Lipopolysaccharide
Lipomannans	Lipid A/Lipid A-associated
Phosphatidylinositol mannosides	proteins (LAP)
Proteins (Purified protein derivative,	Outer membrane proteins (OMP)
Mycobacterial heat shock	Porins/Chaperonins
proteins, Protein A)	
Lipoteichoic acid (LTA)	

Cell wall components of gram-positive and gram-negative species	Extracellular products of gram-positive and gram-negative species
Cell surface proteins	Toxins
Fimbriae and pili	Superantigens
Lipopeptides/Lipoproteins	
Muramyl dipeptide/Peptidolycan	
Polysaccharides	

Source: Ref. 29.

Dozens of microbial compounds have been found to either induce fever or activate host events that may lead to fever, some in combination with endotoxin, but may do so only weakly by themselves or at high doses. See Table 1 for a list of significant host-active microbial components (contaminants). The table does not distinguish the levels of each pyrogen required to bring about a host response or the type of response. LAL activation is considered analogous to the response considered to be pyrogenic but is specific for endotoxin and is capable of detecting host defense activation at subsystemic levels.

15.3. STRUCTURE OVERVIEW

The outer membrane of the gram-negative bacterial (GNB) cell wall is an asymmetrical distribution of various lipids interspersed with proteins (see Figure 1). The membrane is "asymmetrical" in that the outer layer has an inner and outer leaf made up of different constituents. The outer layer

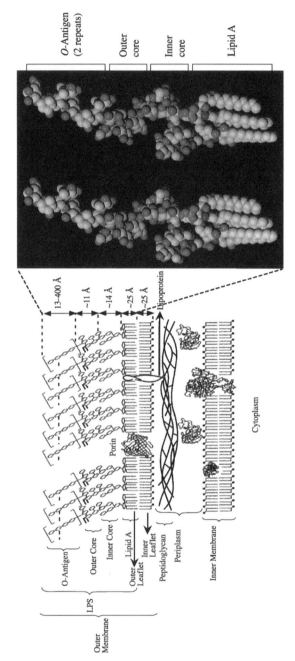

FIGURE 1 Schematic diagram of a portion of the cell envelope in gram-negative bacteria. Stereo view of the model for a single LPS molecule with two repeats of O-antigen is displayed to the right. The model of the single LPS molecule was built using the Sybyl molecular modeling software (Tripos Associates, St. Louis, Mo.), and was energy-minimizing using the Tripos force field available in Sybyl. Reproduced with permission from Langmuir (7).

contains lipopolysaccharide (LPS) and the inner leaf contains phospholipids and no LPS. The outer face is highly charged and interactive with cations; so much so that the anionic groups can bind fine-grained minerals in natural environments (3). LPS contains more charge per unit of surface area than any other phospholipid and is anionic at neutral physiological pH due to exposed ionizable phosphoryl and carboxyl groups (4).

The basic architecture of endotoxin (LPS) is that of a polysaccharide covalently bound to a lipid component, called lipid A. Lipid A is embedded in the outer membrane of the bacterial cell, whereas the highly variable polysaccharide extends into the cell's environment. The long hairlike, protruding polysaccharide chain is responsible for the GNB cell's immunological activity and is known as O-specific side chain (O = Oligosaccharide), or O-antigen, or somatic antigen chain. Endogenous endotoxin (as well as purified LPS, depending on the method of extraction) contains cell membrane–associated phospholipids and proteins as well as nucleic acids and glucans (5). Rietschel and Brade have likened the structure of LPS to that of a set of windchimes (6). The fatty acids resemble the musical pipes and are embedded in the outer membrane parallel to one another and perpendicular to the cellular wall and to the pair of phosphorylated glucosamine sugars, which form the plate from which they dangle. The "plate" is skewed at a 45° angle relative to the membrane. Connected to the plate is the O-specific chain, which, in this analogy, is the long filament from which the windchime hangs (if in fact it did hang rather than protrude from the core sugar plate attached to the lipid A fatty acid "pipes" embedded in the outer cell layer).

The O-specific side chain consists of a polymer of repeating sugars and determines the O-specificity of the parent bacterial strain. The O-chain can be highly variable even within a given GNB species and is responsible for the LPS molecule's ability to escape an effective mammalian antigenic response due to the number of different sugars and combinations of sugars that are presented by different strains. Serological identification of members of the family Enterobacteriaceae utilizes the variation inherent in this region of LPS and is the only means of identifying certain pathogenic strains of *E. coli* (8) such as *E. coli* O157, which has been implicated in recent outbreaks of food-borne illness (9). The O-chain generally (for the most highly studied family: Enterobacteriaceae) contains from 20 to 40 repeating saccharide units that may include up to eight different six-carbon sugars per repeating unit and may occur in rings and other structures. Whereas there are in excess of 2000 O-chain variants in *Salmonella* and 100 in *E. coli*, there are only two closely related core types in the former (10) and five in the latter (11). Strains with identical sugar assembly patterns may be antigenically different due to different polysaccharide linkages (12). For this reason, an immune response evoked for one variant of *Salmonella* may produce antibodies oblivious to 2000 other *Salmonella* invaders.

The O-antigen side chain connects to the core oligosaccharide, which is made up of an outer (proximal to the O-chain) and inner (proximal to lipid A) core. The outer core contains common sugars: D-glucose, D-galactose, N-acetyl-D-glucosamine, and N-acetyl-D-galactosamine (in *E. coli* and *Salmonella*). The inner core contains two uncommon sugars: a seven-carbon heptose and 2-keto 3-deoxy-D-manno-octulosonic acid (Kdo, systematically called 3-deoxy-D-manno-2-octulosonic acid) (12). These residues are usually substituted by charged groups such as phosphate and pyrophosphate, giving the LPS complex an overall negative charge that binds bivalent cations such as Ca^{2+} and Mg^{2+}. Kdo very rarely occurs in nature outside of the LPS molecule. Kdo as a polysaccharide acts to solubilize the lipid portion of LPS in aqueous systems (as does O-antigen when it remains attached).

Westphal and Luderitz first precipitated the lipid-rich hydrolytic fragment of LPS and named it lipid A (13) (and the other more easily separated portion lipid B) (14). Lipid A is a disaccharide of glucosamine, which is highly substituted with amide and ester-linked long chain fatty acids. Lipid A is highly conserved across GNB LPS and varies mainly in the fatty acid types (acyl groups) and numbers attached to the glucosamine backbone. The molecular weight of lipid A has been determined to be approximately 2000 daltons (15) as a monomer, but largely exists in aggregates of 300,000 to 1,000,000 daltons in aqueous (physiological) solutions (16). The structure of lipid A demonstrates the general form of lipid A as seen in the *E. coli* structure and natural variants that occur in the fatty acid part of the molecule. Bacterial LPS inside the family Enterobacteriaceae share the prototypical asymmetrical structure with *E. coli* and *Salmonella*, but other GNB organisms may or may not share the structure. The fatty acid groups (acyl groups) may be in either an asymmetrical or symmetrical repeating series and occur almost exclusively with even-numbered carbon chains. Endotoxic lipid A structures are invariably asymmetrical (14). It is still unknown whether the endotoxic conformation "relates to a single endotoxin molecule or to a particular aggregation state..."(11).

15.4. WHY THE PARENTERAL FOCUS ON ENDOTOXIN?

The importance of endotoxin contamination control in parenteral manufacturing becomes apparent when confronted with four aspects of its existence. The first is its ubiquity in nature, the second is the potent toxicity it displays relative to other pyrogens, the third is its stability or ability to retain its endotoxic properties after being subjected to extreme conditions, and fourth is the relative likelihood of its occurrence in parenteral solutions. The concern for endotoxin from a parenteral manufacturing contamination control perspective has superseeded concerns for guarding against "all pyrogens" that predominated in the first half of almost a century of parenteral testing. The paradigm shift of concern from pyrogens in general to endotoxin specifically[*] began with the testing of pharmaceutical waters and in-process materials and culminated in the availability of the *Limulus* amebocyte lysate (LAL) test for most end-product items as an alternative to the USP Pyrogen test in 1980 (17).

The structure of the endotoxin complex has a number of unique properties tied inseparably to its potent ability to elicit host defense mechanisms. A single bacterial cell has been estimated to contain about 3.5 million LPS molecules occupying an area of 4.9 square micrometers of an estimated 6.7 square micrometers of total outer surface area (4). The outer membrane consists of three-quarters LPS and one-quarter protein. Endotoxin molecules are crucial to the survival of the GNB, providing structural integrity, physiological, pathogenic, immunologic, and nutrient transport functions. No GNB lacking LPS entirely has been found to survive in nature (6). Endotoxin molecules are freed from bacteria by the multiplication, death, and lysis of whole cells and from the constant sloughing off of endotoxin, in a manner analogous to the body shedding skin or hair. It builds up in solution as the viable cells and skeletons of dead bacteria accumulate. When such solutions rich in GNB cellular residues find their way into mammalian blood, they retain their ability to activate host defense mechanisms in nanogram per kilogram amounts. GNB organisms occur in virtually every environment on the Earth, thus making endotoxin one of the most prevalent complex organic molecules in nature. GNB have been isolated [and are being isolated still (18)] wherever man has gone; in soil, fresh and salt water, frigid oceans and hot springs, as well as in significant amounts in ocean sediment. Some GNB organisms are able to grow in the coldest regions known ($< 10°C$) (19). The GNB count of seawater was taken at Woods Hole Oceanographic Institute and found to be in excess of one million organisms per milliliter and the sand from the shore contained almost a billion organisms per gram (20).

Given the ubiquity of endotoxin, one wonders at the mammalian host's exaggerated response to it. It is as though mammalian (and virtually all multicellular organisms (21,22)) and prokaryotic organisms are waging war with the mammals always on the defensive, living in fear, and shouting "bar-

---

[*] And perhaps full-circle in the future to include more host-active bacterial and fungal artifacts.

barian at the gates" at the shadow of this invader. It is as though something larger loomed; as if the body fears another plague or typhoid (GNB invaders) lies ready to threaten the larger society and, therefore, the body reacts accordingly. Viewed in this context, the host response to endotoxin is not as exaggerated as it would seem at first glance. The spectrum of organisms induced to fever by endotoxin is extensive including reptiles, amphibians, fish, and even insects such as cockroaches, grasshoppers, and beetles (23). Some animals that were initially believed to be insensitive to LPS such as rodents have subsequently been shown to respond (24).

Endotoxin achieves greater leverage in eliciting deleterious host effects than any other microbial pyrogen as is seen in the relative amount of endotoxin needed to provoke a response, which is in the nanogram per kilogram range. If endotoxin is an alarm marker for hosts in recognizing microbial invasion (25), then it elicits the loudest and most variable response. The leverage of endotoxin can be seen in the wide variety of endogenous mediators elicited, which are active in the picogram (even femtogram) per kilogram range. Therefore, a miniscule amount of endotoxin generates a huge host response in terms of both severity and variety. The complexity of the host response has frustrated efforts to devise treatments. The complexity arises from the interplay of the various mediators (cytokines) produced that may have proinflammatory and antiinflammatory host effects as well as synergistic effects on their own kind. A few nanograms of endotoxin translate into the production of a myriad of extremely bioactive manufactured endogenous pyrogens.

In the early use of the pyrogen assay, no attempt was made to quantitate the amount of endotoxin needed to produce a pyrogenic response in rabbits. *E. coli* and *Salmonella* were later chosen, as among the most endotoxic of families of bacteria (Enterobacteriaceae), to determine and quantify the amount of endotoxin by weight considered to be pyrogenic. In 1969, Greesman and Hornick (26) performed a study using healthy male inmates (volunteers) and found the threshold pyrogenic response (TPR) level to be about 1 nanogram per kilogram for *E. coli* and *Salmonella typhosa* (approx. 0.1 to 1.0) and 50 to 70 ng/kg for *Pseudomonas*. The same study revealed that the rabbit and human threshold pyrogenic responses are approximately the same. Therefore, the amount of purified *E. coli* needed to initiate pyrogenicity in both man and rabbits is approximately 1 ng/kg, which represents about 25,000 *E. coli* bacterial cells (27). In terms of whole cells, the injection of an estimated 1000 organisms per milliliter (10,000/kg) of *E. coli* causes a pyrogenic reaction in rabbits, as compared to 10^7 to 10^8 organisms per kilogram of gram-positive or fungal organisms (28). The fact that many non-LPS products have been recently identified as macrophage activators and that many are associated with devastating diseases supports an underlying theme that there is a wide variety of potential modulators of adverse host effects (including fever) that are not endotoxin but that may proceed by endotoxin-like mechanisms and with endotoxin-like potencies when presented by infecting organisms (though not necessarily relevant from a parenteral manufacturing perspective) (Table 2).

Peptidoglycan (PGN) is usually described only in association with gram-positive bacterial (GPB) infection but PGN has been found to be released into hosts in several instances of GNB infection (31). PGN is released (by GPB) during infection and can reach the systemic circulation (32). Sensitive methods of quantifying PGN and its subunits in a clinical setting

TABLE 2 The Relative Biological Activity of Cytokine-Inducing Microbial Components Compared to LPS

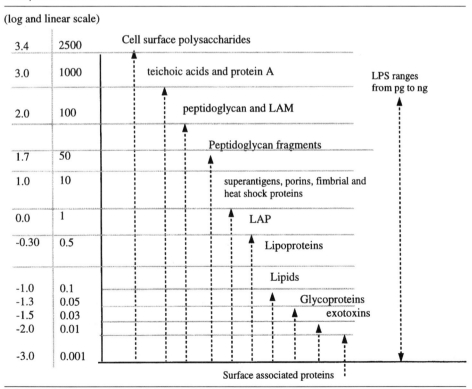

Source: Ref. 30.

have yet to be developed,[*] leaving the levels associated with GPB sepsis largely unknown.[†] The incidence of GPB sepsis in the hospital setting is known to equate to that caused by GNB organisms, though studies have proposed that PGN and LTA act synergistically (34,35).

Given the plethora of evidence for nonendotoxin pyrogens (albeit less potent than endotoxin), it remains to be seen which components will be excluded and which will remain classified as "pyrogens." It does seem intuitive that given the range of prokaryotic cellular debris, endotoxin will not be the only significant pyrogenic (or bioactive) harbinger of bacterial origin.

A relevant note concerning the lack of attention given to nonendotoxin cellular components in pareneteral manufacturing is the degree of difficulty researchers encounter in obtaining the materials in a pure state devoid of endotoxin. The presence of endotoxin overrides many efforts to study non-endotoxin components due to its potency and can affect research study end points at almost undetectable background levels (fg/mL) as compared with the levels necessarily used in the study of non-LPS substances (typically in µg-mg/mL) (Table 2).

Beveridge describes the enduring nature of the GNB cell wall as "strong enough to withstand ~3 atm of turgor pressure, tough enough to endure extreme temperatures and pHs (e.g., *Thiobacillus ferrooxidans* grows at a pH of ~1.5) and elastic enough to be capable of expanding several times their normal surface area. Strong, tough, and elastic..."(3). Endotoxin is extremely heat-stable and remains viable after ordinary steam sterilization, normal desiccation, and easily passes through filters intended to remove whole bacteria from parenteral solutions. Only at dry temperatures exceeding 200°C for up to an hour or at extreme pH do they relent.

The amphiphilic nature of the LPS molecule also serves as a resilient structure in solution with the hydrophobic lipid ends adhering tenaciously to hydrophobic surfaces such as glass, plastic, and charcoal (27) as well as to one another. Many of the most basic properties of LPS are those shared with lipid bilayers in general, which form the universal basis for all cell-membrane structures (36). In aqueous solutions, LPS spontaneously forms bilayers in which the hydrophobic lipid A ends with fatty acid tails are hidden in the interior of the supramolecular aggregate as the opposite hydrophilic poly-saccharide ends are exposed to and subject to solubilization in the aqueous environment. A property adding to the stability of LPS as a lipid bilayer is its

---

[*] Or at least widely accepted as the SLP method is a sensitive detection method for PGN.

[†] Although muramic acid has been used as a sensitive marker for GC-MS detection of GP cellular residues in clinical specimens (septic synovial fluids) at levels of ≥30 ng/mL (33).

propensity to reseal when disrupted, thus preserving the structure's defense against the environment.

A central question that arose with the proposal to replace the rabbit pyrogen test with the *Limulus* Amebocyte Lysate test was (and still is): How can one be sure in testing only for endotoxin that other microbial pyrogens will not be allowed to go undetected in the parenteral manufacturing process? In part, we have answered the question by considering the ubiquity, stability, potency (based on severity of host response), combined with the relative likelihood of endotoxin-bearing GNB as parenteral contaminants. The minimal growth requirements of GNB allow their growth in the cleanest of water. Conversely, the answer can be found by disqualifying from undue concern (a) the environmental predisposition of non-GNB organisms that prevent them from proliferating in largely water-based parenteral manufacturing processes, (b) the relative ease of degradation of their by-products (except heat-stable GPB exotoxins, which derive from microbes having significant growth requirements), and (c) modern aseptic manufacturing procedures required by current Good Manufacturing Practices (cGMP).

15.5. CONTAMINATION CONTROL PHILOSOPHY IN PARENTERAL MANUFACTURING

Endotoxin is a concern for people only when it comes into contact with the circulatory system. The two relevant mechanisms for such contact involve infection and medically invasive techniques including injection or infusion of parenteral solutions. A notable exception to limiting the concern for endotoxin to blood contact is the effect that minute, almost undetectable, quantities of endotoxin may have upon cell cultures used in pharmaceutical manufacturing. The manufacture of biologics makes use of complex cell culture media including the addition of fetal bovine serum (FBS) as a growth factor (which has been associated with microbial contamination[*]) to grow mammalian cells used in recombinant and monoclonal expression systems. Serum has presented manufacturers (and clinicians) difficulties in quantifying and reproducing endotoxin levels due to little-understood interference factors. The regulatory precautions set in place are, in many cases (if not most), due to the poor probabilities associated with finding contamination by quality control sampling techniques. The generally accepted sterility acceptance level

---

[*] Being a bovine blood product subject to temperature abuse and containing gram negative bacteria.

(SAL) has been often repeated to be 10^{-6} (i.e., one possible survivor in a million units), but according to Akers and Agalloco, the value was selected as a convenience (37). They maintain that 10^{-6} is a minimal sterilization expectation and should be linked "to a specific bioburden model and/or particular biological indicator...(otherwise) it is a meaningless number that imparts little knowledge on the actual sterilization process."

Bruch relates that the PSI (probability of a survivor per item) for a can of chicken soup is 10^{-11} whereas the assurance provided by the USP Sterility test alone is not much better than 10^{-2} given a 20-item sampling and is, as Bruch says, due to the rigorous heating cycles developed by the canning industry to prevent the possibility of survival of *Clostridium botulinum* (38). Bruch maintains that the industry has "never relied on a USP-type finished product sterility test to assess the quality of its canned goods...(because) the statistics of detecting survivors are so poor that the public confidence...would be severely compromised through outbreaks of botulism." He cites the generally accepted sterility assurance for a large volume parenteral item as 10^{-9} and 10^{-4} for a small-volume parenteral that has been aseptically filled and sterile filtered as opposed to terminally sterilized. The apparent contradiction in the necessity of more stringent sterility assurance for a can of soup than for a parenteral drug is due to the ability of organisms to grow in soup as opposed to the likelihood of such growth in the parenteral manufacturing environment. See Table 3.

The predominate potential source of endotoxin in a pharmaceutical manufacturing environment is the purified water used as a raw material (also used in component sterile rinse depyrogenation processes). Many different

TABLE 3 Probability of Survivor Estimates for Sterilized Items (38)

Item	Probability of survivor/unit
Canned chicken soup[a]	10^{-11}
Large-volume parenteral fluid	10^{-9}
Intravenous catheter and delivery set[a]	10^{-6}
Syringe and needle[a]	10^{-6}
Urinary catheters[a]	10^{-3}
Surgical drape kit[a]	10^{-3}
Small volume parenteral drug (sterile fill)	10^{-3}
Laparoscopic instruments	10^{-2}
(Processed with liquid chemical sterilants)[b]	

[a] Dosimetric release: no sterility test.
[b] Limits of USP sterility test: $10^{-1.3}$ (with 95% confidence).

grades of water are used and may be variously labeled according to their origin, the treatment they have undergone, quality, or use; and different groups employ different nomenclature (39). The only waters that require endotoxin monitoring are Water for Injection (WFI) and Water for Inhalation and are prepared via a validated distillation or reverse osmosis process. Distillation is the preferred method and results in sterile, endotoxin-free condensate. However, any water may become contaminated via a number of subsequent distribution or storage mechanisms, including the cooling or heating system, storage container, or distribution method such as hoses (39).

15.6. DEVELOPING AN ENDOTOXIN CONTROL STRATEGY FOR DRUG SUBSTANCES/EXCIPIENTS

Finished products often contain ingredients in addition to the active drug substance. Excipients serve as solvents, solubilizing, suspending, thickening, and chelating agents; antioxidants and reducing agents, antimicrobial preservatives, buffers, pH adjusting agents, bulking agents, and special additives (40). Recent endotoxin excipient testing references (41,42) dictate limits for some parenteral excipients and, therefore, require the establishment of endotoxin quality control tests. However, the majority of parenteral excipients still do not have established endotoxin limits. The FDA Guideline on Validation of the LAL Test (43) outlines the determination of limits for "end-product" testing and can be misapplied to drug substance and excipient testing. Relevant activities to be established to gain control over a given drug manufacturing process from an endotoxin control perspective include the following:

Identifying the types of excipients used in various drugs
The relative amounts of those excipients in each drug type
Relevant tolerance limits for drug substances and excipients given the above

This exercise should establish that proposed limits are appropriate and that existing excipient and drug substance limits used in the manufacturing process will not allow an associated drug product to fail its end-product testing. As the cost of drugs derived from biotechnology increase, so do the business-related requirements for ensuring that the raw materials that go into making the intermediates of the manufacturing process as well as end products meet appropriate, relevant, and stringent predetermined specifications.

Every marketed product has a level of endotoxin safely tolerated (i.e., an amount below the tolerance limit), which is defined as TL = K/M, where K is the threshold pyrogenic dose (TPD) constant in endotoxin units (EU) per kilogram and M is the maximum human dose in units per kilogram of body weight [70 kg/hr as per FDA Guideline (43)]. The TPD is the level of endotoxin capable of eliciting a pyrogenic response in a patient. The relevant dose is that administered in an hour. The TPD constant (K) differs depending upon the route of administration (parenteral or intrathecal/radiopharmaceutical). The formula is straightforward except for the units, which vary from product to product depending on the manner in which the product is administered. For drugs administered by weight, the weight to be used is that of the active drug ingredient in milligrams or in units per milliliter. For drugs administered by volume, the potency is equal to 1.0 mL/mL.[*]

The formulas adjust for a product's potency based on either the weight of the active ingredient or the volume of the drug administered; they constitute a package for determining "how much the product can be diluted and still detect the limit endotoxin concentration" (43). An endotoxin control strategy (ECS) is a tool to organize and facilitate laboratory testing of drug substance and excipients at appropriate tolerance limit (and therefore test dilution) levels (44,45). An example strategy is shown in Table 4.

The table allows the user to view TPD in terms of total EUs delivered in a dose. This rationale for drug substances (active ingredients) and excipients has not been described in any guideline (in that only tolerance limit calculations for "end products" are described), but the necessity for relevant testing has become a clear expectation as evidenced by the publication of recent monographs for mannitol and sodium chloride and by ongoing excipient harmonization efforts.

In lieu of using the table, a drug substance tolerance limit adjusted for excipients can be calculated:

TL(drug substance with excipients(ds/e))

$$= \frac{\{350 - ((TLe_1 \times We_1) + (TLe_2 \times We_2)\ldots)\}}{W_A}$$

where TLe_1 is the tolerance limit of excipient 1 and We_1 is the weight of excipient 1 per dose of active drug and W_A is the weight or units of active drug per dose. Note that the formula $((\ldots))$ indicates all relevant excipients without an exclusion rationale should be included in the calculation. Compare the calculated value of 7.48 EU/mg to the end-product tolerance limit calculated in the formula: TL = 5.0 EU/kg/(35 mg/70kg) = 10 EU/mg (Table 4).

---

[*]See Appendix D of the FDA Guideline for exceptions to the general formulas including the use of radiopharmaceutical and intrathecal doses, and the use of pediatric weights.

TABLE 4 Endotoxin Control Strategy Steps

1.	Drug product constituent and weight	
Obtain the unit	API	1.0 mg
formula for a given	Mannitol	2.14 mg
drug product	NaCl	1.43 mg
	Polysorbate	2.5 mg

2.	Constituent	Weight/dose
Determine the relative	API	35 mg
amounts of API and	Mannitol	75 mg
excipients based on the	NaCl[a]	50 mg
dose of API	Polysorbate 80[b]	87.5 mg

3.	Constituent	Proposed or existing TL assigned
Assign existing	API	nmt[d] 7.0 EU/mg
TLs or propose	Mannitol	nmt 0.0025[c] EU/mg
TLs for the	NaCl	nmt 0.005 EU/mg
drug substance	Polysorbate 80	nmt 1.0 EU/mg
and excipients		

4.	Constituent	Weight/dose	Proposed TL	EU's
Ensure that the	API	35 mg	7.0 EU/mg	245 EU
final product	Mannitol	75 mg	0.0025 EU/mg	0.19 EU
cannot exceed	NaCl	50 mg	0.005 EU/mg	0.25 EU
the TPD given	Polysorbate 80	87.5 mg	1.0 EU/mg	87.5 EU
each assigned TL				
			Total EU/dose =	332.94 EU

5. Document both the "control strategy" and any "exclusion rationale(s)" used for excipients deemed not to require endotoxin testing.

[a] See European Pharmacopoeia (3rd Ed. 1997) monograph for Sodium Chloride (p. 1481) (41).
[b] No endotoxin limit in monographs.
[c] See European Pharmacopoeia (3rd Ed. 1997) monograph for Mannitol (p. 1143) (41).
[d] Not more than can be interpreted as less than since a test containing the limit concentration of endotoxin would be positive and hence fail the test.

For the above example, the formula would be filled in as follows:

$$TL(ds/e) = \frac{\{350 \text{ EU} - ((0.0025 \text{ EU/mg} \times 75 \text{ mg}) + (0.005 \text{ EU/mg} \times 50 \text{ mg}) + (1.0 \text{ EU/mg} \times 87.5 \text{ mg}))\}}{35 \text{ mg}}$$

$$TL(ds/e) = \frac{\{350 \text{ EU} - ((0.19 \text{ EU}) + (0.25 \text{ EU}) + (87.5 \text{ EU}))\}}{35 \text{ mg}}$$

$$= \frac{\{350 \text{ EU} - 87.94 \text{ EU}\}}{35 \text{ mg}}$$

$$TL(ds/e) = \frac{262.06 \text{ EU}}{35 \text{ mg}} = 7.48 \text{ EU/mg}$$

An ECS is appropriate for drug products containing the following:

Numerous excipients

Significant (large amounts of one or more) excipients relative to the active

Excipients with tolerance limits set with relatively high limits (perhaps due to difficult/incompatible laboratory tests or ill-conceived historical method of determining its limit)

Drug substances and/or excipients with tolerance limits previously calculated using end-product formulas

Excipients of natural (animal or plant) origin

Conversely, an ECS may be unnecessary for drug products containing the following:

Few or no excipients (drug substance = drug product)

Excipients in miniscule amounts relative to the active

Excipients with very low tolerance limits (i.e. those with compendial requirements)

Excipients incapable of adding appreciable endotoxin because they are antimicrobial and/or inhospitable to microbes due to their method of manufacture, nature or origin, or as a miniscule constituent.

As an example, Cresol (hydroxytoluene) is an antimicrobial excipient obtained from either sulfonation or oxidation of toluene (46). Therefore, it is (a) manufactured from materials inhospitable to microbial growth (b) at temperatures that are depyrogenating and (c) is unlikely to be post-manufacture contaminatable due to the lack of water needed to support microbial growth.

End-product testing provides a test of the total contents of a given vial. See Table 5 for a proof of this. The ECS is concerned with providing in-process testing that demonstrates that when the parts are combined, they

TABLE 5 BET Calculations—Active Versus Total Solids

Calculated by active drug concentration	Calculated by total solids (TS) method (do not use this method, for illustration only)
If active drug is 200 mg and is reconstituted with 20 mL then the solution is 10 mg/mL. The potency, TL, and lambda constitute a "system" to determine the appropriate limit and subsequent dilution (MVD)	If TS of drug is 1 gram (this value is not constant as identical drugs made by different manufactures will differ in excipient use and therefore TS).
$TL = K/M = 5.0$ EU/kg/ $(200 \text{ mg}/70 \text{ kg}) = $ nmt 1.75 EU/mg of drug since $MVD = TL \times PP/\lambda$ $MVD = \dfrac{1.75 \text{ EU/kg} \times 10 \text{ mg/mL}}{0.01 \text{ EU/mL}}$ $= 1{:}1750$ dilution	$TL = 5.0$ EU/kg/(1000 mg/70 kg) $=$ nmt 0.35 EU/mg (TS method) $MVD = \dfrac{0.35 \text{ EU/kg} \times 1000 \text{ mg}/20 \text{ mL}}{0.01 \text{ EU/mL}}$ $= 1{:}1750$ dilution

cannot cause the end-product to fail its specification. The trend in drug development is clearly toward greater complexity. New biologically derived drugs may contain a number of unusual excipients in significant amounts (for example, new sustained-release parenterals contain excipients not traditionally found in non-sustained released drugs (47) and/or present in large quantities). An endotoxin control strategy can provide a frame of reference to determine appropriate drug substance and excipient limits (as opposed to their arbitrary assignment). While there are arguably safety factors included in endotoxin limit calculations [see "Understanding and Setting Endotoxin Limits" (48)], there are also confounding factors such as multiple parenterals given to patients simultaneously). A complete process to account for a drug's entire potential endotoxin contents will aid manufacturers in gaining greater endotoxin control (also see Cooper's method in Chapter 16).

15.7. BET STANDARDIZATION

Tied to the concept of a "standard" endotoxin is the historical determination of a threshold pyrogenic dose for endotoxin. The establishment of a defined, specific threshold pyrogenic response level allowed the concept to be established that a certain amount of endotoxin is allowable and a certain amount of endotoxin should not be delivered into the bloodstream. The advent of LAL allowed the quantification of endotoxin as a contaminant. In turn, quantitation allowed for the creation of specific and relevant endotoxin limits for manufactured drug products, raw materials, active ingredients, devices, components, depyrogenation processes, and in-process sample that constitute the legal requirement for releasing to market products that are not considered "adulterated" by international regulatory bodies.

Today's user of the LAL test rightly views such concepts as the bread and butter of endotoxin testing, but it is good to appreciate the degree to which today's system of endotoxin quantitation has progressed in that:

"Quantitation" in the rabbit assay was limited to a pass/fail response (rabbit response = 0.6°C temp. rise).

The pyrogen test was initially established without attempting to quantitate the amount of endotoxin necessary to produce a febrile response.

Early LAL testing used the weight of dried bacterial endotoxins in nanograms initially with various GNB organisms and then with a specific *E. coli* strain without accounting for the variable potency of a given weight of endotoxin.

None of the early tests could have been used effectively to develop product-specific tolerance limits as they exist today, much less provide the degree of in-process control needed for modern pharmaceutical manufacturing. In some respects the 10–1000-fold greater sensitivity of the LAL test created the "luxury" of controversy on several fronts. A whole new system of relating the new assay to the existing test had to be developed to avoid unnecessary product test failures due to the greater sensitivity of the LAL assay (49). The "system" included the formation of or association with (a) the Endotoxin Unit[*] (EU) as a measure of relative biological activity, (b) the tolerance limit (TL) (endotoxin limit concentration), (c) the maximum valid dilution (MVD) to relate the product dose to the allowable endotoxin content (realizing that a positive LAL response in any given solution as in the Pyrogen assay would be inappropriately stringent), and (d) the lysate sensitivity (lambda [λ] to standardize the relative reactivity of each LAL to each control standard endotoxin (CSE). Prior to this "system" several of the principals of the early LAL assay expressed concern that the greater sensitivity of the assay would end up becoming an apparent disadvantage used by some to confound

---

[*] EU is defined as 1/5th the amount of *E. coli* (EC-2) endotoxin required to bring about the threshold pyrogenic response (as established by Greisman and Hornick as 1 ng/kg).

industry efforts to develop the assay as a replacement for the rabbit pyrogen test. ("I hope that we do not turn the advantage provided by the greater sensitivity of the *Limulus* test into a problem"—Jack Levin) (50).

A number of criticisms were put forward with the use of the first assigned endotoxin standard. The major criticisms included the fact that the standard was not "pure" lipid A for which the chemical formula had been defined and the fact that other, more potent endotoxins, were available. The criticism concerning the purity of the endotoxin was discounted due to the need for a readily soluble standard (lipid A being insoluble). The goal of obtaining a standard endotoxin largely free of biologically active proteins, peptides, polynucleotides, and polysaccharides had been achieved. As for the potency of the new endotoxin reference standard, it was believed that an "average" potency would be more relevant to the testing of a wide range of endotoxins, with a range of potencies, likely to be encountered in real world testing.

As recently as the late 1990s there have been as many as five different official international standards active at once (51). For an international manufacturer, this meant either the construction of a single test designed to overlap all the test requirements, including the use of a control standard calibrated against each official reference standard or the performance of multiple testing of each lot of drug material. An initial International Standard (IS) for endotoxin testing was established by the World Health Organization's (WHO) Expert Committee on Biological Standardization (ECBS) in 1987 (52). The first international standard was calibrated against the US national standard, EC5. However, the potency assignments for the semiquantitative LAL gel clot and photometric tests did not agree. Most of the collaborative data consisted of gel clot testing; therefore, the ECBS of WHO assigned IS-1 as a gel clot standard (53). The assigned potency was 14,000 IU/ampule.

In 1994, the ECBS of WHO acknowledged that the use of the photometric tests (end-point and kinetic chromogenic and turbidimetric) had greatly grown in terms of the number of LAL users since IS-1 was established and recognized the need for a common standard for both gelation and photometric tests (53). The USP made available 4000 vials of a batch of USP-G/EC-6 for the proposed WHO second International collaborative study. The stage was therefore set for a comprehensive study organized by the WHO involving the U.S. European, and Japanese Pharmacopoeias.

Poole, Dawson, and Gaines Das describe the ambitious aims of the study (53):

1. Calibrate the IS as compared to EC5 (USP-F) (although superseded by EC-6 it was the primary calibrant for IS-1 and the JP reference standard) and assign a single IS unit for all endotoxin applications.

2. Compare the current IS (IS-1), EC5, and the candidate standard (CS) using LAL gelation, kinetic, and end-point assays (chromogenic and turbidimetric).
3. Determine the relationship of IU to EU.
4. Compare the CS to the US, European (BRP-2), and Japanese reference standards.

A common lysate (supplied by Associates of Cape Cod) was used in 24 laboratories for each of the gel clot and kinetic assays together with an "in house" lysate (i.e., whatever was already being used in that laboratory). In all, the participants performed a total of 108 gel clot assays; 133 assays were performed using end-point chromogenic (3 labs), kinetic chromogenic (13 labs), and kinetic turbidimetric (12 labs). In the gel clot tests, the geometric mean for the candidate standard sublots (both sublots were therefore considered as a single lot) did not differ significantly from one another, from laboratory to laboratory or from LAL to LAL reagent source (53).

The candidate standard geometric mean result for each assay type obtained in terms of EC-5 is shown in Table 6 (graphically represented in Figure 2).

In January of 2001, USP 25 created the first harmonized microbiological test, the Bacterial Endotoxin Test (BET) concomitant with the formation of IS-2 as an international standard endotoxin. Overall, the newly harmonized test has received high marks industrywide for ease of understanding and practicality when applied to real-world test conditions. Furthermore, to multinational companies that must meet international requirements, the benefits of the harmonized test cannot be overstated. In a nutshell, the benefits of the harmonized test include the following:

> An elevation of the status of non-gel clot tests, including kinetic and end-point chromogenic and turbidimetric tests by including them.

TABLE 6 Results Obtained in WHO IS-2 Collaborative Study

Assay Type	Mean recovery	# Tests (n)
Gelation assay	10,300 EU/vial	103
Kinetic chromogenic assays	11,700 EU/vial	13
Kinetic turbidimetric assays	11,800 EU/vial	11
Chromogenic endpoint assays	11,200 EU/vial	3
All assays (gel and photometric)	10,400 EU/vial	68
IS-2 assigned value	10,000 IU/vial	

Source: Ref. 53.

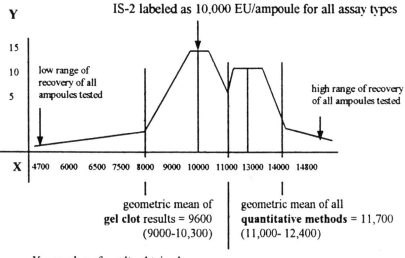

Y = number of results obtained
X = EU of EC5/amp IS-2

FIGURE 2   Graphical representation of the range of geometric means obtained and the grouping of results for all valid gelation and photometric assays as n (number of assays) versus EU of EC5 per ampule of candidate standard.

The gel clot assay has been split into a limit test or an assay, something that is fairly routine but not specified previously and the limit test no longer requires the confirmation of label claim with each block of tubes tested.

The requisite positive product control standard recovery has been widened from 50–150% to 50–200%, which is in effect the recovery associated with the gel clot assay (one two-fold dilution). This change only allows for one's test to overestimate the recovery of endotoxin all the more (200% versus 150% recovery).

An innovative change in the application of kinetic testing (and a novel use of a referenced standard curve) has been put forward by Charles River (Endosafe) with their new PTS®, "a rapid, point-of-use test system that utilizes existing FDA-licensed LAL reagents in a test cartridge with a hand-held spectrophotometer."[*] The reader is miniaturized and the little plastic

---

[*] Refer to *http://www.criver.com/products/invitro/endotoxin/endo_research.html*

cartridge inserts contains all the reagents necessary for testing. The cartridges are coded by lot to reference an archived standard curve. The reader unit is intended to provide "point-of-use" results for research and manufacturing (in-process) testing. Accurate results can be obtained in 15 min with minimal training in endotoxin testing.

15.8. ORIGIN AND IMPORTANCE OF LAL

The rabbit pyrogen assay served as the only official pyrogen test for 37 years. However, during the early 1960s several events occurred that would eventually lead to the development of a seemingly unlikely replacement: a blood product (lysate) derived from the horseshoe crab, *Limulus polyphemus*. The importance of the changes brought about in the pharmaceutical industry by the switch from the in vivo based rabbit pyrogen test to the in vitro bacterial endotoxin test are often underappreciated for a couple of reasons. First, the labor intensity inherent in the rabbit pyrogen assay served as a lid on the amount of in-process testing that could be realistically be expected to be performed (from a cost and resource perspective) to support the manufacture of parenteral lots (100 rabbit pyrogen tests a day would be a colossal effort). The advent of LAL testing has allowed the broad application of current Good Manufacturing Practices (cGMP's) as they relate to the detection of endotoxins across the entire manufacturing process. The quality control testing of only the later forms of a parenteral drug provides a greatly reduced probability of detecting a contaminated unit of that material from a statistical standpoint and would make it impossible to preclude the use of contaminated materials prior to manufacture as a means of precluding the manufacture (and subsequent QC rejection) of an expensive biological lot.

Modern pharmaceutical manufacturing processes include sampling and LAL testing of not only the finished (beginning, middle, and end of lot), bulk and active pharmaceutical ingredient (API) material but also in-process materials, including containers and closures, sterile water, bulk drug materials, and more recently, excipients. The pyrogen assay included the housing of dedicated rabbits and was therefore very expensive and its expansion unlikely given cost and other resource constraints. Second, the inability to quantify endotoxin associated with pyrogen testing acted as a "blind spot" to restrict the improvement of processes that are now readily monitored given the sensitivity and quantification associated with the LAL test. It is difficult to work toward lower specifications when performing an assay that has an inherent invisible pass/fail result. Modern biopharmaceuticals may indeed contain trace amounts of endotoxin or may have activity (i.e., interferon) mimicking endotoxin and in such cases the accurate and reproducible quantification of these minute levels as well as the differentiation of interference and endotoxin content becomes paramount to demonstrating that allowable levels are present.

The first application of the clotting reaction discovered by Levin and Bang was made by Cooper, Levin, and Wagner in their use of the "pre-gel" to determine the endotoxin content in radiopharmaceuticals in 1970 (54). According to Hochstein (55), Cooper was a graduate student at Johns Hopkins in 1970 and worked for the Bureau of Radiological Health. That summer Cooper persuaded the Bureau of Biologics (BoB) group lead by Hochstein that a lysate from the horseshoe crab's blood would be useful in detecting endotoxin in biological products. Given the short half-life and stringent pyrogen requirements associated with radiopharmaceutical drugs, Cooper believed that LAL could be used to accomplish the improved detection of contaminated products. Though Cooper left the BoB to finish his graduate studies, Hochstein continued the Bureau's efforts to explore the use of LAL in the testing of drug products.

The potential for improvement in the area of pharmaceutical contamination control was evident in Cooper, Hochstein, and Seligman's very first application of the LAL test involving a biological (56): the results of 26 influenza virus vaccines included as a subset of a 155 sample test using LAL varied from lot to lot by up to 1000-fold and revealed endotoxin in the 1 μg range in the 1972 study. Cooper later pointed out (57) that newer vaccines used in mass inoculation of Americans for A/Swine virus were subsequently required to contain not more than 6 ng/mL of endotoxin, a level that could not be demonstrated with pyrogen testing. Suspected adverse reactions were reported prior to the inception of the LAL assay and were an expected part of some drug reactions such as that associated with L-asparaginase antileukemic treatment as a product of *E. coli* (58). A third early application (radiopharmaceuticals and biological vaccines mentioned above) involved the detection of endotoxin in intrathecal injections (into the cerebrospinal fluid) of drugs. Cooper and Pearson report (57) that 10 such samples implicated in adverse patient responses were obtained, tested by LAL, and all 10 reacted strongly. The rabbit pyrogen test was negative for all samples when tested on a dose-per-weight basis. They concluded that the rabbit pyrogen test was not sensitive enough for such an application given that endotoxin was determined to be at least 1000 times more toxic when given intrathecally.

15.9. LAL DISCOVERY

In 1956 Frederik Bang, at the Marine Biological Laboratories in Massachusetts, was studying the effects of what he initially believed to be a bacterial disease causing the intravascular coagulation (coagulopathy) of the blood of a horseshoe crab in a group that he was observing. He isolated the bacterium from an ill *Limulus*, believing it to be a marine invertebrate pathogen such as (he cites) the marine bacterium *Gaffkia*, which killed lobsters. He described the basic observation that prompted him to publish the landmark study in A Bacterial Disease of Limulus Polyphemus (59) as follows:

> Bacteria obtained at random from fresh seawater were injected into a series of horseshoe crabs (*Limulus polyphemus*) of varying sizes. One *Limulus* became sluggish and apparently ill. Blood from its heart did not clot when drawn and placed on glass, and yet instant clotting is a characteristic of normal limulus blood...The bacteria caused an active progressive disease marked by extensive intra-vascular clotting and death. Injection of a heat stable derivative of the bacterium also caused intravascular clotting and death. Other gram-negative bacteria or toxins also provoked intravascular clotting in normal limuli. When these same bacteria or toxins were added to sera from normal limuli, a stable gel was formed!

After Bang made his initial observations, he paired up with a hematol-ogist, Jack Levin, at the suggestion of another colleague. Together they explored the requisite coagulate factors of *Limulus* and published a paper entitled "The Role of Endotoxin in the Extracellular Coagulation of Limulus Blood" (60) in an effort to "study the mechanism by which endotoxin affects coagulation in the *Limulus*, and to elucidate the mechanism by which endotoxin exerts its effect in a biological system that may be less complex than that found in mammals." This study contained a number of observations:

> The amebocyte is necessary for clotting.
> Clotting factors are located only in the amebocytes (not in the blood plasma).
> The formation of a gel clot reaction occurs by the conversion of a "pre-gel" material upon addition of gram-negative bacteria.

Levin and Bang demonstrated that extracts of the amebocytes gelled in the presence of GNB endotoxin. In the introduction of that early paper they describe the phenomenon that would later become the basis for the LAL assay:

> *Limulus* blood contains only one type of cell called the amebocyte. When whole blood is withdrawn from the *Limulus*, a clot quickly forms. Thereafter, this clot shrinks spontaneously, and a liquid phase appears. Under appropriate conditions, this liquid material has the

capability of gelling when it is exposed to bacterial endotoxin, and is defined here as pre-gel... The results (of the study that served as the basis for their April 1964 publication) demonstrate that cellular material from the amebocyte is necessary for coagulation of Limulus plasma, and that plasma free of all cellular elements does not clot spontaneously nor gel after addition of endotoxin (60).

Levin and Bang not only used the initial bacterial isolate (they had now identified it as a *Vibrio* species) to bring about gelation, but they also used *E. coli* (Difco) because they now believed that endotoxin common to GNB was bringing about the gelation phenomenon. Their study revealed that agitation of the amebocytes (amebocyte disruption) aided in the production of the pre-gel (i.e., in the production of gel precursor most susceptible to subsequent endotoxin clotting) and that the rate of gelation of pre-gel was directly related to the concentration of endotoxin in the mix. In their third paper, Levin and Bang describe the "striking similarities between *Limulus* amebocytes and mammalian platelets..." during cellular coagulation upon exposure to endotoxin (61).

15.10. HEMOLYMPH COAGULATION IN LIMULUS AND TACHYPLEUS

Invertebrates lack adaptive immune systems and rely on innate immunity to antigens common to pathogenic organisms. Nakamura, Iwanaga, Kawabata, Muta and others have extensively studied the hemolymph (blood) system of the Japanese horseshoe crab (*Tachypleus tridentatus*) and found amebocytes contain two types of granules, large (L) and small (S), which contain the clotting factors, proteins, and antimicrobials that are released via a process called degranulation into the crab's plasma (62). Regardless of the relative simplicity of the crab's defense system (the amebocyte), Nakamura, Morita, and Iwanaga consider it to be "a complex amplification process comparable to the mammalian blood coagulation cascade" and "very similar to those of mammalian monocytes and macrophages..." (63). The ability of *Limulus* and *Tachypleus* blood to clot and form webs of fibrinlike protein serve as a means of entrapping and facilitating the deactivation of both invading organisms and endotoxin by the release of additional anti-endotoxin and antimicrobial factors. The clotting action also serves to prevent leakage of hemolymph at external sites of injury.

The "fibrinogen-like" invertebrate protein is called coagulogen in its soluble form and coagulin in its (post enzyme activated) gelled form (63). The conversion of coagulogen to coagulin is mediated by the sequential activation (cascade) of several zymogens arising from the single blood cell of *Limulus* or *Tachypleus* (the amebocyte or granulocyte). The L-granules contain all the clotting factors for hemolymph coagulation, protease inhibitors, and anti-lipopolysaccharide (LPS) factor, as well as several tacylectins with LPS binding and bacterial agglutinating activities (Figure 3).

Upon GNB invasion of the hemolymph, hemocytes detect LPS on their surface and release their granule contents (degranulate). The known bio-sensors consist of coagulation factor C and factor G, which serve as the triggers for the coagulation cascade that converts soluble coagulogen to the insoluble coagulin gel. These two serine protease zymogens are autocatalytically activated by LPS and (1,3)-β-D-Glucan respectively. The LPS initiated cascade (via activation of the proclotting enzyme) involves three serine protease zymogens: factor B, factor C, and proclotting enzyme. The final step of the clotting reaction involves the creation of coagulin from coagulogen by the excision of the midsection of the protein, called peptide C. Without peptide C, the monomers form AB polymers consisting of the NH2-terminal

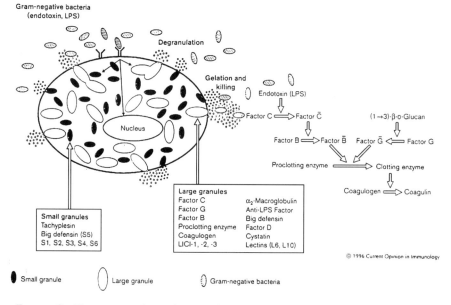

FIGURE 3 The conversion of coagulogen to coagulin is mediated by the sequential activation (cascade) of several zymogens arising from the single blood cell of *Limulus* or *Tachypleus*. (Reprinted from Ref. 64 with permission from Elsevier Science.)

A chain and the COOH-terminal B chain covalently linked via two disulfide bridges (65) (Figure 4).

15.11. PROMINENT LAL TESTS

Early on, Levin and Bang described three critical properties of the gelation of LAL in the presence of LPS that formed the basis for subsequent assays (66):

> Increase in OD that accompanies coagulation is due to the increase of clottable protein.
> The concentration of LPS determines the rate of the OD increase.
> The reaction occurs in the shape of a sigmoid curve (i.e., a plateau, a rapid rise, and a final plateau).

The total amount of clotted protein formed depends upon the initial LAL concentration. An excess of LAL is provided for LAL testing and the amount of clotted protein eventually ends up the same regardless of the amount of

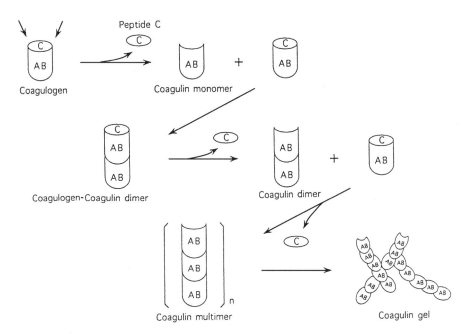

FIGURE 4 Hypothetical mechanism of coagulogen gel formation. Upon gelation of coagulogen by a horseshoe crab clotting enzyme, peptide C is released from the inner portion of the parent molecules. The resulting coagulin monomer may self-assemble to form the dimer, trimer, and multimers. (From Ref. 65.)

endotoxin in the sample. The end result of the enzymatic cascade is the formation of a web of clotted protein. The gel clot and endpoint tests take a single time point reading from the data to determine if the reaction reached an assigned level during the assigned time, whereas the kinetic tests are "watching" (at the appropriate wavelength) throughout the entire course of the reaction. The endotoxin concentration determines the rate of protein clot formation and thus the optical density change over time as determined by measuring the time to reach an assigned mOD value. The rate of OD formation is then related to the standard curve formed using control standard endotoxin. It can be seen from a plate that sits out that all wells containing endotoxin will eventually form a dark colorimetric or turbidimetric solution regardless of the endotoxin concentration, demonstrating that it is the speed of the reaction that correlates to the endotoxin concentration.

Besides the basic gelation of LAL in the presence of LPS, the two methods of observing the assay include the endpoint and kinetic assays. In the endpoint test, the reaction proceeds until it is stopped by the user by the addition of a stop reagent (such as acetic acid) at which point the optical density readings are recorded for all sample and standard curve points. The drawbacks associated with the endpoint method of observing the reaction are (a) necessity of the user attention at the end of data collection (typically 30 min) and (b) the limited standard curve range (a single log). In the kinetic assay, the spectrophotometer records the optical density reading continuously (as determined by the software settings within the manufacturer's recommendations—typically 1:30 to 2:00 minute intervals). Kinetic testing measures the rate of the optical density change, by recording the time it takes to reach a preset optical density setting called the "onset" or "threshold" time. The kinetic assay plots the log of the resulting reaction time in seconds against the log of the endotoxin concentration of the known standards and can span several logs (typically 2–4) and proceeds unattended, thus overcoming the two disadvantages presented by the endpoint tests.

The gel clot test is a simple test not far removed from Levin and Bang's original observations. Until recently it was the most widely used procedure for the detection of endotoxin in solutions. When equal parts of LAL are combined with a dilution of sample containing endotoxin, one can expect to see gelation in the amount equivalent to the endotoxin sensitivity (called lambda, λ) of the given lysate. A series of dilutions will reveal the approximate content of a sample with those samples containing equal to or greater than the given sensitivity being positive and those below the sensitivity not clotting the mixture. The solutions are incubated at a temperature correlating to a physiological temperature (37°C) for one hour and clots are observed by inverting the tubes 180°. The clot must remain in the bottom of 10mm × 75mm depyrogenated test tube when inverted. The method is considered semiquantitative because the true result obtained (indicated by the last gelled sample in the series) is actually somewhere between the two serial dilutions because the result cannot be extrapolated between the (usually twofold) dilution tubes as it is in the kinetic and end-point assays via the use of a mathematical standard curve extrapolated over the entire range of standards.

Because commercial lysates are available with various standardized end points (sensitivities), the assay can be used to quantify the level of endotoxin in a particular solution or product. The level of endotoxin is calculated by multiplying the reciprocal of the highest dilution (the dilution factor) of the test solution giving a positive end point by the sensitivity of the lysate preparation. For example, if the sensitivity of the LAL employed were 0.03 EU/mL and the dilution end point were 1:16, then the endotoxin concentration would be $16 \times 0.03 = 0.48$ EU/mL. For products administered by weight, the result in EU/mL is divided by the initial test solution potency (as reconstituted or as per the liquid in the vial) to give a result in EU/unit (EU/mg, EU/insulin unit, EU/mL of drug, etc.) that can then be compared to the tolerance limit specification. The geometric mean calculation is used for assays as opposed to the pass-fail limit test (that is reported as a "less-than" number if there is no activity).

Given that kinetic assays continue to be the overwhelming area of growth in LAL testing (listed as a primary reason for the harmonization of endotoxin standards in IS-2), it is relevant to discuss details of kinetic testing. The development of the chromogenic assay was largely driven by the desire to accurately determine the endotoxin content for bacteremia (67), endotoxemia (68), and bodily fluids such as blood plasma and cerebrospinal fluids (69). Table 7 lists standard curve parameters obtained for a typical kinetic test standard curve and curve parameter (software) display/printout.

Among the most significant advantages of kinetic and end-point testing over the gel clot assay is that they allow for the quantitative extrapolation of an unknown result between standard points. In the kinetic test, samples are pipetted into a 96-well microtiter plate, layered with LAL, and read spectrophotometrically at 405 or 340 nm (kinetic chromogenic or turbidimetric). The resulting color or turbidity reaction between LAL and endotoxin is recorded in the form of the time in seconds that it takes a sample to reach a threshold optical density reading as defined in the reader's software (OD or mOD). The log of the time obtained for each sample is plotted against the standard curve linear or polynomial regression line formed from the log of the endotoxin content obtained for known standards.

The gel clot quantitative approach, especially for water and in-process testing, has been largely supplanted by kinetic tests due to the ability of kinetic assays to extrapolate accurate results over a wide range of endotoxin concentration. A positive control consisting of a product sample spiked with

TABLE 7 Standard Curve Values from a Kinetic Chromogenic Assay (λ = 0.05 EU/mL) Using a Commercial Reader/Software System

Coefficient of correlation (r):	−0.999	
Y-intercept:	2.943	
Slope (m):	−0.265	
Blank:	**** (no reaction)	average = ****
Standard 1 (0.05 EU/mL):	1984, 1995, 1996, 1984	average = 1989
Standard 2 (0.5 EU/mL):	1007, 997, 999, 1001	average = 1001
Standard 3 (5.0 EU/mL):	594, 591, 593, 575	average = 588

This is the data from which the kinetic reader software uses a linear (or polynomial) regression standard curve to determine result calculations from sample reaction times.

a known concentration of endotoxin and a negative control using non-pyrogenic water is used to ensure the lack of interference in the sample matrix. Although a simple clot end point may be adequate for routine release testing of various pharmaceuticals, the ability to quantify endotoxin is invaluable for troubleshooting production-related pyrogen problems. Daily monitoring of plant water and in-process testing can alert production personnel to potential pyrogen problems before they become critical. Corrective action can be taken to reduce pyrogen loads and levels of endotoxin at this time. Using the gel clot assay, one would not see the increase in activity until the sample forms a clot. Thus, there is little or no warning prior to failing a given lot of water or in-process sample.

The turbidimetric assay gives a quantitative measurement of endotoxin over a range of concentrations. This assay is predicated on the fact that any increase in endotoxin concentration causes a proportional increase in turbidity due to the precipitation of coagulable protein (coagulogen) in lysate (hence forming coagulin). The optical density of various dilutions of the substance to be tested are read against a standard curve obtained that has been spiked with known quantities of endotoxin in sterile water (Table 8).

The chromogenic assay differs from the gel clot and turbidimetric reactions in that the coagulogen (clotting protein) is partially (or wholly) replaced by a chromogenic substrate, which is a short synthetic peptide containing the amino acid sequence at the point of interaction with the clotting enzyme. The end of this peptide is bound to a chromophore, para-nitroanilide (pNA). Japanese workers pioneered the use of chromogenic substrates and lysate (from *Limulus* and *Tachypleus*, the Japanese horseshoe

TABLE 8 Relative Advantages and Disadvantages of Major LAL Test Types

Kinetic and Endpoint Tests Versus Gel Clot Method
- Kinetic quantitative extrapolation of an unknown result between standards via linear or polynomial regression.
- Less prone to variation due to user technique.
- Provides "on board" documentation and calculation capabilities for consumables and products used in the test.
- The mathematical treatment of data allows for the observance of trends and for the setting of numerical system suitability and assay acceptance criteria.
- May have different interference profiles than gel clot assays (useful if the gel clot assay will not give a valid result at a sensitive level).
- Assays may be automated.
- Lambda may be varied by changing the bottom value of the standard curve (within the limits of the given LAL), thus allowing the MVD to be extended for difficult-to-test (interfering) products

Kinetic tests versus endpoint tests
Quantifies a result over a range of several logs (i.e., the difference between the highest and lowest standard curve points) versus a single log.
Tests to completion without user intervention after LAL addition — precision, speed, and accuracy improved.

Chromogenic versus turbidimetric tests (kinetic and end-point)
Calculates a result over a range of several logs (i.e., the difference between the highest and lowest standard curve points) versus a single log.
Tests to completion without user intervention after LAL addition.
Turbidity determinations are made based on the physical blocking of transmitted light (like nephlometry).
Chromogenic methods (end-point and kinetic) are not limited by particulate constraints associated with Beer's Law (absorbance is directly proportional to common parameters such as well depth).
The chromogenic method may be applied to turbid samples.
The turbidimetric method may be applied to samples with a yellow tint.

Recombinant Factor C (fluorescent test)
May provide sample suitability advantages as it does not contain unknown factors associated with the blood of the horseshoe crab (i.e., no glucan pathway).
- Fluorescence associated with emission not absorbance as per kinetic methods.

crab) for the detection of endotoxin (70,71). The chromogenic method takes advantage of the specificity of the endotoxin-activated proclotting enzyme, which exhibits specific amidase activity for carboxyterminal glycine-arginine residues. When such sequences are conjugated to a chromogenic substance, p-nitroanilide (pNA) is released in proportion to increasing concentrations of endotoxin. Thus, it is possible to measure endotoxin concentration by measuring endotoxin-induced amidase activity as release of chromophore. Release of chromogenic substrate is measured by reading absorbance at 405 nm. Testing is conducted with 100 microliters of lysate and an equal amount of sample or diluted sample. The quantitative relationship between the logarithm of the endotoxin concentration and amidase activity can be observed between 5×10^{-6} and 5×10^{-2} ng/mL of endotoxin (72) and, therefore, can be used for the detection of picogram quantities of endotoxin associated with medical device eluates, immersion rinse solutions, and drug products. See Table 10 for a comparison of common methods.

15.12. METHOD DEVELOPMENT AND VALIDATION: THE IMPORTANCE OF A GOOD TEST

Historically, large volume parenteral manufacturers have been foremost in developing tests for bacteria endotoxin assays due to the criticality of even minute endotoxin concentrations in solutions administered in large doses. However, many of today's problems revolve around the recovery of control standard endotoxin spike, the difficulty of which is exacerbated by the chemical nature of the small-volume drug materials being validated rather than their dose, which is often small. Small-volume parenteral drugs often contain high drug concentrations, which interfere both with the physiology of rabbits in the pyrogen assay and with spike recoveries in the LAL assay (73). Some common types of problem compounds encountered in developing endotoxin assays for small-volume parenterals include water-insoluble drugs, drugs containing activity that mimics that of endotoxin, drugs containing endotoxin that must be removed prior to validation, bulk drugs with variable potencies, multiple drugs in a given container, and potent, highly interfering drugs such as chemotherapy drugs. Now that the science of LAL testing has been firmly established, the challenges that remain often reside in difficult, product-specific applications. Perhaps the last great challenge encountered in each parenteral analytical laboratory is the development of, not just an LAL test, but a rugged, reproducible, and perhaps automatable test that will stand the test of time in routine use.

Given all the LAL methods that could be developed the question may be asked: What characteristics must a good LAL test have? A good LAL test from a legal standpoint must meet the appropriate compendial requirements and need not be quantitative except in its ability to demonstrate the detection of the endotoxin limit concentration (gel clot). However, beyond meeting compendial requirements, the best test is the one that provides the most information on the content of the analyte: endotoxin. The regulatory question that must be answered in order to put a drug on the market is: "Does it pass the release test?"* The scientific and business questions that remain to be answered are "How much endotoxin does the sample contain?", "How does the result compare to previous lot measurements?" and "How close to the endotoxin limit concentration is the result?"

Characteristics of a good BET validation test in general terms that cover both the kinetic, endpoint, and gel clot assays, therefore, include the following:

Non-interfering (positive controls are positive and negative controls are negative)

Appropriate product solubility if reconstituted and diluted or as diluted only

Demonstration of that the method chosen does not reduce (destroy) endotoxin that may be present if harsh conditions or solvents are employed[†]

Performed at the appropriate level as determined by the appropriate drug dose (or as per the USP or other compendial monograph tolerance limit assigned for existing drugs), potency, lambda, and proposed or dictated specification requirement

Not subject to significant reagent batch or laboratory test variability

Resolution of a result (well) below the specification to allow manufacturing process contamination problems to be monitored prior to rising to alert levels

---

*Significantly: (a) is the manufacturing process used to produce it compliant with cGMP requirements and (b) does the sampling and testing of precursors to the end-product support the contention that the product is free of endotoxin at the levels required?

[†]Validation via a series of sample dilutions in tubes containing spike demonstrates that the sample spikes endures the harsh treatment. However, if a kinetic or endpoint in-plate spike is used at a significant dilution, then the demonstration that the spike has acceptably endured the entire sample preparation method should be performed in the validation testing. For instance, a sample prepared in dimethyl formimide or other suspected harsh treatment then diluted to 1:1000 in water prior to spike in the plate will not demonstrate that the DMF does not destroy potential endotoxin. This is necessary to mention because of the prevalence today of adding kinetic spikes to only the final dilution of a series in the microtiter plate itself. After all, the goal of validation is to detect, not destroy endotoxin that may be present in the sample.

Demonstration of pH neutrality (6–8) in the I/E sample dilution after combination with LAL

Appropriate laboratory support testing such as labware qualification (endotoxin free and noninterfering), RSE/CSE, LAL label claim (gel clot) or initial qualification (kinetic and endpoint tests), diluent interference tests (i.e., their effect on LAL sensitivity)

Proper documentation of test events

Proper supporting documentation: user training, instrument IQ/OQ's, PM's, computer validation, qualification, data archiving, etc.

Appropriate manufacturing support tests, such as component, excipient, and API testing (i.e., appropriate manufacturing process monitoring)

Some basic information must be gathered prior to developing an endotoxin test for a new chemical entity (NCE) or an established product. A list of questions for the submitting department or developing scientist(s) may be compiled:

1. The Maximum Human Dose, which will typically allow room for the clinic to increase the dose as needed in safety and efficacy studies. The response should be documented in an e-mail or other mechanism for inclusion in the validation documentation

2. The formulation should be documented to establish the appropriate excipient tests (as previously discussed) and because it will likely change.

3. The presentation should be recorded as a critical assay parameter and may be subject to change (i.e., the product potency and volume or weight [for a given indication])

4. The approximate scheduling of the manufacture of the (at least) three lots needed for validation testing (if available)

5. A change notification mechanism to notify the laboratory of potency, dose, and/or presentation changes (who is responsible?)

6. Solubility profile (recommended reconstitution diluent[s]). How water soluble is it? or What is it most soluble in?

7. pH profile. What is the expected sample pH range?

8. Interference related questions:

   Is it a known chelator (such as EDTA)?
   Does it possess enzymatic activity (such as trypsin or serine proteases) likely to interfere with LAL?
   Is the compound likely to be inactivated by heating in a waterbath at 70°C (an enzyme)?
   Is it likely to contain cellulosic material?

. What is the molecular weight of the compound? If there is
endogenous endotoxin, it may be advantageous to remove it
(via filtration) for validation purposes and the MW of the
sample will determine if it may be filtered and still retain the
active compound in the filtrate.

The need for a new bacterial endotoxin test typically begins with a
call from a development scientist with a new compound. Perhaps it is a
compound prepared for an animal toxicology study or perhaps it is a lot
prepared in the development laboratory (a so-called lab-lot). The early lots
of drug substance or drug product will not be used in people, but there is a
need to establish their safety to ensure that the studies being performed are
not skewed in some manner by the presence of endotoxin. Drug development is a costly endeavor and the generation of misleading results can lead
developers down lengthy and costly blind alleys. Typically, compounds
have been handed over to a development team from a discovery research
effort that has been years in arriving. The compound has been formulated
now for parenteral use, perhaps only one of many current or potential
formulations, by combining a drug substance (bulk or active pharmaceutical ingredient [api]) solubilizers, stabilizers, preservatives, emulsifiers, thickening agents, and so forth (74). The compound is in flux and may change
several times in its formulation (excipients), presentation (i.e., potency,
container, size), and application (i.e., dose and perhaps indication). Perhaps, if its prospects seem especially bright, it will spawn a host of sister
compounds that vary in the means of drug delivery (i.e., parenteral, for
inhalation, time-delay parenteral, etc.) and, therefore, in several relevant
parameters required to be defined prior to developing additional suitable
endotoxin tests.

Assay development for the Bacterial Endotoxin Test for a given
compound may be as simple as the following (see Fig. 5 for an overview):

Calculating the new product's proposed tolerance limit (TL) and
maximum valid dilution (MVD) based upon the clinical dose of the
material (or USP monograph listed TL if it is an established drug)
Diluting the material in sterile reagent water
Testing it by the gel clot, end-point, or a kinetic (turbidimetric or
chromogenic) method at a dilution below the MVD.

However, given that early drugs were much less complex than today's drugs, it
seems that the days of simplistic validations that do not require additional
sample treatment(s) have passed. Now one would not realistically expect to
test most drugs in an undiluted fashion. Many compounds have mitigating
factors seemingly designed to frustrate the best assay development efforts as

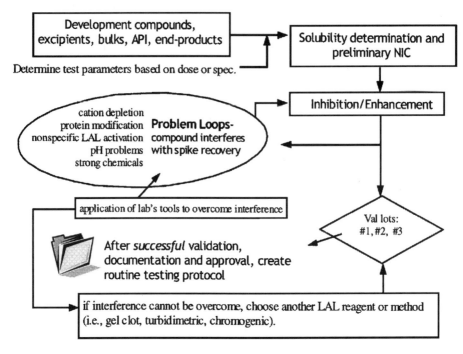

FIGURE 5   Method development—validation process.

previously described. Additional mitigating sample complications include the following:

   Cost: some product candidates are so expensive that product develop-
       ment scientists are reluctant to supply sufficient quantities for pro-
       tracted method development and validation.
   Occurrence of multiple interference properties not overcome by simple
       dilution whereby adjusting one causes a deterioration of another.
   Poorly characterized products: at an early stage of drug development,
       one can expect to see drug products that vary greatly from lot to lot
       (i.e., they are still being adjusted by those charged with establishing
       their formulation).

The types of testing protocols used in developing a new method may include
(a) solubility and pH study protocols, (b) preliminary Non-Inhibitory Concentration (pNIC) protocols, and (c) a validation protocol. The tests performed in this sequence are cumulative. In simple terms, the NIC test varies
the sample concentration while keeping the endotoxin concentration fixed

(none and 2λ for gel clot and the midpoint of the standard curve for kinetic testing), whereas the I/E test varies the endotoxin concentration (to mimic the standard curve) while maintaining a constant product concentration (kinetic I/E uses only the midpoint). The three tests for the gel clot method and subsequent result calculation (which can be applied to the kinetic and endpoint methods with some adjustments) serve to establish parameters upon which to base future routine testing:

(a) Solubility/pH: one cannot perform the pNIC without having a good idea of the solubility and pH characteristics of the material. The gap can be bridged for water-insoluble compounds by dissolving the compound in a suitable solvent that does not destroy endotoxin (DMSO is such a diluent for many water insoluble compounds) but that also is readily diluted with water or buffer. The right proportion will have to be found to keep the compound dissolved, but to allow enough dilution in water to overcome potential interference by both the compound and the solvent. The pH characteristics go hand in hand with the solubility. It may be necessary to acidify a given solution before a compound will go into solution.

(b) The preliminary NIC determines the dilution at which the full validation test may be performed. Typically, at some point in a series of twofold dilutions of both spiked (2λ) and non-spiked sample, a "break-point" will be determined (first positive spike [2λ] recovery of the series coexistent with no recovery in the unspiked sample at the same dilution). If the unspiked twofold dilution is negative and the spiked twofold dilution is positive, then this demonstrates that the observed interference has been overcome by the dilution. Therefore, the noninhibitory concentration is somewhere between the first positive and the negative (2λ) spiked sample test directly preceding it. If it occurs at a level that is compatible with the calculated MVD (MVC for a bulk, excipient or API sample), then one may proceed to the full validation test.

(c) The full validation test typically includes both an NIC confirmation and an inhibition/enhancement curve (I/E), which is simply a standard curve performed in sample solution at the concentration of sample that one will not exceed (validated level). The I/E dilution level must not exceed the MVD (or 1/3 MVD for pooled vial tests) and must exceed the minimum valid concentration (MVC) of sample (or $3 \times$ MVC for pooled vials) needed to detect the endotoxin limit concentration (the tolerance limit amount of endotoxin). The validation test may include a limit test at the proposed routine test dilution, but it is not necessary because that dilution is contained within the NIC and will be greater than or equal to the I/E dilution being tested.

(d) The validation reportable test result will be based on the successful performance of the I/E test. If the I/E test agrees within a twofold dilution

TABLE 9 Overcoming Various Interference Mechanisms

Interference/Ref.	Overcoming interference
(a) Suboptimal pH conditions LAL is a product of a physiological system and many drugs are not. A pH of 6.4 to 8.0 is optimal and a pH requirement of 6.0 to 8.0 taken on a given sample and LAL is referenced by USP (76,77).	Most LAL reagents are buffered either as lyophilized or as reconstituted to overcome minor pH problems. Initial pH adjustment using 0.1N or lower HCl or NaOH may be needed for more acidic or basic samples. Cooper maintains that pH problems "are the most important biochemical cause of LAL-test inhibition." The USP requires the pH of the sample-LAL mixure to fall within the reagent supplier's requirements, which is usually 6.0–8.0. An FDA inspector relates that pH testing is not routinely required for a validated method unless committed to in the firm's NDA. He also says that a failure to study the upper and lower limits of the product pH range (in validation) might necessitate routine testing.
(b) Endotoxin modification is a problem involving the amphiphilic properties of the CSE (78–80).	Strong salts and other solutions causing a large increase in test sample ionic strength will cause endotoxin aggregation and poor spike recovery. Dispersing agents such as Pyrosperse™ (Cambrex) along with dilution ($\leq$MVD) is used to overcome such interference. Adsorption of endotoxin to containers made of polypropylene is avoided in all types of endotoxin testing labware except pipette tips.
(c) Unsuitable cation concentrations. LAL reaction requires cations (81).	Organic chelators (i.e., EDTA) added for the purpose of complexing heavy-metal cations may cause instability in parenteral formulations. 50 mM MgCl2 is routinely used as a test diluent to provide suitable levels of Ca^{+2} and Mg^{+2}. Reagents vary in cation concentration and buffering capacity among those supplied by LAL manufacturers.
(d) Protein or enzyme modification. Enzymes	Alcohols, phenols, oncolytics fall into this category. If the interferring agent is

TABLE 9 Continued

Interference/Ref.	Overcoming interference
needed for LAL gelation reaction[a] are denatured by strong chemicals.	itself an enzyme, it can be denatured by heating a sample or dilution of sample at 70°C for ~10 min prior to or post) dilution before testing. Other offenders may be removed by ion or size filtration, though the validation requirements may be onerous.
(e) Nonspecific LAL activation includes the detection of LAL-reactive material and drugs that mimic endotoxin such as those containing serine proteases	Serine proteases may be heat-inactivated (as above). Products that mimic endotoxin provide a difficult challenge. To show that the activity occurring is not endotoxin, determine the level of activity followed by treatment of the sample to bind endotoxin (if the molecular weight of the product prohibits filtration removal). If the activity is not reduced then it may not be endotoxin. An alternate test method may be needed or one may lower λ to allow sufficient dilution to overcome ("outrun") the enhancement.
(f) Samples containing endotoxin may present a problem similar to (e)	If the levels are relevant to the required test levels, endotoxin must be removed prior to performing the inhibition/ enhancement test (Gel Clot). Methods of removal include filtration (20,000 MW Sartorious filter) when the molecular weight of the sample ingredient(s) do not exceed the cut-off rating of the filter.
(g) *Insoluble drug products*	The lack of a suitable solvent for poorly water-soluble products is problematic. The LAL assay is a water-based test. DMSO has been used successfully. Mallinckrodt[b] described a method of liquid-liquid extraction capable of pulling endotoxin into the aqueous phase, which leaves an inhibitor or difficult to work with sample in the discarded oily phase.

[a] Serine proteases.
[b] Undated product insert.

with the labeled LAL label claim (and the included valid CSE curve), then the sample (test result, TR) can be said to contain:

$$TR = \frac{< \lambda \times DF}{PP}$$

where PP is the product potency of the active ingredient as reconstituted for a weighed sample or as labeled for a liquid sample containing a predetermined potency, DF is the dilution factor, and PF is the pool factor. A geometric mean is not necessary to determine the result calculation here because the I/E is either valid at the given dilution (sample concentration) or is invalid (i.e., does not confirm the label claim) (Figure 5).

15.13. RESOLVING TEST INTERFERENCES

Given that the LAL assay in its many forms is a water-based assay derived from a sensitive physiological environment (blood of the horseshoe crab), it is not too surprising that as one ventures farther from such an aqueous environment the results often correspondingly deteriorate. The Catch-22 of such testing resembles the contradiction presented by endotoxin itself (as an ampiphile) in that an increase in water content of a hydrophobic compound in solution will cause the material to precipitate (and endotoxin to aggregate) but, conversely, as the compound gets away from water, the reaction of LAL and endotoxin will be inhibited. Cooper's paper on interference mechanisms encountered during LAL testing is perhaps the most useful on the subject (75). Cooper lists five major interference mechanisms to be expected when testing various parenteral drugs for BET using the LAL test and points out that often interference mechanisms result from the sample matrix's effect on the aggregation properties of the CSE rather than or as well as on the LPS-LAL reaction itself. The broad mechanisms (a–e) listed by Cooper include (a) suboptimal pH conditions, (b) aggregation or adsorption of control endo-toxin spikes, (c) unsuitable cation concentrations, (d) enzyme or protein modification, (e) nonspecific LAL activation, and (f) sometimes an interfer-ence mechanism cannot be determined. Each broad interference mechanism will be briefly explored along with notable (common or unique) means of overcoming the associated interference in Table 9.

15.14. SETTING ENDOTOXIN SPECIFICATIONS

The group developing the assay plays a key role in verifying that pro-posed specifications set are within the appropriate bounds established by the FDA Guideline calculations and pharmacopoeial requirements. Practically speaking, the lab will determine the informal specification for development testing given the clinician's dose range. At a later date a specification committee will assign an in-house specification. There appear to be two divergent philosophies on setting specifications. The first is to set the most stringent specification that the laboratory can support (i.e., around the limit of detection). The second is to set the specification around the regulatory limit allowed (i.e., the tolerance limit calculated value), which is the highest legal limit.

Concerning the first philosphy, setting the specification too tightly may come back to haunt the participants in the form of a test failure and subsequent destruction of an expensive lot of drug that, scientifically and from a regulatory perspective, does not exceed allowable endotoxin levels. Early clinical doses are often severalfold higher than subsequent marketed drug doses, but companies often are reluctant to ratchet down specifications that have been established as doses decrease in the clinic. When products inevitably go to market, they will do so with a dose that is sometimes significantly lower than that used to establish the endotoxin test. The second philosophy is as poor as the first. If the specifications are set too close to the values allowed by law, then the routine examination of the drugs will not detect changes in endotoxin content until they are at failing levels. Ideally, one wants to "see" the endotoxin content well below the specification to serve as a warning that the manufacturing process is beginning to allow contamination well before it reaches a level relevant to the manufacturing process. If the specification is too high, then there will be no time for corrective action preceding a test failure.

Those unfamiliar with endotoxin limit calculations may see a value and gauge whether it is "high" or "low" simply by how large the number is. However, the specifications is a function of the dose and any specification that is set appropriately will allow less than 350 EU/patient dose/hour. Naturally, a several gram dose may contain less endotoxin on a per milligram basis than will a drug that is delivered in micrograms. The situation may arise in which a limit of nmt 100 EU/mg is set beside another compound with a limit of nmt 0.25 EU/mg, making the 100 EU/mg appear less "stringent" when in fact they both allow the same amount of endotoxin delivery as per their associated dose. The proof of this is in the side-by-side calculation:

$$TL = K/M \quad 5.0 \text{ EU/kg}/(3.5 \text{ mg/70kg/hr}) = 100 \text{ EU/mg} = 350 \text{ EU/dose}^*$$

$$TL = K/M \quad 5.0 \text{ EU/kg}/(1400 \text{ mg/70kg/hr}) = 0.25 \text{ EU/mg} = 350 \text{ EU/dose}^*$$

*By definition, $TL = 350$ EU/dose.

TABLE 10 Time Required to Achieve Multiple Log Reductions Using Different Sources of Endotoxin

Log reduction	Temp°C	Tsuji et al (1978–79)[a] Minutes	Ludwig & Avis (1990)[b] Minutes
3	@210	13.6	7
	@300	0.089	<0.5
5	@210	Infinity*	19
	@300	0.19*	1
6	@300	0.27*	11
		*Extrapolated value.	

Log reduction	Temp°C	Bio Whittaker Minutes	Difco Minutes	ACC Minutes
3	@225	5	5	5
	@250	<0.5	NA	2
5	@225	15	45	45
	@250	5	NA	19

Derived from Ref. 97

[a] Tsuji et al. used aluminum cups.
[b] Ludwig and Avis used glass.
Source: Refs. 86, 96, and 97.

The initial process of validation may be as in flux as the compound itself. Factors subject to change include the product potency, presentation, included excipients, interference factors, containers, and so forth. Factors that are absolutely critical to establishing a test that will detect the endotoxin limit concentration include the maximum human dose (MHD), product potency or concentration (PP), and LAL lambda (λ) to be used in the TL and MVD (or MVC) calculations. An error in calculation or failure to secure a relevant dose for the TL calculation will nullify subsequent efforts to provide an accurate result. The Tolerance Limit is equal to the threshold pyrogenic response (K in EU/kg) divided by the dose in the units by which it is administered (mL, Units, or mg) per 70 kg person per hour. Mistakes in this critical calculation may include the following:

Not adjusting for the body weight (conversion from m^2 may be necessary)

Not clarifying the means of delivery (bolus versus multiple daily doses, etc.)

Basing the dose on a method that is not relevant to the means of administration or is not based on the units of active ingredient (i.e., using mL instead of mg, particularly when the reconstitution may vary)

Not adjusting the MVD formula calculation for a potency change

Having the dose increased in the clinic to a level that exceeds that used as a basis for MVD calculation in the testing laboratory (i.e., poor communication)

The overall process is important in the development of a new LAL assay for a drug to be used in the clinic. Establishing a process that captures all the details is critical to ensuring that the right tasks are performed in the right sequence, the right information is documented, and that the information is correctly applied to the test both in its performance and in the determination of the parameters that govern its proper performance. Such a detailed process may be difficult to capture in a standard operating procedure, and extensive experience will be necessary before an analyst is proficient in all the nuances of developing an LAL assay, particularly for a new drug candidate.

The GMP documentation expectation for any analytical test is that of being able to "recreate" the test including all the materials used in a given assay. For the LAL assay, that can be a daunting task if the right systems are not in place. For any given test there may be dozens of consumables and equipment references (water or other diluent preparation, LAL, CSE, tips, tubes, plates, pipettes, tips, containers, water bath or heating block or kinetic reader, or other equipment, analyst, -etc.) for which lot numbers must be recorded. Preventive maintenance records, training records, product validation documentation, certificates of analysis or other proof (lab test references) that the consumables used are endotoxin-free and do not inhibit or enhance the test, RSE/CSE and/or COA reagent qualification documents used are all part of the items needed to "back up" any given test. Printed laboratory notebooks or worksheets are necessary to collect all the pertinent information in an organized fashion.

15.15. DEPYROGENATION VALIDATION

Integral to the manufacture of sterile and endotoxin-free parenterals is the validation of depyrogenation processes. Endotoxin is notoriously resistant to destruction by heat, desiccation, pH extremes, and chemical treatments. The validation of endotoxin destruction or removal in the manufacture and packaging of parenteral drugs is a critical concern to drug and device manufacturers. LPS requires dry heat treatment of around 250°C for half an hour to achieve destruction and standard autoclaving will not suffice.

Whereas sterilization processes are predictable, depyrogenation procedures are empirical. Many specific instances of applying potent reagents to manufacturing equipment for the purpose of destroying applied endotoxin where one would predict that LPS would be demonstrated to be destroyed have revealed that the LPS has hung on tenaciously, defying preconceived notions of depyrogenation.

Depyrogenation is first thought of as the dry heat incineration of endotoxins from materials able to withstand the protracted dry heat cycle needed to destroy the LPS molecule. Alternatively, the wash/rinse removal of endotoxin from items such as stoppers and plastic vials and alternative vial closures comes readily to mind when heat treatment is not an option. However, there are many additional and hybrid areas of depyrogenation that are less historically entrenched and that are subject to more complex validation support. The two broad classes of depyrogenation processes that may be applied to components, drugs, and articles coming into contact with drugs include inactivation and removal. See Fig. 6.

The past two decades of biotechnology has brought about the concomitant necessity of removing large populations of endotoxin from prod-

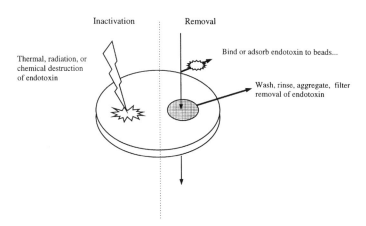

Inactivation

Removal

Thermal, radiation, or chemical destruction of endotoxin

Bind or adsorb endotoxin to beads...

Wash, rinse, aggregate, filter removal of endotoxin

FIGURE 6 Inactivation and removal of bacterial endotoxins. Inactivation: Heat, moist and dry, the use of ionizing radiation of components, chemical inactivation (i.e., strong acid/base solutions), oxidation (i.e., hydrogen peroxide), polymyxin B. Removal: The use of physical size exclusion of endotoxin (ultrafiltration, ion-exchange removal), or aggregation followed by filtration, the use of charge differential (anion exchange), binding treatments (activated charcoal, lipopolysaccharide binding protein products).

ucts due to their manufacture in microbial expression systems (especially *E. coli*). Selected methods of depyrogenation mentioned in Fig. 6 are employed to remove endotoxin from manufactured materials intended for parenteral use. A few of these methods will be examined. The oldest and simplest method of endotoxin removal from solid surfaces is rinsing with a nonpyrogenic solvent, usually Sterile Water for Injection. Low levels of surface endotoxin contamination can be effectively removed from glassware, device components, and stoppers, for example, with an appropriate washing procedure. Rinsewater can be monitored throughout the process with LAL to validate endotoxin removal. An example of such a validation process for large-volume parenteral glass containers was described by Feldstine et al. (82). Distillation is the oldest method known for effectively removing pyrogens from water.

Early investigators studying the thermostability of endotoxin concluded that moist heat supplied in conventional autoclaving was ineffective for depyrogenation. Although autoclave conditions for "normal sterilization" of solutions are ineffective for destruction of endotoxin, Banks (83) was able to demonstrate effective depyrogenation by autoclaving at 20 psi for 5 hr at a pH of 8.2, or for 2 hr at a pH of 3.8. Recent studies show that the action of certain depyrogenating agents can be enhanced by autoclaving. Cherkin (84) found that hydrogen peroxide (H_2O_2) was more effective in destroying pyrogen when the solution was autoclaved. Autoclaving also helped to eliminate residual H_2O_2. Similar findings have been reported for other solutions containing acid or base. Novitsky et al. (85) confirmed that autoclaving following conventional methods (121°C, 15 psi at near neutral pH for 20 min) was not sufficient to eliminate the pyrogenicity of 100 ng/mL of *E. coli* 055:B5. However, autoclaving for longer periods (180 min) successfully reduced endotoxin levels to less than an LAL detectable limit of 0.01 ng/ml. Novitsky et al. also found that activated carbon treatment was more effective in removing endotoxin when solutions containing endotoxin and carbon were autoclaved.

The application of dry heat delivered through convection, conduction, or radiation (infared) ovens has been the method of choice for depyrogenation of heat-resistant materials, such as glassware, metal equipment, and instruments, and of heat-stable chemicals, waxes, and oils. The standard method described in various national and international compendia and reference texts is an exposure of not less than 250°C for not less than 30 min and is based on the studies of Welch et al. (86) on the thermostability of pyrogens as measured with the rabbit pyrogen test. The mechanism of endotoxin inactivation is incineration. The development of the *Limulus* amebocyte lysate (LAL) has provided a more quantitative means of studying dry heat inactivation of endotoxin. Tsuji et al. discovered that the inactivation kinetics of LPS from *E. coli*, *S. typhosa*, *Serratia marcescens*, and *Pseudomonas aeruginosa* was a nonlinear, second-order process in contrast to the inactivation of bacterial spores, which follow first-order kinetics (87–90). They compared the dry heat resistance of intact and purified LPS to that of spores with the greatest heat resistance. Purified LPS was shown to be twice as resistant as the native (whole-cell) endotoxin from which it was derived. Of greater importance was the author's convincing evidence that the general practice of increasing exposure time to compensate for lower process temperature is not supportable for LPS destruction, particularly at 175°C or less. Akers et al. (91,92) confirmed these findings and also determined the F-value requirements for destruction of 10 ng of *E. coli* 055:B5 endotoxin seeded into 50-mL glass vials, using both convection and radiant heat ovens. An F value is the equivalent time at a given temperature delivered to a product to achieve sterilization or, in this case, depyrogenation. There were linear relationships between oven temperatures and the logarithms of the F-values with both treatments.

Before 1978 there were few studies addressing the destruction of endotoxin presumably due to the lack of a suitable quantitative method of measuring endotoxin reductions (93). Along with the LAL assay and the refinement of LPS standardization came a means of applying (as a biological indicator in a manner analogous to the use of sporeforming *Bacillus* species in sterilization studies) and detecting recovered endotoxin for such studies. Methods and mechanisms of proving the depyrogenation of various items have been largely borrowed from sterilization processes and modified to compensate for the thermal and chemical stability of LPS. The two comon types of depyrogenation processes (like sterilization methods) involve (a) the construction of D (death or destruction in the case of endotoxin because it is not alive) values and (b) the use of "bioindicators" as an empirical means of demonstrating that a "worst-case" load of applied pyrogenic residue has been removed by a given proposed depyrogenation process. The definition of the death rate (D value) in sterilization technology is the "time for a 90% reduction in the microbial population exhibiting first-order reaction kinetics" (94,95). The number of organisms decreases during sterilization in a log fashion down to one org (log 0) after which it becomes negative where 10^{-1} is the likelihood of a single survivor per 10 items and 10^{-3} is one survivor in 100 items. Therefore, in theory sterility is never absolutely knowable, but is reduced to a probability (however remote the likelihood of a survivor). Generic procedures (such as that given in the USP) cannot be assumed to work for a given wash or baking process due to the variety of equipment, loading configurations, times and temperatures chosen for different process applications. Validation must include "documented evidence" that the process does hat it purports to do, namely, provides a 3-log reduction of applied endotoxin. Death-rate curves in sterility validation can be constructed by graphing the number of organisms on the Y-axis against the log of either the heating time, exposure time (gas), or radiation dose on the X-axis. Similar destruction curves can be constructed using endotoxin data as shown in Fig. 7.

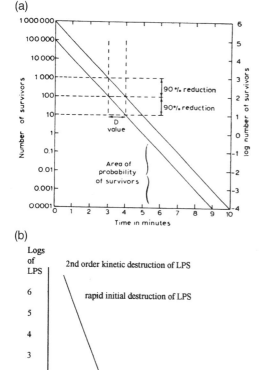

FIGURE 7   Microbial death-rate curves (a) illustrate concept of decimal reduction (D values) and probability of survivors (from Ref. 94) and (b) hypothetically demonstrates the more difficult to achieve reduction of LPS after an initially relatively rapid reduction.

The table above shows the lack of agreement (and thus empirical nature) of depyrogenation processes and hints at the plethora of conditions that can alter the time and temperature needed to bring about adequate depyrogenation (i.e., load and type of material, oven tunnel speed, etc.). Typical parenteral practice involves moving glass vials on a belt through an oven that blasts it with an excess of heat ($\sim 300°C$) at speeds of 5 to 10 min to achieve F values equivalent to or exceeding the targeted half hour at 250°C treatment (F250 = 30).

The requirements for depyrogenation validation processes (from a laboratory perspective) are somewhat vague and interpretive.* A short reference occurs in the USP, Chapter 1211—Sterilization and Sterility Assurance of Compendial articles, Dry-Heat Sterilization section as follows:

> Since dry heat is frequently employed to render glassware or other containers free from pyrogens as well as viable microbes, a pyrogen challenge, where necessary, should be an integral part of the validation program, e.g., by inoculating one or more of the articles to be treated with 1000 or more USP Units of bacterial endotoxin. The test with Limulus lysate could be used to demonstrate the endotoxic substance has been inactivated to not more than 1/1000 of the original amount (3 log cycle reduction). For the test to be valid, both the original amount and, after acceptable inactivation, the remaining amount of endotoxin should be measured.

The only other USP references to depyrogenation are in the Bacterial Endotoxins Test chapter: ⟨ Chapter 85 ⟩ states that one should "treat any containers or utensils employed so as to destroy extraneous surface endotoxins that may be present, such as by heating in an oven at 250°C or above for sufficient time" and then references the above paragraph as a means of validating the oven referred to here. And, "render the syringes, needles, and glassware (to be used in the pyrogen test) free from pyrogens by heating at 250°C for not less than 30 minutes or by any other suitable method" respectively.

The USP/FDA "Guideline on Sterile Drug Products Produced by Aseptic Processing" (98) provides a review of the requirements for container/closure depyrogenation:

> It is critical to the integrity of the final product that containers and closures be rendered sterile and in the case of injectable products,

---

*21 CFR Parts 210 and 211 Subpart E—Control of Components and Drug Product Containers and Closures (211.80, 211.82, 211.84, 211.86, 211.87, 211.89, 211.94) discusses component testing requirements in general terms.

pyrogen-free. The type of processes used to sterilize and depy-rogenate will depend primarily on the nature of the material which comprises the container/closure. Any properly validated process can be acceptable. Whatever depyrogenation method is used, the validation data should demonstrate that the process will reduce the endotoxin content by 3 logs. One method of assessing the adequacy of a depyrogenation process is to simulate the process using containers having known quantities of standardized endotoxins and measure the level of reduction...endotoxin challenges should not be easier to remove from the target surfaces than the endotoxin that may normally be present.

And:

Rubber compound stoppers pose another potential source of microbial and (of concern for products intended to be pyrogen free) pyrogen contamination. They are usually cleaned by multiple cycles of washing and rinsing prior to final steam sterilization. The final rinse should be with USP water for injection. It is also important to minimize the lapsed time between washing and sterilizing because moisture on the stoppers can support microbiological growth and the generation of pyrogens. Because rubber is a poor conductor of heat, proper validation of processes to sterilize rubber stoppers is particularly important.

There should be an awareness on the part of those charged with performing depyrogenation validation that there is a distinct difference between items that may be heat-treated and those that must be washed (inactivation versus removal, respectively). The heat treatment of bottles and vials follows the more easily reasoned path that, given appropriate time and temperature parameters, endotoxins will be destroyed. However, the wash removal of endotoxins is complicated by the tenacity with which endotoxin sticks to rubber and other porous polymers that compose such materials. Entrenched endotoxin's removal is governed by more difficult to assess parameters including agitation and solubility. Thus, with removal there are more variables involved than heat and duration than in the case of incineration.

There is really no perfect way to verify the presence or recovery of low amounts of endotoxin, (i.e., 10 EU/stopper) given the adsorption by porous materials (see Fig. 8). Common methods involve vigorous vortexing, soni-cation, or other means of agitation to dislodge endotoxin prior to testing. The selection of a vigorous method of dislodging of endotoxin is empirical

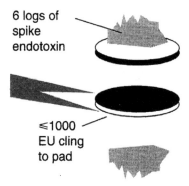

6 logs of
spike
endotoxin

≤1000
EU cling
to pad

FIGURE 8 Is this validation? A mountain of applied spike is turned over (or washed) and the mountain of spike falls off. Has a >3 log reduction transpired? Increasing applied spikes to obtain better percent recovery (rather than developing better removal methods) may result in spikes that are too easily removed, thereby revealing nothing about the depyrogenation process.

(whatever works), and various labs have chosen to use either intense, short duration vortexing or prolonged but less vigorous mixing (such as shaking or sonication), or simply washing with or without added surfactants. James Agolloco has described a theoretical problem associated with cleaning validation studies that relates aptly to depyrogenation validation (endotoxin removal) studies by using a "tar baby" analogy (99):

> The cleanliness of the bath water may not necessarily relate directly to the cleanliness of the baby. If the contamination is not soluble in the cleaning agent, then the contamination will remain on the surface. If the contamination is not soluble in the final rinse, samples of the bath water will not detect the presence of residual contamination. The conclusion will be drawn that the baby is clean, when in fact both the cleaning and evaluation methods are inadequate.

In other words, if one determines the cleanliness of the baby (stopper) by measuring the "tar" (endotoxin) remaining in the bath water (laboratory rinse method), then one has to ensure that the method used does indeed remove the "tar." There must be some validation of the method to serve as a demonstration that the method removes endotoxin from "sticky" surfaces. At least theoretically, endotoxin that clings tenaciously to a stopper (thereby escaping pyroburden detection) can be removed later by the surfactant action of a drug and become available for parenteral administration.

An added step of RSE/CSE characterization of spike solutions to be applied for depyrogenation studies can bring about a greater consistency of recovery given that the potency of the reconstituted solution of concentrated endotoxin (i.e., Difco) used can be highly variable (i.e., may vary from the label and from lab to lab). Additional characterization under laboratory conditions (as opposed to the manufacturer's assigned potency) may aid in "getting back" numerical values that are very close to the theoretical value (i.e., 48,800 EU/component of a 50,000 EU/component spike application).

It is instructive to separate out manufacturing and QC laboratory division of labor in the fragmented depyrogenation validation process. Regardless of how specific companies have bridged the activities, a natural division exists between the manufacturing and QC functions in the depyrogenation validation process. The manufacturing area may have a validation group that runs the studies to document that their processes comply with cGMP requirements including depyrogenation validation. QC laboratories support these efforts by supplying expertise in the endotoxin application. Therefore, the coordination of activities involves manufacturing and lab support. The manufacturing group determines and documents the depyrogenation treatment process (oven [including F values obtained] or washer [settings, rinses]) and the laboratory supplies inoculated components, performs before and after treatment (depyrogenation) LAL testing with accompanying controls, documents and reports the results (as supported by a validated laboratory method). Differences exist in the intention, activities, and requirements of laboratory validation to support pyroburden methods and depyrogenation validation processes (3-log reduction validations) though they are similar in many respects. A significant difference in the two lies in the fact that pyroburden is a release test for components to allow them to be used in marketed products in lieu of (or in addition to when obtained sterile from a vendor) a validated depyrogenation process. As such, the number to be tested should be derived from a statistical (or at least reasoned) sampling of a given lot of components based upon the manufactured component lot size.

ACC intends to publish a procedure to promote the use of LAL to bathe medical devices in situ.[*] Novitsky refers to an in-house study revealing significant LAL reactivity when implants were tested via the LAL in situ bath method versus negative results when tested by traditional extraction in which endotoxin spike recoveries are notoriously difficult to recover. Such a

---

[*] Novitsky, T.J., BET vs. PT Non-Endotoxin Pyrogens, LAL Update, April 2002, Vol. 20, No. 2.

method would overcome, in theory, many of the adsorption issues involved with recovering endotoxin from glass vials and rubber stoppers.

15.16. ENDOTOXIN REMOVAL IN PHARMACEUTICAL MANUFACTURING PROCESSES

Modern techniques used to remove endotoxins from drugs during parenteral manufacturing often involve the combination of several methods. Macromolecules cannot be removed by simple ultrafiltration given that their size may be similar to endotoxin aggregates. Two case studies will be reviewed in which endotoxin removal processes were devised for (a) a 32 Kda enzyme (superoxide dismutase [SOD]), (b) a high MW α-1,6 branched a-1,4 glucan (amylopectin) derived from corn or potato starch and used as an encapsulation matrix for pharmaceutical products.

An endotoxin removal process to meet a proposed specification level of less than 0.25 EU/mg of protein was performed at Sigma Chemical (referred to as case study 1) (100). Held et al. designed the initial purification of the protein to achieve greater than 99% purity using "extraction, heat treatment, clarification, and ammonium sulfate fractionation...." followed by three chromographic steps that removed the majority of endotoxins. Subsequently, the product yielded endotoxin values between 0.16 and 0.72 EU/mg, which provided no consistency in meeting the necessary specification (nmt 0.25 EU/mg). The authors employed a "polishing step" to perform the remaining threefold reduction of endotoxin with an eye on adding only a minimal additional cost to the process. They used a positively charged, 1-ft^2, 0.2 μm disposable Posidyne filter (Pall) to achieve the required endotoxin reduction without product loss. The natural negative charge of LPS above a pH of 2.0 allows the use of ion exchange as a means of binding the endotoxin to the filter matrix while the protein solution passes through.

In case study 2, the same Sigma Chemical group had a more formidable task of reducing endotoxin in amylopectin from approximately 500 EU/gram to less than 20 EU/gram (<0.02 EU/mg). The low solubility and viscosity of the product prevented the filtration removal of endotoxin. They added 400 grams of food-grade amylopectin to 20 liters of 2-mM EDTA to reduce the aggregate size of the endotoxins. They heated to 85–90°C and stirred the mix for an hour. After cooling the mix to 54–56°C, they added NaOH to a final concentration of 0.25 M and stirred for another hour to hydrolyze the endotoxin base labile bonds (i.e., lipid A-KDO). The solution was neutralized using HCl and cooled to room temperature. Repeated ultrafiltration with 300,000 MW cutoff filters removed salts and endotoxin. Upon concentration to 10 liters, the solution was diluted to 30 liters with endotoxin-free water.

This was followed by repeated reconcentration to 10 L followed by redilution in endotoxin-free water a total of nine times. The final solution was filtered through a 0.45 μm Posidyne filter (Pall), frozen, lyophilized and stored overnight under vacuum. Thus, the group combined three different, well-known mechanisms to remove the endotoxin in stages: treatment with moderate heat and alkali, filtration separation by molecular weight cutoff filters, and ion-exchange binding to the 0.45 μm filter. They quantitated the endotoxin removed by each of the processing steps to find that the reduction factors achieved were 20, 5, and 2, respectively. The final filtration resulted in a solution of less than 1 EU/gram. The authors advise "even water with endotoxin levels that are below the detection limit can become a major contributor to endotoxins when large volumes are used for repeated cycles of dilution and concentration of a product." Historical methods of obtaining multiple log reductions in parenteral processing have involved chromatography and adsorption. Particularly problematic is the removal of endotoxin tightly bound to biologicals drug compounds (proteins, polysaccharides, or DNA) (101,102).

15.17. THE FUTURE AND ENDOTOXIN TESTING

Two important reoccurring themes that may help form a view of the future direction of parenteral contamination testing are as follows:

1. Endotoxin is the major microbial cell residue, but it is not the only important cellular artifact (nonliving residue).
2. Endotoxin is the most potent of such artifacts and induces a wide range of deleterious host effects at the cellular and systemic levels, but it is not the only one or the only potent one.

Two general questions form the broad outline for this section: What are some likely paths to future prospective tests for endotoxin and might such prospective tests be expanded to include non-endotoxin parenteral contaminants? Pyrogen testing originated with a fairly insensitive but broadly inclusive method (rabbit pyrogen) to the exquisitely sensitive but narrow LAL method. Characteristics to be desired for a new assay may not only test for bacterial endotoxin but also for other potentially deleterious host-active microbial substances. A futuristic test would be more inclusive than LAL (reminiscent of the pyrogen test) and as sensitive and specific as it. Given the recent advances in molecular biology, the successor to the LAL test may be an LAL test using a recombinant LAL product (now available from Cambrex and soon to be from ACC (103)). The recombinant test merely maintains the status quo of LAL testing although without the need to bleed horseshoe crabs.

There are three likely roads that lie ahead with (a) being the expansion of the current LAL path (including the use of recombinant LAL), (b) the supplementation and perhaps eventual replacement of LAL testing with the Whole blood Test*, or (c) either an increased specificity for the detection of endotoxin† as one of several detected artifacts). The LAL assay is almost entirely specific for endotoxin but has been criticized for both its specificity (i.e., can't detect GPB or viral contamination) and its lack of specificity (some preparations are sensitive to β-glucans)). The road toward greater specificity and broader application to other microbial artifacts has been explored in that several methods are applicable to both endotoxin and non-endotoxin pyrogens (i.e., mononuclear cell assays and the use of GC-MS for the detection of multiple markers).

It may prove desirable to screen drug products for as many microbial contaminants as possible simultaneously with a single test—i.e., supplanting sterility, bioburden, indicator organism recovery (microbial purity), fungi (β-glucan), mycoplasma, endotoxin and other microbial by-product detection, such as enterotoxins and superantigens (many of which are not now analytically precluded)) or, more realistically perhaps, one test for living organisms and another relevant microbial artifacts. The justification for such testing would be driven by (a) product-specific (indication-specific) concerns of non-endotoxin artifact contamination, (b) the potency (relative biological activity) of some non-endotoxin modulins, (c) the emerging technology itself, (d) an increase in the likelihood of non-endotoxin contamination given an increase in manufacturing methods sensitive to alternative (non-GNB) contamination, or (e) necessity, in the case that LAL becomes unavailable and would therefore have to be supplanted with a new technology.

The *PDA Journal of Pharmaceutical Science and Technology* technical report No. 33 (104) describes three broad categories of microbiological testing technologies, including (a) viability-based, (b) artifact-based, and (c) nucleic acid–based technologies. Clearly, the concern for endotoxin as a contaminant lies in its occurrence as an artifact. It is the enduring potent biological activity of endotoxin as an artifact coupled with its almost indestructible nature that separates it from other host artifacts and modulins that are both less biologically active and less resistant to inactivation by heat, chemical and other common pharmaceutical manufacturing treatments. Therefore, the

---

* Though a broad assay, the pyrogen test is hardly sensitive enough to be all-inclusive.
† GC-MS detection of β-hydroxymyristic acid.

viability-based and nucleic acid-based technologies can be viewed as much less relevant as proposed tests to any eventual replacement of LAL although they could and do currently find utility in relevant applications such as clinical detection in blood plasma or the examination of complex media used in cell culture. According to the PDA report, artifact-based technologies that may prove relevant to the detection and quantification of microbial constituents include (a) the use of fatty acid profiles (gathered by GC-MS), (b) fluorescence antibody techniques, (c) ELISA (enzyme-linked immunosorbent assay), and (d) latex agglutination (as well as the continued reliance on LAL).

A testimony to the BET test is the lack of adverse events associated with pharmaceutical or medical device contamination since the use of LAL. The difficulty of replacing LAL lies in its extreme ease of use, sensitivity and specificity, which in turn is also a testament to the crab's defense system. Some non-LAL assays have served in some instances to complement the LAL and pyrogen tests and some may hold potential as eventual alternative tests as they have already served as complementary or confirmatory tests to the use of LAL testing in specific applications. See Table 11.

Some non-LAL assays such as GC-Mass spectrometry or polymyxin B binding may achieve a stoichiometric determination of LPS content that is not a measure of the relative biological responsiveness of a given endotoxin. Although this may seem at first glance to be an ideal advantage in providing a truer means of LPS quantitation, it is the biological responsiveness of the LAL test that provides the current basis for regulatory acceptability and is one that is strictly enforced (and historically is the result of much effort to achieve) through the establishment of reference standards, controls standards, LAL standardization, and the relationship of LPS activity to the threshold pyrogenic response in both humans and rabbits. In other words, the biological responsiveness of LPS as a means of quantification will not only not go away, presumably it will have to be correlated to any truly quantitative nonbiological measure (i.e., non-LAL or nonpyrogen method) developed. Specialized immunological tests (some used in conjunction with LAL) have been developed for clinical applicatins such as the detection of endotoxemia and other investigational applications.

The effect of blood plasma on LAL tests has made the quantification of endotoxin in blood inconsistent. See Hurley's paper for a detailed discussion of methods of endotoxemia detection (Table 12).*

---

*Hurley, J.C., Endotoxemia: Methods of Detection and Clinical Correlates, Microbiol. Rev. 1995, 8(2) p. 268–292.

TABLE 11 Endotoxin and Non-Endotoxin Assays for Mcrobial Contaminants

β-Glucan–Insensitive LAL and Endotoxin-Insensitive LAL
- Factor G biosensor contained within the LAL reagent has been removed to create an endotoxin-specific LAL reagent for both gel clot and kinetic assays (105).
- The factor C pathway enzymes have been removed resulting in reagents insensitive to endotoxin and specific to various β-glucans including curdlan, pachyman, laminaran and lichenan. Kitagawa and coworkers reported that the sensitivity toward curdlan was approximately 10^{-10} g/mL (106).

Enzyme-Linked Immunosorbent Assay (ELISA) with Monoclonal Antibody against Limulus Peptide C (107)

GC-Mass spectrometry of 3-hydroxy fatty acids
- The GC-MS method quantitates endotoxin by relating (integrating) the (triangular) area in the marker fatty acid recovered (β-hydroxymyristic) to the areas obtained for standards recovered. There is a commercial effort to apply the technology to endotoxin detection. Microbial ID GC is coupled to a computer database to reference chromatograms for standard (ATCC) organisms as well as a variety of environmental and clinical isolates. Biochemical and GC methods work side by side now in many microbial ID laboratories.
- Clinical researchers correlated meningococcal endotoxin levels (determined by GC-MS) in septic shock patients with LAL results (108). Brandtzaeg et al. concede that the utility of the LAL assay in measuring plasma LPS activity is still debatable and that in most cases not feasible due to the low levels of endotoxins present. Due to the high endotoxin plasma concentrations associated with patients afflicted with the deadly *Neisseria meningitidis* infection, their studies were successful. They identified 3-hydroxy lauric acid (3-OH-12:0), the neisserial lipid A marker not found in Enterobacteriaceae. *Neisseria meningitidis* LPS is potent from an endotoxin perspective due to its active production of excess outer membrane material called "blebs" (109).
- The suspected false-positive endotoxin reactions occurring in LAL assays have been confirmed using GC-MS. Maitra, Nachum, and Pearson used GC-MS to test hemodialyzer rinses containing up to 4800 ng of endotoxin equivalents per mL to reveal that the solutions did not contain any measurable β(OH) C_{12}, C_{14} or C_{16} fatty acids (110). It is incumbent on users claiming that LAL activity is not due to endotoxin (such as with β-glucans) to have an independent method to prove such a contention.
- GC-MS has been used in the clinical determination of other markers present in septic synovial fluid and septic arthritic joints via the identification of levels of GPB markers, namely muramic acid (111) and has been used to screen out background peaks to allow researchers to detect 30 ng/mL (a sensitivity increase of 1000 × over prior attempts). The GC-MS method may be a valuable investigative tool utilizing multiple markers.

Cultured human mononuclear cells followed by Pyrogen testin (Human Leukocytic Pyrogen Test) (112). Cultured human mononuclear cells followed by thymocyte proliferation assay (113)

TABLE 11 Continued

Silkworm larvae plasma test (SLP) detects petidoglycan. A novel mechanism of detecting specific non-LPS microbial components including β-glucan (βG) and peptidoglycan (PG) (contained in GPB and in lesser amounts in GNB) (114–116) is available commercially (117) for experimental purposes.

- In a method reminiscent of the early LAL test, the SLP test uses another primitive blood-based host defense system, namely that of the silkworm larvae (*Bombyx mori*) plasma. Melanin, a black-pigmented protein, serves as a self-defense molecule in insect hemolymph and is the end-product of a cascade reaction utilizing multiple serine proteases called the prophenoloxidase (proPO) cascade (116). Commercialized by Wako Pure Chemical Industries, Ltd. (Osaka, Japan), the test contains all the components of the proPO cascade.
- Used as a supplementary tool in the detection of bacterial meningitis (which was also one of the first clinical applications of the early LAL test (118)). Rapid determination of infection type is critical to the patient's treatment.
- Used to show that peptidoglycan may be a pyrogen concern in dialysate contamination, as per their measurements made on 54 dialysate samples from nine facilities (119).

PCR test for specific fragments of bacterial DNA (that should not be present in parenterals): Dussurget and Roulland-Dussoix, at the Institut Pasteur, amplified DNA fragments of mycoplasmas to act as probes and detected as little as 10 fg of specific mycoplasma contaminant sequences (120).

Recombinant Factor C test. Utilizing the cascade "biosensor" Factor C produced recombinantly, Cambrex has began marketing this as an LAL substitute, albeit an alternative assay due to the fluorescence method of detection. It may find application in biologics that show interference using traditional LAL. It is glucan nonreactive as well.

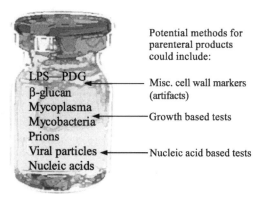

LPS—PDG
β-glucan
Mycoplasma
Mycobacteria
Prions
Viral particles
Nucleic acids

Potential methods for parenteral products could include:

— Misc. cell wall markers (artifacts)

— Growth based tests

— Nucleic acid based tests

FIGURE 9 Ideally, a futuristic test would be both sensitive and specific for as many significant microbial artifacts as possible.

TABLE 12   Microbial Contamination Marker Detection by GC-MS Marker
Indicates

Marker	Indicates Presence of	Non-GC Assays
3-OH fatty acids (lipid A)endotoxin (Gram-negative orgs)		LAL, Pyrogen
β-glucans[a]	yeast & fungi	SLP[b]
ergosterol	yeast & fungi	liquid chromatography
muramic acid	peptidoglycan (gram-positive orgs)	SLP[b]
long chain fatty acids	mycobacteria	acid fast stain
unique lipopeptides	mycoplasma (& other Mollicutes)	broth[c] or agar culture[d], PCR

[a] Detectable by endotoxin insensitive LAL and LC-MS.
[b] SLP = silkworm larvae plasma.
[c] Broth culture: 5% CO_2 up to 6 weeks—sediment and pH change (121).
[d] Agar culture: inverted microscopic observation—"fried egg" appearance (121,122).

15.18.   WHOLE-BLOOD PYROGEN TEST

The concept of an in vitro "human pyrogen test" that utilizes whole blood (and the underlying physiological basis of the fever reaction: the activation of blood monocytes by exogenous pyrogens to produce endogenous pyrogens (cytokines)) has gained support recently with the commission of the Hartung group (University of Konstanz) by the European Commission to investigate the development of such a test with an eye toward eventual compendial inclusion (123–126). The use of isolated monocytes/leukocytes has proved to be highly variable, and therefore Hartung et al. have evaluated tests that employ diluted, fresh, whole blood in a procedure that involves sample incubation and subsequent ELISA detection of immunoreactive mono-phage-secreted cytokines (IL-β, IL-6, and TNF-α). The former two cytokines are largely intracellular as opposed to the latter, which is secreted into the incubated medium (blood) and, therefore, perhaps more amenable to assay. Additionally, IL-6 has been purported to be the principal endogenous percursor to fever and, therefore, the most accurate predictor of the pyrogenic response. Hartung et al. collaborated with the European Centre for the Evaluation of Alternative Methods (ECVAM) beginning in 1999 to propose and perform tests needed to eventually establish such a "human pyrogen test." The test participants summarized their discussions from the ECVAM Workshop 43 (Tables 13 and 14) in ATLA/2001 and claimed a test sensitivity of 0.03–0.1 IU/mL as compared to the BET limit of detection given as 0.03

TABLE 13   Whole Blood Assay (IPT) Claims

Need	Advantage
For non-endotoxin pyrogens	Lists 13 exogenous microbial pyrogen and two exogenous nonmicrobial pyrogen classes (the two nonmicrobial classes are drugs and devices/plastics)
Instances of non-endotoxin contamination	Cites events associated with parenterally manufactured biologicals (most referenced by the group member's own experiences), including immunoglobulins, human serum albumin, hepatitis B vaccine, pertussis vaccine, influenza vaccine, tick-borne-encephalitis vaccine, gentamycin (actually contaminated below or near the limit but given at elevated, off-label dose).
"Comparison of testability"	A range of sample types according to rabbit, LAL, or in vitro pyrogen (IPT) test and lists only recombinant proteins as being questionably tested via the IPT.
"Special problems with biological products..."	Notes that vaccines raise both pyrogen- and LAL-related problems such as when vaccines derived from GNB contain endotoxin as a component, are inherently pyrogenic although LAL nonreactive, or that contain aluminum hydroxide, which interferes with the LAL test; and finally, the fact that many blood products are incompatable with LAL testing.
Medical devices	Adherant pyrogens could be incubated in IPT without the need for elution which is notably inefficient and potentially may affect biocompatibility (i.e. rejection by local inflammatory reaction)
r-DNA used for biologicals	New expression systems (GNB, GPB, fungi, mammalian and insect cells) may be contaminted by expression organisms without LAL detection.

TABLE **14** Materials that Cannot Be Tested with IPT

Drugs that interact with monocytes	IL-1, receptor antagonists, nonphysiological solutions, cytotoxic agents, r-proteins with cytokine activity (i.e., INF-γ), or cytokine detection such as rheumatic factors

IU/mL.* The authors address the "need" for non-endotoxin pyrogen testing in several instances, as shown in Table 13.

Hartung et al. state that the European Pharmacopoeia Commission should examine each monograph individually to determine if replacement of the rabbit pyrogen test requirement should be done by means of LAL or IPT (in vitro pyrogen test). One LAL supplier, Charles Rivers Laboratories (CRL), has marketed a commercial kit for investigational purposes. Some industry debate has begun on the utility of the test and some have called into question the relevance of non-endotoxin pyrogens under any circumstances. Novitsky (Associates of Cape Cod) asserts: "many microbial components once thought to be pyrogens have since been shown to be contaminated with endotoxin. A recent example is lipoteichoic acid (LTA)..."(127). He cites a study by Gao et al. (128) that found contaminating endotoxin in commercial preparations of LTA and another by Morath et al. (129) (that includes Hartung as a co-author) suggesting crude preparations of LTA are not suitable for use as indicators of immune cell activation. However, pointing to the lack of general agreement, Novitsky maintains that β-glucans "represent a clear case of an adulterated (i.e. contaminated product when present in an otherwise cGMP-prepared pharmaceutical drug or device" and suggests differentiating and quantifying such contamination using ACC's glucan-specific LAL products. Elsewhere, he details ACC's current thinking on a particular non-endotoxin "pyrogen": "it has been our policy to treat glucans as 'bioactive' molecules and as 'foreign substances' when present in pharmaceutical preparations" (130).

In the ACC technical report, Novitsky prescribes caution in moving too quickly to IPT and details perceived shortcoming on several fronts:

IPT is not adequately characterized or validated.
There is no valid non-endotoxin pyrogen standard.
The requirement for fresh, whole human blood.

---

* Note that kinetic chromogenic assays can be as sensitive as 0.005 EU/mL.

Variability associated with donor blood in that some contain endotoxin. 12–24 hr incubation for cytokine expression; assay of up to 4 hr for cytokine assay.

Changes in LAL testing probably will not occur until a driving event transpires such as the near extinction of horseshoe crabs on the Atlantic seaboard. If that happens, there will be urgency in looking to cut the use of LAL reagent. In fact, crab populations may have already declined significantly:

Since Hall (of the University of Delaware's Sea Grant College Program) began coordinating an annual springtime census… a decade ago, the number of breeding adults on the shores of Delaware Bay—the center of the species' range and its most important spawning zone—has plummeted from 1.2 million to about 400,000. The main reasons for the decline are the loss of Atlantic beach habitat and—perhaps most significant—the crabs' value as bait for eel and conch fishermen. Though results of this year's census are not yet in, some conservationists already are worried, not just for the crab itself but also for other species, from shorebirds to humans, that depend on this living fossil for their welfare (131).

Tangley goes on to say that the crab's populations have varied sometimes widely in the past, but have always come back. However, the year they don't come back may catch the pharmaceutical industry by surprise, either in the rise in cost of reagents or their lack of availability. Lastly, but perhaps of greatest relevance to parenteral manufacturing in the consideration of potential drivers of change in analytical testing for contamination control, is the exploding knowledge of the interrelationship of microbes, their by-products, and human disease states (see Chapter 1).

Two disease states relevant to such a discussion include systemic fungal infection and sepsis. β-Glucan is a fungal (or cellulosic breakdown) artifact known to the bacterial endotoxin lab due to its LAL reactivity. The substance is not prohibited or excluded by testing from parenteral products and has not been found to be a common contaminant; however, because it is used as a diagnostic marker for systemic fungal infections, it is not hard to envision that those who manufacture parenteral drugs to treat such infections may one day be expected to preclude the possibility of β-glucan contamination. A second, more complex indication and thus a more speculative proposition is the association of minute amounts of non-endotoxin contamination with the occurrence of sepsis. In a similar manner as endotoxin-containing GNB have been correlated with GNB sepsis, GPB have been implicated with GPB sepsis. Indeed, approximately 50% of the instances of sepsis are presumptively caused by GPB infections. What is not known is whether the possibility exists that minute amounts of GPB cellular artifacts introduced from medical devices, infusion solutions, or even parenteral drugs could be relevant contributing factors to this disease state. What is documented is the correlation of the historical rise of sepsis with the use of antibiotics and medical intervention.

REFERENCES

1. Hitchcock, P.J., et al. Lipopolysaccharide nomenclature—past, present, and future. J. Bacteriol. 1986, *166*, 699–701.
2. Mayer, H.; Weckesser, J. The protein component of bacterial endotoxin. In *Handbook of Endotoxin*; Rietschel, E.T., Ed.; Elsevier Science Publishers: NY, 1984; 339 pp.
3. Beverage, T.J. Structures of gram-negative cell walls and their derived membrane vesicles. J. Bacteriol. 1999, *181* (16), 4725–4733.
4. Raetz, C., et al. Gram negative endotoxin: an extraordinary lipid with profound effects on eukaryotic signal transduction. FASEB 1991, *5* (12), 2652–2660.
5. Galanos, C.; Luderitz, O. Lipopolysaccharide: properties of an amphipathic molecule. In *Handbook of Endotoxin, Chemistry of Endotoxin*; Rietschel, E.T., Ed.; Elsevier Science Publishers: NY, 1984; Vol. 1.
6. Rietschel; Brade. Bacterial endotoxins. Sci. Am. 1992, (August), 54–61.
7. Amro, N.A., et al. High-resolution atomic force microscopy studies of the *Escherichia coli* outer membrane: structural basis for permeability. Langmuir 2000, *16* (6), 2789–2796.
8. Kelly, M.T.; Brenner, D.J.; J.J.F. III. Enterobacteriaceae. In *Manual of Clinical Microbiology*; Lennette, E.H., Ed.; American Society for Microbiology: Washington, D.C., 1985; 263–277.
9. Dundas, S.; Todd, W.T.A. *Escherichia coli* O157 and Human Disease. Curr. Opin. Infect. Dis. 1998, *11*, 171–175.
10. Nnalue, A.N. All accessible epitopes in the Salmonella lipopolysaccharide core are associated with branch residues. Infect. Immun. 1999, *67* (2), 998–1003.
11. Rietschel, et al. Bacterial endotoxins: molecular relationship of structure to activity and function. FASEB 1994, *8*, 217–225.
12. Raetz, C. Biochemisty of endotoxins. Ann. Rev. Biochem. 1990, *59*, 129–170.
13. Nowotny, A. Relation of structure to function in bacterial endotoxin. Ann. Rev. Microbiol. 1977, 247–252.
14. Morrison, D.C., et al. Structure-function relationships of bacterial endotoxins, contribution to microbial sepsis. In *Infectious Disease Clinics of North America*; Opal, S.M., Cross, A.S., Eds; Harcourt Brace & Co.: Philadelphia, 1999; 313–340.
15. Reitschel, et al. Chemical structure of lipid A. Microbiology, 1977, 262.

16. Sweadner, et al. Filtration removal of endotoxin (pyrogens) in solution in different states of aggregation. Appl. Envir. Micro. 1977, *34* (4), 285–382.

17. Hochstein, D.H. The LAL test versus the rabbit pyrogen test for endotoxin detection. Pharm. Tech. 1981, *5* (8), 35.

18. Bowman, J.P., et al. Diversity and association of psychrophilic bacteria in Antarctic sea ice. Appl. Environ. Micro. 1997, *63* (8), 217–225.

19. Stanley, J.T.; Gosink, J.J. Poles apart: biodiversity and biogeography of sea ice bacteria. Annu. Rev. Microbiol. 1999, *53* (1), 189–215.

20. Novitsky, T.J. Discovery to commercialization: the blood of the horseshoe crab. Oceanus 1991, *27* (1), 13–18.

21. Dinarello, C.A., et al. New concepts in the pathogenesis of fever. Rev. Infect. Dis. 1988, *10* (1), 168–189.

22. Dinarello, C.A.; Wolff, S.M. Molecular basis of fever in humans. Am. J. Med. 1982, *72*, 799–819.

23. Kluger, M.J., et al. The adaptive value of fever. In *Infectious Disease Clinics of North America*; B.A.C. M.D., Ed.; March 1996; 1–20.

24. Kluger, M.J. Fever: role of pyrogens and cryogens. Physiol. Rev. 1991, *71* (1), 93–127.

25. Horn, D.L., et al. Antibiotics, cytokines, and endotoxin: a complex and evolving relationship in gram-negative sepsis. Scand. J. Infect. Dis. 1996, *101* (9).

26. Greisman; Hornick. Comparative pyrogenic reactivity of rabbit and man to bacterial endotoxin. Proc. Soc. Exp. Biol. Med. 1969.

27. Weary, M.E. Pyrogens and pyrogen testing. In *Encyclopedia of Pharm. Tech.*; Swarbrick, Boyan, Eds.; Marcel Dekker: New York, 1998; 179–205.

28. Braude, et al. Fever from pathogenic fungi. J. Clin. Invest. 1960, *39*, 1266–1276.

29. Wilson, M.; Seymour, R.; Henderson, B. Bacterial perturbation of cytokine networks. Infect. Immunol. 1998, *66* (6), 2401–2409.

30. Henderson, B.; Poole, S.; Wilson, M. Bacterial modulins: a novel class of virulence factors which cause host tissue pathology by inducing cytokine synthesis. Microbiol. Rev. 1996, *60* (2), 316–341.

31. Sriskandan, S.; Cohen, J. Gram-positive sepsis. In *Bacterial Sepsis and Septic Shock*; Opal, S.M.,Cross, A.S., Eds.; W. B. Saunders: Philadelphia, 1999; 397–412.

32. Idanpaan-Heikkila, I., Tuomanen, E. Gram positive organisms and the pathology of sepsis, Chapter 7. In *Sepsis and Multiorgan Failure*; Fein, A.M., et al., Eds; Williams and Wilkins, 1997; 62–73.

33. Fox, A., et al. Absolute identification of muramic acid, at trace levels, in human septic synovial fluids in vivo and absence in aseptic fluids. Infect. Immun. 1996, *64* (9), 3911–3915.

34. van Langevelde, P., et al. Antibiotic-induced release of lipoteichoic acid and peptidoglycan from *Staphylococcus aureus*: quantitative measurements and biological reactivities. Antimicrob. Agents Chemother. 1998, *42* (12), 3072–3078.

35. Hellman, J., et al. Outer membrane protein A, peptidoglycan-associated lipoprotein, and murein lipoprotein are released by *Escherichia coli* bacteria into serum. Infect. Immun. 2000, *68*, 2566–2572.

36. Alberts, B., et al. Chapter 10: Membrane structure. In *Molecular Biology of the Cell*; 3rd Ed.; Garland Publishing, 1994.

37. Akers, J.; Agalloco, J. Sterility and sterility assurance. J. Parent. Sci. Technol. 1997, *51* (2), 72–77.

38. Bruch, C.W. Quality assurance for medical devices. In *Pharmaceutical Dosage Forms: Parenteral Medications*; Avis, K.E. Lieberman, H.A., Lachman, L., Eds; Marcel Dekker: New York, 1993; 487–526.

39. Artiss, D.H. Water, systems validation. In *Valid. Aseptic Pharm. Processes*; Carlton, Agalloco, Eds.; Marcel Dekker: New York, 1998; 207–251.

40. Nema, S.; Washkuhn, R.J.; Brendel, R.J. Excipients and their use in injectable products. PDA J. Pharm. Sci. Technol. 1990, *44* (1).

41. European Pharmacopoeia. *Monograph*; 3rd Ed.; 1997.

42. Croes, R.V. Maltitol solution, mannitol, sorbitol, sorbitol solution, noncrystallizing sorbitol solution-suggested revisions for harmonization of pharmacopeial specifications and procedures. Pharmacopeial Forum 1998, *24* (1).

43. U.S. Dept. of Health and Human Services, FDA Guideline on Validation of the Limulus Amebocyte Lysate Test as an End-Product Test for human and animal Parenteral Drugs, Biological Products, and Medical Devices, Dec. 1987.

44. Williams, K.L. Developing an endotoxin control strategy. Pharm. Tech. 1998, *22* (9), 90–102.

45. Williams, K.L. Developing an endotoxin control strategy for parenteral drug substances and excipients. Pharm. Tech. Asia, Special Issue, 1998, 20–24.

46. Wade, A., Weller, P.J., Eds.; *Handbook of Pharmaceutical Excipients*; American Pharmaceutical Association and The Pharmaceutical Press: Washington and London, 1994.

47. Cleland, J.L., et al. The Stability of Recombinant Human Growth Hormone in Poly(lactic-co-glycolic acid) (PLGA) Microspheres. Pharm. Res. 1997, *14* (4), 420–425.

48. PDA Jour. Sci. Tech. Jan–Feb 1990, *44* (1), 16–18.

49. Cooper, J.F. Formulae for maximum valid dilution. In *Endotoxins and Their Detection with the Limulus Amebocyte Lysate Test*; Watson, S.W., Levin, J., Novitsky, T.J., Eds.; Alan R. Liss: New York, 1982; 55–64.

50. Levin, J. The limulus test and bacterial endotoxins: some perspectives. Endotoxins and Their Detection with the Limulus Amebocyte Lysate Test; Watson, S.W., Levin, J., Novitsky, T.J., Eds; Alan R. Liss: New York, 1982; 7–24.

51. Novitsky, T.J. Selection of the standard. LAL Update 1996, *17* (1), 1–4.

52. World Health Organization. Technical Report Series 1987, *760*, 29.

53. Poole, S.; Das, R.E.G. Report on the Collaborative Study of the Candidate Second International Standard for Endotoxin. Expert Committe on Biological Standardization, 1996.

54. Cooper, J.F.; Levin, J.; Wagner, H.N. New rapid in vitro test for pyrogen in short-lived radiopharmaceuticals. J. Nucl. Med. 1970, *11*, 310.

55. Hochstein, H.D. Review of the Bureau of Biologic's experience with limulus amebocyte lysate and endotoxin. In *Endotoxins and their Detection with the Limulus Amebocyte Lysate Test*; Alan R. Liss: New York, 1982; 141–151.

56. Cooper, J.F.; Hochstein, D.H.; Seligman, E.B. The *limulus* test for endotoxin (pyrogen) in radiopharmaceuticals and biologicals. Bull. Parenteral Drug Assoc. 1972, *26*, 153–162.

57. Cooper, J.F.; Pearson, S.M. Detection of endotoxin in biological products by the *limulus* test. In *International Symposium on Pyrogenicity, Innocuity and Toxicity Test Systems for Biological Products*; Budapest, 1976.

58. Oettgen, H.F., et al. Toxicity of *E. coli* L-asparaginase in man. Cancer 1970, *25*, 253–278.

59. Bang, F.B. A bacterial disease of *Limulus Polyphemus*. Bull. Johns Hopkins Hosp. 1956, *98*, 325–351.

60. Levin, J.; Bang, F.B. The role of endotoxin in the extracellular coagulation of limulus blood. Bull. Johns Hopkins Hosp. 1964, *115*, 265–274.

61. Levin, J.; Bang, F.B. A description of cellular coagulation in the *limulus*. Bull. Johns Hopkins Hosp. 1964, 337–345.

62. Nakamura, S., et al. A clotting enzyme associated with the hemolymph coagulation system of horseshoe crab (*Tachypleus tridentatus*): its purification and characterization. J. Biochem. 1982, *92*, 781.

63. Nakamura, T.; Morita, T.; Iwanaga, S. Lipopolysaccharide-sensitive serine-protease zymogen (factor C) found in Limulus hemocytes. Eur. J. Biochem. 1986, *154*, 511–521.

64. Muta, T.; Iwanaga, S. The role of hemolymph coagulation in innate immunity. Curr. Opin. Immunol. 1996, *8*, 41–47.

65. Iwanaga, S., et al. Hemolymph coagulation system in limulus. In *Microbiology-1985*, Meetings of the American Society for Microbiology; Leive, L., Ed.; *American Society for Microbiology*: Washington, D.C, 1985; 29–32.

66. Levin, J.; Bang, F.B. Clottable protein in Limulus: its localization and kinetics of its coagulation by endotoxin. Thromb. Diath. Haemorrh. 1968, *19*, 186–197.

67. Nachum, R.; Berzofsky, R. Chromogenic limulus amoebocyte lysate assay for rapid detection of gram-negative bacteriuria. J. Clin. Microbiol. 1985, *21* (5), 759–763.

68. Cohen, J.; McConnell, J.S. Observations on the measurement and evaluation of endotoxemia by a quantitative limulus lysate microassay. J. Infect. Dis. 1984, *150* (6), 916–924.

69. Urbaschek, B., et al. Quantification of endotoxin and sample-related inteferences in human plasma and cerebrospinal fluid by using a kinetic limulus amoebocyte lysate microtiter test. In *Endotoxin Detection in Body Fluids*, 39–43.

70. Morita, T., et al. Horseshoe crab (*Tachypleus tridentatus*) clotting enzyme: a new sensitive assay method for bacterial endotoxin. Japan J. Med. Sci. Biol. 1978, *31*, 178.

71. Iwanaga, S., et al. Chromogenic substrates for horseshoe crab clotting enzyme-its application for the assay of bacterial endotoxins. Haemostasis 1978, *7*, 183.
72. Pearson, F.C. III. Pyrogens, endotoxins, LAL testing, and depyrogenation. In *Advances in Parenteral Sciences/2*; Marcel Dekker: New York, 1985.
73. Tsuji, K.; Steindler, K.A.; Harrison, S.J. Limulus amoebocyte lysate assay for detection and quantitation of endotoxin in a small-volume parenteral product. Appl. Environ. Microbiol. 1980, *40* (3), 533–538.
74. Nema, S.; Washkuhn, R.J.; Brendel, R.J. Excipients and their use in injectable products. PDA J. Pharm. Sci. Tech. 1990, *44* (1).
75. Cooper, J. Resolving LAL test interferences. J. Pharm. Sci. Tech. 1990, *44* (1), 13–15.
76. Cooper, J.F. Using validation to reduce LAL pH measurements. LAL Times 1997, *4* (2), 1–3.
77. Motise, P.J. Human drug CGMP notes. PDA Letter, pg. 12, March 1996.
78. Novitsky, T.J.; Schmidt-Gengenbach, J.; Remillard, J.F. Factors affecting recovery of endotoxin adsorbed to container surfaces. J. Pharm. Sci. Tech. 1986, *40* (6), 284–286.
79. McCullough, K.Z. Variability in the LAL Test. J. Pharm. Sci. Tech. 1990, *44* (1), 19–21.
80. Roslansky, P.F.; Dawson, M.E.; Novitsky, T.J. Plastics, endotoxins, and the Limulus amebocyte lysate test. J. Pharm. Sci. Tech. 1991, *45* (2), 83–87.
81. Twohy, C.W., et al. Comparison of Limulus amebocyte lysates from different manufacturers. J. Pharm. Sci. Tech. 1983, *37* (3), 93–96.
82. Feldstine, P., et al. A concept in glassware depyrogenation process validation. Parenter. Drug Assoc. 1979, *33* (3), 125.
83. Banks, H.M. A study of hyperpyrexia reaction following intravenous therapy. Am. J. Clin. Pathol. 1934, *4*, 260.
84. Cherkin, A. Destruction of bacterial endotoxin pyrogenicity by hydrogen peroxide. Biochemistry 1975, *12*, 625.
85. Novitsky, T.J., et al. Depyrogenation by moist heat. PDA Monograph on Depyrogenation, 1985. Technical Report No. 7: 109–116.
86. Welch, H., et al. The thermostability of pyrogens and their removal from penicillin. J. Am. Pharm. Assoc. 1945, *34*, 114.
87. Robertson, J.H.; Gleason, D.; Tsuji, K. Dry-heat destruction of lipopolysaccharide: design and construction of dry-heat destruction apparatus. Appl. Environ. Microbial. 1978, *36*, 705.
88. Tsuji, K.; Lewis, A.R. Dry-heat destruction of lipopolysaccharide: Mathematical approach to process evaluation. Appl. Environ. Microbiol. 1978, *36*, 710.
89. Tsuji, K.; Harrison, S.J. Limulus amebocyte lysate—a means to monitor inactivation of lipopolysaccharide. In *Biomedical applications of the horseshoe crab (Limulidae)*; Cohen, E. Ed.; Alan R. Liss: New York, 1979; 367–378.
90. Groves, F.M.; Groves, M.J. Dry heat sterilization and depyrogenation. In *Encyclopedia of Pharmaceutical Technology*; Swarbrick, J., Boylan, J.C., Eds.; Marcel Dekker: New York, 1991; 447–484.

91. Akers, M.J.; Avis, K.E.; Thompson, B. Validation studies of the Fostoria infrared tunnel sterilizer. J. Parenter. Drug Assoc. 1980, *34*, 330.

92. Akers, M.J.; Ketron, K.M.; Thompson, B.R. F-value requirements for the destruction of endotoxin in the validation of dry heat sterilization/depyrogenation cycles. J. Parenter. Drug Assoc. 1982, *36*, 23.

93. Sweet, B.H., Huxsoll, J.F. Depyrogenation by dry heat. Parenteral Drug Association, Inc., PDA Technical Report No. 7 (Chapter 12), 1985.

94. Bruch, C.W. Quality assurance for medical devices. In *Pharmaceutical Dosage Forms, Parenteral Medications*; Avis, K.E., Lieberman, H.A., Lachman, Eds.; Marcel Dekker: New York, 1993; 487–526.

95. Berger, T.J., et al. Biological indicator comparative analysis in various product formulations and closure sites. PDA J. Pharm. Sci. Tech. 2000, *54* (2), 101–109.

96. Ludwig, J.D.; Avis, K.E. Validation of a heating cell for precisely controlled studies on the thermal destruction of endotoxin in glass. J. Parent. Sci. Tech. 1988, *42* (1), 9–40.

97. Ludwig, J.D.; Avis, K.E. Dry heat inactivation of endotoxin on the surface of glass. J. Parent. Sci. Tech. 1990, *44* (1), 4–12.

98. Sterile Drug Products Produced by Aseptic Processing, CDER, FDA, 1987.

99. Agalloco, J. Points to consider in the validation of equipment cleaning procedures. J. Parent. Sci. Tech. 1992, *46* (5), 163–168.

100. Held, D.D., et al. Endotoxin reduction in macromolecular solutions: two case studies. BioPharm. 1997, 32–37.

101. Wilson, M.J., et al. Removal of tightly bound endotoxin from biological products. J. Biotech. 2001, *88*, 67–75.

102. Brendel-Thimmel, U.; Barenowski, K. Pyrogen removal by adsorption. Pharm. Manufacturing Rev. June 1991, 9–12.

103. Novitsky, T.J. Letter from the president. LAL Update 1997, *15* (2), 1.

104. PDA. Evaluation, validation and implementation of new microbiological testing methods. PDA J. Pharm. Sci. Technol. 2000, *54*, 1–39.

105. Tamura, H., et al. A new test for endotoxin specific assay using recombined Limulus coagulation enzymes. Japan J. Med. Sci. Biol. 1985, *38*, 256–273.

106. Kitagawa, T., et al. Rapid method for preparing a β-glucan-specific sensitive fraction from Limulus (*Tachypleus tridentatus*) amebocyte. J. Chromatography 1991, *567*, 267–273.

107. Zhang, G.H., et al. Sensitive quantitation of endotoxin by enzyme-linked immunosorbant assay with monoclonal antibody against *limulus* peptide C. J. Clin. Microbiol. 1994, *32* (2), 416–422.

108. Brandtzaeg, P., et al. Meningococcal endotoxin in lethal septic shock plasma studied by gas chromatography, mass-spectrometry, ultracentrifugation, and electron microscopy. J. Clin. Invest. 1992, *89*, 816–823.

109. DeVoe, I.W.; Gilchrist, J.E. Release of endotoxin in the form cell wall blebs during in vitro growth of Neisseria meningitidis. J. Exp. Med. 1973, *138*, 1156–1167.

110. Maitra, S.K.; Nachum, R.; Pearson, F.C. Establishment of beta-hydroxy fatty

acids as chemical marker molecules for bacterial endotoxin by gas chromatography-mass spectrometry. Appl. Environ. Microbial. 1986, *52* (3), 510–514.

111. Fox, A., et al. Absolute identification of muramic acid at trace levels, in human septic synovial fluids in vivo and absence in aseptic fluids. Infec. Immun. 1996, *64* (9), 3911–3915.

112. Dinarello, C.A., et al. Human leukocytic pyrogen test for detection of pyrogenic material in growth hormone produced by recombinant *Escherichia coli*. J. Clin. Microbiol. 1984, *20* (3), 323–329.

113. Hansen, E.W.; Christensen, J.D. Comparison of cultured human mononuclear cells, Limulus amebocyte lysate and rabbits in the detection of pyrogens. J. Clin. Pharm. Ther. 1990, *15*, 425–433.

114. Tuchiya, M., et al. Detection of peptidoglycan and β-glucan with silkworm larvae plasma test. FEMS Immunol. Med. Microbiol. 1996, *15*, 129–134.

115. Ashida, M.; Yamazaki, H.I. Molting and metamorphosis. In *Biochemistry of the Phonoloxidase System in Insects: with Special Reference to its Activation*; Onishi, O., Ishizaki, H., Eds; Japan Sci. Soc. Press: Tokyo, 1990; 239–265.

116. Tsuchiya, M., et al. Reactivities of gram-negative bacteria and gram-positive bacteria with Limulus amebocyte lysate and silkworm larvae plasma. J. Endotoxin Res. 1994, *1* (suppl 1), 70.

117. Wako Pure Chemical Industries, L., Package Insert. 1999: Osaka.

118. Kahn, W. New Rapid Test for Diagnosing Bacterial Meningitis. 1996, ASM.

119. Tsuchida, K., et al. Detection of peptidoglycan and endotoxin in dialysate using silkworm larvae plasma and Limulus amebocyte lysate methods. Nephron 1997, *75* (4), 438–443.

120. Dussurget, O.; Roulland-Dussoix, D. Rapid sensitive PCR-based detection of mycoplasmas in simulated samples of animal sera. Appl. Environ. Microbial. 1994, *60* (3), 953–959.

121. Kenny, G.E. Mycoplasmas. In *Manual of Clinical Microbiology*; Lennette, E.H., Ed.; ASM: Washington, D.C, 1985; 407–411.

122. Waris, M.E., et al. Diagnosis of Mycoplasma pneumoniae pneumonia in children. J. Clin. Microbiol. 1998, *36*, 3155–3159.

123. Hartung, T.; Wendel, A. Detection of pyrogens using human whole blood. In Vitro Toxicol. 1996, *9*, 353–359.

124. Hartung, T.; Fennrich, S.; Wendel, A. Detection of endotoxins and other pyrogens by human whole blood. J. Endotoxin Res. 2000, *6*, 184.

125. Hartung, T., et al. Novel pyrogen tests based on the human fever reaction (The Report and Recommendations of ECVAM Workshop 43). ATLA 2001, *29*, 99–123.

126. Hartung, T.; Wendel, A. Detection of pyrogens using human whole blood. In Vitro Toxicol. 1996, *9* (4).

127. Novitsky, T.J. BET vs. PT non-endotoxin pyrogens. LAL Update April 2002, *20* (2).

128. Gao, J.J.; Xue, Q.; Zuvanich, E.G.; Haghi, K.R.; Morrison, D.C. Commercial preparations of lipoteichoic acid contain endotoxin that contributes to activation of mouse macrophages in vitro. Infect. Immun. 2001, *69*, 751–757.

129. Morath, S.; Geyer, A.; Spreitzer, I.; Hermann, C.; Hartung, T. Structural decomposition and heterogeneity of commercial lipoteichoic acid preparations. Infect. Immun. 2002, *70*, 938–944.

130. Novitsky, T. Letter from the president, LAL Update June 2002, *20* (3), Associates of Cape Cod.

131. Tangley, L. The decline of an ancient mariner, A crab's bad tidings for land dwellers. U.S. News Online, 1999.

16

Screening Active Pharmaceutical Ingredients and Excipients for Endotoxin

James F. Cooper
Endotoxin Consulting Services, Greensboro, North Carolina, U.S.A.

16.1. OVERVIEW

Bacterial endotoxin is the most significant pyrogen in the pharmaceutical industry because of its potency and prevalence. Endotoxin was recently implicated in adverse reactions to injectables made by parenteral manufacturers and compounding pharmacies (1–3). Studies of the etiology of these events suggest measures that may prevent future incidents. The compendia prescribe endotoxin limits for finished injectable products, but there are few limits for active pharmaceutical ingredients (APIs) and excipients. This discussion proposes strategies for setting endotoxin specifications and suggests remedies for testing materials that interfere with the Bacterial Endotoxins Test (BET).

16.2. REGULATORY DOCUMENTS FOR THE BET

The FDA's LAL-Test Guideline (1987) was the most influential document to emerge when the pharmaceutical industry converted from the rabbit pyrogen test to the Limulus amebocyte lysate (LAL) reagent test for endotoxin (4,5).

The guide encouraged the industry to take advantage of the new technology by defining requirements for rapid conversion to LAL methods. There was early concern that the new test might miss non-endotoxin pyrogen, but firm evidence to support the existence of non-endotoxin pyrogens or materials-mediated pyrogens in parenteral products did not materialize until recently (6).

The Guideline introduced the concept of the Endotoxin Limit (EL), based on dose, to define a safe level of endotoxin. It also provided formulae for the use of dilution (MVD, maximum valid dilution) or concentration (MVC, minimum valid concentration) to overcome interfering test conditions. It described an assay to qualify analysts and reagents, a validation test to assure the absence of interference factors, and a limit (routine) LAL test to release parenterals by a validated method. Endotoxin test methods are discussed elsewhere in great detail (4,5). Although the Guideline is no longer the principal document for LAL testing, it remains important because procedures in parenteral facilities were written to comply with it. Also, this guide addressed cGMP issues, such as sampling, retests, analyst qualification, and determination of RSE/CSE ratios, that are not found in the compendia.

A harmonized Bacterial Endotoxins Test (HBET) became effective in 2001 that contained sweeping changes (6). This revision was adopted by the International Conference on Harmonization. The LAL-Test Guideline and the new BET are quite similar in requirements for test validation and end-product release. However, the new chapter has simplified procedures, de-scribes all LAL methods and allows for tests that exclude the influence of glucans. The HBET is now the most important regulatory document because it is the minimum standard for LAL testing and harmonizes the BET, globally.

16.3. ENDOTOXIN ALERT LEVEL (EAL) FOR APIs

It is impossible to render materials absolutely pyrogen free because endotoxin is stable, highly potent, and ubiquitous in nature. Therefore, an endotoxin limit (EL) represents the maximum safe amount of endotoxin that is allowed in a dose of a specific parenteral product. When a product contains endotoxin less than its EL, it may be labeled nonpyrogenic. The compendial EL for a product is calculated from the K/M formula where K, the tolerance limit, varies with the type of product and route of administration, as summarized in Table 1; M is the maximum dose in units/kg. The occurrence of aseptic meningitis in patients receiving intrathecal medications led to stringent limits for drugs administered by this route (3). The best sources for product-specific endotoxin limits are drug monographs and chapters in pharmacopoeia.

Only a small number of APIs, such as human insulin and a few antibiotics, have a compendial limit for endotoxin. That leaves the choice of release limits for APIs and excipients to common sense. A strategy for

TABLE 1 Endotoxin Tolerance or Allowable Limit by Type of Parenteral
Material

Parenteral type	Endotoxin tolerance limit (K)
Human or veterinary drugs and biologics	5 EU/kg
Parenterals by intrathecal injection	0.2 EU/kg
Radiopharmaceuticals	175 EU/V[a]
Intrathecal radiopharmaceuticals	14 EU/V[a]
Continuous intraspinal infusion	14 EU/day[a,b]
Large-volume parenterals	0.5 EU/mL
Water for Injection	0.25 EU/mL
Medical devices by extraction	0.5 EU/mL up to 20 EU/device
Medical devices in intrathecal spaces	0.06 EU/mL up to 2.15 EU/device
Multiple ingredient small-volume parenteral	70 EU/V[a,b]
Excipient	3.5 EU per amount in 1 mL of SVP[b]
New chemical entity, preliminary	1 EU/mg[b] until human dose is known

[a] Maximum dose in volume (mL).
[b] Recommended limit by the author, not pharmacopoeia.

assigning endotoxin limits and test methods for noncompendial materials must account for their intended use, origin, and risk for potential endotoxin contamination. High risk for endotoxin is associated with materials that are derived from natural sources or processed in the presence of bioburden. An FDA surveillance study (7) found that 3% of samples had LAL-detectable endotoxin; all were products of natural origin. High risk is also assigned to materials that are intended for injection into a confined site, such as cerebral spinal or intraocular spaces. Low risk is associated with materials that are derived from a synthetic source and are available in pharmaceutical grade.

The assignment of an end-product EL to an API is risky because an additive effect of small amounts of endotoxin from excipients, water, and components might cause a failure. A suitable alert level for an API is a limit that is at least 4 times more stringent than the compendial endotoxin limit. A high endotoxin risk is associated with APIs that are produced by fermentation or recombinant technology, filled by aseptic processing, and intended for intravenous (IV) or intrathecal (IT) administration.

16.4. SYNERGISTIC EFFECT OF ENDOTOXIN WITH OTHER PYROGENS

Thousands of patients were exposed to threshold pyrogenic levels of endotoxin while receiving IV gentamicin therapy in 1998–99 (1). At least 155 pyrogenic reactions were caused by gentamicin produced by two American generic suppliers. The active pharmaceutical ingredient (API) was suspected as the source of the contamination because both suppliers obtained their API from the same third-world vendor. Release tests on the final product by the gentamicin suppliers and retrospective tests by other groups confirmed that the products were released within, but in many cases, near the USP specification. Many patients reacted to doses as low as 2 EU/kg, and in a few cases, reacted to gentamicin lots that had much lower levels of endotoxin by LAL tests.

The need to examine the role of endotoxin and other potential pyrogens led investigators to study gentamicin vials with various tools including kinetic LAL tests, rabbit pyrogen tests, monocyte activation tests (MAT), and assays for peptidoglycan and LAL reactive glucans (LRG) (6). The SLP (silkworm larva) test found no significant levels of peptidoglycan. Tests for LRG were negative. The MAT method in the laboratories of Poole (NIBSC) and Brügger (Novartis) was used to measure cytokine response after isolated peripheral blood mononuclear cells were exposed to gentamicin samples (6). The basis of this test is the exquisite sensitivity of monocytes to exogenous pyrogen. Monocyte activation was measured with IL-6, the most ideal cytokine for this system (8,9). The MAT was positive for all lots of gentamicin that produced pyrogenicity and was negative for batches free of patient reactions. A summary of patient reactions and results of the various pyrogen or endotoxin tests for six representative batches of gentamicin is presented in Table 2.

The root cause of pyrogenic gentamicin was not found. Peptidoglycan from gram-positive microbes was not a factor. However, the FDA investigators cited the API manufacturer with numerous cGMP violations. LAL

TABLE 2 Pyrogenic Reactions and Pyrogen Test Results for Representative Gentamicin Lots

GS/API Lots #	Pyrogenic reactions	Kinetic BET (EU/mg)	MAT NIBSC	MAT Novartis	Rabbit (Δ °C)
G/213	15	0.6	++	++	1.24
D/213	3	0.5	++	+	1.36
I/533	?	0.7	++	++	1.87
F/533	3	0.6	++	++	0.96
E/99	24	<0.25	++	++	NA
A/99	7	<0.25	+	++	NA

GS is IV gentamicin sulfate; MAT is monocyte activation test; + is positive MAT; ++ is highly positive MAT. NA is not available.

endotoxin levels in certain lots related poorly to patient reactions and pyrogenicity by MAT. The lack of full agreement between LAL and test results for pyrogenic gentamicin batches indicated that another pyrogen was present to augment endotoxin activation of the cytokine system. That is, the higher level of impurities in this case may have synergized endotoxicity. The reactions may not have occurred if either endotoxin or the other pyrogen had been absent. A plausible explanation is that unidentified impurities acted synergistically with subpyrogenic levels of endotoxin to induce pyrogenicity in susceptible patients.

These findings suggest that a practice of using finished-product endotoxin limits for the API places a significant number of patients at risk for adverse effects. The following precautions are suggested for an API:

A suitable alert level would correspond to an observed endotoxin level that is out-of-trend with historical data.

A prudent endotoxin action level is 25% of the release limit for an API of a parenteral that is produced by aseptic processing and intended for IV or intrathecal (IT) administration. There is less concern for terminally sterilized drugs, where steam reduces endotoxin and denatures proteins, or for drugs given intramuscularly and subcutaneously, where the risk of pyrogenic reaction is much lower.

The API producer should be audited sufficiently to assure CGMP compliance.

The MAT and SLP tests are indicated should pyrogenicity become problematic for a pre- or post-approval drug. These tests provide a means for detecting and eliminating the pyrogen(s) through process improvement.

The most likely candidates for the MAT are APIs for intravenous and intrathecal drugs that are produced by fermentation or recombinant technology and filled by aseptic processing. These drugs have the highest risk of adventitious impurities that may behave as pyrogens. A MAT is complex, time-consuming, technique dependent and requires a group of human donors. Careful planning and suitable resources are needed to establish a meaningful MAT capability. Poole and Patel recently described a single-plate MAT assay to make the test quicker and simpler (8). The MAT may also be used for complex materials that activate LAL by non-endotoxin pathways.

16.5. ENDOTOXIN LIMITS FOR STERILE PHARMACY COMPOUNDING

Injectables prepared in compounding pharmacies are usually formulated from bulk nonsterile powders, are usually produced individually rather than in batches, and are prepared with oversight from state pharmacy boards. It is difficult to assign an endotoxin limit to compounded injectables because the pharmacy may not know the prescribed dose. There are safeguards that a pharmacy may take to reduce the risk of endotoxin contamination:

Purchase materials from ethical suppliers that provide a Certificate of Analysis (CoA) for purity and endotoxin content, if available.
Screen and qualify incoming lots of drug substances with a validated BET.
Apply aseptic technique and conduct an integrity test on every filter used for membrane sterilization.

Information about valid endotoxin test concentrations is difficult to find. A recent report addressed endotoxin testing of pain medications designed for intraspinal infusion (2). The report gave methods for determining a valid test concentration by gel-clot and kinetic turbidimetric LAL assays. Compounded pain medications were BET-compatible when an individual drug was diluted to 0.5 mg/mL, and the principal drug constituent of mixtures was diluted to 0.25 mg/mL. Table 3 summarizes compatibility and endotoxin limit data. An endotoxin limit of 14 EU/mL was recommended for intraspinal infusion solutions because of the potency of this route of administration and the fact that patient doses seldom exceed 1 mL per day when infused by implanted pump devices.

TABLE 3 Recommended BET Test Concentrations and Safety Data for Intraspinal Infusions Prepared from Bulk Powders

Bulk Powder	LAL-Compatible TC[a] (mg/mL)	EL[b] (EU/mg)	LOD[c] (EU/mg)
Morphine	0.5	0.7	0.12
Baclofen	0.25	7.0	0.3
Bupivacaine	0.5	0.6	0.12
Clonidine	0.25	16.5	0.3
Fentanyl	0.5	14	0.12
Hydromorphone	0.5	1.2	0.12
Morphine mixture[d]	0.25	14 EU/day	0.3

[a] The highest test concentration (TC) that yielded valid recovery of endotoxin positive controls.
[b] EL = The endotoxin limit where 14 EU is divided by the maximum dose per day.
[c] Limit of detection when reagent sensitivity, lambda, is 0.0625 EU/mL. LOD = Lambda/TC.
[d] Morphine mixed with baclofen, bupivacaine, or clonidine.

The procedures described for intraspinal medications are applicable for determining noninterfering BET test conditions, calculating endotoxin limits, and conducting appropriate validation for a broad range of sterile solutions that are compounded in the pharmacy.

16.6. ENDOTOXIN LIMITS FOR EXCIPIENTS

Excipients are essential components of small-volume parenterals (SVP). They serve a variety of functions, including stabilizing, preserving, and buffering. Their concentration varies widely. Mannitol and sodium chloride have both therapeutic and excipient applications. Therefore, an EL calculated for therapeutic use represents a very stringent limit. A calculation for an excipient EL should be based on its use in the industry.

The diversity in the use of excipients makes it a challenge to devise a uniform strategy for selection of limits and test protocols. One could simply set an arbitrary limit or assign limits based on their proportion in an SVP formulation, as proposed by Williams (4). However, excipients have one common attribute to exploit. An SVP is limited to 100 mL; this volume can represent the dose for calculating an endotoxin limit. A compendium of excipients was published that details the range of concentrations for excipients in SVP formulations (10). A uniform way for calculating an excipient endotoxin limit (EL) is proposed that is dependent on the maximum amount of excipient in 100 mL of an SVP:

$$\text{Excipient EL} = \frac{350 \text{ EU(adult endotoxin tolerance limit)}}{\text{Maximum amount of excipient in 100 mL}}$$

$$= \frac{3.5 \text{ EU/mL}}{\text{units/mL}}$$

Table 4 is a list of commonly used excipients as well as a proposed endotoxin alert limit (EAL) and kinetic LAL test parameter for each. The EAL was determined by dividing the tolerance limit by the maximum concentration of an excipient. This number was then divided by 4 and rounded to assure a fourfold margin of safety. The compendial limit is applied for those excipients that also have a therapeutic use, such as mannitol and sodium chloride. A test concentration is provided that is known to be noninterfering with at least one kinetic LAL reagent. Finally, the test sensitivity or limit of detection (LOD) is listed that is derived by dividing lambda by the test concentration. In each case, the LOD is more sensitive (lower value) than the highly conservative EAL calculated by the above formula. The proposed excipient EAL is conservative because the calculation assumes that an SVP is 100 mL, whereas volume of most SVPs is less than

TABLE 4 BET Test Information for Frequently Used Excipients

Pharmaceutical excipient	Concentration[a] (mg/mL)	Endotoxin alert level	LAL test concentration	LOD[b] ($\lambda = 0.05$)
Acetic acid	2–5	0.7 EU/mg	0.1 mg/mL	0.5 EU/mg
Benzyl alcohol	10–30	0.03 EU/mg	2 mg/mL	0.025 EU/mg
Carboxymethylcellulose Na	8	1 EU/mg	1 mg/mL	0.05 EU/mg
Calcium chloride	0.1–1	0.2 EU/mg (USP)	1 mg/mL	0.05 EU/mg
Citric acid	0.1–1	0.5 EU/mg (EP)	0.25 mg/mL	0.2 EU/mg
Dextrose	10–50	10 EU/g (USP)	25 mg/mL	2 EU/g
Disodium EDTA	0.1	0.2 EU/mg (USP)	0.5 mg/mL	0.1 EU/mg
Ethanol	0.1 (v/v)	10 EU/mL	0.05 mL/mL	1 EU/mL
Gelatin	5	0.7 EU/mg	0.5 mg/mL	0.1 EU/mg
Glycerin	150	0.2 EU/mg	1.0 mg/mL	0.05 EU/mg
Glycine	10–24	0.15 EU/mg	2.5 mg/mL	0.02 EU/mg
Hydrochloric acid	Trace	NA	NA	NA
Lactose	10	0.35 EU/mg	1 mg/mL	0.05 EU/mg
Lactic acid	7.5	0.45 EU/mg	2.5 mg/mL	0.02 EU/mg
Magnesium sulfate	100	0.1 EU/mg (USP)	2.5 mg/mL	0.02 EU/mg
Mannitol	100	4 EU/g (EP)	50 mg/mL	1 EU/g
Methylparaben	1.8	1 EU/mg	1 mg/mL	0.05 EU/mg
Phenol	5	0.7 EU/mg	0.25 mg/mL	0.2 EU/mg
Polyethylene glycol	500 (v/v)	0.007 EU/mL	20 mg/mL	0.0025 EU/mg
Polysorbate 80	10	0.1 EU/mg	2.5 mg/mL	0.02 EU/mg
Propylparaben	0.2	4.0 EU/mg	0.5 mg/mL	0.1 EU/mg
Sodium acetate	0.39	2.0 EU/mg	1 mg/mL	0.05 EU/mg
Sodium bisulfite	3.2	0.25 EU/mg	1 mg/mL	0.05 EU/mg
Sodium carbonate	1–33	0.025 EU/mg	2 mg/mL	0.025 EU/mg
Sodium chloride	9	5 EU/g (EP)	10 mg/mL	5 EU/g
Sodium citrate	10–28.5	1.2 EU/mg	2 mg/mL	0.025 EU/mg
Sodium hydroxide	Trace	Depyrogenating	NA	NA
Sodium lactate	10	0.1 EU/mg	1 mg/mL	0.05 EU/mg
Sodium metabisulfite	1–6.6	0.1 EU/mg	1 mg/mL	0.05 EU/mg
Sodium phosphate	1–10	0.1 EU/mg	1 mg/mL	0.05 EU/mg
Sorbitol	48	0.02 EU/mg	5 mg/mL	0.01 EU/mg
Sucrose	50–200	0.004 EU/mg	25 mg/mL	0.002 EU/mg
Thimerosal	0.1	10 EU/mg	0.1 mg/mL	0.5 EU/mg

[a] Excipient concentration source (10).
[b] LOD, Limit of Detection is lambda/TC where $\lambda = 0.05$ EU/mL for a kinetic turbidimetric analysis standard curve of 0.05–5 EU/mL.

10 mL. Test parameters presented in Table 4 were not designed to test an excipient with the greatest sensitivity. Rather, the objective was to propose robust test conditions that were valid with most LAL reagents.

The origin of an excipient is critical for achieving purity. Materials produced from natural sources such as mannitol or sucrose will have LRG and endotoxin as contaminants. In contrast, mannitol produced by electrolytic reduction of mannose or dextrose is free of LRG (LAL reactive glucans). Gelatin is also contaminated with endotoxin to the extent that it may be necessary to screen multiple batches to find one that meets the suggested limit of 0.7 EU/mg. Finally, sodium carboxymethylcellulose is a glucan, so LRG-blocking systems are needed to avoid a false-positive endotoxin result (5).

With a few notable exceptions mentioned above, most excipients are available in a pharmaceutical grade with a CoA for absence of significant endotoxin levels. It is excessive to test all incoming excipients once the reliability of a supplier is established. There is no merit in testing sodium hydroxide pellets that are self-depyrogenating. Sound scientific judgment should be used to establish a meaningful API or excipient BET procedures.

16.7. INTERFERENCE TESTING FOR APIs AND EXCIPIENTS

The validation of BET methods for APIs and excipients is challenging because they are often presented in powder form, have solubility limitations, and may require neutralization (5). A common misconception about pH is that any LAL and sample mixture in the range of pH 6 to 8 is considered noninterfering. Actually, the reaction rate in kinetic BET systems is so pH dependent that recovery of the positive product control (PPC) will be altered if the pH of the LAL reagent and test samples are not within a few tenths of a pH unit. Excipients or APIs that are not pH neutral, such as phenol, acids, and weak bases, may require neutralization with dilute acid or base during the dissolution process, depending on the buffer capacity of the LAL reagent.

Compatibility with LAL reagents is dependent on water solubility. Compounds that are poorly soluble in water may be dissolved in organic solvents that are miscible with water, and then diluted to a suitable test concentration that eliminates solvent interference. Most LAL reagents tolerate up to 5% of ethanol and 2% of dimethyl sulfoxide (DMSO). If a precipitate begins to form in a kinetic BET study, there will be a progressive increase in the optical density that is readily distinguishable from an endotoxin reaction curve; more dilution is indicated in this case.

Endotoxin adsorption problems are often difficult to resolve. An analyst received a sample for qualification from a new API vendor. Even though a validated method was used, PPC recovery was zero. The analysts filtered the sample because it had an uncharacteristic haze. A retest gave normal recovery. It appears that the vendor had failed to filter the API after treating it with silica, a common absorber used to remove impurities.

Finally, it is more efficient and informative to a test new chemical entity (NCE) at a robust, compatible LAL-test concentration than to attempt to develop a test method for an arbitrarily set, interfering test concentration. It is sufficient to test an NCE at a 1 EU/mg until clinical information has progressed to the point that an endotoxin limit calculation becomes realistic.

REFERENCES

1. Fanning, M.M.; Wassel, R.; Piazza-Hepp, T. Pyrogenic reactions in gentamicin therapy. N. Engl. J. Med. 2000, *343*(22), 1658–1659.
2. Cooper, J.F.; Thoma, L.A. Screening extemporaneously compounded intraspinal injections with the bacterial endotoxins test. Am. J. Health Syst. Pharm. 2002, *59*, 2426–2433.
3. Cooper, J.F.; Harbert, J.C. Endotoxin as a cause of aseptic meningitis after radionuclide cisternography. J. Nucl. Med. 1975, *16*, 809–813.
4. Williams, K.L. *Endotoxins: Pyrogens, LAL Testing, and Depyrogenation*, 2nd Ed.; Marcel Dekker: New York, 2001; 193–263.
5. Cooper, J.F. Bacterial endotoxins test. In *Microbiology in Pharmaceutical Manufacturing*; Prince, R., Ed.; Parenteral Drug Assoc: Bethesda, MD, 2001; 537–567.
6. Cooper, J.F.; Brügger, P.; Fanning, M.; Matsuura, S.; Poole, S. Alert and action levels for APIs: A collaborative study of pyrogenic gentamicin. Parenteral Drug Association Annual Meeting; New Orleans, December 2002.
7. Twohy, C.; Duran, A.P.; Munson, T.E. Endotoxin contamination of parenteral drugs as determined by the LAL method. J. Parent. Sci. Tech. 1984, *30*, 190–201.
8. Nakagawa, Y.; Maeda, H.; Murai, T. Evaluation of the in vitro pyrogen test system based on proinflammatory cytokine release from human monocytes: comparison with human whole blood culture test system and with the rabbit pyrogen test. Clin. Diagn. Lab. Immunol. 2002, *9*, 588–597.
9. Poole, S.; Mistry, Y.; Ball, C.; Gaines Das, R.E.; Opie, L.P.; Tucker, G.; Patel, M. A rapid 'one-plate' in vitro test for pyrogens. J. Immunol. Methods 2003, *274*, 209–220.
10. Powell, M.T.; Nguyen, T.; Baloian, L. Compendium of excipients for parenteral formulations. PDA J. Pharm. Sci. Technol. 1998, *52*, 238–311.

17

Viral and Prion Clearance Strategies for Biopharmaceutical Safety

Hazel Aranha
Pall Corporation, New York, U.S.A.

17.1. BIOPHARMACEUTICAL MANUFACTURING: GENERAL CONSIDERATIONS

From the dawn of civilization, the goal of medicine has been to increase longevity while simultaneously enhancing the quality of life. Tribal and tropical medications were the "aspirin" of medieval medicine. The first organized large-scale effort toward the prolongation of life was the use of vaccines to combat infectious diseases. Later, medical intervention included therapeutics derived from mammalian fluids (e.g., plasma-derived coagulation factors or immunoglobulins, hormones derived from human urine, bovine-sourced heparin) and tissues (e.g., human growth hormone [hGH] from the pituitary gland of human cadavers, placenta-derived bovine products such as albumin and collagen). More recently, technologies such as bioinformatics and proteomics created by combining molecular biology techniques with robotics and computers have facilitated the discovery and design of a vast array of biologicals with prophylactic and therapeutic applications.

Vaccinations contributed to the significant decreases in mortality in the late 19th century; however, because the approach was more an art than a science, vaccinations were accompanied, not surprisingly, by occasional catastrophes. Sporadic smallpox infections occurred due to the uncontrolled nature of the source materials; similarly, clinical accidents were traced to the incomplete inactivation of the live attenuated virus used for rabies vaccination. Literature on the benefits and associated hazards of immunization has been reviewed by Wilson (1).

The outbreak of diphtheria following administration of equine-derived diphtheria antitoxin contaminated with tetanus provided the impetus for passage of the Biologics Control Act, in 1902. With close scrutiny of manufacturing operations and adherence to regulations, the safety record of biologicals improved considerably but was not without incident. Incomplete inactivation of the viral immunogen was the causative factor in incidents such as the Cutter incident with polio vaccine (2), Fortaleza accident in Brazil with rabies vaccine (3), and the foot-and-mouth disease (FMD) vaccine (4). More recently, administration of contaminated blood and plasma products has resulted, unfortunately, in viral transmission. Table 1 presents some instances of historical and recent iatrogenic accidents (5–17).

The considerable improvement in the safety record of biologicals is a reflection of the significant scientific strides combined with rigorous and effective process controls. Now, at the beginning of the 21st century, the blood

TABLE 1 Iatrogenic Accidents Associated with the Administration of Biologicals

Source material/production cell line associated

Infected blood donors: transmission of HIV (5,6), hepatitis A, B, C, viruses
    (6–9), parvovirus B19 (10–12)
Cadaver-sourced pituitary glands: Creutzfeld-Jakob disease
    transmission (13)
Infected cell line: SV40 contamination of cell line used for polio
    manufacture (14)

Manufacturing process associated

Incomplete virus inactivation: foot-and-mouth disease (4), polio (2),
    rabies (3)
Incomplete prion inactivation; scrapie transmission with louping-ill
    vaccine (15)
Inadequate virus clearance during manufacturing; low levels of HCV
    present (10)
Contaminated excipients: HAV, B19 transmission via HSA used in
    recombinant coagulation factor; HBV transmission by yellow fever vaccine
    (1,16,17)

supply in the United States is among the safest in the world. This has been facilitated by improvements in donor eligibility criteria and implementation of progressively more sensitive HBV, HIV, HCV, and HTLV screening assays. Nonetheless, our blood supply remains vulnerable to new or reemerging infections. Transmission of viruses such as GBV-C/hepatitis G virus (18,19) and TTV (20,21), has been documented; fortunately, look-back studies have demonstrated no adverse effect on the recipient. The recent reports of West Nile virus transmission via blood and organ donations offer additional corroboration of our vulnerability to viral agents (22,23).

Biopharmaceuticals, in general, have had an excellent virological safety record. Currently, however, it is impossible to establish "absolute" virological safety. Several factors account for this reality. Viral assays lack the sensitivity to detect titers that although low, may be of medical concern. Most viral detection assays are highly specific; consequently, new or reemerging viruses or viral variants may go undetected. Direct testing for the absence of viral contamination from a finished product is not considered sensitive enough for detection of low levels of virus. Multiple approaches are therefore used to minimize and manage virus contamination risks.

This chapter addresses issues related to viral and prion safety of biologicals and biopharmaceuticals. The focus is on safety assurance of continuous cell line-derived (CCL) products, but the safety and procedural considerations in the case of plasma-derived products will also be addressed.

17.2. SOURCES OF VIRAL CONTAMINANTS

Biopharmaceutical processes utilize myriad biologically sourced raw materials, starting with the cell line used and extending to the various manipulations and supplementation undertaken during the production and purification stages. Some of these processes overtly contribute to the potential viral load; others may be less obvious. Thus, for example, the cell line used is a potential source of viral contaminants. Continuous cell lines (CCLs) are extensively characterized, and, consequently, any viral contaminant associated with the CCL will not be cytolytic. Chronic or latent viruses, however, may be present. Endogenous retroviruses and retrovirus-like particles are associated with some CCLs; they are noninfectious but pose a theoretical safety concern. The putative risk stems from their resemblance in morphological, biochemical, or biological terms to tumorigenic retroviruses. The widespread use of murine cell lines in the manufacture of monoclonal antibodies has accentuated the importance of rodent zoonotic agents in viral transmission. Chinese hamster ovary (CHO) cells, a cell line frequently used in monoclonal antibody production, may harbor contaminants such as hantavirus. Monoclonal antibodies produced in human/humanized (human/mouse) cell lines are preferred from an immunological standpoint; however, due to the absence of a species barrier they raise unique viral safety considerations. Humanized cell lines are derived from human B lymphocytes, which can harbor several viruses, including retroviruses, hepatitis viruses, human herpesviruses, cytomegalovirus and human papilloma virus; additionally, cell line establishment or cell transformation is achieved using certain viral agents (e.g., Epstein-Barr virus or Sendai virus). All these contribute to the viral load and the impact of these agents must be assessed.

Viral contaminants may also be introduced adventitiously via the additives used/manipulations undertaken in production. Cell lines are often cultivated in serum-supplemented media (5–10% serum) or reduced serum media (2–4%). Bovine viral diarrhea virus (BVDV) has been identified as the most common contaminant of bovine serum. Other possible contaminants include reovirus, infectious bovine rhinotracheitis virus (IBR), parainfluenza-3 virus (PI-3), bovine leukemia virus, and bovine polyoma virus. Porcine parvovirus is reportedly a common contaminant in preparations of porcine trypsin used for the preparation of cell cultures (24). Serum-free media is the growth medium of choice, but it must be noted that serum-free medium and mammalian supplement–free medium are not synonymous; often media are supplemented with mammalian-derived proteins such as insulin and transferrin. Similarly, a chemically defined medium may be supplemented with recombinant growth factors produced in a serum-supplemented system. A preparation designated protein-free may not contain protein but may contain filtered protein hydrolysates.

TABLE 2 Potential Sources of Adventitious Viral Contamination

Source of viral contamination	Examples of viruses
Virus used for induction of expression of specific genes encoding a desired protein	Epstein-Barr Virus, Sendai virus, other inducing agents
Reagents/additives used during production (e.g., serum, culture media, trypsin, growth factors and other supplements)	Bovine viral diarrhea virus, infectious rhinotracheitis virus, parainfluenza 3
Reagents used during purification (e.g., affinity columns [MAb] for purification)	Viruses from monoclonal antibodies/unknown viruses from large animal polyclonal antibodies
Exicipients used during formulation (e.g., serum)	Human viruses such as hepatitis B
Manufacturing facility/personnel	Rhinovirus, respiratory syncytial virus, rotaviruses

Purification processes may also contribute to the viral load. For example, affinity chromatography using monoclonal antibodies as ligands increases the potential for adventitious virus introduction into the product. Other ancillary sources of viral burden include breach of GMP and consequent virus introduction from manufacturing environments or personnel; these viral contaminants would not be removed by conventional "sterilizing-grade" filters, which are intended for removal of bacterial and microbial contaminants other than viruses. The potential for viral contamination for each of the manufacturing unit operations must be evaluated and its impact on the viral load assessed. Table 2 lists some of the potential sources of adventitious viral contamination.

17.3. VIRUS DETECTION METHODS

In order to assess the effectiveness of the viral clearance process, the ability to quantitate the amount of virus is essential. Virus assays should have adequate sensitivity and reproducibility and should be performed with sufficient replicates and controls to ensure adequate statistical accuracy of the test results. Commonly used detection methods and their advantages and limitations have been summarized elsewhere (25,26).

The effectiveness of any viral clearance method is determined by comparing the virus concentration prior to treatment to the concentration of virus in the sample post treatment. The methods of choice for viral detection in process validation (clearance evaluation) studies and routine monitoring are infectivity and polymerase chain reaction (PCR) assays.

Infectivity assays are the gold standard and essentially involve inoculation of susceptible cell lines with the specific virus, followed by monitoring and observation of cytopathic effects (CPE) (e.g., formation of plaques, focus-forming units or induction of abnormal cellular morphology) as a consequence of the infection. The two types of in vitro infectivity assays commonly used to estimate viral titer are the plaque-(or focus-) forming assay and the 50% tissue culture infectious dose (TCID50) assays. The plaque-forming assay offers extreme sensitivity and is especially useful when the virus is present at extremely low titers. The test essentially involves plating (usually in triplicate) of small volumes of dilutions of the test sample. Following the incubation period, the plaque-forming units (PFU) are scored and the plaque-forming units per milliliter (PFU/mL) for a given sample is calculated. $TCID_{50}$ is defined as that dilution of virus required to infect 50% of a given batch of inoculated cell cultures. The wells in a multititer plate are inoculated with the test sample serially diluted to end point. Post incubation, the wells are scored as positive or negative for a specific CPE (associated with the test virus); statistical analysis is required to determine the dilution of virus that causes CPE in 50% of the inoculated cells. Both the plaque assay and the TCID$_{50}$ assay have been extensively validated for use in process clearance evaluation studies. Although infectivity-based assays are favored due to their extreme sensitivity and specificity, the requirement for a different assay system for each virus (due to the cell culture–specific infectivity) makes biological assays cumbersome.

Molecular probes such as hybridization assays or polymerase chain reaction (PCR) assays are being increasingly used because of their specificity and the rapidity of the results. These methods, in general, detect the presence of nucleic acid (DNA/RNA) but cannot differentiate between infectious or noninfectious particles. Additionally, the method is applicable only when the genomic sequence of the virus is known, as in the case of retroviral genomes. PCR is especially relevant either if the viral agent cannot be grown in vitro (e.g., type A retroviral particles) or for viruses such as hepatitis B and C where there are severe limitations to culturing the agents in vitro. Methods such as polymerase chain reaction (PCR) have provided enhanced assay sensitivity, but a negative PCR result does not prove unequivocally that the preparation is totally free of virions (infectious or not), due to the effect of sample size and assay sensitivity (27). As regards nucleic acid amplification tests (NAT), the Committee for Proprietary Medicinal Products (CPMP), the European Agency for the Evaluation of Medicinal Products (EMEA) cautions, "Validation and standardization of these assays must be unambiguously demonstrated before they are acceptable and extreme caution used in the interpretation of both positive and negative results" (28).

Morphological assays such as electron microscopy (EM), although of limited value to assay viral load in fluids, are used for the examination of the viral load in the production cell line. EM is especially relevant to estimate viral load in cell lines containing noninfectious particles, such as the type A retroviral particles, which are present in several rodent cell lines used in biotherapeutics production.

Biochemical assays such as reverse transcriptase (RT) assays, radiolabel incorporation into nucleic acids, radioimmunoassays, immunofluorescence, and Western blots are also used for virus detection. However, these tests are semiquantitative; also, they detect enzymes with optimal activity under the test conditions and their interpretation may be difficult due to the presence of cellular enzymes or other background material.

Other available viral detection methods are employed as the circumstance warrants. For example, among other tests required for characterization of the murine hybridoma or other rodent lines to establish the master cell bank (MCB), the antibody production test is used. Mouse antibody production (MAP) tests, hamster antibody production (HAP) tests, and rat antibody production (RAP) tests allow detection of viruses that may be associated with the cell line and have the potential for infecting humans and other primates.

It is important to recognize that although positive results are meaningful, negative results are ambiguous. This is because it is not possible to determine whether the negative result reflects inadequate sensitivity of the test for the specific virus, selection of a test system (host) with too narrow a specificity, poor assay precision, limited sample size, or basically, just absence of virus. This is highlighted in cases where limited sensitivity of the screening methods combined with masking of presence of infectious virus by neutralizing antibody in the plasma sample pool have resulted in iatrogenic viral transmission via contaminated plasma products.

17.4. REGULATORY CONSIDERATIONS: A RISK-BASED APPROACH

Since the days when Pasteur introduced attenuation of a pathogenic strain of rabies to induce a protective specific host response, the risk in the administration of biologicals has been inherent, implied, and accepted. Clinical acceptability of biologicals and biopharmaceuticals, must, of necessity, be guided by risk-benefit analysis. Risk assessment involves process analysis to identify sources of risk and their consequence. In view of the unique considerations associated with viral contaminants (i.e., actual versus theoretical presence) and the limitations in the assay methodologies (inability to establish absolute absence of viral presence), regulatory agencies emphasize a holistic approach directed at risk minimization, which, when combined with process monitoring, constitutes an appropriate risk management program.

All guidelines and regulatory documents distinguish between well-characterized biologicals (where viral contamination is often a theoretical concern) and traditional products such as blood derivatives where there is a significant potential for viral presence (e.g., parvovirus B19, hepatitis viruses, HIV). Thus, for example, low levels of infectious virus in plasma products are prohibited and any virus-contaminated source material would be immediately quarantined. However, in the biotechnology industry, cell lines such as Chinese hamster ovary (CHO) cells containing endogenous retrovirus, at levels of 10^6–10^9 particles/mL (as visualized by electron microscopy), are deemed acceptable because the particles are noninfectious and pose primarily a theoretical safety concern.

In general, the major factors influencing the viral safety of biologicals are the following: (a) the species of origin of the starting material (i.e., nonhuman viruses are less likely to initiate infection in humans due to species specificity of these viruses; the species barrier, however, is not absolute), (b) the degree of source variability of starting material (e.g., human plasma–derived products, which are manufactured from pooled donations, versus cell culture products derived from a well-characterized master cell bank) and the possibility of testing the source material for the presence of viral contaminants (feasible for blood donation but not feasible for animal-derived products), (c) the purification and processing steps and their capacity for viral burden reduction, and (d) the existence of specific steps for viral clearance included in the process.

The current risk minimization strategy to guard against inadvertent virus exposure of patients treated with a biological appears to be a combination of three efforts: (a) prevention of access of virus by screening of starting materials (cell banks, tissues, or biological fluids) and raw materials/supplements used in production processes (culture media, serum supplements, transferrin, etc.), (b) monitoring production using a relevant screening assay (adventitious virus testing), and (c) a general evaluation of the manufacturing process, for inactivation and removal of viruses (documentation of viral clearance). Engineering and procedural control over facilities, equipment, and operations are an important component of the safety paradigm.

17.5. HOW MUCH VIRAL CLEARANCE IS 'ENOUGH'?

While the necessity for risk assessment and incorporation of not merely adequate but excess virus clearance capacity is acknowledged, the amount of excess capacity required has not been clearly defined. It has been suggested, "the overall viral reduction should be greater than the maximum possible virus titer which could potentially occur in the source material" (28). The recommendations do not provide specifics with regard to the extent of excess viral clearance that would constitute an acceptable safety margin. The general consensus is that processes must be validated to remove or inactivate 3–5 orders of magnitude more virus than is estimated to be present in the starting material (29). For products derived from CCLs known to harbor endogenous retroviruses, it is necessary to determine the theoretical viral burden per dose equivalent of the biological product and incorporate an appropriate safety margin.

A key factor affecting the overall process clearance factors required for a product is the amount required to produce a single dose of product. The required level of clearance is assessed in relation to the perceived hazard to the target population and is guided by risk benefit analysis. For example, CHO cell lines containing endogenous retroviruses are deemed acceptable if the manufacturing process can be demonstrated to provide adequate retrovirus clearance. The clearance goal is usually chosen based on the product use and the risk to the patient population. The extent of product testing necessary will depend on the source and nature of the product, the stage of product development, and the clinical indication.

Risk calculations to determine retroviral load per dose are shown in Table 3. This example assumes a one-time dose of 1200 mg to the patient. To

TABLE 3   Risk Calculation to Determine the Viral Load per Dose

Retrovirus-like particles/mL:	1.62×10^7 particles/mL
Antibody titer:	0.274 mg/mL
Weight of average person:	80 kg
Dose per mass:	15 mg/kg
One dose:	1200 mg
Viral clearance factor:	Unknown

The total amount of retrovirus-like particles in one dose $= [(1.62 \times 10^7$ particles/mL$)(1200$ mg/dose $\div 0.274$ mg/mL$)] \div 10^{-6}$ particles/dose, $= 7.09 \times 10^{16}$ or 16.85 logs minimum clearance required to achieve a clearance of 1 particle per million doses.

achieve a conservative goal of a probability of a viral contamination event of 1 particle per million doses of product, and assuming a retroviral load of 1.62×10^7 particles/mL in the start material, the purification process for this product would have to demonstrate a minimum log clearance of 16.85 logs to achieve the stated goal of 1 viral particle/10^6 doses.

Biopharmaceutical safety is the result of multiple orthogonal barriers operating in concert. Each approach, individually, may have limitations, yet their use in an integrated manner provides overlapping and complementary levels of protection from putative viruses to recipients of recombinant and monoclonal products. To quote the FDA document: "Confidence that infectious virus is absent from the final product will in many instances not be derived solely from direct testing for their presence but also from a demonstration that the purification regimen is capable of removing and/or inactivation of the viruses."

Multiple orthogonal approaches for virus removal and inactivation are more effective than single steps (30). Similar recommendations have also been made in Europe by the Committee for Proprietary Medicinal Products (CPMP), in the "Note for Guidance on Plasma-Derived Medicinal Products" (31).

17.6.   VIRAL CLEARANCE METHODS

17.6.1.   General Considerations

An ideal clearance method should be robust and have a broad spectrum of clearance (either through inactivation or removal) of viruses (both enveloped and nonenveloped), concomitant with high product recovery. The method should be minimally invasive and noncontaminating: it should not involve addition of stabilizers or other additives that must be removed post treatment and should not alter the biological integrity or reactivity of the product. The mode of action should be well characterized and the method should be scalable and amenable to process validation (clearance evaluation).

Clearance efficiency is evaluated in terms of the $\log_{10}$ titer reduction (LTR), which is the ratio of the viral concentration per unit volume in the pretreatment suspension to the concentration per unit volume in the post-treatment suspension. The unique considerations associated with documentation of viral safety preclude stipulation of a particular test virus for evaluation (validation) or specification of the exact virus load to be used in viral clearance evaluation experiments. Clearance studies are conducted to evaluate the number of logs of viral clearance obtained as opposed to complete absence of virus. In general, any clearance less than 1 log may be due to assay considerations and is not considered significant (30).

17.6.2. Viral Clearance Methods Discussion

Viral clearance may be achieved as a consequence of routine processing, and purification operations or strategies specifically aimed at viral clearance may be incorporated into the manufacturing process. Serendipitous (or fortuitous) viral clearance methods are operations, which are part of the product purification process that offer the added bonus of viral clearance. Methods commonly used in the purification of biopharmaceutical products (clarification, centrifugation, extraction, precipitation and filtration; and affinity, ion-exchange, gel-filtration, hydrophobic interaction, and mixed-mode exchange chromatography) may physically separate virus particles from the product (virus removal) based on size, charge, density, binding affinities, and other differences between the virus and the product. Viral inactivation may occur as a consequence of pH effects during processing (32), use of low pH buffers for elution of proteins from chromatography columns (33), and, inactivation by reagents used in the purification process (e.g., use of acidified solutions of potassium isothiocyanate [KSCN] during the purification of interferon alfa) (34). Similarly, conditions of imunoglobulin manufacture have been reported to inactivate viruses.

Depending on the mode of clearance, virus clearance methodologies are classified as virus removal strategies, which aim at (mechanical) reduction of viral numbers, or virus inactivation methods, in which the objective is irreversible loss of viral infectivity. A multipart comprehensive review by Sofer entitled "Virus inactivation in the 1990s- and into the 21st century" summarizes available information on virus inactivation methods (35–37).

Virus inactivation steps must not compromise a product's stability, potency, biochemistry, or biological activity. The inactivation strategy used will be dictated by the following considerations: lability of the virus, the stability of the biological preparation, and the effect on other components in the preparation (38). Inactivation methods are very effective in decreasing the viral burden; however, there are limitations. Heat treatment can denature certain proteins (39). Stabilizers (sometimes added during inactivation by heat or solvent-detergent, to ensure that the biological activity of the active moiety is not compromised) may be protective, not just to the target protein but to the virus as well. Furthermore, in addition to protein denaturation and the consequent loss of biological activity, the viral inactivation method has the potential to affect functionality (40) or alter/increase the antigenicity of both the active ingredient and other proteins in the product. For example, there is a possibility of production of neo-antigens and thus the induction of antibodies/inhibitors in the recipient (41); also, enhanced thrombogenicity, as a consequence of viral inactivation methods has been reported with Factor IX preparations (42).

One of the important considerations to be addressed in virus inactivation experiments involves evaluation of the kinetics of virus inactivation. This is important because virus inactivation is rarely linear and a persistent fraction can exist that is more resistant to inactivation than the majority of the virus population. Certain process parameters may critically impact viral clearance. Savage et al. (43) reported a minimum threshold moisture level requirement for efficient virus inactivation to occur during dry heat treatment of freeze-dried coagulation factor concentrates; similarly, presence of cations have been demonstrated to contribute to the thermostability of viruses (44).

Viral inactivation methods include heat (43,45), photochemical inactivation (46,47), irradiation (48), inactivation with chemicals such a β-propiolactone and caprylate (49,50) and solvent-detergent inactivation (51,52). Viral removal methods include filtration (53–59), chromatographic separation (60,61), and partitioning into a different fraction (62). Combinations of methods, as, for example, UV irradiation/β-propiolactone (63), and use of psoralens in combination with long-wavelength ultraviolet light (UV-A) (47) have also been evaluated. Additional innovative technologies are constantly being developed, as, for example, inactivation with compounds such as biosurfactants (64) and Inactines™ (V.I. Technologies, Watertown, MA/Pall Corporation, NY)(65) and application of pressure cycling technology (66) for virus inactivation.

Although conventional methods such as heat/solvent-detergent inactivation and viral reduction by fractionation, filtration, and chromatography have been demonstrated and documented over the past several decades to be highly effective in decreasing the viral burden, the efficacy and applicability of all viral clearance procedures has to be demonstrated by conducting appropriate process evaluation (validation) studies.

17.6.3. Process Validation for Viral Clearance

Process validation constitutes an integral part of any manufacturing process and is viewed as a cGMP activity. However, for biotech products, it represents both a regulatory and a compliance activity. The US FDA Center for Drugs and Biologics and Center for Devices and Radiological Health, in its "Guideline on the General Principles of Process Validation" (1987), defines process validation as "establishing documented evidence ... that a specific process will consistently produce a product meeting its pre-determined specifications and quality attributes"(67). Viral clearance validation studies do not meet the above definition to the letter in terms of conforming to a "pre-determined" specification; nonetheless, the objective of the clearance study is documentation of product quality and process specificity in terms of viral safety assurance. These studies are also sometimes referred to as qualification studies or clearance evaluation studies, and these terms are used synonymously with virus validation in this chapter.

17.6.4. Process Analysis and Evaluation of Processes to Validate for Viral Clearance

Strategic decisions with regard to viral inactivation and clearance need to be made at the very beginning of product development when the design of the clinical manufacturing process begins to take shape. The first steps in the validation process involve a critical analysis of the bioprocess to determine likely sources of viral contamination (including pathogenic potential of these contaminants) and process characterization to identify which steps in the manufacturing process have the potential for viral clearance.

Each process step to be tested for viral clearance should be evaluated for the mechanism by which virus clearance occurs (i.e., whether it is by inactivation, removal, or a combination of both). A "robust" step is one in which the viral clearance (inactivation/removal) effectiveness is widely independent of variability in production parameters (68). Both serendipitous methods (those routinely used in the manufacturing process and which have coincidental viral clearance capability—e.g., chromatography and low pH-buffer elution steps) and methods deliberately incorporated for the precise purpose of viral clearance (e.g., filtration and heat inactivation) are usually validated.

Regulatory guidelines (30,31) recommend the incorporation of multiple orthogonal methods for viral clearance—methods that have independent (unrelated) clearance mechanisms. One misconception is that an entire manufacturing process that may include, for example, ion-exchange chromatography, pH inactivation, and detergent inactivation can be tested by challenging with a large spike of virus during the first step and sampling during subsequent steps. Logistically, this is impossible for two reasons: (a) based on the product and possible contaminants, most processes require demonstration of more than 12–15 logs of clearance for individual viruses; it is not possible to grow mammalian viruses to such high concentration, and (b) using a low viral challenge level will result in an initial low viral load, with each successive step in the bioprocess being challenged with fewer viral particles (assuming the previous steps are effective at inactivation/removal of viruses). This study design would also restrict the number of viral clearance steps that can be claimed and reduce the overall claim that can be established for the entire process. The best compromise is to evaluate each of the individual orthogonal steps separately and then sum the amount of clearance obtained for the entire process. Although this method may have some limitations and introduce errors due to overestimation of clearance, it is the only practical approach to a complex problem.

17.6.5. Viral Clearance Studies: Scaling Considerations and Identification of Critical Parameters

Ideally, process validation should be conducted at pilot or full scale; however, of necessity, viral clearance validation studies are conducted on scaled-down models. When evaluating viral clearance strategies, the equivalence of scalability from bench scale to manufacturing scale and vice versa must be demonstrated. Depending on the process, critical operating parameters (e.g., volume, flow rates, contact time and product and/or contaminant load) should be conserved. The composition of the test material should be similar in terms of protein concentration, pH, ionic strength, and so forth; product generated by the large- and small-scale processes should be similar in terms of purity, potency, and yield.

Regulatory guidelines recommend use of virus validation data to set in process limits in critical process parameters (28). Most validation studies are conducted at both process extremes; viral studies, being costly and time consuming, preclude this approach. Instead, testing is performed under "worst case" conditions to demonstrate the minimum clearance a step can provide. Worst case conditions will vary depending on the method and are determined by those factors that influence the clearance mechanism (69). In the case of filtration studies, depending on the filtration mode (direct flow/tangential flow), variables include composition of the solution to be filtered (nature of the target protein, protein size, conformation, concentration; impurity/particulate load), as well as solution characteristics (e.g., pH and ionic strength), pressure, flux, temperature, and processing time requirements (70). Variables in inactivation studies include exposure time, temperature, product concentration, and presence/absence of contaminant protein, volumes, flow rates and container equivalence. General considerations to be borne in mind are the necessity to ensure sample homogeneity prior to the treatment strategy, use of calibrated equipment (e.g., timers, chart recorders), and equipment qualification.

17.7. TECHNICAL ASPECTS OF STUDY DESIGN

17.7.1. Choice of Panel of Test Viruses

There is no single indicator virus to be employed for virus validation studies. Choice of the appropriate panel of viruses to use will depend on the source material (plasma-derived biologicals versus cell line–derived) and on the product phase at which viral clearance testing is conducted. In general, the panel of test viruses used should include relevant viruses (i.e., known/suspected viral contaminants), and model viruses. Relevant viruses are, for example, HIV and hepatitis B and C viruses, which are known contaminants of blood products. Some relevant viruses (e.g., hepatitis B and C viruses) are difficult to propagate in vitro; in these cases specific model viruses may be used. Specific model viruses are viruses that resemble known viral contaminants; for example, murine leukemia virus (MuLV) is often used as a specific model for noninfectious endogenous retroviruses associated with rodent cell lines. Additionally, nonspecific model viruses are also included in the test panel to characterize the theoretical clearance capability of the manufacturing process (i.e., assess the "robustnesss" of the process). These include viruses of different size and varied physicochemical and biophysical characteristics; they are not expected to be associated with the product but are included to address theoretical safety concerns and add confidence that the process can handle unknown or undetected viruses. Examples of viruses that have been used in virus validation studies are provided in Table 4.

In some cases, in view of the cost-prohibitiveness of an entire virus validation package, preliminary testing with surrogates such as bacteriophages can be undertaken. Such testing is, of course, relevant only if removal is size-based, as in filtration; if clearance is dependent on a particular physicochemical or other surface characteristic of the virus, it cannot be used. The applicability of bacteriophages as surrogates for mammalian viruses in filter validation studies has been discussed by Aranha-Creado and Brandwein (71).

Different phases of product may require different approaches concerning virus choice. Prior to Phase I, clearance of known viral contaminants (HIV in the case of plasma-derived products) or specific model viruses is usually assessed. At the Phase II/III level, viral clearance studies should include both specific model and nonspecific model viruses. The entire virus panel evaluated for Phase II/III products should also be evaluated again if

TABLE 4 Viruses Commonly Used in Viral Clearance Evaluation Studies

Virus	Family (-viridae)	Genome	Envelope	Size (nm)	Shape
Vesicular stomatitis virus	Rhabdo	RNA	Yes	70 × 175	Bullet
Parainfluenza virus	Paramyxo	RNA	Yes	100–200 nm +	Pleomorphic/ spherical
Pseudorabies virus	Herpes	DNA	Yes	120–200	Spherical
Herpes simplex virus	Herpes	DNA	Yes	120–200	Spherical
Human immunodeficiency virus (HIV)	Retro	RNA	Yes	80–100	Spherical
Murine leukemia virus (MuLV)	Retro	RNA	Yes	80–110	Spherical
Reovirus 3	Reo	RNA	No	60–80	Spherical
Sindbis virus	Toga	RNA	Yes	60–70	Spherical
SV40	Papova	DNA	No	40–50	Icosahedral
Bovine viral diarrhea virus (BVDV)	Toga	RNA	Yes	50–70	Pleomorphic/ spherical
Encephalomyocarditis virus	Picorna	RNA	No	25–30	Icosahedral
Poliovirus	Picorna	RNA	No	25–30	Icosahedral
Hepatits A	Picorna	RNA	No	25–30	Icosahedral
Parvovirus (canine, murine porcine)	Parvo	DNA	No	18–24	Icosahedral

final manufacturing conditions change or significant scale-up occurs during or after Phase III trials.

17.7.2. Virus Stock–Related Considerations

Although it is unanimously agreed that the quality of the stock preparation has critical bearing on the validation study, there are no standardized methodologies for preparation and purification of virus stocks. Stock preparation varies from vendor to vendor; additionally, there can be intralot variation at the same vendor (69). Another consideration is viral stock titer. In general, starting with a high viral load to challenge a process step will maximize the potential viral clearance claim. The volume of virus spiked into the challenge material and the virus stock titer combine to determine the total virus titer in the spiked product. It is advisable to work with high titer virus stocks, but one must recognize that methods used to concentrate the virus stock and achieve high stock titers may facilitate viral aggregation.

The quality of the virus stocks in terms of presence of viral aggregates, cell debris, or other particulates can influence the results by causing a false enhancement or reduction of viral clearance. Thus, for example, with a chromatography process in a contaminant-binding mode, extra cell debris may compete with the virus for binding sites on the resin, causing a decreased clearance value. In a tangential flow filtration process, use of a virus stock containing high amounts of cell debris would enhance virus retention due to the polarization of the membrane. In direct flow filtration, if the membrane clogs prematurely due to cell debris, the entire load volume cannot be filtered, and, therefore the full log clearance cannot be claimed.

Virus preparation–associated variables (aggregation, debris, stabilizers in stock) can vary depending on the virus and the vendor. Some vendors take precautions to reduce cell debris, whereas other viral stocks are minimally purified and may contain significant amounts of membrane particulates. Virus stock solutions often contain stabilizers such as bovine serum albumin and other additives such as serum, and these may interfere with clearance process evaluation. A small scale run with a spike of virus storage solution (without the virus) provides useful information and should be done when possible.

Prefilters are sometimes used prior to virus removal-filtration to remove any virus aggregates/debris that may falsely increase clearance. Some titer loss may occur over the prefilter that will reduce the amount of virus contacting the test membrane. The amount of loss will vary depending on the method of virus stock preparation and any aggregating effect of the test material on the virus (69).

Viral spike volumes will impact clearance studies (especially if there is a significant amount of debris), and, in general, should be maintained at 10% or less of the final volume to keep the feedstream representative of the manufacturing process (28). Using a 5% virus spike instead of 10% only reduces the number of particles by half or approximately 0.5 logs.

17.7.3. Importance of Adequate Controls in Virus Study Design

To ensure that the decrease in viral titer observed is a consequence of the viral clearance procedure and not an artifact of the test, necessary controls must be incorporated during viral validation studies.

Prior to commencing a viral clearance study, it is necessary to ascertain that the product does not have an inhibitory effect on either the indicator cell line (generalized cytotoxicity control) or the test virus (viral interference studies). Cytotoxicity and interference controls are often conducted considerably in advance of the actual validation study to ensure that the clearance capacity is not overestimated due to product-related considerations. The *cytotoxicity* control is included to ensure that any indicator cell cytopathology observed during the study is the result of the virus alone. The cells are exposed to process components (product intermediates, buffer), in the absence of virus for the length of time the test material will be in contact with the cells; a cytopathic or morphological effect relative to the unexposed control cells is an indication of cytotoxicity. *Viral interference studies* are conducted to determine if process components interfere with the capacity of the test virus to infect the indicator cell line. Essentially, following exposure of the indicator cell line to the process component, the cells are exposed to the virus and evaluated to determine if there is any loss of infectivity and thus viral interference by the product. If either of the above two controls demonstrate positive results, one approach is dilution of the test material (in order to determine a non-inhibitory concentration); another is neutralization or other test solution adjustment. A *media* control consists of virus spiked into the virus cultivation medium at the same ratio as the test material and helps to determine inactivation, if any, by the test material. Media "start" and "end" controls demonstrate the stability of the test virus under the test conditions. The *hold* control is included to ensure that the test virus is stable throughout the test duration and essentially involves virus-spiked starting material held for process time at process temperature. This control essentially demonstrates any inactivation effect that is a consequence of the product (start material). The loss demonstrated by a hold control is not related to the clearance strategy under study and should be evaluated accordingly. Depending on the circumstances of the test, additional controls may be necessary. The *freeze/thaw* control provides information on the thaw of the virus in concentrated form versus effects on the diluted form of the virus from the media controls. Stability and storage issues are primarily a concern if process challenges are performed at a site different from that of the virus vendor. If virus stocks are to be shipped to another location, the stocks are thawed, processed over the manufacturing step to be challenged, and frozen for later shipping. This may differ from challenges performed at the vendor site in that many vendors sometimes assay the test material immediately. This reduction may affect the final clearance claim; freeze/thaw stability should be reviewed. *Shipping* controls determine if temperature changes that may have occurred during shipping affected titers when virus is shipped to a different site.

17.8. CONSIDERATIONS IN DATA INTERPRETATION AND ESTIMATING VIRAL CLEARANCE

Establishing clearance for the entire process (overall clearance value) requires at least two orthogonal robust methods of viral clearance. The individual

TABLE 5 Example of Calculation of Viral Clearance (Log Titer Reduction) for the Overall Process[a] for a Monoclonal Antibody Product

	Log reduction factor
Ion exchange chromatography	>5.39
Nanofiltration	>5.14
Low pH inactivation	>5.85
Detergent inactivation	>6.06
Total clearance	>22.44

[a] Challenge virus: Xenotropic murine leukemia virus (X-MuLV).

steps must possess fundamentally different mechanisms of virus removal or inactivation in order for values to be considered cumulative. Only data for the same model virus is cumulative-because viruses vary greatly with regard to their inactivation or removal profiles. Clearance estimates and their variances are calculated for each orthogonal unit operation; total virus reduction is the sum of individual log reduction factors. In cases of complete clearance, a theoretical titer value is based on a statistical distribution (Poisson distribution). Table 5 provides cumulative virus clearance values for murine leukemia virus.

17.9. VIRAL CLEARANCE VALIDATION STUDIES: PITFALLS AND CAUTIONS

A "good" viral clearance validation study is the consequence of a detailed and well-designed study. Scaled-down studies are, at best, approximations of manufacturing conditions, and the validity of the clearance data is a direct reflection of accurate process modeling and study design. Some of the pitfalls associated with small-scale validation studies are related to the following:

1. Virus-related considerations. Viral spike-related perturbations may make the process nonrepresentative of actual manufacturing conditions. Also, model viruses are used in process validation studies; these are, at best, just that—models—and the wild-type strain may not behave similar to a laboratory strain.
2. Inaccurate process modeling. Conditions in small-scale validations may not always be congruent with process scale conditions (e.g., columns used only once for a validation may not reflect the ability of columns used repeatedly, during manufacture, to remove virus consistently).

3. Sample-related considerations include nonrepresentative sample used in viral validations—for example, either the proper intermediate or actual product sample may not have been used; sample may not be representative in terms of protein concentration, pH or other solution characteristics such as ionic strength; samples may be nonhomogeneous due to inadequate mixing.

4. Assay-related considerations include failure to evaluate buffer toxicity, poor model virus selection, lack of appropriate controls and poor standardization of viral assays. Critical assay performance criteria are accuracy, reproducibility, repeatability, linearity of range, limit of detection (LOD) and limit of quantification (LOQ) and must be validated (26).

Steps that require dilution of the product (e.g., due to viral interference or other toxicity-related considerations) will impact assay results and the ability to make a high viral clearance claim. For example, high salt concentration, pH extremes, or other sample conditions may interfere with the virus titration. Decreasing the actual volume assayed (due to dilution of the sample—e.g., 10×) will result in decreased sensitivity and is especially important when no virus is detected and a theoretical limit titer for the sample is calculated.

There are several similarities in regulatory philosophy in approaches used to ensure virological safety of biopharmaceuticals as well as approaches applied to biologicals contamination by unconventional agents such as prions. The remainder of this chapter addresses biopharmaceutical safety from a transmissible spongiform encephalopathy (prion) safety standpoint.

17.10. PRIONS

Animal- and human-derived materials find applications in biologicals and therapeutic product manufacturing, novel tissue-engineered products, and xenotransplantation. Bovine-derived products are used in the medical devices and biopharmaceutical areas, as well as in the cosmetics and food industries. The bovine spongiform encephalopathy (BSE) epidemic (mad cow disease) in the mid-1980s raised public awareness to the manifestation of these diseases, collectively referred to as transmissible spongiform encephalopathies (TSEs).

Transmissible spongiform encephalopathy (TSE) is a generic term used to describe progressive neurodegenerative diseases caused by unusual/novel infectious agents. The diseases are invariably fatal and are characterized by a long incubation period and a short clinical course of neurological signs. Although TSE diseases in humans were reported in the early part of the 20th century and sheep scrapie has now been identified for over two centuries, the BSE epidemic in the 1980s highlighted the potential threat posed by these diseases.

Creutzfeldt-Jakob disease (CJD) is one of the human manifestations of TSE. It may be sporadic (arise spontaneously at low frequency), be acquired from exogenous exposure (especially iatrogenic sources such as via contaminated surgical instruments, dura mater implants, corneal grafts, and cadaveric pituitary-derived hormones), or familial (due to mutations in a gene on chromosome 20). TSE disease manifestations in humans and animals are tabulated in Table 6.

TABLE 6 Transmissible Spongiform Encephalopathies in Humans and Animals

Disease (susceptible species)	Transmission/mechanism of pathogenesis
Animal diseases	
Bovine spongiform encephalopathy (cattle)	Oral; via ingestion of contaminated MBM
Scrapie (sheep)	Infection in genetically susceptible sheep
Transmissible mink encephalopathy	Infection with prions from sheep or cattle
Feline spongiform encephalopathy (cats)	Ingestion of contaminated MBM
Chronic wasting disease (mule deer, elk)	Unknown
Exotic ungulate encephalopathy (greater kudu nyala, oryx)	Ingestion of contaminated MBM
Human diseases	
Creutzfeldt-Jakob Disease (CJD)	
Sporadic	Somatic mutation in PRNP gene or spontaneous conversion of PrP^c to PrPres
Iatrogenic	Infection via contaminated equipment (EEG electrodes), tissue (corneal) and organ (dura mater) implants, hormone administration (gonadotropin, growth hormone)
Familial	Germline mutation in PRNP gene
Variant	Oral; ingestion of bovine prions in contaminated beef
Kuru	Oral; through ritualistic cannibalism
Fatal familial insomnia	Germline mutation in PRNP gene
Gerstmann-Straussler-Scheinker (GSS) disease	Germline mutation in PRNP gene

MBM: meat and bone meal.

TABLE 7 Properties of Normal and Abnormal Forms of Prion Protein (PrP)

Normal PrP	Abnormal PrP
Referred to as PrP-sen, PrP^c	Referred to as PrP-res, or sometimes PrP^{sc}
Sensitive to proteases	Relatively resistant to proteases
Function only recently elucidated; implicated in homeostasis	Specific in pathology of TSEs; proposed to be congruent with the infectious agent
Found in several tissues and cell types	Found in brain and CNS tissues and other non-neuronal tissues, such as spleen, lymph nodes, reticuloendothelial system
Glycosylphosphatidyl inositol-linked cell surface glycoprotein	Large aggregates and amyloid fibrils
Soluble in mild detergents	Insoluble in mild detergents
Mostly α-helix and loop structure	Mostly β-structure
Native protein	Isoform of a cellular protein conformational differences in the secondary and tertiary structure

A new variant form of CJD (vCJD) believed to have originated from BSE was first reported in 1996 (72). Several lines of experimental evidence (73–77) suggest that the causative agents for BSE and vCJD are indistinguishable. This raises a serious concern of interspecies transmission from bovines to humans via ingestion of contaminated meats. This interspecies transmission could result in intraspecies amplification via exposure to iatrogenic manipulations (blood transfusions, organ transplants, etc.). This has significant implications from a public health standpoint. The reported differences between classic (sporadic) CJD and vCJD are tabulated in Table 7.

This review provides a brief overview of issues related to TSE diseases and pragmatic approaches to addressing risk reduction and risk management concerns.

17.11. ETIOLOGY OF PRION DISEASES

The causative agent of TSEs has yet to be conclusively demonstrated. A feature of all TSEs is conversion of a host-encoded sialoglycoprotein to a protease-resistant isoform as a consequence of infection (78). Agents implicated in the etiology of TSEs include an infectious protein (prion theory); an unconventional/slow virus or a bacterium; also, an autoimmune etiology, where contaminant bacteria/bacterial fragments present in animal feeds could have resulted in molecular mimicry between bacterial components and bovine tissues, has been proposed. The current consensus of opinion is that the etiological agent is devoid of informational nucleic acid and is a protein (prion—*pro*teinaceous *in*fectious agent).

There is no definitive data on the nature of prions. The term prion is used as an operational term for the TSE agent (79). The conversion of a ubiquitous cellular prion protein (PrP^c or PrPsen) to the pathology-associated isoform (PrP^{Sc} or PrPres) is one of the hallmarks of transmissible spongiform encephalopathies such as bovine spongiform encephalopathy (BSE), Creutzfeldt-Jakob disease (CJD), and variant Creutzfeldt-Jakob disease (vCJD).

The prion protein exists as a protease-sensitive, glycosylphosphatidyl inositol-anchored cell surface protein in neurons (designated as PrP^C [i.e., PrPcellular] or PrPsen [i.e., PrP sensitive because it is protease sensitive]). The molecule is believed to be based on a 27–30 kD glycoprotein subunit with both hydrophobic and hydrophilic domains, and it tends to form large amorphous or rod-shaped aggregates in vitro. Because the aggregation process may be complicated and is not well understood, there is currently no straightforward definition of what constitutes a single prion particle. Disease occurs when an abnormal protease-resistant isoform accumulates (PrPres [i.e., PrPresistant because it is relatively resistant to protease K digestion] or PrP^{Sc}, [i. e., PrPscrapie, to denote the sheep disease scrapie, though PrP^{Sc} may not be the appropriate term for the abnormal PrP found in other diseases]). Table 8 lists some salient differences between the cellular and abnormal forms of the prion protein.

17.12. MODE OF PRION TRANSMISSION

Transmissibility of infectivity may be influenced by several factors, including the route of administration, dosage, the strain of agent used, and the presence/absence of a species barrier. Intravenous administration is believed to result in a 10-fold reduction in infectivity compared to the intracerebral route; a species barrier may result in up to 1000-fold reduction in infectivity (80). Prion agents can undergo modification by passage through animal hosts; therefore, a strain of sheep scrapie passaged through cattle might present a risk to humans. The incubation period varies depending on the route of exposure; for example, in cases of iatrogenic transmission of CJD via parenteral transmission, symptoms did not develop until several years (~12 years) after exposure; however, after dura mater grafts and neurosurgical procedures, the incubation period was months rather than years.

Unlike the classic form of CJD that is detected primarily in tissues of the central nervous system, the vCJD strain has a propensity for replication in

TABLE 8 Main Characteristics of Sporadic and Variant CJD Disease

	Sporadic CJD	vCJD
Incidence	~1 case/million population/year	135 cases globally[a]
Distribution	Worldwide	EU (UK primarily, France, Ireland)
Age at onset (mean)	Late middle age, 55–70 years; avg ~60 years, occasionally affects younger people	Relatively young age of onset median ~28 years; range 19–39
Presenting features	Mental deterioration (dementia, myoclonus)	Behavioral abnormalities, ataxia, dysesthesia Psychiatric and/or sensory symptoms
Clinical course	Rapidly progressive	Insidious onset prolonged course
Duration (mean)	Short; ~7 months	Long; ~14 months (range 7–38 months)
EEG	Typical periodic pattern	Nonspecific
PRNP genotype (codon 129)	Predominantly Met/Met homozygous	100% of the cases have Met/Met homozygosity
Neuropathologic features	Synaptic deposits; rarely plaques	Prominent "florid" plaques
PrP-res banding pattern	Type 1 and 2[b]	Type 4 (similar to experimental BSE in mice, macaques, and other species)
PrP positivity in tonsil/spleen/ lymph node	No/No/No	Yes/Yes/Yes Presence of PrP in lymphoreticular tissue

[a] Total number of definite or probable cases (dead and alive) as reported by the National CJD Surveillance Unit, Edinburgh, as of June 4, 2003.
[b] Type 3 is found in iatrogenic cases with intramuscular inoculation.

peripheral lymphoreticular tissue. BSE prions in cows seem confined to the neural compartment. Passage into humans, and consecutive progression to manifest vCJD in humans, likely results in a dramatic shift in the organo-tropism of prions. No cases of iatrogenic transmission of vCJD have yet been reported, but the altered tropism of the infectious agent (manifested in tissues of the tonsils, spleen, and the appendix) (75,81) is disconcerting. Table 9 summarizes the causes for concern.

TABLE 9 Why the Heightened Concern Over vCJD?

vCJD is a different entity from CJD; there are too many unknowns (minimal levels of infectivity required, routes of infection). The number of individuals incubating vCJD is not known. Due to gaps in our information, exclusion criteria cannot be formulated.
The agent withstands conventional sterilization measures.
Currently available detection tests cannot detect infection in the preclinical phase of the disease.
In human-human transfer, due to the absence of a species barrier there is a potential for intra-species amplification and disease transmission.
There is evidence for a higher degree of lymphoid involvement in vCJD compared with sporadic CJD. Therefore, vCJD could be spread by lymphoid cells in the blood and consequently increase probability of iatrogenic transmission.

17.13. PRION DETECTION METHODS

The novel biology of prions makes their detection difficult. Nucleic acid–based methods are not directly applicable for detecting an agent apparently devoid of a nucleic acid component. Because of a poor humoral immune response (presumably because of sequence similarities between the two isoforms— infectious PrPres and cellular PrP^c), no direct serological assays are available. Confirmatory evidence in TSEs is available only postmortem by histopathological examination of brain sections for spongiform changes, accumulations of prion protein (PrP) by immunohistochemistry (IHC), or by visualization of scrapie-associated fibrils (SAF) using electron microscopy (82).

Infectivity assays (either in mice or in hamsters) are the gold standard for prion detection. However, among the logistic considerations associated with infectivity assays are the extended incubation time required (200–500 days in a mouse-adapted system and 90–200 days in the hamster system) and the expensive nature of these assays. Quantitation uses either the end-point titration or incubation time interval assay. The incubation time assay is based on the fact that under well-defined experimental conditions, there is an inverse relationship between infectivity dose and the subsequent incubation period before the onset of clinical disease (83,84). The longer the incubation period, the less the infectivity. Although "incubation period" assays (without the need for titration) are less labor intensive, they may not be able to detect low levels of the infectious agent (85). End-point titrations of infectivity in animal models allow quantitation of low levels of infectivity but involve even more extended incubation periods.

PrPres is the only specific biochemical marker of prion diseases. Because the commercially available antibodies (e.g., Mab3F4) are unable to distinguish between the two isomers (PrPres versus PrPsen), immunocytochemical methods for PrPres detection depend on a certain amount of pretreatment and presumed denaturation of the normal prion protein (PrP^c). The insolubility of PrPres in nonionic detergents (86,87) and its partial resistance to proteinase K (88) allow for distinguishing it from its cellular counterpart (PrP^c). Immunologically based assays for PrP detection include Western blots, dot blots, and ELISA systems (89–91).

The Western blot (WB) assay, which is commonly used in prion clearance evaluation studies, relies on the differential sensitivity of the PrP^c and PrPres proteins to digestion with protease K. The method essentially involves electrophoretic separation of protease K–digested PrP (which provides a distinct protein fingerprint) followed by immunologic detection using an antibody specific for PrP (92). Although the Western blot assay is not as sensitive as a bioassay, it provides a rapid and useful tool for preliminary screening for prion clearance.

A variety of innovative immunoassays have been developed. The Prionics Western blot (PWB) has been reported to detect BSE in cattle with no overt clinical symptoms (93). Detection of pathological prion protein aggregates in cerebrospinal fluid (CSF) by dual-color laser scanning for intensely fluorescent targets has been developed (94). The sensitivity of the Western blot assay can reportedly be enhanced by a few logs by using dissociation-enhanced time-resolved fluoroimmunoassay (DELFIA) (95). The capillary immunoelectrophoresis (CIE) assay is based on competitive binding between the proteinase K–treated PrP and a fluorescein-labeled synthetic PrP in their binding to antibodies raised against the peptides (96,97). A conformation-dependent immunoassay (CDI) identifies the pathologic isoform of PrP without the need for proteinase K digestion; this is one of its advantages, as some forms of PrPres may be less resistant to proteases (and be missed by assays involving PK digestion) (98). Saborio et al. (99) reported a novel approach for detection of low levels of PrPres. The method, called the protein-misfolding cyclic amplification (PMCA), is conceptually analogous to the polymerase chain reaction (PCR) method and would allow detection of low levels of PrPres by amplification of the undetectable PrPres (as in the case of preclinical infections) to a detectable level. This, when combined with any available detection system, would facilitate diagnosis during the asymptomatic phase of the disease.

Detection of surrogate markers is another approach. Patients in the preclinical phase of the CJD have elevated levels of 14-3-3 in the CSF of patients with CJD (100,101). This marker has been shown to be reliable for detection of classical CJD but not vCJD. It is not CJD-specific and cannot distinguish CJD from certain other neurodegenerative disorders (102).

The protease-resistant prion protein (PrPres) is currently used as a proxy marker for measuring infectivity; however, the precise relationship between the infectivity and PrPres concentrations is still not fully understood. While there have been significant strides in developing detection methodologies, there are currently no minimally invasive molecular or serological diagnostic assays that would permit identification in the preclinical phase of the disease.

17.14. PRION CLEARANCE: A RISK-BASED APPROACH

A precise risk assessment cannot be made with human TSEs due to gaps in our information. Significant unknowns include the etiological agent per se, the minimal infectious dose, the exact distribution of infectivity in tissues, and uncertainties in key epidemiological parameters.

Risk assessment to determine the potential for prion contamination of the biological includes evaluation of the source of raw materials, the type of tissue used, and the route of administration; even genotypic susceptibility may be an influencing factor. Also, the potential risk should be evaluated in the context of its use. For example, bovine neural tissue implanted in the central nervous system would pose a much greater risk than a few microliters of highly purified bovine pancreatic trypsin used in a manufacturing process to recover tissue culture cells that are further purified before use.

Considerable emphasis is placed on the source of raw material and type of tissue used. For example, certain tissues and body fluids constitute a high risk (e.g., brain, spinal cord), whereas others have no demonstrable infectivity (blood, saliva, skeletal muscle). The WHO has published a list of infectious tissues for TSEs (103), which provides a reference for risk evaluation (Table 10).

A predisposing factor in humans is the genotype at polymorphic codon 129 of the *PRNP* gene. The encoding alternatives at codon 129 are methionine

TABLE 10 Distribution of Infectivity in the Human Body as Described by the WHO, 1999

Infectivity category	Tissues, secretions, and excretions
High infectivity	Brain, spinalcord ,eye
Low infectivity	CSF, kidney, liver, lung, lymph nodes/spleen, placenta
No detectable infectivity	Adipose tissue, adrenal gland, gingival tissue, heart muscle, intestine, peripheralnerve, prostate, skeletal muscle, testis, thyroid gland, tears, nasal mucosa, saliva, sweat, serous exudates, milk, semen, urine, feces, blood

Source: Ref. 103.

(Met) and Valine (Val). Homozygosity for methionine at this codon appears to be a predisposing factor to CJD (104) and appears to influence suscepti-bility to sporadic (105) and iatrogenic (106–108) CJD in the United Kingdom (108) and France (109).

When applying quantitative risk assessment methods to estimate the risks via potential exposure routes as, for example, from food ingestion and blood transfusion, and from environmental disposal of BSE-infected resi-dues, the results must be interpreted with caution. For example, the risks predicted for consumption of beef-on-the-bone and for drinking water from an aquifer potentially contaminated with effluent from a cattle-rendering plant are similar at 10^9 and 10^8/person/year, respectively (110). However, the risks predicted for beef-on-the-bone are realistic, but the risks through drinking water could be significantly overestimated for a variety of reasons: route of exposure, number of exposures—single versus multiple exposures, and the threshold effect (i.e., whether a minimum number of BSE prions are needed to initiate infection).

17.14.1. Risk Minimization

Risk minimization strategies have included active surveillance strategies, emphasis on appropriate sourcing and import restrictions, severe restrictions on specified bovine and ovine offals entering the human food chain, and a ban on feeding of mammalian proteins to ruminants, and on cattle over 30 months of age entering the human food chain. Further measures in the United Kingdom to minimize the risks of establishing a reservoir of human-to-human infectivity include a ban on the use of U.K.-sourced plasma for the preparation of licensed blood products (e.g., coagulation factors, albumin), leukodepletion of blood donations, increased levels of decontamination of surgical instruments, and use of disposables wherever possible, especially in surgeries associated with tissues considered high risk, such as in neurosurgery, ophthalmic surgery, and tonsillectomies.

17.14.2. Sourcing

Sourcing is the cornerstone of the risk minimization initiative. For biophar-maceuticals using bovine-derived materials, species and source country of the bovine-derived constituent must be identified. Significant emphasis is placed on evaluation of a country's compliance with BSE-related standards of the Office International des Epizooties (OIE) and on evaluating products with regard to the risk posed by the bovine tissue, in addition to the country of origin. As a safeguard, materials sourced from BSE countries should not be used in the manufacture of FDA-regulated products (111), even though no cases of vCJD transmission have been reported in recipients of bovine insulin or other injectable products manufactured in BSE-affected countries.

Master seeds (MCs) or cell banks (MCBs) that have been prepared 30–40 years ago for vaccine production do not need to be reestablished as this would pose a higher risk; for example, there is a potential of altering the vaccines through rederivatization CBER has required licensed vaccine manufacturers to evaluate all bovine-sourced materials used in fermentation/routine production and in the establishment of working seeds and WCBs to ensure full compliance with the TSE guideline (112).

17.14.3. Risk Management

In general, precautionary measures are directed at selection/exclusion criteria for donors and screening tests for donations, processes of removal or inactivation of the agent, recall of batches from which a donor subsequently develops nvCJD (new variant), and substitution with alternative non–plasma-derived products, where possible.

To date, based on data from systematic surveillance of high-risk groups such as patients with hemophilia (113–116), "look back" studies in which recipients of blood transfusions were monitored for up to 20 years post treatment (117), and statistical analyses of the incidence of classical CJD in blood/blood product recipients as compared with the general population (101,118,119), there appears to be a negligible, if any, risk of classical CJD transmissions via blood and blood products. To quote the SCMPMD (Scientific Committee on Medicinal Products and Medical Devices): "the evaluation of all relevant data leads to the opinion that transmission of CJD by blood and blood products either does not occur or does not contribute to the CJD epidemiology. Although a hazard cannot be excluded, a real risk is not recognizable" (120,121). Unfortunately, the same statement cannot be made for vCJD because of significant gaps in our knowledge. Current understanding of vCJD suggests a potentially different pathogenesis.

17.14.4. Prion Clearance Methods

Prions are notoriously hardy to methods that would be considered overkill for most other microbial and viral agents. Part of this resilience could be attributed to their propensity to be associated with organic matter/cell debris that confers a protective effect; another contributing factor is their propensity to form aggregates (122).

Prion clearance methods used should essentially be reliable and robust and preserve product integrity. This becomes a challenge, especially in the face of an agent whose significant hardiness to inactivation conditions has been consistently demonstrated. In view of the high resistance of prions and the lability of biopharmaceuticals, removal methods such as filtration, precipitation, and chromatography appear to be more applicable than inactivation methods.

There is no single method that has been shown to be 100% effective for inactivating prions (123–125). In general, TSE agents have high resistance to a variety of physical and chemical treatment methods: dry heat (160°C, 24 hr) has no effect; infectivity remains even after exposure to 360°C, 1 hr, autoclaving (infectivity detected post autoclaving at 126°C, 1 hr), UV and ionizing radiation (126–128), wide pH range—pH 2–10 (129), alcohols and alkylating agents, phenolic disinfectants, and so forth (130–136). β-Propiolactone is also reportedly ineffective (137).

Caution must be exercised when reviewing the literature pertaining to prion inactivation because the studies vary in the prion strains used (CJD, scrapie, BSE), kinds of different brain preparation methods (dried/macerated preparations, unspun 20% homogenates, 10% tissue supernates, ultracentrifuged material), and kinds of test systems (animal models). Other variables include exposure times, temperature and the type of tissue studied (brain versus reticuloendothelial). Furthermore, any interpretation of experimental data relating to the infectivity or transmission of TSE must be tempered by the knowledge that studies are done under experimentally induced conditions by intracranial injection (high efficiency of the infectious route), of contaminated brain (high titer of infectivity) into nonhuman primates (low species barrier) to provide accelerated disease progression and prognosis. Often, the assumption is made that those procedures effective for partially purified TSE infectivity are equally applicable for dealing with crude tissue contamination, but this is unwarranted (138).

The physicochemical inactivation profile of vCJD differs from classical CJD. In both types of diseases the infectious agent is likely to consist entirely of the PrPres protein, yet the differing conformations of the protein may be the reason why the vCJD agent is more resistant to heat inactivation.

Bioseparation techniques often include protein precipitation based on differential solubility, and adsorptive behavior. Operations such as precipitation, adsorption, and filtration are commonly used in the manufacture of biopharmaceuticals and plasma-derived biologicals. TSE agents have unique physicochemical properties that make possible serendipitous reduction during processing of biological products. For example, PrPres is readily precipitated by ethanol (136), ammonium sulfate, and polyethylene glycol (139). The first step in the manufacture of plasma products involves cryoprecipitation followed by solvent precipitation steps. Despite differences in experimental methods with regard to the kind of spike used (crude brain homogenate versus microsomal fraction), and detection methods (bioassay versus WB), relatively consistent clearance factors have been reported by several investigators (140–143). This suggests that the effects of cold ethanol on PrPsc are relatively robust.

The prion protein is most commonly membrane-associated and has a tendency to aggregate. These attributes facilitate their removal by depth filtration (140,141). Virus removal filters have been demonstrated to provide prion clearance (144,145). Chromatographic purification in the manufacture of coagulation factor concentrates is extensively used; these methods have the potential to effect prion removal (140).

17.15. PROCESS CLEARANCE EVALUATION: CONSIDERATIONS AND DESIGN ISSUES

As in the case of ensuring viral safety of biologicals, a multifaceted approach involving adequate sourcing, incorporation of multiple orthogonal clearance strategies, and process evaluation for prion clearance are vital. Currently, there are no in vitro tests directly applicable for detection of low levels of infectivity either in the raw materials or in the finished product; consequently, adequate sourcing and demonstration of the prion clearance ability of the manufacturing process constitutes a key paradigm to ensure safety.

The first steps in the design of the clearance evaluation study involve a critical analysis of the entire manufacturing process to determine the potential sources of prion contamination, followed by process characterization to evaluate which steps in the manufacturing process possess clearance capability. The principles applied in evaluation of clearance of conventional viruses are applicable to TSE agents (146). CPMP (Committee for Proprietary Medicinal Products) guidance documents acknowledge that several routine processing steps such as precipitation, chromatography, and nanofiltration can contribute to TSE agent removal (147) and require that whenever TSE clearance claims are made for a particular step, the process should be validated (148).

A number of studies have been undertaken to evaluate purification process steps for their potential for prion clearance using either a crude brain homogenate (89,142,149), detergent-solubilized (144,145), or a microsomal fraction (140,141) of the hamster-adapted sheep scrapie (HSc) agent.

Issues to address when designing a validation study to document prion removal are choice of spiking agent, nature or form of the spiking agent (and its relevance), the design of the study (including appropriate scale-down) and the detection method (in vitro or in vivo assay). Validations are often performed using strains of the sheep scrapie agent that have been adapted to either mice or hamsters, by direct intracerebral inoculation of infected sheep brain into mice or hamsters followed by serial passage of the agent in the same species.

17.15.1. Spiking Agents

Experimental TSE studies have been conducted with a variety of TSE agent spikes—mouse/hamster-passaged scrapie agent, mouse-passaged BSE, mouse/hamster/guinea pig–passaged CJD. Prions from different species are sufficiently similar: they share antigenicity (150) and other physicochemical properties such as resistance to proteinase K treatment. Studies have demonstrated that the PrP^c is well conserved between species, with 90% homology in the mammalian amino acid sequences (151). Process clearance evaluation studies are often performed using strains of rodent-adapted sheep scrapie agent; these models have been accepted by regulatory authorities (149).

17.15.2. Relevance of the Spiking Agent

In general, the hamster-adapted 263 K strain of scrapie is regarded as the optimal system because high titers in the brain combined with a short incubation period are a hallmark of this system (152). Spike-related considerations include the "relevance" of the spike and the concentration of the spiking agent. To date, there is no agreement on whether there is a minimum threshold concentration required for infection; and, routinely, process clearance evaluation studies are done using high titer spikes to expedite testing.

Detergent-solubilized scrapie agent is often used in filtration studies (detergent treatment of the crude brain homogenate, which is composed of membrane-bound PrP, results in dissolution of the lipid membrane), as it constitutes a worst-case challenge. TSE agents in blood would likely be cell-associated; consequently, a more representative spike for plasma products may be a microsomal fraction prepared with crude brain homogenate.

17.15.3. Detection Methods Used in Process Clearance Evaluation Studies

Sample analysis is often undertaken either by infectivity assays (89,142,153) or by the Western blot (WB) method (89,140–142). Although the infectivity assay is the assay of choice, the extended incubation period and the cost-prohibitiveness of the assay present logistic limitations. Though less sensitive than a bioasssay, the Western blot provides a rapid and useful tool for detection and preliminary screening of manufacturing methods for prion clearance. Recent developments have enhanced sensitivity of the WB to within 3 logs of a bioassay (154).

17.16. CONSIDERATIONS IN DATA INTERPRETATION AND ESTIMATING PRION CLEARANCE

Estimating prion clearance for the entire process (overall clearance value) is calculated in a similar manner to viral clearance. Studies in the literature have documented (155,156) that significant prion clearance can be achieved by both serendipitous and deliberate steps. Clearance estimates and their variances are calculated for each orthogonal unit operation; total prion reduction is the sum of individual log reduction factors.

17.17. CONCLUSION

Clinical acceptability of any biological or biopharmaceutical is concomitant with risk assessment and guided by risk-benefit analysis. Contemporary serologic screening methods have dramatically reduced the risk of viral iatrogenic accidents to the point where pharmacoeconomics suggests that we have reached a point of diminishing return. The biotechnology industry has made a significant shift toward serum-free media and well-characterized critical raw materials. To date, no confirmed case of CJD/vCJD has been reported in recipients of blood transfusions, plasma-derived biologicals, monoclonals, or other recombinant products manufactured with animal-sourced raw materials.

Current risk management approaches to ensure virological/TSE safety of biologicals have utilized a multifaceted approach that involves screening of source materials and supplements, evaluation of manufacturing processes, and incorporation of viral clearance methodologies, in conjunction with rigorous process controls. Risk-benefit includes a balance between the therapeutic necessity of the biological and the risks posed by the quantity of source material required to produce a daily dose, the number of daily doses, and the route of administration.

Technological advances, demographic and societal changes, and subtle changes to our ecosystems will continue to leave us vulnerable to new or reemerging viruses, as, for example, West Nile virus. Although even a minimal risk is viewed as unacceptable in the case of transfusion-transmitted viral infection and TSEs, the ideal zero-risk goal must be balanced with the consequences of severe shortfalls of treatments that may be lifesaving.

The TSE crisis and the sporadic reports of new or reemerging viral threats epitomize the difficulties in risk management when the nature of the risk and absolute risk levels are unknown. Our only recourse is a constant vigilance combined with pragmatic regulations and guidelines that take cognizance of the latest scientific and technical information.

REFERENCES

1. Wilson, G.S. *The Hazards of Immunization*; The Athlone Press: London, 1967.
2. Nathanson, N.; Langmuir, A.D. The Cutter incident: poliomyelitis following formaldehyde-inactivated poliovirus vaccination in the United States during the spring of 1955. II. Relationship of poliomyelitis to Cutter vaccine. Am. J. Epidemiol. 1963, *142*, 109–140.
3. Para, M. An outbreak of post-vaccinal rabies (rage de laboratorie) in Fortaleza, Brazil, in 1960: Residual fixed virus as the etiological agent. Bull. World Health Organ. 1965, *33*, 177–182.
4. King, A.M.; Underwood, B.O.; McCahon, D.; Newman, J.W.; Brown, F. Biochemical identification of viruses causing the 1981 outbreaks of foot and mouth disease in the U.K. Nature 1981, *293*, 479–480.
5. Darby, S.C.; Ewart, D.W.; Giangrande, P.L.; Dolin, P.J.; Spooner, R.J.D.; Rizza, C.R. Mortality before and after HIV infection in the complete UK population of haemophiliacs. Nature 1995, *377*, 79–82.
6. Schreiber, G.B.; Busch, M.P.; Kleinman, S.H.; Korelitz, J.J. The risk of transfusion-transmitted viral infections. N. Engl. J. Med. 1996, *334*, 1685–1690.
7. Chudey, M.; Budek, I.; Keller-Stanislawski, B.; McCaustland, K.A.; Neidhold, S.; Robertson, B.H.; Nubling, C.M.; Lower, J. A new cluster of hepatitis A infection in hemophiliacs traced to a contaminated plasma pool. J. Med Virol. 1999, *57*, 91–99.
8. Robertson, B.H.; Alter, M.J.; Bell, B.P.; Evatt, B.; McCaustland, K.A.; Shapiro, C.N.; Sinha, S.D.; Souci, J.M. Hepatitis A virus sequence detected in clotting factor concentrates associated with disease transmission. Biologicals 1998, *26*, 95–99.
9. Scheiblauer, H.; Nubling, M.; Willkommen, H.; Lower, J. Prevalence of Hepatitis C virus in plasma pools and the effectiveness of cold ethanol fractionation. Clin. The. 18 (Suppl. B) 59–70.
10. Blumel, J.; Schmidt, I.; Effenberger, W.; Seitz, H.; Willkommen, H.; Brackmann, H.H.; Lower, J.; Eis-Hubinger, A.M. Parvovirus B19 transmission by heat-treated clotting factor concentrates. Transfusion 2002, *42* (11), 1473–1481.
11. Eis-Hubinger, A.M.; Sasowski, U.; Brackmann, H.H. Parvovirus B19 contamination in coagulation factor VIII products. Thromb. Haemost. 1999, *81*, 476–477.
12. Santagostino, E.; Mannucci, P.M.; Gringeri, A.; Azzi, A.; Morfini, M.; Musso, R.; Santoro, R.; Schiavoni, M. Transmission of parvovirus B19 by coagulation factor concentrates exposed to 100°C heat after lyophilization. Transfusion 1997, *37*, 517–522.
13. Bown, P.; Preece, M.A.; Will, R.G. "Friendly fire" in medicine: hormones, homografts, and Creutzfeldt-Jakob Disease. Lancet 1992, *340* (8810), 24–27.
14. Sah, K.; Nathanson, N. Human exposure to SV40: review and comment. Am. J. Epidemiol. 1976, *103*, 1–12.
15. Greig, J.R. Scrapie in sheep. J. Comp. Pathol. 1950, *60*, 263–266.
16. Aygoren-Pursun, E.; Scharrer, I. A multicenter pharmacosurveillance study for the evaluation of the efficacy and safety of recombinant factor VIII in the

treatment of patients with hemophilia A. German Kogenate Study Group. Thromb. Haemost. 1997, *78*, 1352–1356.

17. Fox, J.P.; Manso, C.; Penna, H.A.; Pare, M. Observation on the occurrence of icterus in Brazil following vaccination against yellow fever. Am. J. Hyg. 1942, *36*, 68–116.

18. Alter, H.J.; Nakatsuji, Y.; Melpolder, J.; Wages, J.; Wesley, R.; Shih, J.W.; Kim, J.P. The incidence of transfusion-associated hepatitis G virus infection and its reltion to liver disease. N. Engl. J. Med. 1997, *336*, 747–754.

19. Gutierrez, R.A.; Dawson, G.J.; Knigge, M.F.; Melvin, S.L.; Heynen, C.A.; Kyrk, C.R.; Young, C.E.; Carrick, R.J.; Schlauder, G.G.; Surowy, T.K.; Dille, B.J.; Coleman, P.F.; Thiele, D.L.; Lentino, J.R.; Pachucki, C.; Mushahwar, I.K. Seroprevalence of GB viru-C and persistence of RNA and antibody. J. Med. Virol. 1997, *53*, 167–173.

20. Desai, S.M.; Meurhoff, A.S.; Leary, T.P.; Erker, J.C.; Simons, J.N.; Chalmers, M.L.; Berkenmeyer, L.G.; Pilot-Matias, T.J.; Mushahwar, I.K. Prevalence of TT virus infection in US blood donors and populations at risk for acquiring parenterally transmitted viruses. J. Infect. Dis. 1999, *179*, 1242–1244.

21. Handa, A.; Dickstein, B.; Young, N.S.; Brown, K.E. Prevalence of the newly described human circovirus, TTV, in United States blood donors. Transfusion 2000, *40*, 245–251.

22. Centers for Disease Control and Prevention. Investigations on West Nile virus infections in recipients of blood transfusions. MMWR Morb Mortal Wkly Rep 2002, *51* (43), 973–974.

23. Stephenson, J. Investigation probes risk of contracting West Nile virus via blood transfusion. JAMA 2002; *288* (13):1573–154.

24. Castle, P.; Robertson, J.S. Animal sera, animal sera derivatives and substitutes used in the manufacture of pharmaceuticals. Biologicals 1998, *26*, 365–368.

25. Aranha, H. Viral clearance strategies for biopharmaceutical safety: Viral clearance strategies for biopharmaceutical safety: Part 1: General Considerations. BioPharm. 2001; 28–35.

26. Darling, A.J.; Boose, J.A.; Spaltro, J. Virus assay methods: accuracy and validation. Biologicals 1998, *26*, 105–110.

27. Hilfenhaus, J.; Groner, A.; Nowak, T.; Weimer, T. Analysis of human plasma products: Polymerase chain reaction does not discriminate between live and inactivated virus. Transfusion 1997, *37*, 935–940.

28. *Committee for Proprietary Medicinal Products (CPMP)*. Note for guidance on virus validations: the design, contribution and interpretation of studies validating the inactivation and removal of viruses, Feb 29, 1996.

29. FDA, Center for Biologics Evaluation and Research. Points to consider in the manufacturing and testing of monoclonal antibody products for human use, 1997.

30. FDA, International Conference on Harmonisation. Guidance on viral safety evaluation of biotechnology products derived from cell lines of human or animal origin. Fed. Reg. 1998, *63* (185), 51074–51084.

31. Committee for Proprietary Medicinal Products (CPMP). Note for Guidance on Plasma Derived Medicinal Products, June 27, 1997.

32. Hamatainen, E.; Suomela, H.; Ukkonen, P. Virus inactivation during intravenous immunoglobulin production. Vox. Sang. 1992, *63*, 6–11.

33. Grun, J.B.; White, E.M.; Sito, A.F. Viral removal/inactivation by purification of biopharmaceuticals. Biopharm 5(9): 22–30.

34. Horowitz, B.; Horowitz, M.S. Human leukocyte alpha-interferon preparations: laboratory characterization and evaluation of clinical safety. In Zoon, K.C., Noguchi, P.P., Lie, T.Y., Eds.; *Interferon: Research, Clinical Applications and Regulatory Considerations*; Elsevier Science: New York, 1984; 41–53.

35. Sofer, G. Virus inactivation in the 1990s- and into the 21st century. Part 3a: Plasma and plasma products (heat and solvent-detergent treatments). BioPharm 2002, *15* (9), 28–42.

36. Sofer, G. Virus inactivation in the 1990s- and into the 21st century. Part 3b: Plasma and plasma products (treatments other than heat or solvent (detergent). BioPharm 2002, *15* (10), 42–49, 51.

37. Sofer, G. Virus inactivation in the 1990s- and into the 21st century. Part 4: Culture media, biotech products, and vaccines. BioPharm 2003, *16* (1), 50–57.

38. Barrowcliffe, T.W. Viral inactivation vs biological activity. Dev. Biol. Stand. 1993, *81*, 125–135.

39. Gleeson, M.; Herd, L.; Burns, C. Effect of heat inactivation of HIV on specific serum proteins and tumour markers. Ann. Clin. Biochem. 1990, *27*, 592–594.

40. Suontaka, A.M.; Blomback, M.; Chapman, J. Changes in functional activities of plasma fibrinogen after treatment with methylene blue and red light. Transfusion 2003, *43* (5), 568–575.

41. Peerlinck, K.; Arnout, J.; Gilles, J.G.; Saint-Remy, J.M.; Vermylen, J. A higher than expected incidence of Factor VIII inhibitors in multitransfused haemophilia A patients treated with an intermediate purity pasteurized Factor VIII concentrate. Thromb. Haemost. 1993, *69*, 115–118.

42. Prowse, C.V.; Williams, A.E. A comparison of the in vitro and in vivo thrombogenic activity of factor IX concentrates in stasis (Wessler) and non-stasis rabbit models. Thromb. Haemost. 1980, *44*, 81–86.

43. Savage, M.; Torres, J.; Franks, L.; Masecar, B.; Hotta, J. Determination of adequate moisture content for efficient dry-heat viral inactivation in lyophilized Factor VIII by loss on drying and by near infrared spectroscopy. Biologicals 1998, *26*, 119–124.

44. Melnick, J.L. Virus inactivation: lessons from the past. Dev. Biol. Stand. 1991, *75*, 29–36.

45. Murphy, P.; Nowak, T.; Lemon, S.M.; Hilfenhaus, J. Inactivation of hepatitis A virus by heat treatment in aqueous solution. J. Med. Virol. 1993, *41*, 61–64.

46. Prudouz, K.N.; Fratantoni, J.C. Viral inactivation of blood products. In *Scientific Basis of Transfusion Medicine*; Anderson, K.C., Ness, P.M., Eds., Saunders: Philadelphia, 1994; 852–871.

47. Knutson, F.; Alfonso, R.; Dupuis, K.; Mayaudon, V.; Lin, L.; Corash, L.; Hogman, C.F. Photochemical inactivation of bacteria and HIV in buffy-coat-derived platelet concentrates under conditions that preserve in vitro platelet function. Vox Sang. 2000, *78*, 209–216.

48. Walter, J.K.; Nothelfer, F.; Werz, W. Virus removal and inactivation, a decade of validation studies: critical evaluation of the data set. In *Validation of Biopharmaceutical manufacturing processes*; Kelley, B.D., Ramelmeier, R.A, Eds.; ACS Symposium series 698, ACS: Washington, DC, 1998.

49. Dichtelmuller, H.; Rudnick, D.; Breuer, B.; Ganshirt, K.H. Validation of virus inactivation and removal for the manufacturing procedure of two immuno-globulins and a 5% scrum protein solution treated with β-propiolactone. Biologicals 1993, *21*, 259–268.

50. Lundblad, J.L.; Seng, R.L. Inactivation of lipid enveloped viruses in proteins by caprylate. Vox Sang. 1991, *601*, 75–81.

51. Alonso, W.R.; Trukawinski, S.; Savage, M.; Tenold, R.A.; Hammond, D.J. Viral inactivation of intramuscular serum globulins. Biologicals 2000, *28*, 5–15.

52. Horowitz, B. Investigations into the application of tri(n-butyl) phosphate/detergent mixtures to blood derivatives. In *Virus Inactivation in Plasma Products*; *Curr. Stud. Morgenthaler*, J.J Ed.; Hematol. Blood Transfus. 1989, *56*, 83–96.

53. Aranha-Creado, H.; Oshima, K.; Jafari, S.; Howard, G. Jr.; Brandwein, H. Virus retention by a hydrophilic triple-layer PVDF microporous membrane filter. PDA J. Pharm. Sci. Technol. 1997, *51*, 119–124.

54. Burnouf-Radosevich, M.; Appourchaux, P.; Huart, J.J.; Burnouf, T. Nano-filtration, a new specific virus elimination method applied to high-purity factor IX and factor XI concentrate. Vox Sang. 1994, *67*, 132–138.

55. DiLeo, A.J.; Vacante, D.A.; Deane, E. Size exclusion removal of model mammalian viruses using a unique membrane system, Part I: Membrane qualification. Biologicals 1993, *21*, 275–286.

56. DiLeo, A.J.; Vacante, D.A.; Deane, E. Size exclusion removal of model mammalian viruses using a unique membrane system, Part II: Module qualification and process simulation. Biologicals 1993, *21*, 287–296.

57. Oshima, K.H.; Evans-Strickfaden, T.T.; Highsmith, A.K.; Ades, E.W. The use of a microporous polyvinylidene fluoride (PVDF) membrane filter to separate contaminating viral particles from biologically important proteins. Biologicals 1996, *24*, 137–145.

58. Oshima, K.H.; Evans-Strickfaden, T.T.; Highsmith, A.K. Comparison of filtration properties of Hepatitis B virus (HBV), Hepatitis C virus (HCV) and Simian Virus 40 (SV40) using a polyvinylidene fluoride (PVDF) membrane filter. Vox Sang. 1998, *75*, 185–188.

59. Roberts, P. Efficient removal of viruses by a novel polyvinylidene fluoride membrane filter. J. Virol. Methods 1997, *65*, 27–31.

60. Burnouf, T. Chromatographic removal of viruses from plasma derivatives. Dev. Biol. Stand. 1993, *81*, 199–209.

61. Cameron, R.; Davis, J.; Adcock, W.; MacGregor, A.; Barford, J.P.; Cossart, Y.; Harbour, C. The removal of model viruses, poliovirus type 1 and canine parvovirus, during purification of human albumin using ion-exchange chromatographic procedures. Biologicals 1997, *25*, 391–401.

62. Morgenthaler, J.J.; Omar, A. Partitioning and inactivation of viruses during

isolation of albumin and immunoglobulins by cold ethanol fractionation. Dev. Biol. Stand. 1993, *81*, 185–190.

63. Lawrence, S.A. β-Propiolactone: Viral inactivation in vaccines and plasma products. PDA J. Pharm. Sci. Technol. 2000, *54*, 209–217.

64. Vollenbroich, D.; Ozel, M.; Vater, J.; Kamp, R.M.; Pauli, G. Mechanism of inactivation of enveloped viruses by the biosurfactant surfactin from Bacillus subtilis. Biologicals 1997, *25*, 289–297.

65. Chapman, J. Progress in improving the pathogen safety of red cell concentrates Vox Sang. 2000; *78* (suppl 2) 203–204.

66. Bradley, D.W.; Hess, R.A.; Tao, F.; Sciaba-Lentz, L.; Remaley, A.T.; Laugharn, J.J.; Manak, M. Pressure cycling technology: a novel approach to virus inactivation in plasma. Transfusion 2000, *40*, 193–200.

67. FDA Center for Drugs and Biologics and Center for Devices and Radiological Health. *Guideline on the General Principles of Process Validation*, 1987; 5–9.

68. Willkommen, H.; Schmidt, I.; Lower, J. Safety issues for plasma derivatives and benefit from NAT testing. Biologicals 1999, *27*, 325–331.

69. Aranha, H.; Forbes, S. Viral clearance strategies for biopharmaceutical safety: a multifaceted approach to process validation. BioPharm. 2001, *14* (5), 42–54.

70. Aranha, H. Viral clearance strategies for biopharmaceutical safety: Part 2: Filtration for virus removal. BioPharm. 2001, *14* (2), 1–8.

71. Aranha-Creado, H.; Brandwein, H. Application of bacteriophages as surrogates for mammalian viruses: a case for use in filter validation based on precedents and current practices in medical and environmental virology. PDA J. Pharm. Sci. Technol. 1999, *53*, 75–82.

72. Will, R.G.; Ironside, J.W.; Zeidler, M.; Cousens, S.N.; Estibeiro, K.; Alperovitch, A. A new variant of Creutzfeldt-Jakob syndrome in the UK. Lancet 1996, *347*, 921–925.

73. Bruce, M.E.; Will, R.G.; Ironside, J.W.; McConnell, L.; Drummond, D.; Suttie, A.; McCardle, L.; Chree, A.; Hope, J.; Birkett, C.; Cousens, S.; Fraser, H.; Bostock, C.J. Transmissions to mice indicate that 'new variant' CJD is caused by the BSE agent. Nature 1997, *389*, 498–501.

74. Collinge, J.; Sidle, K.C.; Meads, J.; Ironside, J.; Hill, A.F. Molecular analysis of prion strain variation and the aetiology of 'new variant' CJD. Nature 1996, *383*, 685–690.

75. Hill, A.F.; Zeidler, M.; Ironside, J.; Collinge, J. Diagnosis of new variant Cruetzfeldt-Jakob disease by tonsil biopsy. Lancet 1997, *349*, 99–100.

76. Lasmezas, C.I.; Cesbron, J.Y.; Deslys, J.P.; Demaimay, R.; Adjou, K.T.; Rioux, R.; Lemaire, C.; Locht, C.; Dormont, D. Immune system-dependent and -independent replication of the scrapie agent. J. Virol. 1996, *70*, 1292–1295.

77. Scott, M.R.; Will, R.; Ironside, J.; Nguyen, H.-O.B.; Tremblay, P.; DeArmond, S.J.; Prusiner, S.B. Compelling transgenetic evidence for transmission of bovine spongiform encephalopathy prions to humans. Proc. Natl. Acad. Sci. USA 1999, *96*, 15137–15142.

78. Carp, R.I.; Meeker, H.; Sersen, E. Scrapie strains retain their distinctive

characteristics following passages of homogenates from different brain regions and spleen. J. Gen. Virol. 1997, *78*, 283–290.

79. Aguzzi, A.; Weissman, C. Prion research: the next frontiers. Nature 1997, *389*, 795–798.
80. Bader, F.; Davis, G.; Dinowitz, M.; Garfinkle, B.; Harvey, J.; Kozak, R.; Lubiniecki, A.; Rubino, M.; Schubert, D.; Wiebe, M.; Woollett, G. Assessment of risk of bovine spongiogorm encephalopathy in pharmaceutical products, part 1. Biopharm 1998, *11*, 20–31.
81. Schreuder, B.E.C.; van Keulen, L.J.M.; Vromans, M.E.W.; Langeveld, J.P.M.; Smits, M.A. Preclinical test for prion diseases. Nature B, 1996; 381–563.
82. Hope, J.; Reekie, L.J.; Hunter, N.; Multhaup, G.; Beyreuther, K.; White, H.; Scott, A.C.; Stack, M.J.; Dawson, M.; Wells, G.A. Fibrils from brains of cows with new cattle disease contain scrapie-associated protein. Nature 1988, *24, 336* (6197), 390–392.
83. Outram, G.W. The pathogenesis of scrapie in mice. In Amstedam: *Slow Virus Diseases of Animals and Man*; Kimberlin, R.H., Eds.; North Holland, 1976; 325–327
84. Prusiner, S.B.; Groth, D.F.; Cochran, S.P.; Masiarz, F.R.; McKinley, M.P.; Martinez, H.M. Molecular properties, partial purification, and assay by incubation period measurements of the hamster scrapie agent. Biochemistry 1980, *19*, 4883–4891.
85. Lax, A.J.; Millson, G.C.; Manning, E.J. Can scrapie titers be calculated accurately from incubation periods? J. Gen. Virol. 1983, *64*, 971–973.
86. Meyer, R.K.; Oesch, B.; Fatzer, R.; Zurbriggen, A.; Vandevelde, M. Detection of bovine spongiform encephalopathy-specific PrP(Sc) by treatment with heat and guanidine thiocyanate. J. Virol. 1999, *73*, 9386–9392.
87. Bolton, D.C.; Bendheim, P.E.; Marmorstein, A.D.; Potempska, A. Isolation and structural studies of the intact scrapie agent protein. Arch. Biochem. Biophys. 1987, *258*, 579–590.
88. Prusiner, S.B.; McKinley, M.P.; Groth, D.F.; Bowman, K.A.; Mock, N.I.; Cochran, S.P.; Masiarz, F.R. Scrapie agent contains a hydrophobic protein. Proc. Nat. Acad. Sci. USA 1981, *78*, 6675–6679.
89. Lee, D.C.; Stenland, C.J.; Hartwell, R.C.; Ford, E.K.; Cai, K.; Miller, J.L.C.; Gilligan, K.J.; Rubenstein, R.; Fournel, M.; Petteway, S.R. Jr. Monitoring plasma processing steps with a sensitive Western blot assay for detection of the prion protein. J. Virol. Methods 2000, *84*, 77–89.
90. Serban, D.; Taraboulos, A.; DeArmond, S.J.; Prusiner, S. Rapid detection of Creutzfeldt-Jakob disease and scrapie prion protein. Neurology 1990, *40*, 110–117.
91. Grathwohl, K.-U.; Horiuchi, M.; Ishiguro, N.; Shinagawa, M. Sensitive enzyme-linked immunosorbent assay for the detection of PrPsc in crude extracts from scrapie-affected mice. J. Virol. Methods 1997, *65*, 205–216.
92. Rubenstein, R.; Kascsak, R.J.; Merz, P.A.; Papini, M.C.; Carp, R.I.; Robakis, N.K.; Wisniewski, H.M. Detection of scrapie-associated fibril (SAF) proteins using anti-SAF antibody in non-purified tissue preparations. J. Gen. Virol. 1986, *67*, 671–681.

93. Schaller, O.; Fatzer, R.; Stack, M.; Clark, J.; Cooley, W.; Biffiger, K.; Egli, S.; Doherr, M.; Vandevelde, M.; Heim, D.; Oesch, B.; Moser, M. Validation of a Western immunoblotting procedure for bovine PrPSc detection and its use as a rapid surveillance method for the diagnosis of bovine spongiform encephalopathy (BSE). Acta Neuropathological 1999, *98*, 437–443.

94. Bieschke, J.; Giese, A.; Schulz-Schaeffer, W.; Zerr, I.; Poser, S.; Eigen, M.; Kretzschmar, H. Ultrasensitive detection of pathological prion protein aggregates by dual-color scanning for intensely fluorescent targets. Proc. Natl. Acad. Sci. USA 2000, *97*, 5468–5473.

95. MacGregor, I.; Hope, J.; Barnard, G.; Kirby, I.; Drummond, O.; Pepper, D.; Hornsey, V.; Barclay, R.; Bessos, H.; Turner, M.; Prowse, C. Application of a time resolved fluoroimmunoassay for the analysis of normal prion protein in human blood and its components. Vox Sang. 1999, *77*, 88–96.

96. Schmerr, M.J.; Jenny, A.L.; Bulgin, M.S., et al. Use of capillary electrophoresis and fluorescent labeled peptides to detect the abnormal prion protein in the blood of animals that are infected with a transmissible spongiform encephalopathy. J. Chromatogr. A. 1999, *853*, 207–214.

97. Schmerr, M.J.; Jenny, A. A diagnostic test for scrapie-infected sheep using a capillary electrophoresis immunoassay with fluorescent-labeled peptides. Electrophoresis 1998, *19*, 409–414.

98. Safar, J.; Wille, H.; Itrri, V.; Groth, D.; Serban, H.; Torchia, M.; Cohen, F.E.; Prusiner, S.B. Eight prion strains have PrPSc molecules with different conformations. Nat. Med. 1998, *4*, 1157–1165.

99. Saborio, G.P.; Permanne, B.; Soto, C. Sensitive detection of pathological prion protein by cyclic amplification of protein misfolding. Nature 2001, *41*, 810–813.

100. Hsich, G.; Kenney, K.; Gibbs, C.J.; Lee, K.H.; Harrington, M.G. The 14-3-3 brain protein in cerebrospinal fluid as a marker for transmissible spongiform encephalopathies. N. Engl. J. Med. 1996, *335*, 924–930.

101. Zerr, I.; Brandel, J.P.; Masullo, C.; Wientjens, D.; deSilva, R.; Zeidler, M.; Granieri, E.; Sampaolo, S.; vanDejin, V.; Delasnerie-Lauprete, N.; Will, R.; Poser, S. European surveillance on Creutzfeldt-Jakob disease: a case control study for medical risk factors. J. Clin. Epidemiol. 2000, *53*, 747–754.

102. Green, A.J.; Thompson, E.J.; Stewart, G.F.; Zeidler, M.; McKenzie, J.M.; MacLeod, M.A.; Ironside, J.W.; Will, R.G.; Knight, R.S. Use of 14-3-3 and other brain-specific proteins in CSF in the diagnosis of variant Creutzfeldt-Jakob disease. J. Neurol. Neurosurg. Psychiatry 2001, *70*, 744–748.

103. WHO Working Group on International Reference Materials for Diagnosis and Study of Transmissible Spongiform Encephalopathies. Geneva, September, 1999.

104. Laplanche, J.L.; Laupretre Delasnerie, N.; Brandel, J.P.; Chatelain, J.; Beaudry, P.; Alperovitch, A.; Launay, J.M. Molecular genetics of prion diseases in France. French Research Group on Epidemiology of Human Spongiform Encephalopathies. Neurology. 1994, *44*, 23471–23451.

105. Palmer, M.S.; Dryden, A.J.; Hughes, J.T.; Collinge, J. Homozygous prion protein genotype predisposes to sporadic Creutzfeldt-Jakob disease. Nature 1991, *352* (6333), 340–342.

106. Collinge, J.; Palmer, M.S.; Dryden, A.J. Genetic predisposition to iatrogenic Creutzfeldt-Jakob disease. Lancet 1991, *337* (8755), 1441–1442.

107. Brown, P.; Cervenakova, L.; Goldfarb, L.G.; McCombie, W.R.; Rubenstein, R.; Will, R.G.; Pocchiari, M.; Martinez-Lage, J.F.; Scalici, C.; Masullo, C. Iatrogenic Creutzfeldt-Jakob disease: an example of the interplay between ancient genes and modern medicine. Neurology 1994, *44*, 291–293.

108. Zeidler, M.; Estibeiro, K.; Will, R.G. The genetics of Creutzfeldt-Jakob disease in the United Kingdom. J. Neurol. Neurosurg. Psychiatry 1997, *62*, 206.

109. Deslys, J.-P.; Jaegly, A.; d'Aignaux, J.H.; Mouthon, F.; Billette de Villemeur, T.; Dormont, D. Genotype at codon 129 and susceptibility to Creutzfeldt-Jakob disease. Lancet 1998, *351*, 1251.

110. Gale, P. Quantitative BSE risk assessment: Relating exposure to risk. Lett. Appl. Microbiol. 1998, *27* (5), 239–242.

111. CDRH. Guidance for FDA reviewers and Industry: Medical devices containing materials derived from animal sources (except for in vitro diagnostic devices). CDRH BSE Working Group, Center for Devices and Radiological Health, Nov 6, 1998.

112. CPMP/CVMP note for guidance on minimizing the risk of transmitting animal spongiform encephalopathy agents via human and veterinary medicinal products. Explanatory note for medicinal products for human use on the scope of the guideline. EMEA/CPMP/BWP/498/01, Feb 28, 2001.

113. Brown, P. Can Creutzfeldt-Jakob disease be transmitted by transfusion. Curr. Opin. Haematol. 1995, *2*, 472–477.

114. Evatt, B.L. Prions and hemophilia: Assessment of risk. Haemophilia 1998, *4*, 628–633.

115. Evatt, B.L.; Austin, H.; Barnhart, E.; Schonberger, L.; Sharer, L.; Jones, R.; DeArmond, S. Surveillance for Creutzfeldt-Jakob disease among people with hemophilia. Transfusion 1998, *38*, 817–820.

116. Lee, C.A.; Ironside, J.W.; Bell, J.E.; Giangrande, P.; Ludlam, C.; Esiri, M.M., et al. Retrospective neuropathological review of prion diseases in UK haemophilia patients. Thromb. Haemost. 1998, *80*, 909–911.

117. Heye, N.; Hensen, S.; Muller, N. Creutzfeldt-Jakob disease and blood transfusion. Lancet 1994, *343*, 298–299.

118. Holman, R.C.; Khan, A.S.; Belay, E.D.; Schonberger, L.B. Creutzfeldt-Jakob disease in the United States, 1979–1994: using national mortality data to assess the possible occurrence of variant cases. Emerg. Infect. Dis. 1996, *2*, 333–337.

119. Wientjens, D.P.; Davanipour, Z.; Hofman, A.; Kondo, K.; Matthews, W.B.; Will, R.G.; van Duijn, C.M. Risk factors for Creutzfeldt-Jakob disease: a reanalysis of case-control studies. Neurology 1996, *46*, 1287–1291.

120. EC Scientific Committee on Medicinal Products and Medical Devices (SCMPMD): Opinion on the risk quantification for CJD transmission via substances of human origin. 21st October, 1998.

121. EC Scientific Steering Committee. Opinion: Oral exposure of humans to the BSE agent: infective dose and species barrier, 13–14 April, 2000.

122. Masel, J.; Jansen, V.A.A. The measured level of prion infectivity varies in a

predictable way according to the aggregation state of the infectious agent. Biochim. Biophys. Acta 2001, *1535* (2), 164–173.

123. Antloga, K.; Meszaros, J.; Malchesky, P.S.; McDonnell, G.E. Prion diseases and medical devices. ASAIO J. 2000, *46*, S69–S72.

124. Steelman, V.M. Prion diseases—An evidence-based protocol for infection control. AORN J, *69*, 946–967.

125. Taylor, D.M. Inactivation of unconventional agents of the transmissible degenerative encephalopathies. In *Principles and Practice of Disinfection, Preservation and Sterilization*; Blackwell Science: Oxford, 1999; 222–236.

126. Bellinger-Kawahara, C.G.; Cleaver, J.E.; Diener, T.O.; Prusiner, S.B. Purified scrapie prion resist inactivation by UV irradiation. J. Virol. 1987, *61*, 159–166.

127. Latarjet, R. Inactivation of the agents of scrapie, Creutzfeldt-Jakob Disease, and kuru by radiations. In *Slow Transmissible Disease of the Nervous System*; Prusiner, S.B., Hadlow, W.J., Eds.; Academic Press: London, 1979; Vol. 2, 387–407.

128. Taylor, D.M.; Diprose, M.F. The response of the 22 A strain to microwave irradiation compared with boiling. Neuropathol. Appl. Neurobiol. 1996, *22*, 256–258.

129. Mould, D.L.; Dawson, A.M.; Smith, W. Scrapie in mice. The stability of the agent to various suspending media, pH and solvent extraction. Res. Vet. Sci. 1965, *6*, 151–154.

130. Brown, P.; Gibbs, C.J.; Amyx, H.L.; Kingsbury, D.T.; Rohwer, R.G.; Sulima, M.P.; Gajdusek, D.C. Chemical disinfection of Creutzfeldt-Jakob disease virus. N. Engl. J. Med. 1982, *306*, 1279–1282.

131. Brown, P.; Rohwer, R.G.; Green, E.M.; Gajdusek, D.C. Effects of chemicals, heat and histopathological processing on high-infectivity hamster-adapted scrapie virus. J. Infect. Dis. 1982, *145*, 683–687.

132. Brown, P.; Liberski, P.P.; Wolff, A.; Gajdusek, D.C. Resistance of scrapie agent to steam autoclaving after formaldehyde fixation and limited survival after washing at 360°C: Practical and theoretical implications. J. Infect. Dis. 1990, *161*, 467–472.

133. Dickinson, A.G.; Taylor, D.M. Resistance of scrapie agent to decontamination. N. Engl. J. Med. 1978, *229*, 1413–1414.

134. Ernst, D.R.; Race, R.E. Comparative analysis of scrapie agent inactivation methods. J. Virol. Methods. 1993, *41*, 193–202.

135. Fraser, H.; Bruce, M.E.; Chree, A.; McConnell, I.; Wells, G.A.H. Transmission of bovine spongiform encephalopathy and scrapie to mice. J. Gen. Virol. 1992, *173*, 1891–1897.

136. Prusiner, S.B. Prions. Proc. Natl. Acad. Sci. USA 1998, *95*, 13363–13383.

137. Haig, D.A.; Clarke, M.C. The effects of β-propiolactone on the scrapie agent. J. Gen. Virol. 1968, *3*, 281–283.

138. Taylor, D.M. Decontamination of Creutzfeldt-Jakob disease agent. Ann. Neurol. 1986, *20*, 749.

139. Turk, E.; Teplow, D.B.; Hood, L.E.; Prusiner, S.B. Purification and properties of the cellular and scrapie hamster prion proteins. Eur. J. Biochem. *176*, 21.

140. Foster, P.R.; Welch, A.G.; McLean, C.; Griffin, B.D.; Hardy, J.C.; Bartley, A.; MacDonald, S.; Bailey, A.C. Studies on the removal of abnormal prion protein by processes used in the manufacture of human plasma products. Vox Sang. 2000, *78*, 86–95.

141. Foster, P.R.; McLean, C.; Welch, A.G.; Griffin, B.D.; Hardy, J.C.; Bartley, A.; McDonald, S.; Bailey, A. Removal of abnormal prion protein by plasma fractionation. Transfusion Science 2000, *22*, 53–56.

142. Lee, D.C.; Stenland, C.J.; Miller, J.L.C.; Cai, K.; Ford, E.K.; Gilligan, K.J.; Hartwell, R.C.; Terry, J.C.; Rubestein, R.; Fournel, M.; Petteway, S.R. Jr. A direct relationship between the partitioning of the pathogenic prion protein and transmissible spongiform encephalopathy infectivity during the purification of plasma products. Transfusion 2001, *41*, 449–455.

143. Lee-, H.-S.; Brown, P.; Cervenakova, L.; Garruto, R.M.; Alpers, M.P.; Gajdusek, D.C.; Goldfarb, L.G. Increased susceptibility to kuru of carriers of the PRNP 129 methionine/methionine genotype. J. Infect. Dis. 2001, *183*, 192–196.

144. Aranha, H.; Martin, J. Potential prion risks and clearance by filtration. Genetic Engineering News 2001, *21* (11), 58–59.

145. Pocchiari, M.; Peano, S.; Conz, A.; Eshkol, A.; Maillard, F.; Brown, P.; Gibbs, C.I. Jr; Xi, Y.G.; Tenham-Fisher, E.; Macchi, G. Combination ultrafiltration and 6 M urea treatment of human Creutzfeldt-Jakob disease virus contamination. Horm. Res. 1991, *35*, 161–166.

146. Hellman, K.B.; Asher, D.M. International Workshop on Clearance of TSE Agents from Blood Products and Implanted Tissues, Meeting Report. Biologicals 2000, *28*, 189–192.

147. EMEA Expert Workshop on Human TSEs and Plasma-derived Medicinal Products. July 27, 2000. CPMP/BWP/1244/00.

148. CPMP Working Party on Biotechnology: Note for guidance on minimising the risk of transmitting animal spongiform encephalopathy agents via medicinal products (CPMP/BWP/877/96).

149. Bader, F.; Davis, G.; Dinowitz, M.; Garfinkle, B.; Harvey, J.; Kozak, R.; Lubiniecki, A.; Rubino, M.; Schubert, O.; Wiebe, M.; Woollett, G. Assessment of risk of bovine spongiogorm encephalopathy in pharmaceutical products, part 1. Biopharm 1998, *11* (20–31), 56.

150. Bendheim, P.E.; Bockman, J.M.; McKinley, M.P.; Kingsbury, D.T.; Prusiner, S.B. Scrapie and Creutzfeldt-Jakob disease prion proteins share physical properties and antigenic.

151. Harmeyer, S.; Pfaff, E.; Groschup, M.H. Synthetic peptide vaccines yield monoclonal antibodies to cellular and pathological prion proteins of ruminants. J. Gen. Virol. 1998, *79*, 937–945.

152. Rosenberg, R.N.; White, C.L.; Brown, P., et al. Precautions in handling tissues, fluids, and other contaminated materials from patients with documented or suspected Creutzfeldt-Jakob disease. Ann. Neuro. 1986, *19*, 75–77.

153. Brown, P.; Rohwer, R.G.; Dunstan, B.C.; MacAuley, C.; Gajdusek, D.C.; Drohan, W. The distribution of infectivity in blood components and plasma

derivatives in experimental models of transmissible spongiform encephalopathy. Transfusion 1998, *38*, 810–816.

154. Lee, D.; Stenland, C.; Gilligan, K.; Ford, E.; Hartwell, R.; Cai, K., et al. PrPsc partitioning during plasma fractionation. Development and application of a sensitive Western blot assay. Thromb Haemost. 1999, *82* (suppl.), 757.

155. Peano, S.; Reiner, G.; Carbonatto, M.; Bodenbender, L.; Boland, P.; Abl, K.J. Determination of the clearance factor for transmissible spongiform encephalopathy during the manufacturing process of polygeline. Intens. Care Med. 2000, *26*, 608–612.

156. Golker, C.F.; Whiteman, M.D.; Gugel, K.H.; Gilles, R.; Stadler, P.; Kovatch, R.M.; Lister, D.; Wisher, M.H.; Calcagni, C.; Hubner, G.E. Reduction of the infectivity of scrapie agent as a model for BSE in the manufacturing process of Trasylol. Biologicals 1996, *24*, 103–111.

18

Statistical Sampling Concepts

Hewa Saranadasa
Pharmaceutical Group Americas, A Division of Ortho-McNeil
Pharmaceuticals, Raritan, New Jersey, U.S.A.

18.1. INTRODUCTION

The objective of statistics is the use of sample information to infer the nature of a population. For example, laboratory diagnoses about the state of our health are made from a few drops of blood. This procedure is based on the assumption that the circulating blood is always well mixed and that one drop tells the same story as another. In this regard, selecting a sample is not very important, because the material from which we are sampling is uniform and any sample gives similar results. But that is not the case always: sometimes the material is far from uniform. In those instances, it is important to know how to choose a representative sample that exhibits characteristics similar to those possessed by the population about which we wish to make inferences. The scientific way to meet this goal is to select a sample in such a way that every sample of same size has an equal probability of being selected. This statistical procedure based on probability is called the random sampling. In theory, population items are assumed to be well mixed before choosing a representative random sample.

In some populations the sampling units in certain groups are close to each other. The population can be divided into similar groups called strata before choosing a sample. Selecting random samples from each of the strata has the major advantage of reducing the variance of the estimation procedure over selecting a random sample from entire population. This sampling procedure is called stratified sampling. For pharmaceutical industry applications, this sampling is chosen at different stages of the production process. For example, this kind of sampling is used in estimating bulk product characteristics of a lot. The raw and/or processed materials are often stored in drums. Typically, one first selects a number of drums randomly and then obtains random samples from the top, middle, and bottom parts of the sampled drums. One then tests these samples to determine whether the material meets some in-house quality assurance specifications. The number of drums should be decided based on the statistical justification regarding the precision of the estimate.

In particular, in an application relevant to contamination control of parenteral product, random samples are chosen, representing beginning, middle, and end from a lot for testing bacterial endotoxin level of the end products (vials or syringes). In a production process of a lot, we may have seen the differences of the items due to unavoidable causes of nature. Therefore, it is desirable to choose random samples from each stage or stratum. This sampling procedure has an additional advantage of obtaining separate estimate potency in each stratum from the same sample data and testing for the homogeneity of the product in the process.

In microbiological and parenteral manufacturing contamination, it is evident that the level of concentrations of contaminants is very low. Therefore, microbiologists often need to determine the amount of sample that must be drawn in order to have a high degree of confidence that contamination of microorganisms will be detected at the level that is of concern to the company. The Poisson distribution would be well suited to model the microbial contamination in these situations.

18.2. POISSON PROBABILITY DISTRIBUTION

In microbiological applications, suppose a very low bacteria concentration of a population of V liter. The number of occurrences of the event, say number of microbes, (r) in a volume of v approximately obeys a Poisson probability law: the approximation is quite accurate if v is much smaller than the population, with parameter μv, more precisely the probability that exactly r events (P_r) occur in a volume of v is equal to

$$P_r = \frac{e^{-\mu v}(\mu v)^r}{r!},$$

where $r = 0, 1 \ldots \infty$, The μ is the mean rate of occurrence of events per unit volume—i.e., V(liter) is the ovearall material to be tested, v(liter) is the volume of the sample, and n is the absolute number of bacteria distributed in V so that $\mu = n/V$. Let the probabilities that this sample does contain none, less than one, and less than two viruses be denoted as P_0, P_1, P_2. Then

$$P_0 = e^{-\mu v}$$

$$P_1 = e^{-\mu v} + (\mu v)e^{-\mu v}$$

$$P_2 = e^{-\mu v} + (\mu v)e^{-\mu v} + \frac{1}{2}(\mu v)^2 e^{-\mu v}$$

The probability of detecting at least one, two, and three events are given by $(1 - P_0)$, $(1 - P_1)$ and $(1 - P_2)$.

In microbiological testing, the underlying assumption is that only a single organism is required for detectable growth to occur. However, it is believed that more than one organism is usually required to show growth in broth (1). Suppose a solution contains 1000 microbes in a liter: it does not mean that each milliliter will contain exactly a single organism. However, the chance of capturing 50 organisms in a 500 mL sample is much higher than capturing one organism in each milliter volume. Therefore, it is more informative to calculate the volume that should be taken from a preparation to have a certain level of confidence that an organism will be detected given a certain level of contamination of the preparation or product.

18.2.1. Example

Suppose that a microbiologist wants to know how much material should be tested to declare a $(1 - \alpha)$ 100% confidence that the product is sterilized. The α is the chance of the preparation's lack of sterility. From past experience, the biologist knows that the contamination rate was 2 per liter. What would be the sample size that needs to be tested, if he wants to have 95% and 99% confidence the preparation is sterilized?

The absolute sterility cannot be practically achieved without testing the entire preparation. Therefore, the sterility of the preparation is claimed only in probabilistic term by testing V amount of the preparation. Let P be the probability of detecting an organism in V, and we also know the mean rate of contamination, $\mu = 0.022$ per mL. Then we have $P = (1 - e^{-\mu v})$ and v can be expressed as

$$v = -\frac{\ln(1 - P)}{\mu}.$$

If $P = 0.95$, then $v = 1498$ mL, and if $P = 0.99$, $v = 2303$ mL needs to be tested if material was present as a single bulk preparation. On the other hand,

TABLE 1 Volume (mL) of Sample for Detecting at Least 1–3 Organisms with a Given Probability

μ/mL	At least one			At least two			At least three		
	.90	.95	.99	.90	.95	.99	.90	.95	.99
100	.023	.030	.046	.040	.047	.067	.053	.063	.084
50	.046	.060	.092	.078	.095	.133	.106	.126	.168
25	.092	.120	.184	.156	.190	.266	.213	.252	.336
15	.154	.200	.307	.259	.316	.443	.355	.420	.561
10	.230	.300	.461	.389	.475	.664	.532	.630	.840
5	.461	.600	.922	.778	.949	1.33	1.06	1.26	1.68
2	1.15	1.50	2.30	1.94	2.37	3.32	2.66	3.15	4.21
1	2.30	3.00	4.60	3.89	4.74	6.64	5.32	6.30	8.41
.75	3.07	4.00	6.15	5.19	6.33	8.85	7.10	8.40	11.20
.50	4.61	5.99	9.20	7.78	9.49	13.29	10.65	12.59	16.83
.25	9.21	11.98	18.41	15.56	18.98	26.58	21.29	25.19	33.66
.10	23.03	29.96	46.05	38.89	47.45	66.42	53.22	62.97	84.07
.075	30.71	39.94	61.41	51.86	63.25	88.58	70.97	83.95	112.09
.05	46.05	59.92	92.09	77.80	94.89	132.83	106.45	125.94	168.14
.025	92.10	119.83	184.18	155.59	189.78	265.67	212.89	251.88	336.28
.01	230.35	299.65	460.01	388.97	474.37	663.15	532.17	629.45	841.65
.0075	307.07	399.44	614.07	518.63	632.75	884.20	709.67	839.27	1120.9
.005	460.69	599.30	920.02	777.94	948.73	1326.3	1064.3	1258.9	1683.3
.0025	921.39	1198.6	1840.0	1555.9	1897.5	2652.6	2128.7	2517.8	3366.6
.001	2302.7	2997.2	4604.1	3890.1	4743.9	6638.9	5322.3	6295.9	8407.8

if the material was contained in 10 mL final production vials, the micro-biologist would need to collect 150 vials to have 95% confidence and 231 vials to have a 99% high degree of confidence.

Table 1 gives the volume of sample for detecting at least one to three organisms with a given probability. Any other desired level of confidence level, for the mean rate not given in the table, can be calculated from the formula given above.

18.3. THE LIMIT OF QUANTITATION

In analytical method development, the limit of quantitation (LOQ) of an assay is an important performance characteristic that is fundamentally im-portant to the interpretation of assay results. It refers to the ability of the assay method to reliably detect differences as the concentration of analyte approaches zero.

In microbiological applications, the similar concept appears as how capable the method is of detecting an organism, if it is present by the sterility testing procedure. In this regard, the LOQ is often based on the sample size. For example, if the sample is 0.6 mL, and an organism is accurately detected by the test, then the limit of detection is two organisms/mL. In making this statement, a strong assumption of even distribution is made and ignores the statistical nature of distribution of the organisms in the preparation and dis-regards the low confidence of detecting an organism in the sterility test. If we use the probability distribution of Poisson (as we discussed earlier), we have:

$$P = 1 - e^{-\mu v} \quad \text{and} \quad \mu = -\frac{\ln(1 - P)}{v}$$

with $P = .95$ and $v = 0.6$ mL we have $\mu = 3/0.6$ and 5 organisms per milliliter and 8 organism per mL for 99% detecting probability of a single organism. The correct statement is to say 8 organisms/mL is the limit of detection of 99% chance of detecting contamination or 5 organisms/mL is the limit of detection of 95% chance of detecting contamination. Table 2 gives the orga-nisms per milliliter required to detect at least one, two, or three organisms in the sample volume with the probability given.

18.4. LIKELIHOOD OF NOT DETECTING AN ORGANISM

The probability of not detecting a microorganism for a selected sample volume (v mL) heavily depends on the density or concentration of micro-

TABLE 2 Organisms per Milliliter (µL) Required to Detect at Least 1–3 Organisms in the Sample Volume Shown with the Probability Given

v(mL)	At least one			At least two			At least three		
	.90	.95	.99	.90	.95	.99	.90	.95	.99
0.01	230.5	299.7	460.0	389.0	474.4	663.2	532.2	629.5	841.7
0.02	115.7	149.8	230.0	194.5	237.2	331.6	266.1	314.7	420.8
0.03	76.8	99.9	153.5	129.7	158.2	221.1	177.4	209.8	280.2
0.04	57.6	74.9	115.0	97.2	118.6	165.8	133.0	157.4	210.4
0.05	46.1	60.0	92.1	77.8	94.9	132.8	106.5	126.0	168.1
0.06	38.4	50.0	76.7	64.8	79.1	110.5	88.7	105.0	140.1
0.07	32.9	42.8	65.8	55.6	67.8	94.9	76.4	90.0	120.1
0.08	28.8	37.5	57.5	48.6	59.3	83.0	66.5	78.7	105.2
0.09	25.6	33.3	51.1	43.2	52.7	73.8	59.1	70.0	93.4
0.1	23.0	30.0	46.1	38.9	47.4	66.4	53.2	63.0	84.1
0.2	11.5	15.0	23.0	19.5	23.7	33.2	26.6	31.5	42.0
0.3	7.7	10.0	15.4	13.0	15.8	22.1	17.7	21.0	28.0
0.4	5.8	7.5	11.5	9.7	11.9	16.6	13.3	15.7	21.0
0.5	4.6	6.0	9.2	7.8	9.5	13.3	10.7	12.6	16.8
0.6	3.8	5.0	7.7	6.5	7.9	11.1	9.0	10.5	14.0
0.7	3.3	4.3	6.6	5.6	6.8	9.5	7.6	9.0	12.0
0.8	2.9	3.7	5.8	4.8	5.9	8.3	6.7	7.8	10.5
0.9	2.6	3.3	5.1	4.3	5.3	7.4	5.9	7.0	9.4
1	2.3	3.0	4.6	4.0	4.7	6.6	5.3	6.3	8.4
2	1.2	1.5	2.3	2.0	2.4	3.3	2.7	3.2	4.2
5	0.46	0.6	0.9	.8	0.9	1.3	1.1	1.3	1.7
10	0.23	0.3	0.5	.4	0.5	0.7	0.5	0.6	0.8
15	0.15	0.2	0.3	.3	0.3	0.4	0.4	0.4	0.6

organisms in the target preparation. Table 2 shows that if the sample volume is 0.6 mL and if the microbe concentration varies from 8 organism/mL to 4 organism/mL, the probability of not detecting an organism (risk) increases from 1% to 10%. That is, we have a higher associated risk of non-detection in the same volume sample of two different decreasing contamination levels. Therefore, it is important to the researcher to know the risk in advance for different samples of known contamination ranges. For example, in order to ensure that tapwater meets minimum federal safety drinking water standards, a township municipal authority tested 100 mL sample for microbiological contaminants and found no coliform bacteria in the sample and released it for public usage. Table 3 shows that the associated risk of non-detection of the bacteria in the water system would be 90% and 37% for one per liter and 10 per liter contaminated water systems and there is no risk if the contamination rate is more than 100 per liter water systems. If more than 100 per liter

TABLE 3 A Probability of Not Detecting at Least One Organism

	Number of organisms per liter (μ)				
v (mL)	1	10	100	1000	10000
0.1	99.99	99.90	99.00	90.48	36.80
0.2	99.98	99.80	98.02	81.87	13.53
0.3	99.97	99.70	97.05	74.08	4.98
0.4	99.96	99.60	96.08	67.03	1.83
0.5	99.95	99.50	95.12	60.65	0.67
0.6	99.94	99.40	94.18	54.88	0.25
0.7	99.93	99.30	93.24	49.66	0.09
0.8	99.92	99.20	92.31	44.93	0.03
0.9	99.91	99.10	91.40	40.66	0.01
1	99.90	99.00	90.48	36.80	.005
2	99.80	98.02	81.87	13.53	0
5	99.50	95.12	60.65	0.67	0
10	99.01	90.48	36.80	.004	0
15	98.51	86.07	22.31	0	0
20	98.02	81.87	13.53	0	0
25	97.53	77.88	8.21	0	0
30	97.05	74.08	4.98	0	0
40	96.08	67.03	1.83	0	0
50	95.12	60.65	0.67	0	0
75	92.77	47.24	0.06	0	0
100	90.48	36.80	0	0	0

TABLE 4 Probability of Not Detecting at Least Two Organisms

	Number of organisms per liter (μ)				
v (mL)	1	10	100	1000	10000
0.1	100.00	100.0	99.99	99.53	73.58
0.2	100.00	100.0	99.98	98.25	40.60
0.3	100.00	100.0	99.96	96.31	19.92
0.4	100.00	99.99	99.92	93.85	9.16
0.5	100.00	99.99	99.88	90.98	4.04
0.6	100.00	99.99	99.83	87.81	1.74
0.7	100.00	99.99	99.77	84.42	0.73
0.8	100.00	99.99	99.70	80.88	0.30
0.9	100.00	99.99	99.62	77.25	0.12
1	100.00	99.99	99.53	73.58	0.05
2	100.00	99.98	98.25	40.60	0
5	99.99	99.87	90.98	4.04	0
10	99.99	99.53	73.58	0.05	0
15	99.99	98.98	55.78	0	0
20	99.98	98.25	40.60	0	0
25	99.97	97.35	28.73	0	0
30	99.96	96.31	19.92	0	0
40	99.92	93.85	9.16	0	0
50	99.88	90.98	4.04	0	0
75	99.73	82.66	0.47	0	0
100	99.53	73.58	0.05	0	0

contamination rate were critical, the authority should test more than 75 mL sample to release the water system with more than 99% confidence (see Table 3). Tables 3 through 5 provide the risk level associated with known sample volume and the selected contamination levels for number of detection of bacteria (0,1,2).

In the following sections I discuss the lot-to-lot acceptance sampling by attributes and variables type. These sampling plans deal with the evaluation of the finished products before they are released to the market. In these sampling plans, hypergeometric and normal distributions find applications.

18.5. HYPERGEOMETRIC DISTRIBUTION

Let us assume N, p, and n to be lot size, proportion of defective items in the lot, and sample size, respectively. The number of defective items in the sam-

TABLE 5 Probability of Not Detecting at Least Three Organisms

V (mL)	Number of organisms per liter (μ)				
	1	10	100	1000	10000
0.1	100.00	100.00	100.00	99.98	91.97
0.2	100.00	100.00	100.00	99.89	67.67
0.3	100.00	100.00	100.00	99.64	42.32
0.4	100.00	100.00	99.99	99.21	23.81
0.5	100.00	100.00	99.99	98.56	12.47
0.6	100.00	100.00	99.99	97.68	6.19
0.7	100.00	100.00	99.99	96.59	2.96
0.8	100.00	100.00	99.99	95.26	1.37
0.9	100.00	100.00	99.99	93.71	0.62
1	100.00	100.00	99.98	91.97	0.28
2	100.00	100.00	99.89	67.67	0
5	100.00	99.99	98.56	12.47	0
10	100.00	99.99	91.97	0.28	0
15	100.00	99.95	80.88	0.00	0
20	100.00	99.89	67.67	0	0
25	100.00	99.78	54.38	0	0
30	100.00	99.64	42.32	0	0
40	99.99	99.21	23.81	0	0
50	99.99	98.56	12.47	0	0
75	99.99	95.95	2.02	0	0
100	99.98	91.97	0.28	0	0

ple is a random number between 0 and n. Each item has equal probability, p, of being a defective item. The probability of the number of defective items in the sample, d, follows a distribution called hypergeometric. The probability is given by the following distribution function:

$$Pr(D = d) = \frac{\binom{Np}{d}\binom{N(1-p)}{n-d}}{\binom{N}{n}}, 0 \leq d \leq n$$

where the $\binom{N}{n}$ notation indicates the number of different samples of size n drawn from a population or lot size N. That is,

$$\binom{N}{n} = \frac{N!}{(N-n)!n!}$$

where $n! = n(n - 1)(n - 2)\ldots1$ (for example, $4! = 4 \times 3 \times 2 \times 1 = 24$). When the lot size is sufficiently large, the binomial distribution gives reasonably accurate probabilities of the above. If n is sufficiently large (≥ 10) and p is close to zero (≤ 0.10), the Poisson distribution could be used as an approximation to calculate above probabilities.

Binomial approximation when N is large:

$$Pr(D = d) \cong \binom{n}{d} p^d (1 - p)^{n-d}.$$

Poisson approximation when $n \geq 10$ and $p \leq 0.10$:

$$Pr(D = d) \cong \frac{e^{-np}(np)^d}{d!}.$$

18.6. LOT-BY-LOT ACCEPTANCE SAMPLING ATTRIBUTES

There are three types of attribute sampling plans: single, double, and multiple. In the single sampling plan, one sample is taken from a lot and a decision to reject or accept the lot is made on the inspection results of that sample. Double and multiple sampling plans are somewhat more complicated and are not discussed in this chapter. The single sampling sampling plan is the most common and easiest to use of the sampling plans: a predetermined number of units (sample) from each lot are inspected by attributes. In this type of attribute sampling scheme, the items are classified as conforming/acceptable or nonconforming/not acceptable.

For example, consider a media fill trial of evaluating an aseptic filling process. If the number of contaminated units is greater than the acceptance number of units in the sampling plan given, we reject the lot due to its potential risk of susceptibility to microbiological contamination, even though after passing the lots from the sampling inspection plan there would be an expected minimum risk of susceptibility to microbial contamination. The maximum acceptable level of microbial contaminated rate is called acceptable quality level (AQL).

For example, consider a plan of 4750 media fill units with one as the maximum tolerable number of contaminated units given in ISO/DIS 13408-1 plan (2). The contaminated level with 95% probability of accepting the batch from this plan is tabulated as 0.0017%. This rate of contamination is called the AQL. This level is dependent on safety and regulatory requirements. It is important to keep in mind that any sampling does not involve certainty. When we draw samples, we assume some risks. We assume the risk of rejecting a lot that should have been accepted (producer risk) and accepting a lot that should have been rejected (consumer risk). On other occasions, by chance alone the sample may contain a disproportionately small number of defectives (biased sample) and the lot may be accepted when in fact the quality level was such that it should have been rejected. The question then arises as to what probabilities would exist in the plan that we have devised. We can find this probability/percent defective (quality level) relationship for any sampling plan by constructing its operating characteristic (OC) curve.

18.7. OPERATING CHARACTERISTIC CURVE

The operating characteristic (OC) curve expresses the discriminating power of a given sampling plan; in other words, its ability to distinguish between product of conforming and nonconforming quality.

The OC curve decreases from 0 to 1 as the proportion of nonconforming units increases from 0 to 1.

Determining the probability of acceptance for each of several values of incoming quality can develop an OC curve. There are three types of attribute probability distributions that can be used to find the probability of acceptance: hypergeometric, binomial, and Poisson distributions. When the defective rate is 10% or less and the sample size is relatively large, the Poisson distribution is preferable because of the ease of table use. The Poisson formula as applied to acceptance sampling is given below.

$$P_r = \frac{e^{-np}(np)^r}{r!}, r = 0, 1, \ldots \alpha$$

The probability of exactly r defectives in a sample of n denoted by P_r and np is the expected average rate of defectives and p is the proportion of defectives in the population. The above distribution can be used to find cumulative probability of the number of defectives less than or equal to the number of defectives (Ac) allowed by the plan. That is,

$$Pr \ (\# \text{ of defectives} \leq Ac) = \sum_{r=0}^{Ac} \frac{e^{-np}(np)^r}{r!}.$$

18.7.1. Example

Consider a sampling plan with the following parameters: $n = 150$ acceptance numbers 2, 3 and 4. Construct the table for acceptance for quality level ranging from 0.33% to 6% fraction of nonconforming units.

Probability of Acceptance for the Single Sampling Plan: n = 150

p	np	Acceptance probability for # defectives		
		≤2	≤3	≤4
0.33%	0.5	.986	.999	1
1%	1.5	.809	.935	.982
2%	3.0	.423	.647	.815
3%	4.5	.174	.343	.533
4%	6.0	.062	.151	.285
5%	7.5	.020	.060	.132
6%	9.0	.006	.021	.055

18.7.2. Example

Consider the earlier example with acceptance number = 3. Examine the OC curve if sample size increases to 50, 100, 150, and 300.

Probability of Acceptance for the Single Sampling Plan: AC = 3

p	n = 50		n = 100		n = 150		n = 300	
	np	AP	np	AP	np	AP	np	AP
1%	0.5	.999	1.0	.981	1.5	.935	3.0	.647
2%	1.0	.980	2.0	.857	3.0	.647	6.0	.151
3%	1.5	.935	3.0	.647	4.5	.343	9.0	.021
4%	2.0	.857	4.0	.433	6.0	.151	12.0	.002
5%	2.5	.757	5.0	.265	7.5	.060	15.0	.000
6%	3.0	.647	6.0	.151	9.0	.021	18.0	.000

AP: Acceptance Probability.

Examples 18.7.1 and 18.7.2 showed that OC curves become steeper when n is increased or acceptance number is decreased. The steeper OC curves reject the batch at a much faster rate at given AQL and it discriminates between acceptable and unacceptable lots. Figures 1 and 2 illustrate the OC curves graphically.

18.8. METHOD OF DESIGNING A SAMPLING PLAN

We have demonstrated that the amount of quality protection provided by sampling scheme depends on its operating characteristic curve. One way to

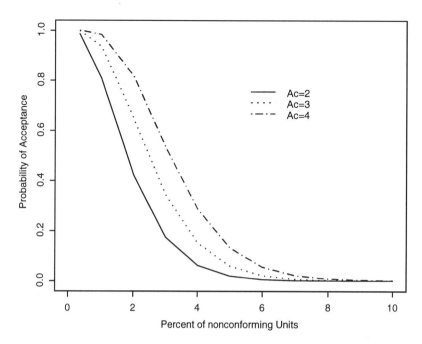

FIGURE 1 Comparison of operating characteristic curves of three sampling plans with $n = 150$.

establish such a plan is to choose the parameters Ac (acceptance number) and n (sample size) in such a way as to satisfy two required characteristics on the OC curve. In the design we select two quality levels, p_1 (in Acceptable Quality Level, AQL) and p_2 (in Rejectable Quality Level, RQL). Select values for n and Ac in such a way that the producer risk and the consumer risk are maintained at a desired level at the two points selected.

18.8.1. Computation Strategy

Let us assume the acceptance probabilities of a lot at the two quality levels selected on the OC curve are $(1 - \alpha)$ and β respectively for p_1 and p_2. We need to solve for sample size n and acceptance number (Ac) from the two equations obtained from cumulative binomial distribution with parameters (n, p_1) and (n, p_2). It may be easy to solve them using binomial to chi squared (χ^2) approximation. Using the approximation, we have the following 2 equations that need to be solved for n and Ac.

$$\int_{2np_1}^{\alpha} \chi^2_{2A_c+2} \, dx \geq 1 - \alpha$$

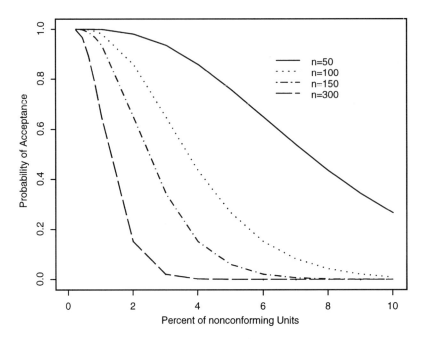

FIGURE 2 Comparison of operating characteristic curves of four sampling plans with Ac = 3.

$$\int_{2np_2}^{\alpha} \chi^2_{2Ac+2} \, dx \le \beta$$

The solutions for n and Ac can be obtained by χ^2 tables, which are available in any standard statistics textbook. The following steps help to find the required sampling plan.

Step 1: Look at the 2 columns of χ^2 table given right areas of $(1 - \alpha)$ and β and compare the corresponding variate values $(\chi^2_{(1)}, \chi^2_{(2)})$ given in the two columns for various values of degrees of freedom (v).

Step 2: Locate the position where $\chi^2_{(1)} / \chi^2_{(2)} = p_1/p_2$ and note down corresponding degree of freedom.

Step 3: Using the equation $2Ac + 2 = v$ calculate acceptance number Ac.

Step 4: Obtain the value n using the equation $2np_1 = \chi^2_{(1)}$ or $2np_2 = \chi^2_{(2)}$

18.8.2. Example

Set aside for a moment traditional retest requirements for the bacterial endo-toxin test so we may compare a orthogonal, statistical method to the traditional retest criteria (refer to FDA Guideline on Validation of the LAL Test as an End-Product Endotoxin test for Human and animal parenteral Drugs, Biological Products and Medical Devices for official retest method) for information purposes.

In a certain pharmaceutical company, a lot of parenteral drug product is rejected due to one syringe of 20 samples that had endotoxin level above the reference level. For extra precautions the quality assurance department wants to test a certain number of syringes from each lot produced in the same week but passed the release test for endotoxin. For the sake of explanation, assume AQL level 0.025% and RQL level is 1.12%. The probability of acceptance of the lot at the above two levels is 95% and 10%, respectively. What would be the sample size and the acceptance number?

AQL and RQL ratio is 0.0223 and at steps 1 and 2 we see that at the degree of freedom 2, two chi squared variates ratio is 0.103/4.61 and it is very close to AQL/RQL ratio. From step 3 we get Ac = 0, and step 4 directs us to have required sample size. That is $n = 206$. If we found none of syringes in the sample had the endotoxin level above the specification, then the lot is accepted; otherwise the lot should be rejected.

Some of the shortcomings in the above method are as follows: (a) the distribution of the proportion of defectives in the lot is not considered at all; (b) the plan becomes unrealistic if two points on the OC curve are not selected due to consideration of the process curve; (c) the method usually yields a large sample size, thereby creating practical difficulties; and (d) in practice it is desirable to reduce the sampling risk when dealing with large lots, but the method does not permit such an adjustment directly since the sample size is not related to lot size. In the industry, concerns associated with the latter point are very critical.

The MIL-STD-105E/ANSI/ASQ Z1.4 scheme (3) addresses the latter. This scheme was originally designed by a team of experts from the United States, the United Kingdom, and Canada. The following are the main features of the scheme:

1. The plans consist of tables listing sample size code letter.
2. Several inspection levels are recommended (Normal, Reduced, and Tightened) for using different situations.
3. Under each inspection level, the lot size is divided into number of ranges and the particular sample size is recommended for each range. This provides a method of deciding on a suitable sample size based on practical consideration.

4. For each selected value of n, the table gives the series of Ac each of which is categorized under specified value of AQL.
5. Schemes obtained as above have the property that the probability of acceptance at AQL takes some value in the interval 88–99% and AQL is the only reference point governing behavior of the plan.
6. There is provision for switching over to a different level of inspection on the basis of the results of the past experience.

18.8.3. Inspection Levels

Three general inspection levels (I, II, III) are given in ANSI/ASQC Z-1.4 tables. The different levels of inspection provide approximately the same protection to the producer, but different protections to the consumer. Inspection level II is the norm, with level I providing about half the amount of protection (less steep OC curves) and level III providing about twice the amount of protection (steeper OC curves and consequently more discrimination and increased inspection cost). The decision on the inspection level is also a function of the type of the product. For inexpensive items wherein destructive testing is performed, inspection level II should be considered. When product costs are higher or when the items are complex and expensive, inspection level III may be applicable.

The special inspection levels S-1, S-2, S-3, and S-4 are also provided in the plans. These plans use smaller numbers of samples, so the sampling risk must be unavoidable and should be used when one can tolerate the associated risk due to small sampling situation. When the produced goods are submitted in a continuing series of lots, the following switching rules are applicable.

Reduced inspection. Under reduced inspection, the plan allows a smaller sample to be taken than under normal inspection. The reduced inspection may be implemented when it is evident that quality is running unusually well.

Normal Inspection. Normal inspection is used when there is no evidence that the quality of product being submitted is better or poorer than the specific quality level. Normal inspection is usually used at the start of inspection.

Tightened Inspection. Under tightened inspection, the inspection plan requires much stringent acceptance criteria. Such a plan is used when it becomes evident the quality is deteriorating.

Let us discuss how we use MIL-STD tables to obtain the required sample size for the following example.

18.8.4. Example

A production lot of 2000 units is subjected to sampling by attributes. Determine the scheme of single sampling for general inspection level II, normal inspection, and AQL = 1%.

Using lot size N = 2000 and inspection level II, sample size letter code K is obtained from Table 6; given the AQL level 1%, the table will show the required sample size. It is n = 125, AQL 1%, acceptance number (Ac) = 3 and rejection no (Re) = 4. Thus, from a lot of 2000, a random sample of 125 should be inspected. If fewer than four nonconforming units are found, the lot is accepted; otherwise, the lot is rejected. If the lot sizes are 800, 400, and 80 and for the same AQL and the same sampling level inspection, what would be the sampling plans for each of lot size?

For N = 800, the sample letter J and n = 80, Ac = 2 and Re = 3
For N = 400, the sample letter H and n = 50, Ac = 1 and Re = 2
And for N = 80, sample letter E, and n = 13, Ac = 0 and Re = 1

TABLE 6 Inspection by Attributes According to Military Standard 105E at General Inspection Level II, Normal Inspection

Lot size	Sample letter	Number of sample	AQL % at pass/fail			
			0/1	1/2	2/3	3/4
2–8	A	2	6.5	25	—	—
9–15	B	3	4.0	15	25	—
16–25	C	5	2.5	10	15	25
26–50	D	8	1.5	6.5	10	15
51–90	E	13	1.0	4.0	6.5	10
91–150	F	20	0.65	2.5	4.0	6.5
151–280	G	32	0.40	1.5	2.5	4.0
281–500	H	50	0.25	1.0	1.5	2.5
501–1200	J	80	0.15	.65	1.0	1.5
1201–3200	K	125	0.10	.40	.65	1.0
3201–10000	L	200	0.065	.25	.40	.65
10001–35000	M	315	0.040	.15	.25	.40
35001–150000	N	500	0.025	.10	.15	.25
150001–500000	P	800	0.015	.065	.10	.15
>500000	Q	1250	0.010	.040	.065	.10

Source: Refs. 1 and 3.

18.9. VARIABLE TYPE SAMPLING PLANS

All attribute sampling plans are based on the data that can be counted. Each inspected item is classified as either conforming or nonconforming and an accept/reject decision is based on a previously selected sample risk. The variable sampling plans require unit measurements. The sample data is recorded and processed to yield statistics such as sample average and standard deviation. This type of sampling plan uses actual measurements and therefore requires an assumption regarding the distribution of the observations. The measurements obtained in most of the practical situations approximately follow a normal distribution.

The concern here is whether the lot meets the lower or upper specification limit. Also, we may not have prior knowledge of process variability. Without loss of generality we assume the upper specification such as no more than desired level of bacteria counts. The acceptance criteria, if $\bar{x} + ks < U$, where $\bar{x}$ and s are the sample mean and sample standard deviation k is a constant and U is the upper specification. We have the following distribution:

$$\bar{x} + ks \approx N\left(\mu + k\sigma, \frac{\sigma^2}{n} + \frac{k^2\sigma^2}{2n}\right)$$

(i.e., normal distribution with mean, $\mu + k\sigma$ and variance),

$$\frac{\sigma^2}{n} + \frac{k^2\sigma^2}{2n}$$

To design a sampling plan, two conditions on the OC curve are needed. Let the probability of acceptance of the lot be $1 - \alpha$ if the nonconforming rate is p_1 and the probability of acceptance of the lot be β if the nonconforming rate is p_2. Using the normal distribution above we have the following:

$$k + Z_{p_1} = Z_\alpha \sqrt{\frac{1}{n} + \frac{k^2}{2n}} \quad \text{and}$$

$$k + Z_{p_2} = -Z_\beta \sqrt{\frac{1}{n} + \frac{k^2}{2n}},$$

where Z_p is the Z-value corresponding to $100p\%$ percentile of the standard normal distribution.

Solving above two equations, we have expressions for k and n:

$$k = -\frac{(Z_\alpha Z_{p_2} + Z_{p_1} Z_\beta)}{Z_\alpha + Z_\beta} \quad \text{and} \quad n = \left(1 + \frac{k^2}{2}\right)\left(\frac{Z_\alpha + Z_\beta}{Z_{p_1} - Z_{p_2}}\right)^2.$$

If we know the process variability we need only

$$n = \left(\frac{Z_\alpha + Z_\beta}{Z_{p_1} - Z_{p_2}} \right)^2$$

samples.

18.9.1. Example

The production line of syringe filling of particular product needs to be tested for fill weight accuracy. The upper specification of the fill weight is 0.856 gram. The AQL of the overfilled product is 0.025% and the worst quality of the lot would be considered if the proportion of overfilled syringes were 1.12%. The probability of accepting such lots would be 95% and 10%, respectively. What would be the sampling plan? If syringe-to-syringe standard deviation were known to be 0.014, what would be the sampling plan?

$$p_1 = 0.00025, \alpha = 0.05 \quad \text{and} \quad p_2 = 0.0112, \beta = 0.10$$
$$Z_{p_1} = -3.48, Z_\alpha = -1.64 \quad \text{and} \quad Z_{p_1} = -2.28, Z_\beta = -1.28$$

Plugging the above values for expressions n and k above, we get $k = 2.80$ and $n = 30$.

Suppose the fill weight of 30 random sample was obtained and the mean and standard deviation of the sample were .814 g and .0072 g, respectively. Then we calculate the upper quality index,

$$Q_u = \frac{(U - \bar{x})}{s}$$

which is in this example 5.78. Since the $Q_u > 2.80$ (k, obtained above), we accept the lot. If we know the syringe-to-syringe variability (i.e., $\sigma = 0.014$), the sample of six syringes needs to be tested. If

$$Q_u = \frac{(U - \bar{x})}{\sigma} > 2.80$$

then we accept the lot.

18.10. MIL STANDARD PLANS FOR VARIABLE TYPE

Similar to the attribute-type sampling plans, the procedure and the criteria for the inspection by variables are described in the MIL-STD 414/ANSI/ASQZ1.9 document (4). These plans are designed according to the lot size and the AQL level. Provision is made for normal, tightened, and reduced inspection and four general inspection levels, which are based on the consumer risk level. A letter is attached to the lot size interval given in the table and it leads to the required sample size and the AQL level leads to the acceptability constant (k). The most common type of inspection plan is the general inspection at level IV. The Table 7 gives the sample size and the acceptability constant for given AQL.

Based on the lot size and the AQL level, we can get the plan from Table 7. Then we have the sample information about the variable. The quality index Q_u and Q_L for upper and lower specification are defined as follows:

$$Q_u = \frac{(U - \bar{x})}{s} \quad \text{and} \quad Q_L = \frac{-(L - \bar{x})}{s}$$

where $\bar{x}$ and s are the sample mean and the sample standard deviation. If this index is greater than the k, then we accept the lot; otherwise, reject the lot.

18.10.1. Example

Let us find a sampling plan for a production lot of size 1000 units that is subjected to sample by variable type. Let AQL be 0.25%. Since the AQL =

TABLE 7   Inspection by Variable According to MIL-STD 414/ANSI/ASQZ1.9 at Normal Inspection at Level IV

Lot size and letter code		Sample size	AQL % for Inspection Level IV (normal)					
			0.04	0.10	.25	1.0	1.5	2.5
3–15	B	3	—	—	—	—	—	1.12
16–25	C	4	—	—	—	1.45	1.34	1.17
26–40	D	5	—	—	—	1.53	1.40	1.24
41–65	E	7	—	—	2.00	1.62	1.50	1.33
66–110	F	10	—	—	2.11	1.72	1.58	1.41
111–180	G	15	2.64	2.42	2.20	1.79	1.65	1.47
181–300	H	20	2.69	2.47	2.24	1.82	1.69	1.51
301–500	I	25	2.72	2.50	2.26	1.85	1.72	1.53
501–800	J	30	2.73	2.51	2.28	1.86	1.73	1.55
801–1300	K	35	2.77	2.54	2.31	1.89	1.76	1.57
1301–3200	L	40	2.77	2.55	2.31	1.89	1.76	1.58
3201–8000	M	50	2.83	2.60	2.35	1.93	1.80	1.61
8001–22000	N	75	2.90	2.66	2.41	1.98	1.84	1.65
22001–110000	O	100	2.92	2.69	2.43	2.00	1.86	1.67
110001–550000	P	150	2.96	2.73	2.47	2.03	1.89	1.70
550001 & over	Q	200	2.97	2.73	2.47	2.04	1.89	1.70

Source: Refs. 1 and 4.

0.25% and lot size = 1000, the above table gives the sample letter K and the sample size would be 35 and k = 2.31 for 0.25% AQL level.

Note that in general the variable type of sampling needs fewer number of sampling units compared to the attribute type. For example, the attribute-type example presented in 18.8.2 needs 206 sampling units, whereas the variable-type example (18.9.1) described above with the same level of protections as the example in 18.8.2 needs only 30 sampling units.

18.11. COMMON TYPES OF STATISTICS

Let us discuss the common types of statistics used in parental manufacturing and what they are used to do routinely in the industry.

I. Control Charts (Shewhart) are used to identify trending of potency of finished production lots over time. The control limits are set at average $\pm 3\sigma$, where σ is the standard deviation of the potency of the lot.

II. T-test is used to compare the average values of two methods—for example, the rabbit pyrogen test and Limulus Amebocyte Lysate (LAL) tests for the sensitive indicator of the presence of bacterial endotoxins (pyrogens) level in a biological product.

III. Correlation coefficient measures the strength of the association of two variables ($-1 \leq r^2 \leq 1$, closer to ± 1 indicates strong linear correlation). For example, the in vitro bioassay potency of a finished product to its in vitro bioassay potency of the bulk product used in the production.

IV. Analysis of variance (ANOVA) is used to identify the effect of independent variables (X's) to the dependent or Y variable; for example, to compare laboratory, concentration levels, manufacturing site, and their interaction to potency estimates of a bioassay.

V. F-Test is the test for homogeneity; the statistic is the variance ratio of the two methods or two laboratories. Suppose one wants to compare the equality of variances of two methods for example indicated in (II) above. The variance is the measure of the closeness of the values to their average. The method with smaller variance is more precise compared to the other method. The observed F-ratio would be compared with critical F-value at 5% significant level.

VI. Linear regression is used to demonstrate the linearity of two methods tested on different concentrations.

VII. 7.95% confidence interval of the ratio of averages is used to demonstrate the equivalency of two methods. For example, consider the results obtained by either 100 mL or 25 mL sample chamber of the MicroCount digital system or to demonstrate the equivalency of results obtained at MicroCount digital system and USP recommended method (5).

We know that the bacteria count data for a given volume follows a Poisson distribution. We used conditional binomial distribution to find the 95% confidence interval of the ratio of two averages of Poisson data of (7). Let us discuss an example (18.11.1) to show the calculations of 95% confidence interval for small trials.

18.11.1. Example

Suppose we have three trials. Method A obtains results of bacteria counts 1, 2, and 1 CFU per mL; method B, 0, 0, 1 CFU per mL. What would be the 95% confidence interval for the ratio of the average counts of method B to A? Suppose the two methods are claimed to be equivalent if the lower and upper 95% confidence limit on the ratio falls between 0.55 and 1.45. Could we claim method A and B are equivalent with 95% confidence?

Let S_A, S_B, and S_T be the total of A, B, and both. The upper and the lower bounds on S_A/S_B can be obtained by calculating the binomial parameter P_L and P_U for the cumulative binomial distributions with parameters P_L and P_U and $S_T = 5$ and $S_B = 1$. That is we solve for

$$(1 - P_L)^5 + 5P_L(1 - P_L)^4 = .975$$

and

$$(1 - P_U)^5 + 5P_U(1 - P_U)^4 = .025$$

for P_L and P_U for the above equations. These give $P_L = 0.053$ and $P_U = 0.716$.

The 95% lower limit $= 1/P_U - 1$ and the 95% upper limit $= 1/P_L - 1$ (i.e., the lower and upper limits on the ratios) were 0.396 and 18.87, respectively. The 95% confidence limits were outside of the interval 0.55 to 1.45; therefore, we could conclude that the two methods are not equivalent with 95% confidence.

ACKNOWLEDGMENT

I express my thanks and appreciation to my colleagues Drs. Stan Altan, and Tom Geneva, Centocor Pharmaceutical Research and Development,

Johnson & Johnson, who read this chapter and made useful comments and suggestions.

REFERENCES

1. Kuwahara, S.S. Microbiological based statistical sampling. In *Microbiology in Pharmaceutical Manufacturing*; Prince, R., Ed.; Davis Horwood: Godalming, Surrey, 2001; 485–504.
2. Bernuzzi, M.; Halls, N.; Raddi, P. Application of statistical models to action limits for media fill trials. Eur. J. of Parent. Sci. 1997, *2* (1), 3–11.
3. Military Standard 105E. Sampling Procedures and Tables for Inspection by Attributes. U.S. Army Armament Research, Development and Engineering Center. U.S. Department of Defense, 1989.
4. Military Standard 414. Sampling Procedures and Tables for Inspection by Variables for Percent Defective. Standardization Division, U.S. Department of Defense, 1957.
5. Mariano, G.; Maier, C.; Cundell, A.M. A comparison of the MicroCount Digital System to plate count and membrane filtration methods for enumeration of microorganisms in water for pharmaceutical purposes. PDA J. Pharm. Sci. Tech. 2000, *54* (3), 172–192.

19

Environmental Monitoring

John Emerson
Aventis Pharma, Holmes Chapel, United Kingdom

Petra Esswein and Ulrich Pflugmacher
Aventis Pharma, Frankfurt/Main, Germany

Lothar Gail
Siemens Axiva, Frankfurt/Main, Germany

19.1. INTRODUCTION

The purpose of environmental monitoring is to assure that the conditions in the area are adequate to protect the process that is being performed. For the critical areas involved in aseptic processing these environmental conditions are very exacting, whereas for support areas they may be less stringent. In order to define exactly what is required, the regulatory bodies have indicated their expectations. The following is a brief review of the current state of the art.

In the USA, Section 211.42 of the 21 Code of Federal Regulations, which sets down the legal requirements (1) requires for aseptic processes the establishment of a *"system for monitoring environmental conditions."* Advice on interpretation of this is given in the FDA (Food & Drug Administration) guidance documents.

In Europe, EC GMP (2) requires the setting of *"appropriate alert and action limits for the results of particulate and microbial monitoring"* (Annex 1, cl. 6).

In terms of international standardization, ISO 13408-1 (3) requires under cl. 14.3.1.1 that "*the aseptic processing area shall be routinely monitored for the presence of microorganisms, i.e., environmental flora/isolates. Periodic monitoring shall include methods for yeast, moulds and other microorganisms*". Cl.14.4 of the same standard states: "*Particulate monitoring programs shall be implemented for areas or equipment where product quality, personnel safety or testing accuracy can be affected by particulates or environmental conditions.*"

It is possible to define requirements either for the 'at-rest' state or for the 'in operation' state (the latter being referred to as 'dynamic state' by US FDA). Generally, qualification programs, which are performed in the 'at rest'-state, are dedicated to the qualification of systems and processes (Operational Qualification, OQ). Such practices (e.g., HEPA filter leak testing, determination of air flow patterns or air change rates) have been described in Chapter 9, "Airborne Contamination Control."

Most relevant to the manufacturing process is the 'in-operation' state, i.e., with all personnel present and machines operating. The FDA Guideline on Aseptic Processing (4) states that environmental monitoring programs have to "*provide meaningful information on the quality of the aseptic processing environment when a given batch is being manufactured as well as environmental trends of the manufacturing area.*"

Thus Environmental Monitoring (EM) programs are based on operation-related data including the collection of short-term variations and long term trends, "*...allowing for implementation of corrections before product contamination occurs*" (1,4).

This chapter focuses on the environmental monitoring of clean rooms and filling areas with respect to particulate and microbial contamination during performance qualification (PQ) and the subsequent routine environmental monitoring. Other relevant systems that may be considered as either directly or indirectly relevant to environmental monitoring (e.g., utilities including water and gas; bioburden control) are briefly addressed.

19.2. MICROBIOLOGICAL MONITORING

19.2.1. Requirements and Procedures

19.2.1.1. Requirements

The environmental monitoring program should cover at least the following potential sources of viable contaminants:

> Air (especially in critical locations, e.g., filling and adjacent operations)
> > Contaminants suspended in air
> > Contaminants settling from air

Room and equipment surfaces (especially in critical locations, e.g., filling and adjacent operations or from where contamination could potentially enter the product)

Personnel/clean area clothing
    Gloves
    Gown surface
Water
Facility cleaning
Handwashing (where used)
Cleaning agents and disinfectants
Compressed gases
Material entering the area

19.2.1.2. Methods of Monitoring

The following methods are commonly used for microbiological monitoring:

Air	Volumetric samples and/or settle plates
Surfaces	Contact plates for sufficiently large flat surfaces and swabs for smaller or intricate surfaces
Clothing	Fingertip imprints on Petri plates for gloved hands plus contact plates for gown locations—typically wrists and front of gown (chest area)

19.2.1.3. Media for Microbiological Environmental Monitoring

Media used in environmental monitoring should meet the criteria for growth promotion testing. It is also vital to assure that microbes cannot be introduced into a manufacturing area on contaminated media plates. Controls should be in place to avoid this possibility; the use of γ-irradiated plates, supplied in appropriate sealed containers, is strongly recommended—particularly for the critical and sterile processing zones.

a. Settle Plates

Microorganisms settle out of air under gravity and it should be borne in mind that contaminants from the air are often attached to other larger particles such as dust or skin flakes. The settling rate will depend partly on the particle characteristics and partly on air flows. Larger (or denser) particles tend to settle more effectively, and settling is facilitated by still air conditions. Smaller (or less dense) particles have a greater tendency to be prevented from settling by air resistance and air currents. It should be remembered that aseptic areas are specified in a way that make still air unlikely. There may also be a possibility of detecting such contamination by impingement onto the surface from an airstream. Clearly, such techniques are not quantitatively related to the level of contamination per unit volume of air. However, settle plates do give a method of continuously sampling in a location (e.g., close to filling) over a long period allowing a general assessment of contamination risks (5).

The method recommended for measuring microorganisms settling out of air involves exposure of an agar plate for a fixed period of time. The recommended time, consistent with the EC Guide to GMP (2) is 4 hr. However, the exposure time should take into account the possibility of plate drying if air flow is excessive and/or humidity is low and should be validated. Standard size petri dishes are suitable for settle plates. They should be exposed with the surface of the agar horizontal; floor level is suitable for plates exposed in changing rooms and laundries. Data should be available to indicate that microbial recovery is not adversely affected by desiccation. If exposed in areas where antibiotic dust is present, the plates should contain neutralizing substances.

b. Air Sampling

These techniques sample a defined quantity of air and, after suitable incubation, the colony forming units (cfu) from that volume can be counted. In practice these are best described as semi-quantitative as there are limitations on accuracy and recovery. The results given by different techniques are not generally directly comparable, but provided that the same procedure is used consistently, a useful measure of air quality can be obtained and trends can be established. The equipment used must be qualified before use and, where applicable, regularly calibrated. Examples of typical methods that may be used are shown in Table 1.

c. Surface Contact Plates

Use of contact plates (usually 55 mm diameter) is a common test for surface contamination. Plates are pressed lightly once against the surface to be tested and the imprinted plate subsequently incubated. Contact plates are suitable only for flat surfaces. Because the plates leave traces of nutrients on the surfaces sampled, the tested surface should be carefully swabbed with a suitable sterile, nonshedding disinfectant swab or wipe (e.g., with sterile isopropyl alcohol) immediately after taking the imprint. Otherwise the residual nutrient could create a focus for microbial growth. As an alternative to contact plates commercially available systems based on plastic strips or similar formats may be used. If the surface area covered by such devices differs from that of a standard plate, different alert/action-limits may be applied.

d. Surface Swab Testing

This may be done using appropriate sterile swabs moistened with sterile water or sterile saline. It is recommended that swabbing should concentrate on

TABLE 1 Methods for Collecting Environmental Microbial Samples

Principle	Description	Example
Sieve impactor	The apparatus consists of a container designed to accommodate a petri dish containing a nutrient agar. The cover of the unit is perforated, with the perforations of predetermined size. A vacuum pump draws a known volume of air through the cover, and the particles in the air holding microorganisms impact onto the agar medium in the petri dish. Some samplers are available with a cascaded series of containers containing perforations of decreasing size. These units allow for the determination of the distribution of the size ranges of particulates containing viable microorganisms, based on which size perforations admit the particles onto the agar plates.	MAS 100 (Merck) MAirT (Millipore)
Slit-to-agar air sampler	This sampler is the instrument upon which are based the microbial guidelines given in USP 26 (6) (for the various controlled environments). The unit is powered by an attached source of controllable vacuum. The air intake is obtained through a standardized slit, below which is placed a slowly revolving petri dish containing a nutrient agar. Particles in the air that have sufficient mass impact on the agar surface and viable organisms are allowed to grow out on subsequent incubation of the plate. A remote air intake is often used to minimize disturbance of the laminar flow field. If used, the dimensions of this should be specified, and technically acceptable.	Cassella, Mattson-Garvin
Gelatin filter sampler	The unit consists of a vacuum pump with an extension hose terminating in a filter holder that can be located remotely in the critical space. The filter is made from random fibers of gelatin, capable of retaining airborne microorganisms. After a specified exposure time, the filter is aseptically removed and placed on a petri dish.	MD 8 (Sartorius)
Centrifugal sampler	The unit consists of a propeller or turbine that pulls a known volume of air into the unit and then propels the air outward to impact on a tangentially placed nutrient agar strip set on a flexible plastic based.	RCS plus/RCS high flow (Biotest)

surfaces that are difficult to test using contact plates (for example, due to intricate shape) and/or which pose cleaning problems, for instance parts of machinery that are difficult to access but which could lead to product contamination. The moistened swabs should be used to sample the surface and then rolled over the surface of a sterile agar plate, which is then incubated appropriately.

A variant of the swab test is to use soluble, sterile alginate swabs. After the surface swab has been taken, these are dissolved in a suitable medium such as sterile saline solution, which is then membrane-filtered and the membrane transferred to a suitable nutrient agar plate and incubated.

For isolator testing, where swabs are the better alternative (no media residues) and the limit is less than 1, the swab can be put directly in broth and then incubated. (The disadvantage of this is that is gives a $+/-$ result rather than a quantitative estimate of the numbers of cfu's present) The swab techniques are, at best, only semiquantitative. Results may be reported per defined area (where a specific area is swabbed) or per swab (e.g., where small items, cracks, or crevices are swabbed).

19.2.1.4. Detection of Viable Contaminants on Gloves and Clothing

The method recommended for detection of microorganisms on the (gloved) hands of operators working in the clean areas is the finger imprint (finger dab). The operator touches the surface of an agar plate with the tips of all fingers and thumb. The plate is then incubated appropriately. The finger dab test should be carried out at times defined in the site procedure, typically; this is at the end of a work period or after critical interventions. Since the objective is to measure the potential for product contamination, staff should not be permitted to 'disinfect' their gloves immediately before the test is performed (staff tested immediately before leaving the area should discard their gloves without touching any surfaces). Petri plates are recommended for finger dabs because contact plates are too small for all 5 fingers. Clothing may be tested by contact plates either for specific training applications or at the end of a shift. This method should not be used during operations due to the difficulty or removing media residues from the fabric. Typically, sleeve cuffs and/or front of gown are tested.

19.2.1.5. Suitable Media and Incubation Conditions

For routine monitoring, a general purpose medium such as casein soya bean digest* agar should be used. For area qualification exercises, or when

---

* Casein soya bean digest (CSD) agar is the Pharm. Eur. terminology. In the USP and JP, the same medium is referred to as soybean casein digest agar (often abbreviated to SCD). The formulae in the three compendia are identical.

problems with yeast and molds occur monitoring may include a medium for yeasts and moulds such as Sabouraud dextrose agar. Recommended media and incubation conditions for environmental microbial samples are given in Table 2.

19.2.1.6. Media Growth Promotion Testing and Method Validation

Environmental monitoring media are mostly bought ready to use from suppliers, double- or triple-packed, and γ-irradiated. These media are delivered with a certificate. If there are uncertainties with respect to transport conditions (e.g. no cooling during transport), in-house media growth pro-

TABLE 2 Media and Incubation Conditions for Environmental Microbial Samples

Monitoring	Type of media	Incubation period
For routine monitoring:		
Surface monitoring	Casein soya bean digest agar (CSD)	30–35°C for 3 days when using for bacterial counts or 20–25°C for 3 days, followed by 30–35°C for 2 days when using CSD agar for both bacteria and fungi
Settle plate		
Quantitative air sample		
Finger dab		
For specific monitoring for fungi only:		
Tests for mold and yeast (where used for any technique)	Sabouraud dextrose agar (SDA)	20–25°C for 5 days
For specialized anaerobic monitoring:		
Tests for anaerobes (where used for any technique)	Casein soya bean digest agar or suitable alternative medium designed for use in culture of anaerobes	30–35°C for 5 days under anaerobic conditions in an anaerobe jar or suitable alternative system

Notes: Consideration should be given to reading the plates during incubation in order to give rapid detection of potential contamination problems. There should be an assessment of the possible risks of transmissible spongiform encephalopathies (TSE) in the choice of media supplies.

motion tests need to be done. The growth promotion testing should be completed before release for use.

Consideration should be given to performing additional growth promotion tests on representative media after exposure in the monitoring because media may have been affected by inhibitory agents or conditions. The same procedures apply to media prepared in-house. Sterilization processes used to prepare growth media must be validated and, in addition, media must be examined for sterility as indicated in the monographs for the sterility test. The following test organisms should be used for growth promotion testing:

> *Staphylococcus aureus* such as ATCC 6538
> *Pseudomonas aeruginosa* such as ATCC 9027
> *Bacillus subtilis* such as ATCC 6633
> *Candida albicans* such as ATCC 10231
> *Aspergillus niger* such as ATCC 16404
> *Escherichia coli* (such as ATCC 8739) may also be used.

In addition, representative microflora isolated from the controlled environment may also be used (or preparations of the same species as these isolates, obtained from recognized culture collections such as the ATCC [American Type Culture Collection]). Media must be able to support growth when inoculated with 10–100 cfu of the challenge organisms. If a qualified or certified supplier is used, testing could be reduced, for example not all strains or media batches would need to be tested. The medium should be supplemented with additives to overcome or to minimize the effect of sanitizing agents or antibiotics if used or processed in these environments. Whatever methods are chosen to prepare the media, there should be strict controls to avoid contamination from the test items.

Where the preparation details are important for the function of the test plate, it is important to control these parameters in the routine preparation of media. For example, petri plates used for settle-plate monitoring must contain a standard volume of agar that is sufficient to avoid excessive drying (as validated) and contact plates require a consistent volume of medium to give the correct surface meniscus profile.

19.2.2. Test Procedures, Test Program, Limits, and Data Evaluation

19.2.2.1. Test Procedures

The test procedures must define the operational requirements such as the following:

> Who is to prepare the materials
> Preparation of medium (including records of preparation, lot number, etc.)

Where applicable, how items are to be sterilized (e.g., critical parts of
   sampling devices)
How the materials are to be labelled
Who is allowed to perform the test (training aspect)
How to protect and transport the test materials securely
How test materials are to enter the area
Guidance on times between exposure and incubation
Incubation conditions and times
Instructions for reading and recording the results

19.2.2.2. Validation of the Environmental Monitoring Testing Schedule

The EM testing program should be based on a validation exercise. The
validation protocol should include (as a minimum):

Clear rational for the setting of test locations, taking into consideration:

Contamination risks for the product
Contamination risks for the area

This should preferably be done by descriptive text and graphic representation

Frequency of testing
The detailed test procedures, including all the necessary test parameters,
   for example:

The test methods/equipment/media to be employed at each
   location
How test methods/equipment/media should be used (e.g., sampling volumes for active air sampling, exposure time for settle
   plates)
Time or process stage at which finger checks are to be performed
The incubation temperatures and times required

The requirements (action limits, rationales for further identification of
   microbial isolates)
Responsibilities

19.2.2.3. Evaluation of the Validation Exercise and Setting of the Routine Environmental Monitoring Program

The analysis of data obtained from the validation study should comply with
the requirements of the validation protocol for the specific clean room class.
Excursions of the action limits must trigger a failure investigation. Based on
the data obtained during the validation study and taking into consideration

the specific contamination risks, the sampling locations and the alert limits are set for the routine monitoring program. The use of fixed locations facilitates trend analysis. However, it is possible to include some additional variable locations if deemed appropriate in specific instances.

19.2.2.4. Guidance on Setting Action and Alert Levels

The alert and action levels for the evaluation of environmental monitoring samples must be defined. A course of action (and corresponding responsibilities) must be defined in the event of levels being exceeded. There should be clear guidance on what isolates are to be identified and to what level (e.g., general type, family, genus, or species).

a. Action Levels

Despite differences between methods, general standards have been adopted by the regulatory authorities in terms of cfu per air volume or surface area for the different zone categories. Recommended action limits for environmental microbial contamination according to EC GMP and USP classification (2,6) are given in Table 3.

EC GMP permits averaging. There is no clear guidance on which basis averaging is accepted. Evidently, it is not adequate to obscure individual high results by averaging. For example, for class A, where <1 count is

TABLE 3 Action Limits for Environmental Microbial Contamination

Grade (a)		Active air sample, [cfu/m^3]		Settle plates [cfu/4 hours] (c)	Contact plates [cfu/plate]		Glove print 5 fingers [cfu/glove]	
EC	USP	EC (b)	USP	EC (b)	EC (b)	USP	EC (b)	USP
A	M3.5	<1	<3	<1	<1	3 (including floor) gown:5	<1	3
B	M5.5	10	<20	5	5	5 gown: 20	5	10
C	M6.5	100	<100	50	25	10 (floor)	(d)	
D	—	200	—	100	50	—	—	—

Notes:
(a) The correlation of EC and USP classification is made on the basis of in-operation data.
(b) These are average values.
(c) Individual settle plates may be exposed for less than 4 hours.
(d) For the monitoring of gloves in zone C, there is no guidance in current GMPs or compendia and a level of 25 cfu per 5 fingers is suggested for the action level.

expected, a set of 10 data from a distinct test location in a series could be used, whereas each limit transgression has to be carefully investigated.

b. Alert Levels

The alert level is set lower than the action level and should be based on validation and subsequent trend review. It should be designed to be above normal levels but to give a sufficient margin to pick up abnormal trends before action level is reached. For the calculation of alert levels from aerial and surface counts, see (7,8) for possible approaches.

19.2.3. Routine Environmental Monitoring Program

The overall testing schedule and the detailed test procedures to be used must be clearly defined in the SOP for environmental monitoring. In addition to the above-mentioned aspects (see 1.2.2) an action plan in case of limit transgressions, trends, or special occurrences must be included.

19.2.4. Water Samples

All pharmaceutical water systems should be validated, maintained and monitored. Adequate alert- and action limits should be set. Whenever the action levels are exceeded or adverse trends are detected, the main species present should be identified. Representative isolates from the general typical population of the water system should be periodically identified to establish the normal flora. Changes in flora or abnormal isolates should be investigated.

19.2.5. Bioburden Samples

Adequate limits for the bioburden should be set, e.g., for raw materials or pre-filtration bulk-solutions. Isolates obtained should be identified periodically or in case of out-of-specification counts.

19.2.6. Gas Samples

Samples from gasses (e.g., compressed air, nitrogen), which could come into contact with products should be taken by adequate means. Adequate limits should be set. An action plan should be in place if limits are transgressed, if trends are detected or organisms are found which are able to persist in gas distribution systems (e.g., spore former).

19.2.7. Identification of Contaminants

The identification or characterization of isolates is particularly important where isolates are recovered from critical processing operations, particularly those involved in aseptic processing. Microbial identification should be performed by suitably qualified, trained, and experienced staff. Initially it is based on morphology and classical microbiological techniques such as the Gram's stain, etc. Following initial characterization, some form of advanced procedure may be used to facilitate this and to make it more efficient and consistent (examples include API, Vitek and MIDI systems). The results of such techniques may be limited by the nature of the database used as reference (for example whether it is largely based on clinical microbiology isolates rather than typical environmental isolates). Interpretation should be done in the context of the overall information.

The extent to which isolates need to be identified is often debated and on this point the following guidance is suggested. As a minimum, the following environmental isolates and related IPC isolates should be identified to species level wherever possible:

> Isolates from critical raw materials or bulk solutions (incl. water)
> Isolates from the vicinity of aseptic fill locations (zone A)
> Isolates from the vicinity of sterile filtration operations (zone A)
> Isolates from the vicinity of exposed aseptic processing applications (zone A)
> Isolates from finger and gown checks in aseptic processing areas
> Isolates from the zone B support areas in aseptic processing/filling applications

In other controlled zones isolates should be characterized, but not necessarily identified to species level. The characterization should be sufficient to reveal any trends or shifts in typical flora and to detect objectionable organisms where appropriate.

19.2.8. Quality Assessment of Microbiological Data

The monitoring program should facilitate the reconstruction of possible contamination paths and to assess the significance of the contamination with respect to the product, process, or system. An evaluation of microbiological environmental monitoring data should be performed on a routine basis. Particular attention should be paid to any excursions outside the action/alert levels and to identification results that indicate presence of unusual, atypical, or objectionable organisms. Trends towards higher counts or uncommon identification results indicate failures in clean-room performance resulting from personnel or technical errors. The microbiological data should preferably be evaluated in conjunction with the physical environmental data to detect possible root causes of contamination.

19.3. PARTICULATE MONITORING AND OTHER PARAMETERS

19.3.1. Purpose of Particulate Monitoring

A fixed correlation between particle and microbiological contamination has not been determined. Interdependencies between both parameters, however, have been frequently reported (9,10). Owing to the fact that the number of physically measured particles normally greatly exceeds the number of cfu's isolated in viable monitoring, particulate monitoring offers a considerably better resolution than the determination of viable particles. For sterile/aseptic cleanroom environments, where the limit values of viable contamination may be close to the detection limits, this proves to be an important advantage. Hence, results of particulate monitoring are being used to demonstrate the overall effectiveness in maintaining the air quality of a cleanroom environment. For demonstrating compliance with process-specific requirements, however, microbiological monitoring cannot be dispensed with. By offering real-time measurement, particulate monitoring contributes a particularly valuable feature to state-of-the-art pharmaceutical process control.

19.3.2. Test Methods and Equipment for Particulate Monitoring

Methods for particulate monitoring are dealt with in ISO 14644-1 (11) and -2 (12). ISO/DIS 14644-3 Metrology and Test Methods (13) specifies further details on test performance and requirements of test equipment. The cleanliness classification of ISO 14644-1 may also be used for specifying environmental monitoring limits. It should be borne in mind, however, that this classification has been prepared for specifying and controlling cleanroom installations and not for the purposes of process control. For process control purposes, smaller or even greater tolerances than those of the ISO 14644-1 classification may be suitable. Validation of test methods is required as far as nonstandardized methods are employed.

Where remote sampling installations are used (e.g. for controlling the aseptic processing environment), rules have to be specified, covering, e.g., probe position relative to air flow, sensor calibration, and tube length. Recommendations dealing with probe orientation are given in Ref. (4). The limits for particulate monitoring may be defined and tested as part of the performance qualification (PQ).

Some operations (e.g., aseptic powder filling) can generate sterile particles, causing a process-related background particle level that may prohibit monitoring by standard particle counting procedures. GMP guides, such as Refs. (2) and (4) take that fact into account. Alternative methods are to use the closest representative position offering a sufficiently low background particle level and/or to validate with the machine operating but without powder in place. This should be periodically repeated.

19.3.3. Results and Deviations

All results should be examined for compliance with the standards of the zone in question; the limits should be defined following relevant guidelines and Pharmacopeias.* All out-of-limit results must be investigated and appropriate corrective actions have to be implemented and reported (or documented).

19.3.4. Other Environmental Parameters

19.3.4.1. Cleanroom Segregation by Differential Pressure or Airflow

Differential pressures between zones of different cleanliness levels are established for controlling the efficiency of cleanroom segregation against airborne contamination. Critical applications require continuous differential pressure control by automated systems. Differential pressure monitoring shall give full and reliable information on the function of a pressure cascade system. When zones of different cleanliness levels are segregated by exchanging a high airflow volume, airflow control may be more suitable for the control of small differential pressures; see ISO 14644-4 (14).

Appropriate limits must be set based on the minimum and maximum values required to maintain the function of the segregation system. Where nonautomated systems are used, the readings should be checked and recorded regularly and at a frequency appropriate to the criticality, see Table 1.

19.3.4.2. Temperature and Humidity

Temperature and/or humidity assessment may be included as part of EM programs. For standard applications, such measurement is required for collecting information on environmental trends of the manufacturing area; for specific processing, it is needed for collecting batch-related information.

19.4. OVERALL PERFORMANCE OF ENVIRONMENTAL MONITORING

Routine performance is carried out to ensure that operational conditions of the processing environment as well as environmental trends of the manufac-

---

* For air cleanliness classification data see Chapter 9, "Airborne Contamination Control."

turing area remain within validated conditions. A suggested test schedule for routine performance is given in Table 4.

19.5. OVERALL QUALITY ASSESSMENT OF ENVIRONMENTAL MONITORING DATA

Excursions of limits and uncommon occurrences call for decisions affecting the release of product and/or changes or adjustments to the operating procedures, processing equipment or control equipment. The results of all monitoring tests should be analyzed in an overall approach for any problems or trends. This may be by either graphical representation or by appropriate statistical means so that progressive or sudden changes in the results may be

TABLE 4 Minimum Testing Requirements in Operational Conditions

Parameters	Operational test schedules for different zones		
	Critical zone	Processing zone	Support zone[b]
	Grade A[a]	Grade B[a]/ Grade C[a]	Grade C[a]/ Grade D[a]
Viable Counts			
Air monitoring	Each shift	Each shift—weekly	Monthly
Surface monitoring	Each shift	Each shift—weekly	Monthly
Personnel monitoring	Each shift	Each shift—depending on the process	
Physical parameters			
Particulate monitoring	Ind, aRec, Alarm[1]	Ind, aRec, Alarm	mRec
Differential pressure	Ind, aRec, Alarm[2]	Ind, aRec, Alarm	Ind, mRec
Temperature	—	aRec, Alarm	mRec
Humidity	—	aRec, Alarm	mRec

Notes

[a] according to EU GMP Guide to Good Manufacturing Practices, Annex 1.

[b] see (3,4).

[1] For certain applications (e.g., blow-fill-seal and aseptic powder processing) this monitoring may not be possible in operation.

[2] Differential pressure measurement between grade A and the background area is only applicable to contained processing environments, e.g. isolators.

Ind = Continuous indication.

aRec = Continuous automatic recording.

mRec = Manual recording.

observed. All data should be regularly reviewed and the possible sources of both microbial and particulate contamination should be critically assessed and, where necessary, appropriate action taken to eliminate the source.

ACKNOWLEDGMENTS

We thank Hans Amelung for reviewing the manuscript and all members of the Aventis IO/GET for Sterile Manufacturing contributing to the Aventis' Environmental Monitoring principles.

REFERENCES

1. Code of Federal Regulations (21 CFR Ch.1, Part 211).
2. EC GMP Guide to Good Manufacturing Practice. Revised Annex 1: Manufacture of sterile medicinal products. 2003.
3. ISO 13408-1: 1998-08: Aseptic processing of health care products—Part 1: General requirements; ISO: Geneva.
4. Guideline on Sterile Drug Products Produced by Aseptic Processing. Center for Drugs and Biologics and Office of Regulatory Affairs, Food and Drug Administration, Rockville, Maryland 20857, June 1987.
5. Whyte, W. In support of settle plates. PDA J. Pharm. Sci. Technol. 1996, *50*, 201–204.
6. USP 26 ⟨1116⟩: The United States Pharmacopeia. United States Pharmacopeial Convention, Rockville, MD, USA, Jan 1, 2003.
7. Wilson, J.D. Setting Alert/Action Limits for Environmental Monitoring Programs. PDA J. Pharm. Sci. Technol. 1997, *51*, 161.
8. PDA Technical Report No. 13, Fundamentals of an Environmental Monitoring Program, PDA J. Pharm. Sci. Technol. 2001, 55, 5, Suppl. TR13.
9. NASA Biological Handbook of Engineers, Marshall Space Flight Center, 1976.
10. Wada, J. The Relationship between the Number of Microorganism and Environmental Condition (Cleanliness). Proc. 12th ICCCS Symp. Cont. Control, Yokohama, 87–92.
11. ISO 14644-1: 1999: Cleanrooms and associated controlled environments—Part 1: Classification of air cleanliness; ISO: Geneva.
12. ISO 14644-2: 2000: Cleanrooms and associated controlled environments—Part 2: Specifications for testing and monitoring to prove continued compliance with ISO 14644-1; ISO: Geneva.
13. ISO/DIS 14644-3: 2002: Cleanrooms and associated controlled environments—Part 3: Metrology and test methods: ISO: Geneva.
14. ISO 14644-4: 2000: Cleanrooms and associated controlled environments—Part 4: Design, construction and start up; ISO: Geneva.

20

Prevention and Troubleshooting of Microbial Excursions

Elaine Kopis Sartain
Steris Corporation, St. Louis, Missouri, U.S.A.

20.1. INTRODUCTION

One of the most frustrating and time-consuming, not to mention costly, aspects of managing pharmaceutical and biotechnology production facilities is the investigation of environmental monitoring excursions. For purposes of this chapter, an excursion is defined as environmental monitoring data that fall at or above the action levels for the area in question. The regulatory agencies stipulate that environmental monitoring data should be trended, in order to better predict, and possibly prevent, the onset of an excursion. However, as most statisticians who have studied this subject know, in most areas, there are not significant data with which to predict such an occurrence. All too often, excursions appear seemingly out of nowhere. It is only upon closer examination of all operational aspects of the facility that a predictor of these excursions may be identified. All too often, these predictors have little to do with the environmental monitoring data, which was collected for trending purposes. These predictors are related to three key areas: facilities design, personnel management, and cleaning/sanitization programs. Problems with these areas cannot necessarily tell you when problems will occur, but it is almost assured that they will occur if these problems exist.

20.2. FACILITIES DESIGN

I have toured manufacturing facilities on three continents, and one thing that I have observed is that there is no universal standard for the age or general condition of facilities in which drug manufacturing takes place. The facilities that I have seen range from brand-new, state-of-the-art facilities, the design of which includes high-speed filling lines encased in isolators, to WPA projects built as a result of President Roosevelt's New Deal during the Depression, which have endemic mold problems and too little HEPA filtration. In terms of microbial excursions, the latter situations clearly present more challenges for several reasons.

There are several ways to address microbial problems once they are identified—but one of the best ways is to keep them out of the facility. Most modern manufacturing facilities have been designed to accomplish this goal through the use of positive pressure, HEPA filtration, temperature and humidity controls, materials selection, and design features that allow for more effective cleaning. They have also been designed to physically buffer the manufacturing area from uncontrolled areas—minimizing introduction of contamination through traffic. The recent introduction of antimicrobial polymeric flooring and wall coverings may provide an adjunct to other control measures. Those who have the advantage of the newer, state-of-the-art facilities will find on careful examination that any environmental excursions that are experienced are usually due to the other two key areas listed above, which will be discussed in more detail later. Those who are not so fortunate should take a critical look at their surroundings and start a to-do list.

In terms of troubleshooting excursions, a good first step is to complete an objective evaluation of the facility, including a review of the HVAC system. I have worked with facilities that cannot control temperature spikes and that must use supplemental dehumidifiers during the summer months in order to maintain a reasonable degree of control. The problem with this approach is that often the dehumidifiers are not put into use until after a humidity spike has occurred. This is the environmental equivalent of closing the barn door after the horse has escaped. Temperature and humidity control—consistent control—are paramount to preventing microbial excursions. Temporary fixes should be avoided because all too often, the "fix" is implemented after the damage has occurred and typically is not rugged enough to consistently maintain the environment under manufacturing conditions. One example of this is the use of flexible duct material as a permanent alternative to rigid ductwork. This may lead to leaks in the system and may cause airflow balance problems.

The condition of flooring, walls, and ceilings is another key element of managing excursions. Water leaks, uncontrolled humidity, and frequent cleaning can take a toll on the materials with which these surfaces are constructed. Recently I was working with a facility that continually experienced high mold counts in the master seed room. The onset of the problem coincided with damage to the roof—which allowed leakage. Although the roof was repaired, the interstitial area was not examined for the presence of water, nor was it cleaned and sanitized for mold removal. Subsequently, the mold in this area, nourished by the water, grew to uncontrollable levels. When the problem was discovered, the entire area was cleaned and fogged with an oxidizing chemical. However, this area was not designed for cleanability, and several treatments were required to bring the mold levels under control.

Flooring is particularly problematic because of the high level of direct contact with other potentially contaminated surfaces, such as shoes, wheels, and equipment. I find it interesting that some facilities with which I work do not wish to monitor the floor, particularly in Grade D (Class 100,000) areas. The argument presented is that this is not a USP requirement, so why do it? The answer is that the floor may be a barometer of the condition of the rest of the facility. If the floor is not kept under control, then other areas may be more difficult to control due to the spread of contamination via traffic. Of course, this may not always be the case. It depends on the condition of the floor and whether or not floor drains are present. Damaged flooring—or example, epoxy surfaces that have visible signs of erosion or physical damage from chemicals or heavy equipment—is particularly difficult to decontaminate. The challenge stems from the surface irregularities allowing both liquids to pool and microorganisms to find protection from the disinfectants that are applied during the cleaning process.

Floor drains can be a tremendous source of contamination for two key reasons: they allow nearly unfettered egress of microorganisms into the room and they are very difficult to effectively decontaminate. Floor drains are probably underestimated in their potential to cause problems. One facility with which I work has noted that on occasion, when another area of the facility is discharging heavy amounts of effluent, liquid from the floor drain in their manufacturing suite literally shoots out of the floor drain, as high as approximately 12 feet. This "geyser" effect spreads organisms to the walls and potentially to the ceiling—and certainly onto manufacturing equipment. It is best to remove or cap floor drains whenever possible. In the event that this is not possible, floor drains should be sanitized regularly with a strong oxidizing chemical, and floors in these areas should be cleaned more frequently than other areas. Additionally, problems with water flow and backup must be treated as a priority and resolved.

A facility designed with unidirectional traffic flow will usually have fewer problems with environmental excursions—this is an extension of the "clean to dirty" principle, pervasive in cleanroom practices. Unfortunately, many facilities must contend with multidirectional traffic—due to design limitations or to ever-increasing space demands that require that areas be used in a manner inconsistent with design—in order to meet production needs. Even in these cases, the environment can be adequately controlled. However, this usually requires more frequent cleaning and strict adherence to both gowning and entry/exit procedures. This scenario presents potential problems due to the heavy reliance on compliance with SOPs to control the environment.

20.3. PERSONNEL MANAGEMENT

Systems that rely on human behavior are difficult to control—in part because often we are unaware of our behavior. At other times, we are fully aware of what we are doing, but we are fully ignorant of the consequences of our actions. I have seen numerous examples of this ranging from operators removing goggles (ostensibly to see better), scratching their heads underneath their hoods, kneeling on the floor with an operations manual—then placing the manual onto a work station with no decontamination step. These are examples of thoughtless behavior that can lead to problems.

I have also observed "thoughtful" behavior—such as deliberate deviations from SOPs. During a training session several years ago, an operator told me that the disinfectant that they were using was not working well because it did not foam enough. She stated that she had resolved the problem by adding dish detergent to the disinfectant solution. This is a clear example of an operator deliberately deviating from an SOP. One could argue that her motives were pure. However, she clearly did not understand the potential consequences of her actions—or that deviating from the SOP alone was a serious problem from a regulatory compliance perspective.

More recently, a supervisor was observed to direct his operators to apply disinfectant directly to the floor, and then to distribute the concentrated disinfectant with a water-dampened mop. This was a significant deviation from the SOP, which directed that the product be diluted with water prior to application, as well as from the safe and approved use of the disinfectant. In this case, the supervisor was quite willing to deviate from the SOP and potentially create compliance, safety, and performance problems. Training and qualification for his role as a supervisor should have been sufficient to prevent this deliberate behavior from occurring.

In order to achieve greater compliance from our personnel, we need training programs that focus on more than just execution of the SOP. Additionally, these programs should seek to do more than just define cGMP—they should seek to engender a cGMP mindset. This creates a philosophical shift that focuses not only on execution of an SOP but also on compliance to cGMP as an operating principle—as an absolute requirement of the job. In other words, the emphasis should be not only on the outcome but also on the consistent execution of the job according to standard procedures.

Of course, cGMP training alone will not ensure compliance. Personnel require training in the specific SOPs and in topics peripheral to proper execution of the SOPs. For example, a training program for cleanroom cleaning should include basic concepts in microbiology, chemistry, and cleaning principles, as well as a tutorial on relevant regulatory standards and guidelines.

In terms of microbiology, the training should provide information on the different classes of microorganisms—for example, bacterial endospores, fungi, bacteria, and viruses. This training should emphasize the sources of these organisms in a manner that reinforces cleanroom behavior, including traffic flow and gowning requirements. It is axiomatic that people are the chief contributors of contamination in cleanrooms. Training that is rooted in an understanding of the sources of microorganisms and in the mechanisms by which they are transported to controlled areas, via personnel, is essential to compliance.

The fact that all microorganisms do not succumb to all disinfectants equally should also be emphasized. This is the true rationale for using multiple disinfectants, including extremely harsh materials such as oxidizers. A basic understanding of the chemistry of antimicrobial agents, such as phenolics, quaternary ammonium chloride compounds, and oxidizers (bleach and hydrogen peroxide/peracetic acid blends) can be useful in reinforcing compliance and safe handling practices. This training should include an elementary overview of product formulation (i.e., the function of various ingredients), toxicity data, and safe handling practices.

A comprehensive training approach should address decontamination measures required to control viable and nonviable particulates. In particular, this training should emphasize the relative difficulty involved in effectively removing particulate from surfaces, and in not recontaminating areas during the cleaning process. As someone who frequently trains cleanroom personnel, I often ask questions to gauge the relative sophistication of the group in terms of cleanroom control issues. I find it interesting that many do not know or appreciate the size of a 0.5 μm particle. By comparison, the diameter of a human hair is approximately 100 μm. Yet this principle is very important in enforcing compliance. When the expectation is that one will control "nonvisible" particulate, then visual indicators may not be sufficient alone to determine whether or not the objective is being met. In other words, consistent compliance to good cleanroom practices is critical to achieving consistent control, because what is not seen can be harmful. Examples of this are "clean to dirty" application practices and two-bucket cleaning, both of which minimize recontamination of surfaces through the spread of nonvisible particulate.

Prevention should be emphasized during training, including:

Prevention of adulterated product
Prevention of injuries
Prevention of contamination
Prevention of damage to surfaces
Prevention of microbial excursions
Prevention of noncompliance

It goes without saying that the training program should include SOPs covering the scope of the job function. Before SOP training is engaged in the documents should be reviewed for clarity and accuracy, and updated if required. Often during training programs I have been informed that SOPs are not being followed because they are difficult to understand or outdated. This problem may be prevented by developing SOPs through the use of multi-disciplinary committees. These committees should include the operators who will actually be performing the procedures. SOPs should be reviewed periodically and updated as required. I have reviewed hundreds of SOPs from a number of different manufacturers, and I often find significant inaccuracies within these documents, including inaccurate instructions for cleaning agent dilution and incorrect product name.

SOP training should include the following steps:

Read it
See it
Do it

Reading comprehension levels vary tremendously from one individual to another. Therefore, merely reading a document does not ensure understanding of the document—or proper execution of the instructions, the ultimate goal. Observation of an experienced operator performing the SOP is an important step in bridging the gap between reading the instructions and understanding the instructions. The operator-in-training (OIT) should be encouraged to ask questions of the experienced operator during the observation period. If this is not convenient, then the operator should be given an opportunity following observation. Finally, the OIT should follow the SOP, unassisted, while being observed by an experienced operator or supervisor who is responsible for execution of the SOP. Submission of a written test to determine the effectiveness of SOP training is another option. However, because reading comprehension varies among individuals, a written test may not be the best means to determine the effectiveness of SOP training.

It would be remiss to discuss personnel management issues and not to mention the role of the supervisor in maintaining compliance and control. In many ways the supervisor sets the tone for compliance and control by his actions or lack thereof. Recently I worked with a facility in which the supervisor crawled under the filling line as a shortcut to get to the exit. He directed that I follow suit—neither of us was able to do this without some part of our gowning touching the floor. This was a completely unnecessary and risky practice that no doubt left our garb vulnerable to tears and contamination. It was no surprise that during the audit I observed several significant examples of noncompliance to SOPs from his personnel. The title of supervisor, or manager, or director does not convey with it an exemption from following the rules. It is the supervisor's responsibility to engender compliance to SOPs and cGMPs, through behavior as well as words.

Over the years, I have noted an increasing trend among middle management and front-line supervisors: All too often, instead of supervising or managing, these personnel are attending meetings, conducting noncompliance-related investigations, and taking part in the paper chase that is anathema to, but a significant requirement for, this industry. In the face of reduced supervision, effective training is more important than ever in ensuring compliance.

20.4. CLEANING AND SANITIZATION PROGRAMS

As a technical support specialist who works for a supplier of contamination control products, I frequently take part in discussions with end users regarding the design and execution of cleaning and sanitization programs. One of the most important elements of designing an effective cleaning and sanitization program is understanding the features and the limitations of the products that are used for these purposes. No one product is a panacea for all contamination control problems within the manufacturing environment.

Overuse of certain products can lead to significant substrate damage, which in turn will lead to more difficulties in controlling microbial and particulate contamination. Underutilization of certain products can lead to poor microbial control. Effective contamination control programs must be designed to strike a balance between eradication of microorganisms and damage to the surfaces being treated. Additionally, most products designed for microbial control applications will leave residues on surfaces. These residues may further facilitate corrosion of surfaces, or may build up to levels that affect the functionality or esthetic properties of surfaces. There is also a possibility of product and component contamination, depending on the surfaces treated with the product. Therefore, it is important to develop a rinsing strategy as part of the overall cleaning and sanitization program.

It is also important to understand that cleaning and sanitization are two different functions. Cleaning may be defined as a process that results in the removal of particulate from surfaces. Sanitization is defined as a process that results in a significant reduction of viable microorganisms from surfaces. It is difficult to achieve consistent, effective microbial reduction without achieving effective particulate removal. Antimicrobial products do not necessarily have the properties necessary to ensure effective particulate removal from surfaces. However, there are many products available that are formulated with the proper additives, such as surfactants and dispersants, which aid in particulate removal from surfaces. This is especially important in the case of small particulate where surface attractions are quite strong.

Any tool is only as effective as the skill with which it is applied. In my view, application problems account for a large percentage of performance issues encountered in controlled environments. Sometimes these problems are due to poor SOP development and poor training, and sometimes they are due to poor compliance. For example, recently while touring a facility I witnessed an area supervisor direct the operators to prepare disinfectant and then pour the disinfectant onto a pool of cell culture suspension on the floor. The disinfectant was literally poured from a bucket onto the floor, mixing with an unknown quantity of cell suspension—the material was then pushed to the floor drain with a squeegee. This procedure was not part of an approved SOP for cleaning the floor—and certainly was not the most efficacious manner in which to do so. Clearly, even the best designed cleaning program will not work in the face of deliberate noncompliance.

SOPs should include clear instructions for preparation and storage of the disinfectant solution, including expiration dating of the use-dilution and the opened container. Other considerations include type and temperature of water (WFI vs. USP purified) to be used, filtration, and bioburden monitoring. Often the supplier can provide supporting information on the product.

Application instructions for the disinfectant should include area-to-volume ratio of the disinfectant. I have witnessed disinfectant applications during which a mop has been dipped into the solution, wrung out, and then applied to an extremely large area to the point where the surface was not being effectively wetted by the mop. Instead of following a well-defined cleaning plan to re-wet the mop after a certain area had been covered, the operator would re-wet the mop whenever the notion occurred to him. SOPs may include diagrams indicating when it would be appropriate to wet the mop.

Understandably there are some challenges with this approach—but it is not a one size fits all world, and some guidance is valuable to ensure that proper wetting occurs.

Wetting is critical to disinfectant performance. Many products contain volatile ingredients that are key elements of overall performance. One cannot depend on nonvolatile residues left behind on surfaces to provide antimicrobial activity. In other words, disinfectants must be in the wet state to provide effective control on surfaces. I have reviewed hundreds of SOPs, and most of them stipulate that the disinfectant should be left on a surface in the wet state for a contact time of at least 10 min. Yet most SOPs and training programs do not address how this is to be accomplished. The wet dwell time will depend on the temperature and humidity of the room, and the solution volume to area ratio being used. The surface type and condition, room activity level, and several other factors may also influence dwell time. Physical coverage of the target areas is also an important element. I have witnessed cleaning being performed during which large gaps of wall and flooring were not treated with the disinfectant solution due to poor application techniques.

Contamination control problems arise from a number of directions. Sometimes these problems call for capital expenditures and significant downtime, such as replacement of flooring or airhandling units. In many cases, the facility has to implement a fast, short-term fix in order to keep manufacturing until a scheduled maintenance shutdown. One way in which this can be accomplished is to increase cleaning frequency, or to increase the frequency of sporicide application. Most sporicides designed for hard surface applications are either toxic (formaldehyde, glutaraldehyde) or corrosive (bleach, peracetic acid). Therefore, overdependence on these products is not desirable. Sporicides should be used periodically to address bacterial endospores, such as *Bacillus cereus*, and particularly tenacious molds, such as *Aspergillus niger*. Together with a broad-spectrum routine disinfectant, such as a quaternary ammonium compound or a phenolic compound, this is an effective strategy. However, I have witnessed facilities implement routine overuse of sporicidal agents to combat systemic problems with personnel management, traffic flow, and facility design problems. Inevitably, this leads to additional control problems due primarily to damage arising from overdependence on the corrosive agents.

20.5. CONCLUSION

Effective contamination control is a balancing act. It requires a thorough understanding of both the key contributors and the key barriers to contamination: facilities design, personnel management, and cleaning/sanitization programs. Failure to address these key issues from both a tactical and a strategic standpoint leads to the time and expense invested in investigating excursions. Any time there is an excursion, there is generally a question of whether or not laboratory error has occurred. Although this should be investigated as a possible cause, the other areas discussed within this chapter are more likely to be the causes and should be used as a guide during microbial excursion investigations.

In terms of investigating an excursion there are several pieces of information that may be required. Knowing the identification of the microorganism detected is a key place to start. By knowing the microorganism, at least to the genus level, you can determine several things—including whether or not your sanitization program has failed due to inappropriate product selection. This is, of course, predicated on having either a validation study conducted on the product that involves the isolated organism or a scientifically valid presumption of performance. For example, it is well established that bacterial endospores are susceptible to oxidizing chemicals and are relatively unaffected by exposure to phenolics, quaternary ammonium compounds, and alcohol. If you have in vitro data demonstrating that the product is capable of killing the organism, then the question becomes, was the product used correctly? Determining proper application (time, temperature, concentration, coverage) is a matter of reviewing SOPs, interviewing personnel responsible for application, and observation.

The question may also arise whether the area is being recontaminated with the organism due to personnel management issues or facility design limitations. As discussed previously, cleaning frequency may be increased as a way to ameliorate these issues. However, there should be a plan to address the root causes of the contamination, rather than merely establishing an over-dependence on a chemical solution. Key to developing such a plan is reviewing data in the context of activities. In this way a cause-and-effect relationship may be isolated, which leads to sound strategy and better contamination control.

Establishment of cause-and-effect relationships is easy to suggest but may be challenging to implement, in part because the cause may be hidden from observation. For example, damage behind walls and in interstitial spaces may lead to endemic mold problems but is difficult to detect. However, as this chapter demonstrates, there are many obvious causes of contamination control failures, such as facility deficiencies and personnel behavior problems, which occur every day with both well-meaning and well-trained personnel. The key to success in combating these problems is first to understand them. Observation, documentation, retraining, and facility upgrades will subsequently lead to better compliance and fewer microbial excursions.

21

Simulation of Aseptic Manufacture

Nigel A. Halls
Chorleywood, Herts, U.K.

21.1. INTRODUCTION

The focus of sterile drug manufacture by aseptic manufacture is the avoidance of microbiological contamination. This is achieved through careful design of the facilities, cleanrooms, and work stations where aseptic manufacture takes place, through good (and preferably fail-safe) engineering practices governing their construction and operation, and through highly disciplined personnel work practices in their day-to-day practices.

Nonetheless, aseptic manufacture involves bringing together, with the avoidance of microbiological contamination, the several components of a final product that have been separately sterilized. As a consequence of the ubiquity of microorganisms in nature, aseptic manufacture must incur some tangible risks of microbiological contamination of the dosage form, the aseptic processing equipment, and of the surroundings. Despite these risks, aseptic manufacture is a "success story"; many millions of aseptically manufactured sterile units are used daily to treat patients throughout the world, yet the incidence of actual occurrences of nonsterility is extremely low.

With the high standards of aseptic practices that are being achieved in the pharmaceutical industry supported by rigorous enforcement (some might say escalation) by the regulatory bodies, the probability of finding a genuinely nonsterile unit in batch-by-batch release testing is extremely low. This is mainly due to the extremely poor statistical power of the sampling plans used in the pharmacopoeial Test for Sterility. How then can we be confident for a new aseptic process or a newly built aseptic manufacturing facility that its operation is safe and its risks of microbiological contamination are acceptably low? How then can we be confident that for an existing aseptic process that some key element or subsystem or working practice has not deteriorated to an extent that an unacceptably high risk of microbiological contamination has arisen? Clearly, we cannot ethically allow the effect on the patient to be the arbiter.

Simulation of the process with a placebo for which every filled unit can be scored as contaminated or not contaminated is the means by which the pharmaceutical industry obtains an index of the risk of microbiological contamination occurring in aseptic manufacturing processes. Although the information provided through simulation is itself limited by various factors, it provides at least a benchmark against which aseptic processes can be compared.

21.2. SIMULATION: ITS PURPOSE

Simulation trials, media fills, broth fills, and so on are all synonymous names for an exercise undertaken as part of the validation of a new aseptic process, and as part of a frequent validation review thereafter. Simulation has long been a familiar part of regulatory compliance in the aseptic manufacture of sterile dosage forms and since the 1990s has become necessary for aseptic manufacture of sterile drug substances.

The purpose of simulation is to provide a measure of the probability of microbiological contamination arising in particular aseptic processes. A placebo is substituted for the product and is processed in a manner identical to that in which the product is processed.

In its simplest form, an aqueous liquid microbiological growth medium is substituted for an aqueous liquid dosage form. The medium is filled into ampules or vials. The filled units are incubated and the number of contaminated units are scored versus the number of uncontaminated units, thereby providing an index of the probability of contamination (the proportion contaminated) arising from the aseptic process.

It should be emphasized that simulation trial results do not provide an index of the Sterility Assurance Level (SAL) or PNSU (probability of a non-sterile unit) being achieved for any particular aseptically filled product. This conceptual difference between the proportion contaminated in a simulation trial and the SAL of sterile products is generally poorly understood.

In properly conducted simulation trials the aseptic process is carried out exactly as it would be carried out routinely. The only difference is the use of a placebo to replace the dosage form; an aqueous placebo is used to simulate aqueous liquid dosage forms, a solid placebo is used to simulate sterile solid dosage forms, and something with similar rheological characteristics to an ointment is used to simulate ointments.

Table 1 compares the composition of the placebo most commonly used for aqueous liquid media fills (Tryptone Soy Broth, TSB), with the formulation of a particular but not atypical aqueous injection. There are few similarities except that they are both aqueous.

TSB is widely used as a placebo for simulation trials because it is a general-purpose microbiological growth medium formulated to support the survival and growth of a broad spectrum of types of microorganisms. On the other hand, the injection described in Table 1 has been formulated for medicinal purposes; most significantly, it contains a preservative (0.5% phenol) for the express purpose of inhibiting the survival and growth of microorganisms.

The proportion of contaminated units found in a simulation trial may well be an index of the probability of a nonsterile unit (PNSU, SAL) for TSB being filled aseptically in the particular process because it can be expected that every microorganism arriving in TSB will survive and increase in numbers to discernable levels. This would not be the case, however, for most pharmaceutical products—certainly not for the injection described in Table 1 for which it would be expected that most microorganisms arriving in the product would die. Therefore, the proportion contaminated found in simulation trials should be understood to be an index of the potential for contamination

TABLE 1 Comparison of Tryptone Soy Broth (TSB) with an Aqueous Injection

	TSB (g/L)	Aqueous injection (g/L)
Drug substance		28
Casein	17	
Soybean meal	3	
Dextrose	2.5	
Phenol		5
NaCl	5	
K_2HPO_4	2.5	
KH_2PO_4		1
Na_2HPO_4		2.4

associated with a particular process, not the SAL for all products filled in that process.

To further emphasize this point—that simulation trials provide an index of process contamination and not product SALs—consideration may be given to different dosage forms with different formulations, different drug substances, preserved and nonpreserved, and so forth being aseptically manufactured in the same filling process. It is not logical that they should all have the same SAL, because of the effects of their formulations on contaminant survival—but it is only usual to perform one set of simulation trials for each process and to obtain only one index of the probability of contamination in the process.

The proportion of contaminated units found in simulation trials is an index of the potential for contamination arising from a particular process, and is arguably the worst SAL for any product filled in the process. Frankly, however, this is not likely to bear a major resemblance to the real probability of finding a nonsterile unit in a manufactured population, batch or lot. This is something that ought to be considered when "routine" simulation trials "fail."

21.3. PLACEBOS

The principle of the simulation trial is to replace the product with a placebo. The choice of placebo is critical. The most commonly used placebo is TSB replacing aqueous injections. TSB is not, however, the only placebo that may be used, and therefore the general case merits some discussion.

Aside from the basic requirement that liquids should be used to replace liquids and solids to replace solids, the one essential requirement of placebos used in simulation is that they should not be inhibitory to microorganisms. When a microorganism alights on an object or material, one of several things may happen—it may die, it may survive without growing, or it may survive, grow, and multiply. TSB is a liquid medium that allows for survival, growth, and multiplication of a wide range of bacteria, yeasts, and fungi. Thus, it is a very convenient placebo for liquid injections because not only are contaminating microorganisms not inhibited when they "fall into" a container of TSB, but they are also very likely to multiply to discernable numbers. In fact, TSB serves a dual function in the simulation of aqueous liquid products.

A better general case can be drawn from the use of solid dry powder placebos for simulation of solid dosage forms, such as antibiotics. Microorganisms do not typically multiply to discernable numbers on dry substrates. Therefore, the simulation of solid dosage forms always requires a second aseptic process for the addition of a liquid in which the solid placebo is dissolved to allow any contaminating microorganisms to multiply.

In principle, dehydrated TSB could be used as a placebo (followed by addition of water) for simulating solid dosage forms—but in practice it rarely is. This is principally because it has poor flow characteristics (it is "sticky"). Other microbiologically "neutral" solids such as polyethylene glycol, mannitol, or lactose are used in preference, followed by the addition of liquid TSB.

In the case of simulation of solid dosage forms, we can distinguish the essential function of the placebo for recovering contaminating microorganisms from the secondary process by which we detect the microorganisms as a result of their multiplication.

This difference can become important when consideration is given to simulation of aseptic manufacture of sterile active pharmaceutical ingredients (APIs). Sterile APIs are most often manufactured in massive plant that far more closely resembles that of the chemical industry than of the "light engineering" equipment used to aseptically fill sterile dosage forms. TSB is rarely used to simulate manufacture of sterile APIs because of its potential to contaminate the equipment behind gaskets, in valves, in complex pipe-work runs, and so forth. Even minute traces of TSB can lead to foci of contamination for growth of microorganisms, which may prove difficult or even impossible to clean and sterilize without major equipment strip down. Therefore, for practical reasons it makes sense to simulate liquid stages of such processes with sterile water, saline, Ringer's solution and solid stages with polyethylene glycol, mannitol, lactose, and so forth, with subsequent use of TSB to encourage multiplication in the end products. A comparison can be drawn between the placebo used in simulation and the recovery fluids used in the Test for Sterility and the Microbial Limit Test.

For disclosure of the presence of contaminants, TSB is a reasonably good all-round, general-purpose microbiological medium that can support growth of aerobic bacteria when incubated at temperatures in the range of 20–35°C. Equally, it is a reasonably good medium for supporting the growth of yeasts and fungi, when incubated in the range 20–25°C. It is recommended for the Test for Sterility in all of the major pharmacopoeias.

It is, however, worth noting that many microorganisms will not grow readily in TSB; and some will not grow in TSB at all. It is a good recovery medium for gram-positive and human commensal–type bacteria, but not the best recovery medium for gram-negative bacteria. The latter grow better with lower nutrient concentrations, and at lower incubation temperatures, than those recommended in the pharmacopoeias for the Test for Sterility.

TSB is not the best recovery media for yeasts and fungi. No mycology specialist would dream of using TSB as first choice for surveying an environment for yeasts and fungi. Neither is it the best recovery medium for anaerobic and microaerophilic microorganisms such as the common skin commensal, *Propionibacterium acnes*.

Why is TSB used so widely, displaying as it does so many limitations? The answer is, quite simply, that it is a compromise medium. It is available commercially. It is uncomplicated and robust. It has the reflected authority of the pharmacopoeias to support it. Most important of all, it has become the industry standard.

There is no doubt that by using other media in addition to TSB (e.g., those with lower nutrient concentrations, which favor gram-negative microorganisms) and/or other incubation conditions (e.g., anaerobic incubation) and/or temperatures (e.g., lower temperatures and longer incubation periods), we could increase the "range" of the simulation trial. The simulation trial fill is not, however, an exhaustive search for every microorganism that could possibly be contaminating an aseptic process—it is a "snapshot" in time with a recognized and limited "focal range."

Occasionally, a case is put forward for anaerobic simulation trials, particularly for processes in which the dosage form may in routine manufacture be sparged with nitrogen and/or held under a nitrogen or carbon dioxide head space. The simulation trial is not itself a validation of the sterility of these gases—this is done by other means. Obligate anaerobes do not survive in air (oxygen is toxic to them)—in fact, they are difficult to recover even from those environments where they are capable of persisting (mud at the bottom of ponds, the cecum of the goose, etc.). It is difficult to see what value may result from the additional effort. Simulation trials are done at a cost to industry, the cost of unsaleable goods (media-filled containers), labor, production capacity, and so forth, which in the long run is passed on to society. For simulations, a line must be drawn somewhere; it is the view of the author of this chapter that as far as media are concerned, TSB is sufficient.

21.4. PROCESS SIMULATION

The general principle of simulation is that the process should be simulated in a way that addresses *every* risk of microbiological contamination that could occur in practice. In other words, the process must be conducted exactly as it would in routine operation. In reality, some differences are made specifically for simulation. Some of these differences arise for reasons of practicality and others in order to achieve regulatory compliance.

Most, if not every, aseptic process is unique. Even in the same factory, two lines set up for the simplest process such as filling liquid products into ampules could differ significantly one from the other. Therefore the treatment given below can at best only be a generalization that may (or may not) be helpful to deciding how specific processes and specific problems of simulation may be resolved.

21.4.1. Simulation of Aqueous Liquid Aseptic Filling Processes

Filling liquid dosage forms into ampules, vials, syringes, and so forth is on the face of it the simplest type of aseptic process simulation, and that which is most often given as the typical case in regulatory documentation. Figure 1 schematically represents two broadly similar (but technically different) aseptic filling processes for liquid dosage forms.

In the first technology the empty ampules are depyrogenated in a double-door oven and loaded onto the filling machine; in the second, the empty ampules are depyrogenated in a tunnel linked to the filling machine. Other than that, the processes are the same: the bulk sterile dosage form is passed through bacteria-retentive filters from a nonsterile to a sterile bulk tank in the filling room, and personnel enter the filling room to service and operate the processes via an air-locked change room. Filled ampules leave the room on a conveyor. A double-door autoclave is represented for bringing sterile equipment into the filling room. Were this to have represented a vial-filling line, the rubber stoppers would have been introduced via the autoclave; otherwise, liquid filling of vials and ampules is broadly similar.

For simulation, the placebo is TSB. Traditionally, aqueous liquid simulation trials were done by taking vessels of autoclave presterilized TSB into the filling room, connecting them one by one to the filling line, and then filling the ampules or vials, and so forth. Some manufacturers, however, have been interpreting regulatory pressure to "simulate the whole process" to mean that they must take dehydrated culture medium as their starting point, make it up in the controlled but nonsterile manufacturing areas, pass it through the process sterilizing filters, and then connect to the filling room and fill ampules or vials. Both approaches have some merit and some disadvantages.

The origins of the "traditional approach" lie in older slow-speed technology and in the days when often a regulatory-satisfactory simulation could be achieved by filling as few as 1000 units (see below). Generally, there would be sufficient laboratory autoclave capacity to sterilize sufficient media in aspirators, or large vessels, which could then be brought to the filling machine.

One aseptic connection would have to be made between the media vessel and the filler. However, in routine operation, there would most likely be other additional aseptic connections—for example, between the downstream side of the sterilizing filter and the sterile holding vessel. Very few older processes would have SIP (sterilize-in-place) systems addressing the whole line from filters to filling needles, now developed for newer processes. Inevitably, therefore, aseptic connections that would be required in practice would not

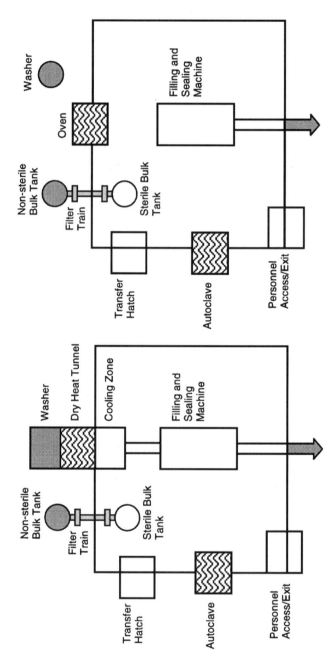

FIGURE 1 Schematic representation of ampule filling technologies.

be *required* in the "traditional" approach to simulation. Conscientious manufacturers might simulate these aseptic connections separately to the media fill, others would ignore them.

With the advent of high-speed filling lines and the need for larger numbers of filled units, laboratory autoclave capacity often became a limiting factor to complying with regulatory requirements on numbers of units filled. The question was inevitably asked as to why autoclave sterilization was necessary for media, when there was perfectly good sterilizing process (filtration) being used for product. So filtration through the production filtration setup became a fairly commonplace practice. Nonetheless, the perceived advantages of doing this merit some discussion.

1. *Taking dehydrated media through all the stages of dispensing and compounding in production vessels takes account of every potential for contamination arising from the process.* There is some confusing logic in this contention.

Dehydrated microbiological media is most usually heavily contaminated with microorganisms not unusually reaching levels of around 10^4 colony-forming units (CFU)/g. Raw materials for aseptic manufacture are invariably specified to be within standards of contamination of no more than 10^3 CFU/g and rarely ever approach those limits. Compounding areas are required to be restricted and microbiologically controlled—they are a medium-level cleanroom; operators in compounding areas are required to wear dedicated footwear, clean overalls, head-covers, and gloves and to conduct themselves and their work to high but not aseptic standards of hygiene. At least twice a year, in the name of QA and regulatory requirement, the notion of bringing nonsterile, dehydrated microbiological media through these areas makes a mockery of the other enforced controls. If simulation of the filtration process is thought to be valuable to the media fill it is sensible to have the dehydrated media sterilized by gamma radiation or for prepared media to have been autoclaved before it is brought into the compounding areas.

The simulation trial is intended to detect weaknesses in *aseptic* processing. Compounding is intended to be sufficiently clean to prevent there being increases in numbers of contaminants or of their by-products (e.g., endotoxins) as a result of conditions in the manufacturer's premises, but it is not an aseptic process. The simulation trial should not be seen as an instrument for detection of problems in nonaseptic manufacture—there are simpler and more straightforward methods to achieve that end.

2. *The media follows exactly the same route as the product and is therefore an exact simulation of the process including the risks associated with sterile filtration.* Indeed, there is some contention that the simulation trial validates sterile filtration—it does not.

There is a totally independent regulatory requirement for sterile filtration to be validated by a bacterial challenge test that is specified in detail and relates to the way filters, particular microorganisms in particular concentrations, and specific products interact. Sterilizing filters are not intended to retain microorganisms from microbiological media at particularly high challenge levels and at the viscosity of microbiological media.

Probably the newer approach to simulating the challenges to product is a fairer one than the "traditional" approach but it is important that its limitations are recognized.

Whichever approach is taken to bringing the TSB on line, the filling process is then run as identically as possible to that used in routine practice, the following exceptions generally being necessary.

1. *Any inert gas (e.g., CO_2 or nitrogen) used to fill or sparge the ampule headspace should be disconnected, or compressed air should be substituted for the gas.* The principle of the use of placebos is to create conditions where there is as great a likelihood as possible of recovering any contaminants present. Most contaminants likely to be present in pharmaceutical manufacturing environments metabolize aerobically (either obligatorily or in preference to other methods of metabolism) and the creation of anaerobic conditions in the headspace above the media would decrease the probability of recovering the majority of potential contaminants.

2. *The volume of TSB added to each ampule need not necessarily be the same volume as the volume of product.* Typically they are identical for small fills but with larger fills it is not always usual practice to replicate the exact volume of product as long as the filling speed is adjusted to leave the ampules open under the filling heads for the same time as they would be in routine filling and as long as the volume of medium is sufficient to at least half-fill the container. The principal reasons for doing this are in connection with the logistics of media preparation. In a 1996 Parenteral Drug Association (PDA) survey of aseptic manufacturers, some 34% respondents did not fill the same volume of media as they filled of product in routine production (1).

3. *All of the contaminating events permitted in a specific routine process must be conducted in the time that the simulation is running even though some of them may be infrequent events in practice.* In modern, high-speed, tunnel-linked ampule filling lines, this often results in as many as 10,000 or 20,000 units being filled just to give enough time to simulate everything. Before simulation trials are run to validate a new process, and perhaps where there is little past experience, the operational Standard Operating Procedure (SOP) should be carefully scrutinized and the process "brainstormed" to prepare a list of potential contaminating events that can be checked off during the simulation trial at the time they take place.

With existing processes, where personnel or wear and tear may have introduced informal changes to the process, it is sensible to repeat the "brainstorm" with the operational personnel periodically, and to have the operational process observed closely over several shifts, noting what happens and how often it happens. Typical contaminating events are greater for vial filling than for ampule filling: they include, but are not restricted to, the following:

> Replenishment of stoppers in the stopper-hopper. This is also generally a manual process.
> Replenishment of vials in the vial-feed if this is a manual operation from a depyrogenating oven—this is not an issue with tunnel depyrogenation.
> Set up and adjustment of filling machines at the beginning of the process. This is often a complex and difficult job. This is reflected in the operating practices of many companies in which there is an intensive disinfection after "setup" and before running, plus a routine rejection of the first however-many filled product units.
> Filling machine adjustment that might be necessary in routine operation in response to, for example, volume checks. These in-process machine adjustment must be simulated even though they may not be necessary in the actual media fill.
> Filling machine stoppages.
> Removal of vials that have fallen over, etc.
> Off-loading of stoppers from autoclaves.
> Personnel shift changes and other occasions where personnel may leave or enter the filling room.
> Microbiological monitoring.

4. *As it is generally thought that the most potent source of contamination in aseptic processes is personal, it is important that any potentially contaminating events associated with manual intervention is addressed through each of the human variables.* In other words, each aseptic operator should be required to actually perform or simulate the performance of each potentially contaminating event in each simulation trial. In order to do this reasonably, it is customary to split human intervention potentially contaminating events into two or three categories: major and standard, or critical, major, and standard. It should be ensured that each aseptic operator performs all of those within the most serious category for each media fill, but less serious interventions need only be addressed by the "team," as distinct from each member of the "team."

5. *The media fill need not be required to be run over exactly the same length of time for which the process may be run in practice.* It has to be run for

long enough to fill a statistically significant minimum number of units (see below); it needs to be run for long enough to be able to simulate all of the potentially contaminating events; and it needs to be run for long enough to address the potential for contamination to build up over time.

The contents of each ampoule, vial or syringe are only likely to be contaminated while they are open and their contents unprotected; this will be for a matter of seconds only in most aseptic processes. Irrespective of whether a shift is 6, 8, or 12 hours long, each unit is still only open for a few seconds. Admittedly there is a *possibility* of the concentration of contaminants increasing in a cleanroom over the time it is manned and operational but this is addressed in routine liquid media fills by doing them at the end of a normal production run with the personnel who have been working in the area.

21.4.2. Simulation of Lyophilization Processes

Those sterile dosage forms that are stable only for a short time in solution are frequently marketed in lyophilized presentations. The lyophilization process is represented in a simplistic way in Fig. 2.

The process is more complicated than standard liquid vial filling although it may involve many items of common equipment. Basically, vials are aseptically filled in the normal way but the stoppers (which are of a special design) are not fully inserted.

The filled, partially stoppered vials are "trayed," taken and loaded into a lyophilizer. The "traying" and transfer of the vials from the filling machine to the lyophophilizer may be done manually or by an automated means (e.g., by robotics and automatic goods vehicles), but irrespective of which means this is done by, there is some vulnerability of the contents of the vials to contamination while they are only partially stoppered.

Within the lyophilizer the liquid in the vial is frozen and a vacuum is drawn. The water from the solid (frozen) phase sublimes directly to vapor, and the dosage form dehydrates. At the end of the cycle the vacuum is broken and the closures are automatically rammed home. The main vulnerability to microbiological contamination within the lyophilizer is clearly at the point where the vacuum is broken. Replacement air must be filtered sterile but other undiscovered means of air contamination from leaks, bypasses, and so forth cannot be discounted.

21.4.3. What Should and What Should Not Be Simulated?

1. *The aseptic filling process should be simulated exactly as any other vial-filling process.* However, the closures will differ and the vials may differ, so

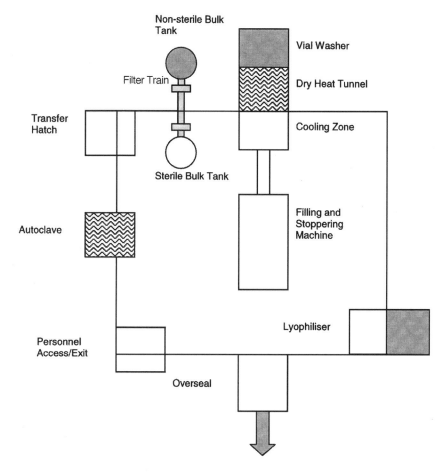

FIGURE 2 Schematic representation of lyophilization in vials.

attention should be given to simulating any activities that are peculiar to filling lyophilized vials as distinct from liquid-filled vials. For instance, there may be a greater frequency of intrusion to, for example, free blocked closure chutes or to remove vials which have fallen over. Any such difference will be unique to the particular process and have to be determined empirically.

2. *The "traying" and transfer process should be simulated exactly.*

3. *The lyophilization process itself need not, and indeed must not, be simulated exactly.*

Freezing should not be simulated. The formation of ice crystals in the freezing process kills microorganisms by cellular disruption, 24 of 26 manufacturers using lyophilization who responded to the PDA's 1996 survey of aseptic manufacture claimed not to freeze the contents of vials filled for simulation (1).

A complete vacuum as specified for the lyophilization process should not be drawn. In addition to the technical difficulties of foaming, which would happen if a complete vacuum were to be drawn over the medium in its liquid phase, consideration should also be given to any fluid loss from the media and its effect on the viability of microorganisms and the ability of the media to support microbial growth. These are two issues, not one. The media after some concentration may still be able to support the growth of microorganisms, but injured microorganisms may have died as concentration took place. Typically, a partial vacuum of say 20–28 inches Hg is drawn, held for about 2 hr and "broken." Conscientious simulators of worst-case conditions may repeat this process although it is not typical of routine practices.

If there is danger of unfrozen media foaming when the vacuum is applied, and thus contaminating the lyophilizer, it may be necessary to perform the simulation in two stages; simulating all of the risks up to and including loading of the freeze-dryer in stage one of the trial, which is then incubated without freezing, and then simulating the subsequent risks in stage two with the rest of the filled vials, which are passed through the complete process including freezing. This is not commonly thought necessary.

Some companies perform complete simulation of the lyophilization process from filling, through transfer, to lyophilization. Others may split the process into three simulations to help provide a clearer focus on what might have gone wrong if contaminated units result from the media fill. The decision as to which approach to take or how to develop a responsible combination of the two approaches is a matter of judgment. A balance has to be struck between regulatory pressure to simulate the process as closely as possible, and the need (also pursued rigorously by regulatory inspectors) to diagnose the source of contamination accurately enough to implement satisfactory corrective/preventive actions.

21.4.4. Simulation of Solid Dosage Form Aseptic Filling Processes

Figure 3 describes aseptic filling of a sterile solid dosage form (e.g., an antibiotic) into vials, in the same simplistic way as was used to describe aseptic filling of liquid dosage forms.

Again, the two different technologies—a batch process in which the vials are depyrogenated in a double-door oven and a continuous process in which depyrogenation takes place in a tunnel linked to the filling machine.

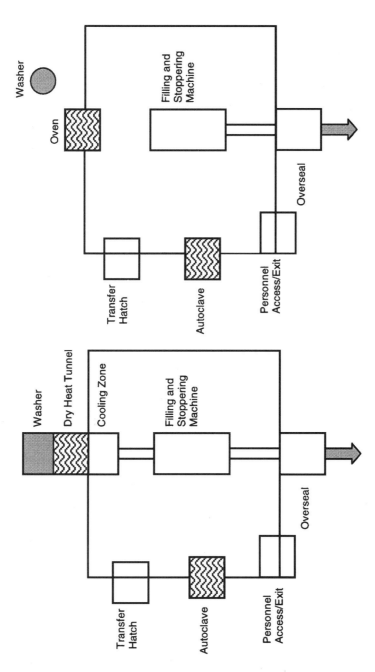

FIGURE 3 Schematic representation of aseptic filling of solid dosage forms in vials.

A wider variety of placebos is used for solid dosage forms. Generally speaking, the placebo is filled into the unit containers and then TSB is added, either on-line or off-line. It is possible to add the TSB before the placebo, but it is not the general practice. The placebo is dissolved in the TSB and incubated.

The chosen placebo should have similar flow characteristics to the product or products that it has been chosen to represent—if it does not have these similar characteristics, it might be effectively impossible to simulate the intended process.

It must be sterilizable. Gamma irradiation is the method of choice for sterilizing solids because, provided they have a low moisture content, this method is unlikely to induce chemical or physical changes. Irradiation is reliable and also penetrative through bulk quantities.

The placebo must be soluble in TSB and must not be inhibitory to the growth of microorganisms. This requirement is always understood to mean that the placebo is not inhibitory to microorganisms when dissolved in TSB, at whatever concentration is compatible with the amount of placebo necessary to simulate the process, and with whatever limitations of remaining-available volume in the unit container are chosen to determine the amount of TSB that can be added.

For simulation trials, the placebo is substituted for the bulk sterile dosage form in exactly the same type of container. It is brought into the filling room through the hatch and taken and connected to the filling machine and filled. TSB is then added to each vial. This may be done using an on-line liquid filler, bearing in mind that it adds an extra aseptic stage to the filling process, or off-line. If the medium is added off-line, the time interval between filling the placebo and adding the medium becomes critical because irrespective of the "neutrality" of the placebo, microbial viability is always adversely affected by desiccation and time.

The filling process is then run as identically as possible in routine practice. There is no doubt that simulation of aseptic filling of solids is more complicated than simulation of liquid dosage forms. The following considerations apply.

1. *Any inert gas (e.g., CO_2 or nitrogen) used to fill or sparge the vial headspace should be disconnected, or compressed air should be substituted for the gas.*

2. *The weight of placebo added to each vial need not necessarily be the same weight as the weight of product.* This is similar to the situation with liquid fills but there is a reason in addition to the logistics reason.

The final concentration of placebo after the media is added must not be so high as to be inhibitory to microbial growth; therefore, the smaller the weight of placebo present per vial, the easier it is to achieve this. Polyethylene glycol is not inhibitory to microbial growth in TSB in concentrations of up to 100 g/L.

3. *All of the contaminating events permitted in a specific process must be simulated.* Specifically, with simulation of aseptic filling of solid dosage forms it is important to ensure that placebo-container changes are simulated. This may prove more complicated than it would appear at first sight. Routine aseptic filling runs are generally longer than simulation trials and fill weights are generally larger; therefore, the frequency of placebo-container changes may be much fewer in simulation trials unless provision is made for this. One means of doing this is to use partially filled placebo-containers for simulation, changing them at a similar frequency to the frequency of routine production. The downside of this is that the lighter partially filled placebo-containers may not present the operators with the same difficulties of maintaining asepsis when changing them.

4. *The media fill need not be required to be run over a complete shift.* It has to be run for long enough to fill a statistically significant minimum number of units (see below); it needs to be run for long enough to be able to simulate all of the potentially contaminating events; and it needs to be run for long enough to address the potential for contamination to build up over time.

The contents of each vial are only likely to be contaminated while the vial is open and its contents unprotected; this will be for a matter of seconds only in most aseptic processes. Irrespective of whether a shift is 6, 8, or 12 hours long, each vial still is open for only a few seconds. Admittedly, there is a possibility of the concentration of contaminants increasing in a cleanroom over the time it is manned and operational but this is addressed in routine liquid media fills by doing them at the end of a normal production run with the personnel who have been working in the area. The only exception to this practice is for antibiotic filling, where it is important that all antibiotic traces are cleaned out of the filling equipment and the filling room before the placebo is filled. This is to prevent the antibiotics inhibiting recovery of microorganisms in the medium. If it is possible, it is advisable to use personnel who have completed or come close to the end of a shift on another filling line to simulate antibiotic filling to simulate any "sloppiness" in aseptic technique that may arise from tiredness.

A more rigorous approach may be demanded to the validation of the time a sterile "setup" may be left on a filling machine, especially if filling is done on a campaign basis over more than one day. There are several possible approaches to this:

> Several thousand units may be filled with placebo and medium after start-up. Unless the filling machine is sterilize-in-place (SIP)-equipped to point-of-fill, machine setup and aseptic assembly of

presterilized product contact parts is surely one of the times of greatest contamination risk. Thereafter the machine may be "held sterile" for a period of hours or even days, and then several thousand more vials filled with media, with all interventions included or simulated. Thus the three major risks—setup, interventions, and time-related factors—are all taken into account.

Alternatively, several thousand units may be filled with placebo and medium after start-up, and then the machine may be "run dry" (i.e., with no addition of placebo or TSB for however long as is necessary), with operators freeing jams and simulating sample removal and so forth as usual. The several thousand more vials may be filled with placebo and medium as before.

The third alternative is for the machine to run placebo for the whole of the campaign length that is to be validated. Medium is filled however, only for the first and last several thousand and after any serious interventions during the "placebo-only" period.

21.4.5. Simulation of Processes Involving Aseptic Bulk Compounding Before Filling

Some sterile products are required to be compounded aseptically. Suspensions (e.g., ophthalmic ointments) are one example. Some antibiotic solid dosage forms require to be blended with a carrier. Each particular case is likely to be different. There may be compounding of two liquid phases, both of which have been passed through bacteria-retentive filters; compounding of two solid phases, both of which enter the filling room through pass-through hatches; or compounding of liquid and solid phases.

The general case applying to media fills that needs to be drawn is that the aseptic compounding needs to be included in the simulation.

Performing simulation trials fills on ophthalmic ointments is a nightmare. Placebos are based on TSB made viscous by addition of a substance such as carboxymethyl cellulose at about 65 g/L, although this concentration may differ according to the process settings that are applicable to the range of ointments being simulated.

The nightmarish aspects of ointment simulation trials are threefold as detailed below.

First, there is cleaning up behind them. Actual ointments generally present a sticky mess, which is difficult but obviously not impossible to clean from production equipment. But, add a microbiological growth medium to that sticky mess, and cleaning becomes highly critical, especially if the creation of foci for microbiological growth in the equipment and in the facility is to be avoided—as it must. Good cleanroom practices are difficult to maintain in ointment simulation trials.

Second, ointment tubes are rarely transparent; therefore, inspection of thousands of placebo-filled tubes for growth after incubation is difficult. Generally the tubes are opened and squeezed out, although some users of plastic tubes have special orders of transparent tubes purchased solely for simulation trials.

Third, microorganisms grow as colonies in carboxymethyl cellulose–thickened TSB rather than as general opacity, and carboxymethyl cellulose–thickened TSB is not a clear transparent medium in which colonies can be easily discerned. This is generally addressed by inclusion of a metabolic indicator such as 2,3-tri-phenyltetrazolium chloride in the medium at or around 0.0025%. Tetrazolium chloride is a metabolic indicator that changes to a red/purple color when microorganisms respire.

21.4.6. Simulation of Aseptic Manufacture of Sterile Solid Active Pharmaceutical Ingredients

There is now a regulatory expectation that aseptic processes for the manufacture of sterile active pharmaceutical ingredients (APIs) should be simulated. These processes are quite different in their scale, their equipment, and their methods of operation to aseptic filling processes for sterile dosage forms. This has posed many problems for the pharmaceutical industry in terms of how to go about their simulation. A very detailed account of these difficulties and proposals to resolve them is contained in PDA Technical Report No 28 (2).

Figure 4 is a gross oversimplification of "typical" sterile API manufacture. First of all, the starting materials are dissolved in one or more dissolution vessels; they are then passed through bacteria-retentive filters into a presterilized reaction vessel where typically the API crystallizes or precipitates out. The API is dried by filtration or centrifugation and is then dispensed (off-loaded) into its final containers.

Ideally, this whole operation is hard-piped and sterilized-in-place from start to finish. Most often the greatest vulnerability is at off-load, where large quantities of solid material are aseptically transferred to manageable-sized containers—it is the current expectation for this to be done under the protection of isolation technology, but in fact there is a diverse range of aseptic technologies still in current use. Other vulnerabilities to contamination may be from vacuum driers which operate at lower pressures than the external environment, and from leaks in pipe-work, valves, gaskets, and so forth.

Resolving the technical problems of API simulation is not easy:

Should the placebo be liquid or solid? The combination of "wet" subprocesses where the reactants are in solution, and "dry" subprocesses where the API is being handled pose obvious problems for simulation. Some

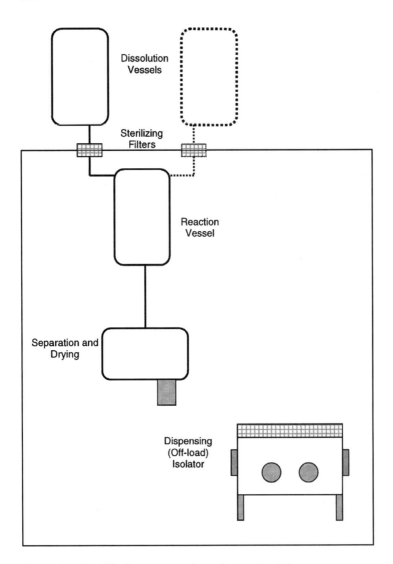

FIGURE 4 Simplified representation of a sterile API process.

API processes may allow for a microbiologically "neutral" placebo to precipitate without addition of antimicrobial reactant—for example, if precipitation/crystallization is achievable by concentration and temperature reduction, then something like lactose could be passed through the whole process. Most often this is not possible, and unit operation liquid stages have to be simulated using a microbiologically "neutral" liquid such as water, saline, or Ringer's solution; and unit operation solid stages simulated by using polyethylene glycol, mannitol, lactose, and so forth. As mentioned above, it is unwise to risk contaminating a large-scale complex chemical plant with TSB. Where the process is dealt with in separate unit operation simulations, it is wise to document a formal analysis of the risks of contamination at the different stages of the process in order to justify the compromises undertaken.

How should the large volumes of placebo required to fill the plant be addressed versus restricted incubation capacity? Some companies have attempted to address this by using TSB as the placebo and incubating it within the plant, but this should be avoided at all costs—such an approach is a de facto conversion of a sterile chemical plant into a fermentation plant. Once "seeded" with microorganisms from heavy contamination arising from a "failed" simulation, it is very difficult and time-consuming to restore the plant to sterile operational conditions. There are two other possible approaches: concentration and sampling.

Large volumes of liquid may be passed through bacteria-retentive filters and the filter either incubated in TSB or plated on TSA. Plating is a poorer means of confirming absence of microorganisms than incubation in liquid media but may permit acceptance criteria (e.g., not more than n colonies recovered) which are to some extent tolerant of the occasional laboratory or test-induced contaminant. The complexity of transferring filters to agar plates should also not be underestimated; it is likely that the volumes requiring filtration require filters of much larger diameters than the standard 90 mm petri dish.

Solids cannot be concentrated; generally, TSB must be added to the placebo, ensuring that the concentration of the placebo in TSB is not inhibitory to microorganisms. This can be done in the final containers, but they must be only partially filled to allow for addition of the TSB. Generally such containers are not transparent, and indications of growth can be obtained only by opening them; other than risking incidental contamination of the trial, this can practically be done only at the end of the incubation period.

If it is not possible to incubate the whole of the solid phase, then some recourse to sampling must be made. This may be very difficult to justify but may be the only practical approach to simulation of very large scale processes.

21.5. MICROBIOLOGICAL CONTROLS

The "ownership" of simulation trials should properly lie with the management of the aseptic process. For greatest efficiency, simulation trials fills should be scheduled into the manufacturing program in the same way as a routine filling activity except that the "product" is units filled with media. In practice, the ownership of simulation trial fills tends to be held jointly between production and microbiological QA because of the numerous additional microbiological considerations and controls that must be complied with for a simulation trial to be fit for its intended purpose within the QA program.

21.5.1. Sterility and Growth Support Controls

Nonsterile materials must not be taken into aseptic manufacturing areas. Sterility of media is best verified by preincubation outside the aseptic manufacturing areas. The prospect of having, for example, 50 liters of microbiological media becoming more and more heavily contaminated through each hour of preincubation within an aseptic filling room is something that should be avoided.

A more subtle but far more serious concern is that the media must be capable of supporting growth. The first responsibility in any microbiological exercise for which "no growth" is the favorable condition is to ensure that the batch of media used is actually capable of supporting growth.

When the placebo is a growth medium (usually TSB) or where TSB is added to some other placebo, growth supportiveness must be verified after it has been in contact with the filling equipment and the product containers. This is to ensure that traces of product, antibiotic, detergent, disinfectant, and so forth in antimicrobial concentrations have not been passed into the media from any one of these or other production-related sources. Additionally, it is often advisable to check the media for growth supportiveness before use to be able to distinguish at the earliest stage if there is any growth support problem associated with the media (or its preparation) itself or with the interaction of the media with manufacturing equipment.

When prepared media is autoclaved for use in simulation, it is usual to aseptically withdraw a sample for growth support checks, which can then be conducted simultaneously with preincubation of the media in a laboratory incubator to verify its sterility. Both results are obtainable before the media need be taken into the aseptic areas.

If the only sterilization of media is on-line filtration, an aseptic sample may be taken from the sterile holding vessel for growth support checks and sterility. The media should conservatively be held until these results are obtained, but the risk of contaminating the filling room by preincubation in these areas is something that is most often thought to present a greater risk than the benefit offered by knowing that the media is growth supportive and sterile before taking it on line—if it has been sourced from an approved supplier and sterilized according to a validated process, the risk of it not being growth supportive and sterile should be minimal.

A second growth support check should be done on filled units. In principle these may be taken and tested at the start or at the end of the incubation of the simulation trial. In terms of managing and scheduling, it is best to take them at the beginning because unsatisfactory media would be revealed at the earliest opportunity. However, current FDA thinking (3) has it that all media-filled units must be incubated, even those that might normally be discarded in routine practice, because (allegedly) the removal of units for controls may result in taking the very units that might be contaminated out of the trial. As a consequence, growth support on filled vials is most usually done at the end of the incubation period of the complete set of filled units, and the results are not available until several days after incubation of the trial.

The medium that is almost always used for media fills is Tryptone Soy Broth (TSB, Casein Soya Digest Medium) because it is used in the pharmacopoeial Test for Sterility. The pharmacopoeias are therefore the usual points of reference for the microorganisms and the conditions that should be applied to growth support checks.

Table 2 shows the current *United States Pharmacopeia* (*USP*) and *European Pharmacopoeia* (*PhEur*) requirements for TSB medium growth support when used for the Test for Sterility; however, the control cultures applying to TSB are cited only for the 20–25°C incubation condition. Simulation trials may, however, be incubated at two temperatures (see later), 20–25°C and 30–35°C. It is therefore good sense to replicate the growth support test across the two temperature ranges. All of the pharmacopoeially recommended microorganisms listed in Table 2 should grow profusely in both temperature ranges within 7 days' incubation from an initial inoculum of 10–100 CFU. Separate media samples should be inoculated with each culture.

In addition to the pharmacopoeial media growth support control cultures, many regulatory agencies insist on at least one isolate from the manufacturing environment being used as a media control. The logic is that if the TSB is intended to recover microorganisms that inhabit the manufactur-

TABLE 2 Microorganisms Required for Sterility Test (Media Growth Support Checks in *USP* XXV and *PhEur* 3rd ed. (1997) (ATCC numbers only are shown for convenience)

Medium	*USP* XXV	*PhEur* 3rd ed. (1997)
TSB at 20–25°C	*Bacillus subtilis* (ATCC 6633)	*Bacillus subtilis* (ATCC 6633)
	Candida albicans (ATCC 10231)	*Candida albicans* (ATCC 2091)
	Aspergillus niger (ATCC 16404)	*Staphylococcus aureus* (ATCC 6538P)

ing environment, it should be shown to have the ability to support the growth of those environmental microorganisms. Local microorganisms could be frail, injured, disinfectant-damaged, and so forth and therefore could be more difficult to recover in TSB than the pampered, well-nourished subcultures from the culture collection.

Conversely, the local environmental isolates used for media controls will have most likely been maintained in a local culture collection for several months at least and will probably have recovered from any physiological damage associated with stressful local conditions.

Irrespective of these doubts and compromises, local environmental isolates are recommended for media control. It is recommended that the chosen isolate is changed periodically such that it can be related to the current rather than the historic microflora of the manufacturing environment.

Where antibiotic filling processes are being simulated it is also important to ensure that at least one of the growth support control cultures is very sensitive to the antibiotic. This is to provide the most sensitive information on the success or otherwise of the cleanup process.

The preparation of control cultures should be clearly specified in laboratory documentation, and records of subculturing should be maintained. The FDA prefers that working control cultures are no more than five generations separated from their national or international culture collection origins. This is to limit the potential for mutation.

Small inocula must be used in media control: the intention is to recover microorganisms when they are present only in low numbers. The pharmacopoeias interpret low numbers to mean between 10 and 100 CFU per inoculum. The reference condition for this is surface culture on Tryptone Soy Agar incubated at 30–35°C for at least 48 hr.

21.5.2. Controls Intended to Expose Incidental or Laboratory Contamination

The microbiological controls required in relation to simulation trials are not restricted to the qualification of the trial. They should also be able to reveal problems of the laboratory's own creation. Microbiological QA exists to identify production problems and to assist in their resolution—microbiological QA should always be wary of creating problems of its own making. Any activities associated with microbiological control of media and any laboratory manipulations that do not exist in routine manufacturing practice should be examined critically. This is best done by detailed analysis of the ways in which simulation trials fills are organized. Some examples are given below.

1. *Aseptic sampling from bulk media is a serious vulnerability*. It is all too easy for the bulk to be contaminated when growth support samples are

being withdrawn. It is not inconceivable that the outcome would be for the simulation trial to be contaminated, probably over several filled units, as a result of the contaminants being distributed throughout the bulk and possibly increasing in numbers before all of the media is filled. If possible, the bulk vessel should be incubated at the same time as the filled units. Contamination of the bulk invalidates the simulation trial and places a lot of pressure on attempting to diagnose production problems that are not of production's making. Regulatory agencies would always expect a simulation trial in which the incubated bulk was found to be contaminated to be repeated irrespective of the quality of the results from the filled containers.

2. *In some cases, solid-dosage form simulation trials require the recovery medium to be added to the placebo-filled containers (usually vials) off-line in a laboratory.* Typically this requires media in bulk, and some apparatus, most often an automatic or repeating syringe, to transfer it to the placebo-filled units. The vulnerabilities are for the bulk to be contaminated when the microbiologist aseptically assembles the transfer apparatus, for the transfer apparatus to be contaminated when the microbiologist assembles it, and/or for the transfer apparatus to become contaminated over the period in which it is used. In this type of simulation trial it is usually easy for the bulk container to be retained and incubated. A sample of the first media passed through the transfer apparatus before any placebo-filled units are filled should be injected into a sterile vessel and incubated. A sample of the last media passed through the transfer apparatus after all of the placebo-filled units have been filled should be injected into a sterile vessel and incubated. It is also usual to intersperse the placebo-filled units with sterile sealed containers at regular intervals. The sterile sealed containers are filled with TSB in the same way as the placebo-filled containers and are intended to disclose any transfer apparatus contamination as close as possible to the stage in media transfer when it happened. The frequency of interspersion of sterile containers is a matter of judgment; it may be every third, fifth or tenth unit according to the degree of confidence in the skills of the microbiologists adding the media. Irradiation is the recommended method of sterilization because sealed empty vials are quite difficult to sterilize by autoclaving. The discoloration obtained in most grades of glass as a result of exposure to gamma radiation is a convenient feature for distinguishing sterilized from placebo-filled units.

21.5.3. Environmental Monitoring and Media Fill Observation

Microbiological monitoring is itself a potential source of contamination. It must therefore be included in simulation trials. It is also one of the best ways of diagnosing the source(s) of contamination when they arise. It should always be assumed that management is anxious to know what, where, and why simulation trial contamination has arisen in order to decide on appropriate corrective and preventive actions and to improve their processes.

For this reason it is advisable to have intensive microbiological monitoring over the period of the simulation trial. The potential advantages outweigh the disadvantages. Microbiological environmental monitoring should be intensive; where it may be routine practice for microbiological monitoring to be applied over a number of locations on a matrix basis (see later), the practice during media fills should be for all locations to be monitored.

It is also advisable that the simulation trial is observed by a person trained in asepsis and familiar with the filling process. Detailed notes should be taken describing what is happening, particularly anything unusual, and when it is happening. The observer may provide an independent verification that the listed contaminating events have been simulated. Whereas production personnel may appear best qualified for this task, it is usual to have microbiological QA doing it to ensure independence.

It may be useful if the "traying" of filled units can be related to the times of filling, although it should not be thought that all contamination of product units happens at the moment and at the point of fill. A contaminated unit may as easily have arisen from microorganisms carried on rubber stoppers contaminated when the autoclave was off-loaded, perhaps some hours or even longer before the time of stopper insertion.

Some regulatory agencies are indicating a preference for simulation trials to be recorded on videotape. This would appear to be the best way of proving that fraudulent claims regarding the conduct of the simulation trials are not being made, but conversely the video camera rarely has the peripheral vision and/or the variability of focus of the human observer. In other words, for information purposes the video recording has to be done intelligently, but with the risk that a regulatory investigator may become more interested in what the cameraperson may not have recorded than what he/she has focused on. A fixed camera focused at point-of-fill gives no information about the risks attendant upon, for example, unloading autoclaves, replenishing stopper bowls, and so forth.

21.6. INCUBATION OF SIMULATION TRIALS

Filled units must be incubated as soon as possible after filling. Regulators, the FDA in particular, are anxious that all units are incubated (with the exception of those without caps, obvious cracks, etc.). This is intended to include all of those "perfect" units that may in practice never be released (e.g., units cleared off the line after a stoppage or similar event). These normally rejectable units should be incubated "for information" separately from the trial units.

Contamination in these units should not be included in assessing the success or failure of the simulation trial.

When media is added in the laboratory after a solid placebo is filled, it is critical that the interval between filling the placebo and adding the medium is as short as possible to prevent contaminating microorganisms dying off in the placebo. The argument against this, based on the premise of genuine product contaminants dying off within the 7 or 14 days' Sterility Test quarantine before release, is not valid. The simulation trial is intended to disclose process contamination—not the probability of nonsterility in product (see earlier) within its marketed shelf life. The maximum interval between filling and media addition should be validated by inoculation of the placebo and then tracking recoverable survivors over time.

Incubation of simulation trials is done over 14 days. This probably originates in the pharmacopoeial Sterility Tests where, over many decades, none of the major pharmacopoeias have asked for any longer period of incubation than 14 days. If the simulation trial were to be considered as an exhaustive search for potential viable microbial contaminants, then the duration of incubation would be potentially limitless. It is well known, for instance, that some coryneform bacteria require 28 days or more incubation to produce visible turbidity in TSB. Some companies incubate validation simulation trials beyond the 14-day period and justify future routine trial incubation at 14 days or whenever the last contaminant was detected in the extended validation exercise, whichever is the longer.

There has been some controversy over the temperature of incubation for simulation—20–25°C or 30–35°C. If a choice of one or the other is made, it will always be open to criticism, and in reality both temperature ranges (and probably some others too) can be reasonably justified. Incubation at both temperatures is widely used, but this still leaves the decision over which temperature should be used in the first 7 days and which should be used in the last 7 days of incubation (or indeed should there be another pattern?). Once again, both options are justifiable, and neither is worth an acrimonious argument with a regulatory inspector and none is likely to arise as long as there is some justification and it is documented. It is probably as well to incubate at 20–25°C for the first 7 days to encourage the recovery of any gram-negative types, which tend to be favored by lower temperatures and longer periods of incubation.

It is normal to incubate the filled units for 7 days in their normal orientation and for 7 days upside down. The principle is to ensure that all of the internal surfaces of the container and closure are bathed in media for long enough to allow any adherent contaminants to be resuscitated, recover, and grow. Almost always incubation in the correct orientation takes place over the first 7 days, and upside-down incubation in the second 7 days. The opposite approach could be as well justified. Some companies are satisfied that they can justify incubation without inversion because they initially swirl the media around to wet all internal surfaces.

The amount of media filling each container should be sufficient to reach halfway up the height of the container so that every internal surface is bathed by the medium for at least 7 days. This is not always done. This is a factor that should be taken into account when determining how the simulation trial is to be conducted (see earlier).

It is advantageous to know if there are any contaminants in simulation trials as early as possible. Visual inspection without disturbing the units is normal on a daily or every second day basis; a thorough visual inspection should be conducted at 7 days when the units are inverted and 14 days when incubation is complete. Damaged or cracked units may be excluded from the results. The total number of units checked at the end of incubation plus any removed for reasons of damage should reconcile exactly with the numbers filled and presented for incubation. Reconciliation limits such as plus or minus 5% used in other aspects of pharmaceutical manufacture are not acceptable.

Visual inspection should be done in good daylight or artificial light by personnel who have good eyesight and are subject to periodic sight tests. Turbidity is the typical indication of microbiological growth, but personnel assigned to this task should also be alert to the possibility of pellicle formation on the surface of liquid media and other forms of microbial growth. Visual inspection becomes more difficult with tinted glass containers, but it is certainly most difficult for ophthalmic ointments where the contents have to be squeezed out usually onto white paper and examined for growth as indicated by the red coloration produced from the oxidation of tetrazolium chloride, or by the presence of bubbles.

The microorganisms from every contaminated unit obtained in any simulation trial should be subcultured, purified, and identified to species level. Where possible the tray number/time of filling of every contaminated unit should be retained. The identity of any microbial contaminants is a major part of the information content of the media fill, and where possible the identified microorganisms should be related to the events that were happening when the contaminated unit was filled. This view appears to contradict the apparent obsession of many pharmaceutical manufacturers, microbiologists, regulators, and standards writers to place the emphasis of contaminated media fills on the numbers of contaminated units or on the proportion of contaminated to uncontaminated units. There is practically no information content in knowing that there were two contaminated units in a media fill of say 4000 units. Conversely, knowing that the two contaminants were, for example, pseudomonads or micrococci points the experienced microbiologist to the most likely source(s) of contamination and allows intelligent diagnosis of the problem and focused corrective/preventive actions.

21.7. APPLICATIONS

Simulation trials fills are used in validation of aseptic processes as one of the final stages of Performance Qualification. They are also repeated periodically in routine operation of aseptic processes. It is arguable whether this latter application should be categorized as part of validation review or as part of environmental monitoring—either way the outcome is the same; the simulation trial is a method of gathering information about microbiological contamination.

21.7.1. Simulation Trials in Validation of Aseptic Processes

The 1987 FDA *Guideline on Sterile Drug Products Produced by Aseptic Processing* (4) refers to simulation trials (media fills) as "an acceptable method of validating the aseptic assembly process." By 1994, the *Guideline to Industry for the Submission Documentation for Sterilization Process Validation in Applications for Human and Veterinary Drug Products* (5) says that specifications for simulation trials fills *should* be among the information submitted in support of sterility assurance for products manufactured by aseptic processing. In 2002 the draft *Concept Paper on Sterile Drug Products Produced by Aseptic Processing* (3) includes the word *should* in relation to process simulations for validation of aseptic processes in line with the 1994 publication. "*Should*" is a directive verb in these publications.

On the other side of the Atlantic, the 1983 "Orange" Guide (Department of Health and Social Security, 1983) gives simulation trials as *an example* (albeit the only example provided) of how the "efficacy of aseptic procedures should be validated." This has been succeeded by the 1992 and 1997 and 2002 editions of the Commission of the European Communities' *Good Manufacturing Practice for Medicinal Products* (6), which state that "validation of aseptic processing *should* include simulating the process using a nutrient medium," "*Should*" is again a strongly directive verb in the language of these requirements.

The point to be drawn is that in the past 15 years, simulation trials have, in the eyes of the regulatory bodies, developed from being a reasonably good way of validating aseptic processes through being the *preferred* way of validating aseptic processes, to now being an *essential requirement* of a properly validated aseptic process. It is now remotely unlikely that any regulatory submission for a new aseptically filled sterile pharmaceutical product or sterile active pharmaceutical ingredient (API) would be acceptable without supportive data from simulation trials, nor is it likely that a manufacturer of an existing aseptically filled sterile product would escape severe regulatory criticism if simulation data were not available.

It is now well accepted in the pharmaceutical manufacturing industry that validation is an exercise that is intended to confirm that a process is capable of operating consistently. As far as asepsis is concerned, the consistency of the contamination control "engineering" of a process is qualified by three successive replicate simulation trials done on separate days. Satisfactory completion of simulation is usually the factor that dictates the time of handover of a process from validation or development into routine usage.

New aseptic processes require validation by simulation. Any process (irrespective of the equipment being old or new) beginning in a new cleanroom requires simulation fills as part of validation. A new filling machine in an established cleanroom requires validation by simulation.

Although this is on the face of it simple, it is quite probable, however, that a range of container sizes may be filled on the same filling machine. The question then arises over the necessity to perform simulation fills on all sizes, and in validation in particular whether it is necessary to replicate each size through three simulations. The simplistic answer is that simulations need only be necessary for the container size presenting the greatest potential for contamination (the worst case). Frequently this is justified to be the container size that takes longest to fill and that has the widest neck diameter, therefore addressing the contamination potential for all smaller sizes. However this is not necessarily true. Wide-necked containers may be more stable than narrower-necked ones, and therefore the wide-necked filling process may be arguably less susceptible to contamination because there are fewer personnel intrusions necessary for rectifying fallen containers. Where there are multiple container sizes, there is probably no secure way of rationalizing simulation to fewer than two sizes. In the long run the decision over what and how many sizes to include in a simulation validation protocol is a local decision, and for regulatory purposes the reasons for taking the particular decisions must be justified and documented. If the rationale for performing simulation on more than one container size is based on the risks of contamination arising from different sources, or from a different balance of sources (e.g., ampules and vials being run on the same machine), rather than from a scale-up of risks from the same sources (e.g., vials of different capacity but with the same neck diameters), then it is logical that the three replicate simulations thought necessary to verify consistency of control must be performed on each container size.

The significant formality of validation simulation trials is the Protocol. There are three principles that must be borne in mind:

21.7.1.1. The First Principle

The first principle of the Protocol is that the process that is to be validated has to have been already defined and documented; in other words, draft operating SOPs have been prepared and personnel have been trained in them. It may be that some trial simulations have been done as part of process development or in OQ or in training to establish the process, but validation follows on only after the process has been established.

21.7.1.2. The Second Principle

The second principle is that the test methods, in this case the simulation conditions, have been defined and documented. The Protocol must define the differences between the simulation trial and the routine process (the simulation will probably fill fewer units and must take account of all permitted personnel interventions, irrespective of whether or not they arise as a matter of course) and the number of units that are to be filled is important.

The minimum number of units is expected to be 3000 units. The origins of this figure are worth justifying. In principle, it is an expression of the minimum number of units for which a contamination rate of no more than one contaminated unit in 1000 units (1 in 10^3 or 0.1%) can be demonstrated with 95% confidence. But, why a contamination rate of no more than one contaminated unit in 1000 units (0.1%)? And why with 95% confidence?

In 1971, Tallentire, Dwyer, and Ley (7) wrote that "sterility testing has several serious defects, not least amongst them being the high frequency of spurious results, sometimes called "false positives," due to contamination during testing. When measured using a population of items known to be sterile under best known test conditions, this frequency is approximately 1 in 10^3" (7).

As far as is known to the author of this chapter, this 1971 paper is the origin of the view that processes involving aseptic manipulation are limited by test-related contamination at or around a frequency of 1 in 1000. The 1 in 1000 level also ties in with the regulatory expectation of Sterility Test failures within any particular laboratory being no greater than 0.5% of all tests conducted. This assertion is based on the typical Sterility Test involving aseptic transfer from 20 product units; therefore, 0.5% test failure represents aseptic transfer from 1000 units.

However, Tallentire's contention (7) was based on the technology of the 1970s. Asepsis has moved on considerably since then, but the 1 in 1000 limit seems to have become permanently attached to simulation trials, probably because it is a practical benchmark for the number of units that can be filled, incubated, and so forth.

The PDA (8) supported a limit of no more than 0.1% contamination for media fills in its 1981 monograph, adding that this should be demonstrated with 95% confidence and that at least 3000 filled units are required to achieve this (8). No reason for choosing 95% confidence rather than 99% confidence or 90% confidence is given.

The idea of 3000 units and 95% confidence reappeared in the FDA 1987 Guide (4) and has become part of the regulatory industry and expectation of media fills. On the other hand, the FDA's 2002 draft *Concept Paper on Sterile Drug Products Produced by Aseptic Processing* (3) has no mention of 3000 units nor 0.1% contamination rates. Instead, it says that "For example, a single contaminated unit in a 10,000 unit media fill ... is not normally considered on its own to be sufficient cause for line revalidation."

The association of 3000 units with 95% confidence of assuring a contamination rate of no more than 0.1% has been elaborated upon by Halls (9). Its support from two different mathematical positions is summarized here.

The PDA (8) references the following equation of an "operating characteristic" curve to describe the probability of detecting one or more contaminated items in a sample size N taken from a population with a contamination rate of 0.1%:

$$P_{(x>0)} = 1 - e^{-NP},$$

when $P_{(x > 0)}$ is made equal to 95%, this equation describes how large a sample size, N, needs to be taken from a universe in which there is 0.1% of contaminated units to find at least one contaminated unit on at least 95% of occasions when samples are taken. In practice, 95% confidence cannot be achieved with a sample size of less than 2996.

Alternatively, the measured contamination rate in a simulation trial may be regarded as an estimate of the true contamination rate (P) in the underlying population, which may be higher or lower than the measured rate (P_{est}). The reliability with which Pest can be claimed to be a true reflection of P can be calculated from the confidence limits of P_{est}.

The 95% confidence limits around Pest may be calculated from the expression:

$$P_{est} - hP_{est}Q_{est}/N < P < P_{est} + P_{est}Q_{est}/N,$$

where h is the number of standard deviations appropriate to particular confidence limits (1.96 for 95% confidence).

If 0.01% is regarded as the upper 95% confidence limit of the lowest measurable number of contaminants obtainable in a simulation trial (one contaminated unit), the lowest value of N can be calculated to be close to 3000 units.

The number of units (if any) in excess of the 3000 requiring to be filled is a further important decision, and there are several views on how it should correctly be made.

One view holds that the number of units filled should be related to the product batch size. This is difficult to reconcile with the fact that the media fill is a process test and should not logically therefore be related to product batching. Different products filled into the same containers on the same filling machine could easily have different batch sizes, perhaps dictated by some complexity of compounding. To which of these batch sizes should the number of simulation units be related? If this approach is taken, the pragmatic answer is usually the largest of the batch sizes. Guidance on media fill dimensions in relation to product batch size given in ISO/IS 13408 *Aseptic Processing of Health Care Products* (ISO, 1997) is summarized as Table 3. In practical terms, this guidance applies only to small batch sizes; for normal production batch sizes, ISO supports only a minimum media fill size of 3000 units.

A second view is that the simulation trial should be run over the same time as an operating shift (see earlier). In many cases this amount of elapsed time may be necessary to simulate all of the potential contaminating events arising in a process. In other cases (e.g., with high-speed ampule filling lines) it could result in vast numbers of units being filled. As long as there are no contaminated units present, this approach to filling ampules gives good assurance of asepsis. Its logic breaks down when contaminated units are identified—perhaps three or four contaminated units would seem to pale into insignificance in comparison to the overall large numbers filled. Regardless of this, they may be significant to contaminating events that occurred during filling but the effect of the large dimensions of the simulation is to dilute their impact.

TABLE 3 Minimum Numbers of Simulation Trial Units Related to Production Batch Size from ISO/IS 13408 *Aseptic Processing of Health Care Products* (ISO, 1997)

Number of units in production batch	Minimum number of units for validation media fills	Minimum number of units for periodic media fills
< 500	5000 in 10 or more runs	Maximum batch size per run
≥ 500–2999	5000 in three or more runs	Maximum batch size per run
≥ 3000	9000 in three runs	3000 per run

The third view is that the dimension of the simulation trial should be dictated by the time necessary to allow simulation of *all* of the potential contaminating events. Exactly what constitutes all potential contaminating events is in the long run a matter of opinion. Nonetheless, there are techniques such as Failures Modes and Effects Analysis that can be used to create a documented structure around the development of these opinions. This type of approach adds to the knowledge of the process.

21.7.1.3. The Third Principle

The third principle of the Protocol is that acceptance criteria must be predetermined. In the case of simulation trials, a maximum number of contaminated units must be specified for each simulation, and indeed the underlying aseptic process, to be acceptable. If the acceptance number is exceeded in any one of the three validation simulations, appropriate action must be taken and the simulation(s) repeated until three successive successful simulation trials are obtained. In an ideal world the appropriate action is preventive—in other words, action appropriate to preventing a further recurrence should be taken, probably involving some change in working practice and to the operating SOP.

In the real world the action is most often corrective—something like a redisinfection of the filling room, retraining of personnel, and so forth. This is because it is not usually easy to accurately diagnose the source of contamination in a simulation, and this difficulty is greatest for a new process.

The question arising out of the predetermination of acceptance criteria is, exactly how many contaminated units are tolerable? This is not an easy question to answer.

If the statistic of no more than one contaminated unit in 1000 units being the acceptance limit is taken as a starting point, and 3000 units as the minimum number of units in a simulation trial, then it might reasonably be expected that zero, one, two, or three contaminated units in 3000 would be acceptable and four or more contaminated units would be unacceptable. Up to four contaminated units would be acceptable in 4000, up to five in 5000, and so forth.

This approach was overtaken by the PDA recommendation (1981) that the no more than one contaminated unit in 1000 limit should be met with 95% confidence. In relation to 3000 units filled, compliance with this modification to the 1 in 1000 limit would be acceptable only with zero or one contaminated units.

Slightly different mathematical treatments result in recommendations of "pass zero, fail one or more contaminated units in 3000" or "pass one or fewer contaminated units, fail two or more contaminated units in 3000."

When simulation trials require numbers of units larger than 3000, it might be considered reasonable to increase the number of contaminated units permissible beyond zero or one. The guidance in ISO/IS 13408 allows maximum numbers of contaminated units ranging from 1 in a 3000 unit simulation to 11 in a 17,000 unit (approximately) simulation (ISO, 1997).

Bernuzzi et al. (10) examined these recommended limits and concluded that they were becoming weaker as the total number of filled units increased (one positive in a 5000 unit media fill does not have the same meaning as 10 units positive when 16970 units are filled).

These authors attempted to develop an alternative set of limits for simulation trials but in all cases found the same statistical frailty as numbers of units filled increased. The limits from ISO/IS 13408 (ISO, 1997) and from the most rigorous plan of Bernuzzi et al. (1997) are summarized in Table 4.

The value of these mathematical treatments merits questioning in light of the practicalities, both technical and regulatory, of simulation data. In practice it is not realistic that, for example, a manufacturer of aseptically filled ampules would repeatedly tolerate (or be allowed by the regulatory agencies

TABLE 4   Maximum Permissible Numbers of Contaminated Units in Simulation Trials According to ISO/IS 13408 *Aseptic Processing of Health Care Products* (ISO, 1997) and to the More Rigorous Scheme of Bernuzzi et al. (1997)

Number of units filled	Maximum permissible number of contaminated units	
	ISO/IS 13408 (1997)	Bernuzzi et al. (1997)
3000	1	0
4750	2	—
6300	3	—
7200	—	1
7760	4	—
9160	5	—
10,520	6	—
11,500	—	2
11,850	7	—
13,150	8	—
14,440	9	—
15,710	10	—
15,800	—	3
16,970	11	—
20,200	—	4

to tolerate) six contaminated units in simulation trials of 10,000 units as the mathematical treatments would appear to suggest. Nor is it realistic that a manufacturer of blow-fill-sealed ampules would repeatedly tolerate even four contaminated units in media fills of 20,000 units. This is because it is known by industry and by the regulatory bodies that these processes are well capable of yielding tens of thousands of uncontaminated media-filled units.

In practice, any conscientious pharmaceutical manufacturer would respond to contaminated units in excess of zero or one irrespective of the overall dimensions of the simulation trial. This is particularly true in validation in which there would inevitably be serious doubts over the approval of any aseptic process in which three or more contaminated units were appearing in validation simulation trials.

Contaminated units should be the stimulus for process improvement. The practical limit in all simulations is that there should be no more than one, possibly two contaminated units. Larger numbers of contaminated units than these must elicit preventive action and improved control.

In summary, it makes best sense that validation simulation trials should be composed of a number of units in excess of 3000 to allow for sufficient elapsed time to simulate all predicted potential contaminating events, and no more than one contaminated unit should be allowed in any single run no matter how many units are filled in total.

21.7.2. Simulation Trials in Validation of Aseptic Processes for Manufacture of Sterile APIs

The statistics of sterility have to date fairly well avoided addressing massive systems. Sterilize-in-place (SIP) systems are but one example for which the probability of a unit within a population of units being nonsterile after treatment is clearly unrealistic, and the determination of the probability of one microorganism surviving in the whole SIP system is impractical. API manufacture is another such example.

The final API product is a single homogeneous bulk. This is more than likely split into smaller units after dispensing (off-loading), but these units have no relationship to the final dosage form units that the patient receives after further processing in secondary manufacture. This creates vast difficulties in terms of setting acceptance criteria that relate to the probability of contamination. Regardless of whether the whole API manufacturing process is simulated as a whole or in a series of simulations, each focusing on one unit operation, the output of the simulation must either be a contaminated or a noncontaminated broth, or a filter with or without colonies growing on it.

PDA *Technical Report* No 28 gives some guidance, but only for simulations in which the output can be evaluated by counting colonies on membranes. Basically, it proposes that quantitative acceptance criteria should be based on a contamination rate of not more than one colony-forming unit in 5000 finished sterile dosage form units. This of course assumes that the sterile API manufacturer knows the future use of the product, which may not always be the case but could be included in technical contracts at the request of the finished dosage form filler. To make this approach conservative, it is necessary to know the largest finished dosage form and divide that into the smallest bulk API batch size.

The author of this chapter has had no involvement with sterile API simulations taking this approach but on the face of it, it would appear to have rather a large number of variables to be able to offer a simple straightforward interpretation of whether the criteria have been complied with. In the long run, a satisfactory sterile API simulation is one in which no contaminants have been recovered by whatever means used. The difficulty lies in sterile API simulations where contaminants are recovered as a consequence of the technical and manipulative differences rather than as a consequence of the process itself—how are we to determine and correct a root cause of contamination within the *real* process when the actual root cause may have originated from some unique problem of simulation?

21.7.3. Periodic Simulation Trials in Routine Operation

It is unlikely that any responsible regulatory body would currently be tolerant of a frequency of less than twice a year for periodic simulation trials of aseptically filled sterile dosage forms nor less than once every two years for sterile APIs. Simulation trials are probably the most sensitive method of detecting unexpected sources of process contamination. Routine environmental monitoring of aseptic processes should generally be biased toward the locations where contamination would be expected (i.e., to known process vulnerabilities), so as a consequence mainly tells the manufacturer only what he already knows. Simulation trials have the capability of revealing new and previously unknown vulnerabilities to contamination.

The regulatory standpoint coming from the principle of protection of the patient is that if unexpected process contamination occurs in a simulation trial and is considered sufficient to compromise the sterility of past product, they would expect market withdrawal. Following this logic, the greater the frequency of periodic simulation the lower the risk to the patient and the lower the commercial risk to the manufacturer.

Simulation is generally done on every filling line at least twice a year (88.5% of the respondents to the PDA's 1996 survey performed media fills at least twice a year). Within this program it is sensible to ensure that on multicontainer filling lines every container size has been filled at least once in a reasonable time frame, say over 2 years. Otherwise, the possibility of unexpected contamination as it relates to a particular size of container may never be addressed.

It is also arguable that at least one of the sizes that were identified in the validation protocols as presenting the worst risks of contamination should be tested in every simulation. If the assumption of the "worst case" is correct, it should surely be evaluated at the greatest frequency. This is usually achieved by setting up some sort of matrix approach to periodic simulations on multicontainer filling lines. Of course, it is much easier on a single container size, single volume filling line.

Periodic simulations are required to be done at the end of a routine production operation. Care should be taken to run a few liters of sterile water through the filling setup, to flush out any product-related inhibitory substances, before filling the placebo. This is intended to address the two possibilities of contamination having built up in a filling room over a period of manned operation between cleanups, and of operator discipline having lapsed as a result of tiredness.

The exception to this is for aseptic filling of sterile antibiotics. In this case the filling room must be cleaned up and all antibiotic traces removed before the placebo is brought in and filled. This is to ensure that recovery of contaminants is not inhibited. This presents more force to the argument concerning the purpose of simulation trials being related to process contamination rather than to *product* sterility. The simulation trial is *intended* to disclose process contamination regardless of whether the contaminants would survive or die in the product.

ISO/IS 13408 recommends a two-tier approach of alert and action to limits for periodic simulations. It does not, however, elaborate on how the two levels should be applied. The action limits are those listed under ISO/IS 13408 in Table 3, the alert limits are lower.

The divergence between the ISO/IS 13408 action limits and acceptable reality has been discussed previously in relation to validation simulations (11). Reality for a valid aseptic process that has been transferred to routine control is that there will have been no more than one contaminated unit per simulation run irrespective of the total number of units filled.

It is axiomatic that periodic revalidation simulations should not generate significantly worse results than the original validation simulations without some appropriate action being taken. Limits for periodic media fills should be related to the results obtained in validation simulations (Table 5).

The numbers of permissible contaminated units given in Table 5 are presented in a two-tier approach for which both levels demand action. The difference in the approach is in the consequences of the actions to production, to scheduling, and to past product.

TABLE 5 Recommended Action Limits for Periodic Simulation Trials Irrespective of Total Numbers of Units Filled and Related to Results from Validation Media Fills

Numbers of contaminated units actually occurring in three successive validation media fills	Action limit (numbers of contaminated units) for marginal failures in periodic media fills	Action limit (numbers of contaminated units) for significant failures in periodic media fills
0, 0, 0	≥ 1 but < 3	≥ 3
0, 0, 1	≥ 1 but < 3	≥ 3
0, 1, 1	≥ 2 but < 4	≥ 4
1, 1, 1	≥ 2 but < 4	≥ 4

The term "action limits for marginal failures" is used here rather than "alert" limit because all contaminants found in simulation trials merit some action, and the use of the "alert" term may detract from this.

The action limits for marginal failures allow for the extremes of statistical variation from the validation trials' results that might be expected without there being any significant change in the real contamination rate. Contaminants are rarely found in well-controlled facilities, but when they occur most fall into this marginal category.

Identification and investigation are essential. The possibility that they may not be a statistical phenomenon should not be discounted. *Bacillus* spp. should, for instance, be treated with extreme suspicion in relation to the possibility of there being some systematic problem with nonsporicidal disinfection, or of there being residual air in autoclave loads, and so forth. Actions from marginal failures that do not appear to have arisen from a systematic failure of one of the systems necessary for the maintenance of asepsis are best dealt with by counseling, retraining, and improved supervision of operators. The simulation should be repeated as soon as possible and a further simulation on the container size implicated should be scheduled into the next periodic simulation trial in addition to those sizes defined by the predetermined matrix.

Successive marginal failures on the same container size should be treated as a consequential failure, as also should marginal failures on three or more successive media fills on the same filling line irrespective of container size. Other circumstances of repeated failures within the marginal range may also be indicative of process conditions that have deteriorated from the validated condition and should be treated as infringements of the action limits.

Table 5 gives action limits that are described as significant. These limits are well beyond the expected variation seen in validation and must therefore be interpreted as indicators of genuine loss or genuinely deteriorating levels of control. It is reasonable to expect that the potential for any patient risk should be minimized while these failures are being resolved. Product manufactured on the filling line after the date of the media fill, and product still in the company's warehouse(s), should be quarantined until the failure investigation is completed.

In an ideal world, production on the line in question should be suspended pending the outcome of the investigation. In practice it may be advantageous to the investigation for production to continue, but this decision should not be taken lightly in view of the commercial risk of possibly having to reject the product made in that period.

The most important factor in the failure investigation is the identification of the contaminants. Any microbiologist should be able to categorize identified contaminants within their most likely sources to the environment (air, dust, etc.), to water, or to human sources. An experienced QA microbiologist may be able to pinpoint the contaminants to their origins in the facility (e.g., nonsterile disinfectants, water leakage, worn-out garments) or to general weaknesses in control.

Environmental microbiology is not an exact science—a weakness in control will always be a weakness in control (a systematic weakness) even though it may not manifest itself every time it is tested by simulation or in environmental monitoring. If a specific problem is diagnosed it should be traced, if possible, to the time it began.

The identified contaminants should be considered for their ability to survive in the products filled on the line in which the significant simulation trial failure occurred. The importance of this information is to determine if product sterility has been compromised. In a multiproduct filling line, the decision might be different for different products. If product sterility is compromised, then product must be withdrawn from the market.

Once the failure investigation is complete and corrective/preventive action implemented, it is customary to repeat the simulation trial. Some would argue for revalidation of the line by repeated media fill, but this decision should be contingent on the extent of the corrective/preventive action implemented.

In some instances of significant failure it may not be possible to pinpoint the exact cause, and corrective/preventive action cannot therefore be targeted. In such cases it is normal to clean, disinfect, fumigate, counsel, train, and improve supervision overall. In these cases three repeat simulation trials should be done to counterbalance the uncertainty of the diagnosis.

Where simulation trials have to be done, the date of recommencement of production is a business that requires account to be taken of any uncertainty surrounding the diagnosis of the cause of the problem. In the event of the source of the simulation failure being quite clear and the preventive action being self-evident, it is probably a reasonable risk to recommence production before the incubation of the repeat simulation trial(s) is complete. The commercial risk is greater where diagnosis of the problem is unclear.

The major practical issue of periodic simulation trials is what to do in response to the results. This is a lesser problem (in principle if not always in practice where deadlines have to be met) in validation than in periodic simulations. The major questions are as follows:

Should production on a particular filling line be allowed to continue if simulation trial results are unfavorable?

Simulation trial results are not available until 14 days after the simulation has been conducted. What should be done with the product manufactured between these dates when results are unfavorable?

Simulation trials are done only every 6 months. What should be done with the product manufactured since the last successful simulation when results are unfavorable?

If a company has the luxury of running terminally sterilized products on the same filling line as aseptically filled products, the opportunity is there to run the line and investigate a simulation trial failure while fully operational. If only aseptically filled products are filled on the line, filling should be suspended until investigations are complete and repeat simulations are satisfactory. However, it is exceedingly difficult to conduct an investigation that requires observation of practical conditions unless the line is running; and although running a series of repeat simulations is essential, the hope is usually that they will pass rather than fail. The information content in a simulation with no contaminated units is low with respect to the diagnosis of the cause of previous failure.

It is normally recommended to "freeze" all of the product still in company control aseptically filled on a line on which a simulation has failed until investigations are complete and repeat simulations have given the line the "go ahead." This strategy, although fine in principle, usually raises significant pressure from marketing and distribution over stock-outs or impending stock-outs.

If a simulation trial fails and it is traceable to a failure in one of the subsystems that make up the sterility assurance system, there is little choice but to reject and recall back to the date it commenced, unless the regulatory bodies can be convinced otherwise. A responsible recall initiative from a company is generally less harmful than a recall requested by an inspector who discovers the matter later on. An example of this could be a tear in a HEPA filter—recall back to the last satisfactory in situ integrity test.

The outcome of the investigation of most marginal media simulation trial failures is inconclusive: as often as not the source is some human commensal microorganism shed by an operator, not necessarily on point-of-fill, possibly even when unloading stoppers from an autoclave. It would not be sensible to recall for this type of phenomenon, and in mitigation there could be some work done on the potential for the particular microorganism to survive and grow in particular products filled on that line.

REFERENCES

1. PDA Technical Report No. 24. Current practices in the validation of aseptic processing. PDA J. Pharm. Sci. Technol. 1996, *51* (Suppl. S2), Parenteral Drug Association Inc.: Bethesda, MD.
2. PDA Technical Report No 28. Process simulation testing for sterile bulk pharmaceutical chemicals. PDA J. Pharm. Sci. Tech. 1998, *52* (5), Supplement, Parenteral Drug Association Inc.: Bethesda, MD.
3. FDA (2002) Draft Concept Paper on Production of Sterile Products by Aseptic Processing, CDER.
4. FDA (1987) *Guideline on Sterile Drug Products Produced by Aseptic Processing*, CDER.
5. FDA (1994) *Guideline to Industry for the Submission Documentation for Sterilization Process Validation in Applications for Human and Veterinary Drug Products*, CDER.
6. CEC, (1992, 1997, 2002) *Good Manufacturing Practice for Medicinal Products*.
7. Tallentire, A.; Dwyer, J.; Ley, F.J. Microbiological quality control of sterilized products: evaluation of a model relating frequency of contaminated items with increasing radiation treatment. J. Appl. Bacteriol. 1971, *34*, 521–534.
8. PDA. *Validation of Aseptic Filling for Solution Drug Products, Technical Monograph No 2*; Bethesda, MD: Parenteral Drug Association Inc., 1981.
9. Halls, N.A. *Achieving Sterility in Medical and Pharmaceutical Products*; Marcel Dekker: New York, 1994.
10. Bernuzzi, M.; Halls, N.A.; Raggi, P. Application of statistical models to action limits for media fill trials. Eur. J. Parent. Sci. 1997, *2*, 3–11.
11. Halls, N.A. *Microbiological Media Fills Explained*; Sue Horwood Publishing Ltd: Storrington, UK, 2002.

22

Standard Methods of Microbial Identification

Myron Sasser

MIDI Labs, Newark, Delaware, U.S.A.

22.1. INTRODUCTION

Identification of bacteria, yeast, molds, mycoplasma, and viruses by methods that are not DNA-based involves several technologies. Molds (filamentous fungi) are typically identified by traditional microscopic techniques but may also be identified by biochemical, fatty acid–based, or mass spectrometric methods. Molds will also be discussed later. Identification of bacteria and yeast will be covered as if the methods were the same and differences noted where applicable.

Traditional taxonomy of bacteria was based on colony morphology, microscopic observations (usually by Gram stain), and by a limited number of biochemical (enzymatic) tests. As newer tools such as fatty acid analysis and DNA sequencing were developed, the taxonomy began to change rapidly and microbiologists found it difficult to keep pace with the names of the new genera and species. Similarly, manufacturers of identification systems had to evolve techniques and databases to keep pace with the evolving taxonomy. The earliest and most widespread systems for identification of bacteria have been based on biochemical tests similar to those initially used to define genera and species. These systems include manual, semiautomated, and automated analysis; have databases for comparison of test results; and while using yes/no results, have evolved to include statistical calculations to help compensate for aberrant reactions by some strains of species. Fatty acid methyl ester (FAME) analysis by automated gas chromatography began to be used in pharmaceutical microbial identification in the 1980s and has evolved into a widely used technique.

22.2. BIOCHEMICAL TEST–BASED IDENTIFICATION

22.2.1. Basic Premises

1. Bacteria of a species all have the same substrate utilization patterns. It is assumed that major enzyme groups within a species should be constant for utilization of certain substrates. Thus, if one strain uses mannitol or histidine, the others will also utilize the substrate.
2. Bacteria in many species can be differentiated if a sufficient number of utilization tests are included. If this is not the case for some related organisms, they can be listed as a "species group."
3. The enzymes for substrate utilization are always expressed under the test conditions. If this is not the case, a statistical algorithm can calculate the probability of the name being correct when some strains of a population do not use one or more of the substrates.
4. Plasmid exchange and/or minor mutations will not seriously affect the utilization of substrates chosen.

22.2.2. Methodology

1. A suspension of bacteria is introduced into miniature tubes (or wells in a microplate) containing the test substrate and either a pH indicator or a redox indicator. The suspension of bacteria must be from a pure culture and typically made to match a specific McFarland level. Certain manual methods use an inoculating needle or wire to introduce part of a bacterial colony into the substrate containers.
2. Typically, within 4 to 24 hr the bacteria have utilized the substrate and caused a color change that may be read visually or by an automated reader.
3. The miniature tubes are contained on a "card" or microplate in a series, which relates to the type of organism being tested. This decision is usually made by observation of the Gram stain but may include other preliminary observations as well.

22.2.3. Manual Biochemical-Test Systems

Manual biochemical-test systems assign a number to the response for each test and this "code" is then compared to information in reference material, which allows assigning a name to the organism. Advantages of manual systems are that these systems have a very low cost of entry as there is no instrument or computer to purchase and they take up very little bench space in the laboratory. Manual systems are primarily of value for pharmaceutical companies having relatively low volumes of bacterial isolates to be identified. As no electronic data are generated, 21 CFR part 11 (Electronic Records and Signatures, ERS) is not a factor. Disadvantages are that the tests are somewhat labor intensive and may be prone to technician error in judgment of reaction and in transcription of code numbers. The organisms are also still living when being tested and thus pose a small health hazard despite the fact they are primarily environmental isolates (although many are from human skin).

22.2.4. Automated Biochemical-Test Systems

Biochemical-test systems that have been automated are commonly inoculated with multitip pipettes or by vacuum introduction of the inoculum into the tubes. The utilization reaction in the tubes is read by a light beam that detects the color change due to a pH shift or a redox reaction (e.g., from reduction of a tetrazolium compound). The utilization pattern is then passed to the computer that compares the pattern to a stored database of reactions of organisms previously tested on that battery of compounds (i.e., that card or plate). Primary advantages over manual systems are that automated reading and database comparison reduces potential operator error and reduces labor cost. The disadvantages compared to the manual system are the initial cost of the instrument and computer equipment, the laboratory space occupied, and the necessary maintenance of the equipment.

Database creation for biochemical-test systems is typically accomplished by obtaining the utilization data for the type strain of the species and then gathering data from users. This allows for determination of variations within a species or species-group and creation of a "confidence level" applied to the naming. As the early systems evolved from clinical to pharmaceutical microbiology Quality Control applications, new cards or plates and new variations of existing cards were developed. For example, in the genus *Bacillus*, there is little clinical relevance to most species and thus a new card was designed for this genus. Databases among the systems vary in size from a few hundred to almost two thousand entries.

Identification of yeast by biochemical tests is similar to that of bacteria; however, identification of filamentous fungi requires an additional plate and some of the methodology for inoculation is different. As this is a relatively recent addition, performance is still an open question but any help in this area is desirable.

22.3. FATTY ACID–BASED IDENTIFICATION

22.3.1. Basic Premises

1. Bacteria make more than 300 fatty acids and related compounds.
2. Gas chromatography of fatty acid methyl esters (FAME) yields profiles that are characteristic for species.
3. FAME compositions may vary depending on growth temperature, the medium on which grown, and the physiological stage of growth, thus requiring standardization of these conditions.
4. Profile matches to a database must consider quantitative data rather than yes/no decision tests as in biochemical test systems.
5. Plasmid loss or gain and/or minor mutations will not seriously affect the FAME profiles.

22.3.2. Methodology

1. Bacteria are grown on Trypticase Soy Broth ™ (BBL etc.) with agar at 28°C for 24 hr. A quadrant streak allows harvesting of approximately 40 mg of cells from a quadrant not heavily overgrown (to achieve the correct physiological stage).
2. Whole cell fatty acids are saponified, methylated, and extracted into an organic solvent.
3. Automated gas chromatographic analysis names and quantifies the FAMEs.
4. The FAME profiles are computer matched to a database by algorithms including covariance, principal component analysis, ratios of quantities of all fatty acids to each other (cross terms) and pattern recognition.

Database construction for fatty acid matching depends on analysis of multiple strains of each species so that a "normal distribution" (bell-shaped curve) can be plotted. An unknown matched to the mean percent FAME composition for the species results in a "Similarity Index" which is related to its distance from the species mean percent. The lipopolysaccharide of gram-negative bacteria contain short-chain fatty acids that have a hydroxyl moiety at the 3-position (e.g., 3 OH 14:0 is a 14-carbon saturated fatty acid with a hydroxyl at the third carbon counting from the carboxyl end), whereas this is not found in appreciable quantities in gram-positive bacteria. Most gram-positive bacteria (and a few genera of gram-negatives) have predominately branched chain fatty acids (e.g., iso and anteiso compounds, which have methyl groups at the penultimate [iso] and ante-penultimate [anteiso] carbons from the end of the chain opposite the carboxyl group [omega end of the carbon chain]). Sporeformers (*Bacillus* and *Clostridium*) contain membranes around the spores that cause changes in the FAME composition as the spores or other inclusions (e.g., crystals in *B. thuringiensis*) are formed. These changes in composition require consideration in building the database but also offer great discrimination between closely related species (e.g., qualitative differences between *B. cereus* and *B. anthracis*).

Users may construct a customized database that will automatically be searched with each analysis. Trend analysis is obtained through exporting the data to a commercial database and using queries within the database tools to find trends in appearance of organisms. Strain tracking is automated so that each analysis may be compared to all previous analyses and strain-level matching will be performed within seconds. The current databases contain about two thousand entries.

22.4. FACTORS TO CONSIDER IN CHOICE OF SYSTEM FOR IDENTIFICATION OF BACTERIA

The features of the various identification systems change rather rapidly so no attempt will be made to differentiate among them in a "snapshot" at this time. Instead, an attempt will be made to briefly discuss features that are currently available in some of the systems and that seem desirable. A trade-off may be made between labor costs and initial purchase price in the degree of automation. Consumables costs vary widely across instruments and may be a major factor in high-volume laboratories. Space considerations and technician comfort level with the technology are also factors to be considered. A key to acceptance of any system should be that it obtains the correct identification of a high percent of the isolates tested. An absolute requirement should be conformance to Food and Drug Administration (FDA) mandated regulations. Some specific factors to be considered include:

1. Volume of samples to be identified daily. High throughput systems often are required in parenteral drug manufacturing, whereas nonsterile products may require less volume of samples.

2. The type of product manufactured may suggest the organisms of greatest interest and thus the system most likely to be strong in identification of these organisms. For example, anaerobes may be of little interest to the manufacturer of topical ointments (with the exception of *Propionibacterium acnes* and similar organisms), but

Staphylococci and Micrococci may take on added significance. A manufacturer of disinfectants or of inhalation devices will be especially interested in *Burkholderia cepacia*, *B. gladioli*, and *Pseudomonas aeruginosa*.

3. Validation of the instrument should not consider only the U. S. Pharmacopoeia strains of *Escherichia coli*, *Pseudomonas aeruginosa*, *Staphylococcus aureus*, and *Salmonella* spp., the technique should be validated with organisms most likely to be found in the manufacturing environment. Thus, the technique must be able to correctly identify these organisms.

4. Compliance with FDA regulations is crucial. The supplier of the technology must provide IQ, OQ, and PQ documentation and may (upon request) suggest a validation protocol that can be customized to fit the manufacturing requirement. The use of computer-stored data leads to the requirement for the system to be compliant with 21 CFR Part 11 (ERS).

5. Database size and breadth of coverage is a crucial factor in obtaining correct identification of many environmental organisms. Numbers of species in the databases of various systems range from a few hundred to about two thousand.

6. Accuracy of naming of organisms is obviously important. As pharmaceutical companies do not publish scientific journal articles about organisms found in their products or facilities (for obvious reasons), the scientific literature on environmental isolate identification and the journals covering bacterial taxonomy are useful sources of how well the technique may work in identification of environmental isolates in a pharmaceutical facility.

7. Highly valuable is the ability of the system to perform "trend analysis." This is the ability to search historic data for trends of occurrence of certain species in relation to time of year, source (e.g., water, air, personnel, or raw material), particular product, room of the facility, and so forth. Trending should help in prevention of contamination by tying the contaminant to a particular source or location and allowing for reduction of the contamination.

8. "Strain tracking" is another useful tool available in some systems. This allows matching the current isolate with individual strains in stored data. This may enable specifically targeting the source of contamination (e.g., it may even point out the raw material source or the individual from whose body the organism came).

9. Initial cost of the system vs. cost of the consumables (applying knowledge of the likely volume of isolates to be identified). The

biochemical test systems use minitube cards or microplates that constitute a cash stream for the manufacturer. If the system requires specialized media sold only by the manufacturer, this may be an added cost. The primary manufacturer of the fatty acid system sells a "calibration" standard made only by the system manufacturer.

10. Reliability of the instrumentation and technical support provided by the manufacturer are major factors once the system is in operation. Check with current users to determine these aspects.

11. Updates of the databases are customarily free of charge by most manufacturers.

22.5. CONCLUSION

Standard techniques for identification vary in ability to identify organisms, with one perhaps superior in identification of enterics, another better at cocci, and a third better at *Bacillus*. The same may be said for DNA sequencing, which is weak in separation of some species easily identified by biochemical tests or by fatty acid profiling but works quite well in some groups not done well by the standard techniques. A recent introduction that may prove interesting is Matrix Assisted Laser Desorption Ionization Time Of Flight (MALDITOF) analysis. The use of bacterial samples without requirement for sample preparation is an advantage. The detection is by mass spectrometry that distinguishes bacterial fragments by size through the time required to fly to the detector. Other methodologies such as pulsed field gel electrophoresis and ribotyping (modeled after RFLP analysis) are useful for strain tracking but require additional labor and expense. Some standard tests such as hemolysis, motility, catalase, oxidase, coagulase, are not discussed above but are frequently used and may add much confirmatory information to the techniques being used. Where possible, use of more than one orthogonal technology (e.g., fatty acid and biochemical) may provide optimal solution. No one technique is totally optimal and the choice of identification technique will often depend on the needs of the specific laboratory.

REFERENCE

For more information: *www.midilabs.com*

23

Rapid Methods of Microbial Identification

Luis Jimenez

Genomic Profiling Systems, Bedford, Massachusetts, U.S.A.

23.1. INTRODUCTION

Parenteral pharmaceuticals are sterile products analyzed for the absence of bacteria, yeast, and mold. Standard microbiological procedures for detecting and identifying microbial contamination rely on the cultivation and isolation of microorganisms from pharmaceutical samples (1,2). Because parenteral pharmaceuticals are manufactured under stringent aseptic conditions, to guarantee the absence of microorganisms from the production facility and finished product, several microbiological procedures are performed. Environmental monitoring of production facilities requires the sampling of analysts, water, air, and surfaces, and finished product testing using compendial methods is performed on production and in-process samples (3–5). Although standard methods are routinely used for quality control analysis, they require long incubation times, continuous manipulation, and time-consuming procedures (6). Furthermore, it has been recently reported that standard methods underestimate the microbial communities present in pharmaceutical environments (7–9). This is because standard methods were developed for isolating and detecting microorganisms from clinical samples. Microorganisms living

685

under high organic carbon habitats such as clinical samples respond to those conditions with totally different metabolic and survival strategies. However, when confronted with low organic carbon and other environmental fluctuations (e.g., changes in temperature, pH, pressure, presence of antimicrobial ingredients), microorganisms exhibit several survival strategies that will make detection and identification by standard methods more complicated (4,8,10,11).

Despite regulatory guidelines and recommended methods, microbial contamination is still one of the major causes for product recalls worldwide (6,12,13). Contamination is due to the presence of objectionable microorganisms in raw materials, finished products, and water, or from questionable practices during product manufacturing. When products are not sterile, microbial growth will have a negative impact on product integrity and create serious health threats to consumers.

Over the past 30 years, implementation of good manufacturing practices (GMP) has been the foundation for improving industrial quality analysis and process control. In the 21st century, one of the ways to improve quality control and good manufacturing practices is to apply rapid microbiological methods that will optimize sample analysis and product release. These procedures have been shown to be sensitive, accurate, and robust, and they provide faster results that might indicate problems in processes and systems used in pharmaceutical environments (6,14). Earlier detection of microbial contamination allows rapid implementation of corrective actions, resulting in the minimization of manufacturing losses and optimization of risk assessment. This chapter discusses the different technologies available to pharmaceutical scientists working on parenteral pharmaceuticals for the rapid detection of microbial contamination that might lead to the optimization of process control and quality.

23.2. ATP BIOLUMINESCENCE

Of all the molecules present in a microbial cell, adenosine triphosphate (ATP) is the most important high-energy phosphate compound (15). ATP serves as the prime energy carrier for essential functions in the viability and growth of microorganisms. ATP bioluminescence technology is based on the reaction of the enzyme complex luciferase-luciferin, in the presence of oxygen and magnesium, with ATP released from microbial cells resulting in the production of light (Table 1). The light emitted is proportional to the amount of ATP released.

Several studies have demonstrated the applicability of ATP bioluminescence to pharmaceutical quality control. The first reported studies have relied on laborious sample preparation for ATP extraction from microbial

TABLE 1 ATP Bioluminescence Reaction

Firefly Luciferase
Magnesium
↓
$ATP + D\text{-Luciferin} + Oxygen \rightarrow Light\ AMP + PPi + Oxyluciferin + CO_2$

cells and manual addition of reagents (16). Once the ATP is extracted and reacted with the enzyme, the samples are added to a luminometer to detect the production of light. These studies were used as an alternative to the visual end point used in standard sterility testing by determining the total microbial biomass present in the sample in a shorter time period. For instance, standard sterility testing relies on the addition of product samples to different types of enrichment media. Because of the chemical composition of some pharmaceutical products, the addition of the product to the media results in a turbid broth that does not indicate the presence of microbial growth. However, after incubation, ATP bioluminescence has indicated that although the broth was turbid there was no microbial growth.

During the 1990s, technological improvements in instrumentation provided for the complete automation and processing of multiple samples, cell lysis, and reagent addition, allowing minimization of sample handling and time-consuming extraction procedures. Some instruments have developed quantitative information but others only indicated the presence or absence of microbial cells in samples after an incubation step (17,18). ATP bioluminescence assays have been shown to detect microbial contamination in pharmaceutical waters and finished products.

A qualitative ATP bioluminescence system has been shown to allow high throughput screening of more than 180 samples/day. Furthermore, faster detection times for finished product samples range from 24 to 48 hr (17,19). Because of the need for an enrichment-incubation step, assay optimization requires the development of different enrichment media to overcome the antimicrobial nature of the different pharmaceutical actives (Table 2). For instance, for optimal recovery of bacteria, yeast, and mold (e.g., 24–27 hr detection time) from pharmaceutical products containing halogenated compounds, it was necessary to add sodium thiosulfate to the enrichment media (R, MR, and MR2 broth) (Table 3). Furthermore, different nutrients are also added to optimize recovery for *Staphylococcus aureus* (e.g., glycine) and mold (e.g., sodium acetate and glycerol) (Table 3).

A wide variety of pharmaceutical formulations have been validated using ATP bioluminescence. Different types of pharmaceutical drug delivery systems such as capsules, tablets, liquid, solids, and emulsions were found to

TABLE 2   Enrichment Media for ATP Bioluminescence Analysis

R broth	*R2 broth*
TAT broth	TAT broth
4% Tween 20	4% Tween 20
1% Dextrose	1% Dextrose
1% Neopeptone	1% Neopeptone
0.25% Sodium Thiosulfate	0.25% Sodium Thiosulfate
	0.5% Sodium Acetate
MR broth	1% Glycerol
TAT broth	1% Sucrose
10% Tween 20	
1% Dextrose	
1% Neopeptone	*MR2 broth*
1% Glycine	TAT broth
1% Triton X-100	10% Tween 20
0.5% Sodium Phosphate Dibasic	1.2% Dextrose
0.5% Sodium Thiosulfate	1.2% Neopeptone
	1% MgSO4
Letheen Broth with 1.5% Lecithin	0.25% KH2PO4
	0.25% Sodium Thiosulfate
TAT Broth with 4% Tween 20	

be compatible with the system (Table 4). When 1% product suspensions in enrichment media are analyzed, no indigenous ATP concentration is found; the reaction is neither enhanced nor inhibited. The product response to ATP ranges from 25% to 200%, which is within the specifications recommended (17,19). After the samples are spiked with different types of microorganisms, detection times range between 24 and 27 hours (Table 5). The criteria for

TABLE 3   Detecton Times (Hours) of Microbial Contamination by ATP Bioluminescence

	Product A		Product B	
Enrichment Media	R broth	MR broth	TAT broth	R broth
P. aeruginosa	24	24	24	24
S. aureus	48	24	24	24
E. coli	24	24	24	24
S. typhimurium	24	24	24	24
C. albicans	24	24	24	24
A. niger	24	27	48	27

TABLE 4 ATP Concentration Values and Inhibition/Enhancement Effects of Pharmaceutical Samples Using 1% Sample Suspensions

Test sample	Broth	ATP concentration of broth (nanomolar)	ATP concentration sample suspensions (nanomolar)	% Product suspension response to ATP
Sample A	R	0.053	0.055	101.9
Sample B	R	0.020	0.030	44.0
Sample C	R	0.017	0.021	101.2
Sample D	R	0.019	0.025	69.8

passing or failing a sample are simple. A positive sample is indicated when the relative light units (RLU) of the contaminated samples in the enrichment broth are 2 times the values of the sample in the broth. When product suspensions inoculated with different concentrations of microorganisms are incubated, a positive response is detected in all the samples exhibiting 2 times the values of the control (Table 5). As shown in Table 5, all microorganisms spiked into pharmaceutical product A have been shown to grow on standard media and exhibited bioluminescence values twice the values of the control sample. Bacteria and yeast are easily detected after a 24-hr incubation period, but mold detection requires 27 hr. Table 6 shows pharmaceutical product suspensions containing 1% and 2.6% product suspensions exhibiting similar responses when spiked with ATP. Evidently, increasing the product suspension from 1% to 2.6% does not inhibit the reaction or add additional ATP.

Pharmaceutical waters have also been analyzed using ATP bioluminescence. Standard methods for water testing comprise membrane filtration and incubation times ranging from 48 hr, with Plate Count Agar (PCA), to 72 hr with R2A media. Water is extremely susceptible to microbial contamination. Microbial contamination of pharmaceutical water systems can create major manufacturing problems and product recalls. Therefore, microbiological analysis of process water is a critical control point in pharmaceutical manufacturing.

After a 4-month performance evaluation, a quantitative ATP bioluminescence assay has been shown to provide a 24-hour total count of bacteria present in water samples taken from a reverse osmosis/ultra-filtration water system, hot water circulating system, and cold tapwater (20). The overall correlation between the assay and standard methods is greater than 82%. After membrane filtration by the analyst, the system simultaneously lyses the microbial cells on the filters, adds the reagents, and quantitatively determines the number of cells in a given sample. Water samples with microbial numbers from 1 to 75 colony-forming units (CFU)/100 mL are accurately quantitated.

TABLE 5 Detection of Different Levels of Spiked Microorganisms in a Pharmaceutical Product by ATP Bioluminescence and Standard Methods

P. aeruginosa
ATCC 9027   Detection time = 24 hr

Mean RLU of R broth	Mean RLU of 1 gram sample in R broth	CFU per 10 μL	Mean RLU of 1 g sample in R broth + 10 μL of inoculum	Growth on agar
1306	1627	25.0	22,500,000	+
		3.0	22,400,000	+
		1.0	22,400,000	+
		0.3	2173	−

S. aureus
ATCC 6538   Detection time − 24 hr

Mean RLU of R broth	Mean RLU of 1 gram sample in R broth	CFU per 10 μL	Mean RLU of 1 g sample in R broth + 10 μL of inoculum	Growth on agar
1306	1627	27.0	1,516,276	+
		4.0	7,640,774	+
		1.0	16,321,052	+
		0.3	2440	−

E. coli
ATCC 8739   Detection time = 24 hr

Mean RLU of R broth	Mean RLU of 1 gram sample in R broth	CFU per 10 μL	Mean RLU of 1g sample in R broth + 10 μL of inoculum	Growth on agar
1799	1752	43.0	392,736	+
		5.0	295,350	+
		2.0	1805	−
		0.2	2054	−

S. typhimurium
ATCC 13311   Detection time = 24 hr

Mean RLU of R broth	Mean RLU of 1 gram sample in R broth	CFU per 10 μL	Mean RLU of 1g sample in R broth + 10 μL of inoculum	Growth on agar
1549	1523	31.0	864,608	+
		7.0	1,960,812	+
		2.0	1,628,083	+
		0.1	1802	−

TABLE 5   Continued

C. albicans
ATCC 10231   Detection time = 24 hr

Mean RLU of R broth	Mean RLU of 1 gram sample in R broth		CFU per 10 µL	Mean RLU of 1g sample in R broth + 10 µL of inoculum	Growth on agar
1549	1523	3,250	33.0	1,110,004	+
		575	6.0	254,703	+
		185	2.00	39,269	+
		35	0.4	1879	−

A. niger
ATCC 16404   Detection time = 27 hr

Mean RLU of R broth	Mean RLU of 1 gram sample in R broth	CFU per 10 µL	Mean RLU of 1g sample in R broth + 10 µL of inoculum	Growth on agar
1853	1953	12.0	38,369	+
		3.0	49,387	+
		1.0	2341	−
		0.30	2243	−

TABLE 6   Sample Effects of Pharmaceutical Products Using 1% and 10% Sample Suspensions

Product	Response to ATP 25–200%	ATP picomolar
A		
1g	100	10
10g	97	10
B		
1g	107	9
10g	85	7
C		
1g	109	10
10g	102	11

However, accurate quantitation is not possible with water samples containing more than 75 CFU/100 mL. The linearity between the bioluminescence assay and standard methods is demonstrated when the system is challenged with water samples artificially contaminated with *Pseudomonas aeruginosa* ATCC 9027.

A different quantitative ATP bioluminescence system has been shown to be effective for monitoring purified water and water for injection in a pharmaceutical plant. After a one-month evaluation, comparable counts are obtained with the system and standard methods (18). Microbial counts are obtained within 24 hr. The system combines a specialized membrane filtration assay with ATP bioluminescence and enhanced image analysis for quantitation purposes. The linearity, accuracy, and reproducibility of the system are demonstrated by analyzing water samples artificially contaminated with *Burkholderia cepacia* ATCC. Similar responses are demonstrated with water samples artificially contaminated with *Pseudomonas aeruginosa* and *Bacillus subtilis* (21). Replica plates of microbial colonies enumerated with the ATP bioluminescence system are identified and compared to the microorganisms found using standard methods. Bacterial species such as *Ralstonia pickettii, Bacillus sphaericus, Stenotrophomonas maltophilia,* and *Staphylococcus* species have been isolated using both methods (18).

In conclusion, two different ATP bioluminescence systems are available: one is based on the presence and absence of microorganisms in a given pharmaceutical sample after enrichment, and the second one enumerates the microbial colonies grown on the media plates after incubation. Both methods require incubation steps for detecting microorganisms in samples. One is qualitative, the other one is quantitative. Neither allows the immediate detection of microorganisms from samples. However, they do reduce testing time and labor and allow faster sample release.

23.3. DIRECT VIABLE COUNTS (DVC)

Microbial enumeration in pharmaceutical samples can be performed using plate counts and direct microscopy along with viability dyes. Direct counting of individual microbial cells using epifluorescence microscopy has been shown to detect physiologically active bacteria in purified water used in manufacturing processes (9). The samples have been processed through a 0.45 μm filter to retain the bacteria. The bacteria on the filter are then stained with different types of dyes. The dyes are specific for different types of metabolic reactions in the microbial cell. Fluorescent staining with 5-cyano-2,3-ditolyl tetrazolium chloride (CTC) and 6-carboxyfluorescein diacetate (6CFDA) has detected bacterial cells with respiration and esterase-activity, respectively. The CTC and 6CFDA results have indicated that large number of bacteria in purified water retained physiological activity, whereas a large percentage could not form colonies on conventional media. Therefore, microbial counts using DVC are always higher than standard plate counts. However, epifluorescence microscopy analysis is a time-consuming procedure at the time and does not allow the rapid screening of multiple samples.

23.4. FLOW CYTOMETRY

Several studies have shown the applicability of using "viability markers" and flow cytometry for the rapid enumeration of microorganisms in pharmaceutical grade water (22–24). The viability maker most commonly used is based on the reaction of bacteria with the ChemChrome B (CB) dye. This dye, a fluorescein-type ester, is converted to a fluorescent product, a free fluorescein derivative, by intracellular esterase activity after being taken up by microbial cells previously captured by membrane filtration. Microbial cells with an intact cell membrane retain only the fluorescein derivative. The bacteria are then enumerated using a laser scanning instrument, which has been shown to be sensitive down to one cell in a sample within 90 min. and demonstrated a substantially wider linear range than the conventional heterotrophic plate count method. Similar results have been found by fluorescent staining using the DAPI, membrane filtration with TSA and R2A as growth media, and flow cytometry. An ion-exchange system, reverse-osmosis system, and purified water in a hot loop have been sampled and processed. Fluorescence microscopy analysis of water samples using DAPI has resulted in higher microbial counts because DAPI stained all cells containing DNA, including dead cells. Of the two growth media used for membrane filtration, R2A has shown higher microbial numbers than TSA due to the longer incubation time. However, the laser scanning instrument generally has demonstrated a cell recovery closer to R2A. Rapid and accurate enumeration of labeled microorganisms is completed within 90 min. Bacterial numbers obtained by the laser scanning instrument appear to be higher than standard plate counts by an order of magnitude. Analysis of tapwater, purified water, and water for injection (WFI) at several pharmaceutical sites has also shown that flow cytometry is equivalent to the conventional method. Recovery studies in pure cultures demonstrate a good correlation between methods, with a coefficient of correlation of greater than 0.97 for all organisms tested (vegetative bacteria, spores, yeast, and mold). However, none of the studies provide for the multiple processing of water samples.

Furthermore, the assay does not provide accurate quantitation when samples exhibit more than 10^4 cells/membrane. The scanning of the filters is interrupted due to the agglomeration of cells, resulting in a high fluorescence background. Nevertheless, because of recent modifications to the instrument, a higher accuracy can be achieved with 10^5 cells/membrane for bacteria and 10^4 for yeast and mold (25).

Additional studies have recently been performed on the macrolide antibiotic spiramycin, using solid-phase cytometry (26). Artificially contaminated samples of the antibiotic have been analyzed. The solid-phase cytometry has been found to detect all microbes regardless of their sensitivity to the bacteriostatic activity of the drug. With the conventional heterotrophic plate method run in parallel, complete recovery has been obtained only for spiramycin-resistant organisms. The spiked microorganisms that were sensitive to the antibiotic have remained inhibited or stressed by the action of the spiramycin and do not grow on the plate but are detected by flow cytometry. These results further indicate the inadequacy of standard methods to recover injured microorganisms.

Bioburden of in-process samples of recombinant mammalian cell cultures have also been performed using flow cytometry (27). Instead of the 7 days incubation time required for standard bioburden testing, analyses are completed within 4 hr. The assay is sensitive enough to detect from 5 to 15 CFU/mL after 4 hr. Bioburden results are known before a batch is pooled or processed. However, to optimize the detection of bacteria from a background of mammalian cells, different lysis procedures and modification of the original protocol are needed. Residual fluorescence appears to be a problem when detection limits go down to 1 cell/filter.

23.5. IMPEDANCE

When microorganisms grow in enrichment media, some of the substrates are converted into highly charged end products. These substrates are generally uncharged or weakly charged but are transformed during microbial growth. Because of their nature, the end products increase the conductivity of the media, causing a decrease in impedance.

Impedance is the resistance to flow of an alternating current as it passes through a conducting material. Impedance detection time (Td) is when the resistance to the flow of an alternating current indicates the growth of a particular microorganism as a result of changes in the growth media. Several studies have shown the applicability of direct impedance for detecting microbial activity in pharmaceutical products. Because impedance is a growth-dependent technology, a medium must be chosen that will support the growth of microorganisms and also can be optimized for electrical signal. Substrates for this kind of media will be uncharged or weakly charged—such as glucose, which when converted to lactic acid will increase the conductivity of the media. However, a current modification called indirect impedance monitors microbial metabolism by measuring the production of carbon dioxide. The carbon dioxide removed from the growth media results in a decrease in conductivity. The use of indirect impedance allows the use of media that might not generate an optimal electrical response by using the direct method.

A good correlation between direct impedance detection time (Td) and total colony counts has been obtained for untreated suspensions of *S. aureus* ATCC 6538, *Candida albicans* ATCC 10231, *Aspergillus niger* ATCC 16404, and *Pseudomonas aeruginosa* ATCC 9027 in phosphate-buffered saline (PBS) (28). Similar results have been found with suspensions of test microorganisms treated for varying contact periods with selected concentrations of antimicrobial agents. The only difference found is that the detection time for treated cells is extended.

Impedance has been compared to the direct epifluorescence technique (DEFT-MEM) and ATP bioluminescence (ATP-B) for detecting microbial contamination in cells exposed to different antimicrobial agents (29). ATP-B, impedance, and DEFT-MEM have shown a strong correlation between the rapid method response and total colony counts for bacteria and yeast. However, for mold, impedance has been the only rapid method that showed a strong correlation between colony counts and the rapid method. When chlorhexidine-treated suspensions of *S. aureus* ATCC 6538 and *C. albicans* ATCC 10231 have been analyzed by impedance, a good dose-response curve was obtained. Different results have been found with ATP-B and DEFT-MEM methods, which underestimate the kill by the order of 1–6 logs. Impedance application to pharmaceutical screening requires the development of growth curves for different microorganisms. Furthermore, the systems available do not provide high throughput screening.

23.6. PCR TECHNOLOGY

Deoxyribonucleic acid (DNA) contains the genetic information that encodes for the development of a microbial cell. With the latest advances in genomics where scores of microbial genomes have been sequenced, the potential to use genetic information for the detection and discrimination of microorganisms is endless. Genetic technologies can increase the resolution and specificity of microbial detection and identification in pharmaceutical environments. DNA-based technologies are used in clinical, food, and environmental samples providing valuable information on the survival, distribution, and function of microorganisms in those habitats (30,31). One of the technologies based on DNA analysis is the polymerase chain reaction.

The polymerase chain reaction (PCR) amplifies specific DNA sequences along the microbial genome. For example, a set of DNA primers is used to target the specific sequence to be amplified (Table 7). The PCR reaction takes place in three different steps.

TABLE 7 PCR Assay Reaction Steps

1. Double helix denatured by heating.

```
5'                      3'
A T C G C A G G G A T C        95°C        5'                      3'
T A G C G T C C C T A G                     A T C G C A G G G A T C
3'                      5'         →
                                            T A G C G T C C C T A G
                                            3'                      5'
```

2. Primers are bound to complementary sequences on template strands.

Template Strand

```
5'                      3'
A T C G C A G G G A T C
T A G
        |  >   >   >   >  |
        Target Region
                    |
  |   <    <    <    A T C
T A G C G T C C C T A G
3'                      5'
```

Template Strand

3. Primers are extended by DNA polymerase, resulting in two DNA strands.

```
5'                      3'
A T C G C A G G G A T C
T A G C G T C C C T A G
3'                      5'


5'                      3'
A T C G C A G G G A T C
T A G C G T C C C T A G
3'                      5'
```

First, the target sequence is denatured by heating. Second, the primers anneal to complementary sequences on the target DNA strands. Third, the primers are extended by the DNA polymerase enzyme, resulting in two strands. The three steps are repeated again for a given number of cycles (e.g., 30 to 35). Once the target is amplified, the products are detected by gel electrophoresis. However, new systems have been developed that rely on fluorescence detection of amplified products. PCR-based assays are used routinely in the food industry and clinical laboratories to detect and identify bacteria, yeast, and mold (30,31). In pharmaceutical laboratories, PCR-based assays have been shown to be capable of detecting *Salmonella typhimurium*, *Escherichia coli*, *Pseudomonas aeruginosa*, *Staphylococcus aureus, Burkholderia cepacia, Aspergillus niger,* and eubacterial sequences after an incubation period (32–38). Analysts, raw materials, equipment, or water contamination introduces some of these microorganisms into pharmaceutical environments. Furthermore, when analysts do not follow good laboratory practices they become major sources of microbial contamination in cleanrooms. Rapid detection of objectionable microorganisms results in faster implementation of corrective actions. Detection times using PCR range from 24 to 27 hours (Table 8). This is a significant reduction when compared to the standard 5–7 days detection time (6,14). Furthermore, high throughput screening of samples is possible by using a 96-well format.

The simplification of PCR analysis for pharmaceutical quality control is achieved by using a tablet and PCR bead formats. The PCR reagents, including DNA primers, are combined in a tablet form; the beads provide the necessary reagents for the PCR reaction but without the DNA primers. Time-consuming preparations and handling of individual PCR reagents are not required due to the tablet and bead formats incorporated in the assay. DNA extraction from sample enrichments is performed in single-step assays. For bacteria and yeast, a sample preparation using Tris-EDTA-Tween 20

TABLE 8   Pharmaceutical Samples Analyzed by PCR

Inhibitory	Reaction	Detection	Dilution	Time (hr)
Neobee oil	No	Yes	1/10	24–27
Simethicone	No	Yes	1/10	24–27
CMC	No	Yes	1/10	24–27
Sodium alginate	No	Yes	1/10	24–27
Rasberry flavor	No	Yes	1/10	24–27
Hydroxymethylcellulose	No	Yes	1/10	24–27
Xantham gum	No	Yes	1/10	24–27
Silica calcinated	No	Yes	1/10	24–27
Guar gum	No	Yes	1/10	24–27
Starch	No	Yes	1/10	24–27
Lactose monohydrate	No	Yes	1/10	24–27
Diatomaceous earth	No	Yes	1/10	24–27
Tablets	No	Yes	1/10	24–27
Medicated skin cream	No	Yes	1/10	24–27
Ointment	No	Yes	1/10	24–27
Antiflatulent drops	No	Yes	1/10	24–27
Medical device	No	Yes	1/10	24–27
Laxative tablets	No	Yes	1/10	24–27

Source: Refs. 21, 23, and 27.

buffer with proteinase K at 3°C resulted in high-quality DNA; boiling the samples in SDS for 1 hr is required for efficient mold DNA extraction. None of the product suspensions show PCR inhibition, allowing rapid determination of sample quality (Table 8). The development of new PCR formats allows for the simplification of PCR protocols where only sample addition and primers are needed to perform the assay. With the latest advances in microbial genomics, the availability of DNA primer sequences are limitless, allowing the development of universal primers for bacteria, yeast, and mold. A recent study has shown the applicability of detecting bacterial contamination by using a simple PCR assay. The study is based on the universal nature of the DNA sequences coding for bacterial ribosomal genes. DNA primers targeting these common bacterial sequences are capable of rapidly screening samples for bacteria contamination.

All the studies previously discussed have been performed using a single PCR amplification format where a specific microorganism DNA sequences was targeted. However, simultaneous detection of bacteria and mold DNA sequences has been recently reported using a gradient thermocycler (39). The gradient thermocycler allows the use of primers with annealing temperatures ranging from 54° to 65°C, leading to the detection of different microorganisms in a single PCR run. This allows the immediate screening of a pharmaceutical sample for bacteria, yeast, and mold. PCR has also been used for the monitoring of pharmaceutical water samples in manufacturing processes (7). Ribosomal DNA sequences are amplified with universal bacterial primers. After amplification, the samples are loaded onto polyacrylamide gels (denaturing gradient gel electrophoresis (DGEE)) to detect the amplified products. This will allow the separation of DNA fragments of the same length but different pair sequences. After the separation, the gels are scanned to generate a densitometric profile. The sequencing of the amplified fragments has revealed that the dominant bacteria in the water samples are not culturable on standard media. Most of the culturable bacterial species have been found to be related to *Bradyrhizobium* sp., *Xanthomonas* sp., and *Stenotrophomonas* sp.; the dominant bacterial species have not been characterized.

23.7. IMMUNOASSAYS

Although enzyme-linked immunosorbent assays (ELISA) are widely used in clinical and food analyses, it was not until recently that these methods were applied to pharmaceutical quality control. ELISA tests are performed using different formats. The most common format to pharmaceutical quality control analysis is based on the immobilization of high-affinity antibodies, specific for different types of microorganisms, on the surface of microtiter wells. The sample is then applied to the well and incubated. If there is a microorganism in the sample, it is captured by the immobilized antibody. An enzyme-conjugate antibody is then added to react with the captured microorganism. This will result in the formation of an antibody-microorganisms conjugate "sandwich." To develop a detection signal, a chemical substrate is added to react with the enzyme in the conjugate. If there is a microorganism in the sample, a color reaction will develop (Fig. 1). Absence of a specific microbial target is indicated by the absence of color.

Pharmaceutical samples contaminated with pure and mixed cultures have been shown to detect microbial contamination by *S. aureus* within 24 hr (40). These results indicated that the assays are specific enough to detect the target microorganisms in the presence of other microbial species. When compared to the 4–5 days detection time using standard methods, the ELISA method is found to be more effective in reducing detection time and labor.

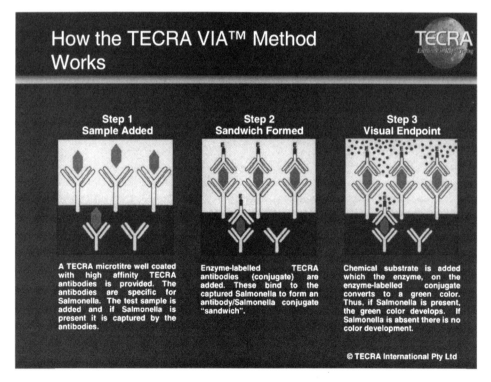

FIGURE 1 Immunoassay format used for detecting microbial contamination in pharmaceutical samples. (Courtesy of TECRA INTERNATIONAL.)

TABLE 9   Protocol for Analysis of Pharmaceutical Samples
Using Immunoassays

A. Heat treatment
   Salmonella spp.
   P. aeruginosa
   1 mL of broth is heated for 15 min in boiling water.
   S. aureus
   5 mL of broth, add 50 μL of ADDITIVE (8), then boil for 15 min.
B. Addition of samples
   Add 200 μL of samples and controls to Remova wells.
   S. aureus
   P. aeruginosa
   60 min 35–37°C
   Salmonella spp.
   30 min 35–37°C
C. First wash
   5X *P. aeruginosa*
     S. aureus
   3X *Salmonella* spp.
D. Conjugate addition
   P. aeruginosa
   Salmonella spp.
   Add 200 μL of conjugate (4) to each well.
   Incubate for 30 min at 35–37°C.
   S. aureus Add 200 μL of conjugate (4) to each well.
   Incubate 10 min at 35–37°C.
E. Second wash
   4X
   P. aeruginosa
   Salmonella spp.
   S. aureus
F. Substrate addition:
   P. aeruginosa
   Salmonella spp.
   Add 200 μL of substrate (6) to each well.
   S. aureus
   Add 200 μL of substrate (5)
   Incubate for 10 min at 25C
G. (Optional) Add 20 μL of STOP solution.

Furthermore, multiple processing and analysis of samples has been possible due to the 96-well microtiter format. Another validation study has been undertaken to compare ELISA assays with standard methods. Other products tested included a range of pharmaceuticals such as cough mixtures, laxatives, ulcer treatments, infant formulas, antiseptic cream, as well as some pharmaceutical ingredients (41).

A recent study in the Consumer Product Testing laboratory ascertained the applicability of three different types of ELISA assays for rapid detection of pathogens. Product suspensions are inoculated with 10 colony-forming units/mL of *P. aeruginosa*, *S. aureus*, and *S. typhimurium*. Samples are then incubated for 24 hours at 35°C. After incubation, samples are analyzed as described in Table 9. Table 10 shows the results of the analysis of pharmaceutical products by using three different types of ELISA methods. Results demonstrated that standard methods and the immunoassays exhibit a 100% correlation. No interferences, false negatives or positives were found by any of the products. However, the immunoassays detected the bacteria in 24 hr while standard methods required 4 to 5 days. Using the 96-well plate format, sample output is 48 samples every 2 hr, counting two positives and two negatives controls simultaneously run with each plate. In an 8-hr laboratory shift, a total of approximately 176 samples can be screened for *P. aeruginosa*, *S. aureus*, and *S. typhimurium*.

TABLE 10 Detection of Microbial Contamination Using Immunoassays

Product	ELISA pharmaceutical bacteria	Dilution method	Standard (days)	Detection time
A	*S. aureus*	1:100	4–5	1
	P. aeruginosa	1:100	4–5	1
	S. typhimurium	1:100	4–5	1
B	*S. aureus*	1:10	4–5	1
	P. aeruginosa	1:10	4–5	1
	S. typhimurium	1:10	4–5	1
C	*S. aureus*	1:10	4–5	1
	P. aeruginosa	1:10	4–5	1
	S. typhimurium	1:10	4–5	1
D	*S. aureus*	1:10	4–5	1
	P. aeruginosa	1:10	4–5	1
	S. typhimurium	1:10	4–5	1
E	*S. aureus*	1:10	4–5	1
	P. aeruginosa	1:10	4–5	1
	S. typhimurium	1:10	4–5	1

23.8. CRITERIA FOR VALIDATING RAPID METHODS

A recent technical report by the PDA has described the different requirements and guidelines to validate and support the implementation of rapid microbiological methods (42). For instance, systems installation, operation, and performance must be verified before validation studies are performed. The Installation Qualification (IQ) of the instrument indicates that the system is correctly installed based on manufacturer's specification. All operational specifications and parts are analyzed to determine if they comply with the manufacturer's specification. For example, in the ATP bioluminescence system, the instrument is turned on and off to determine the operability of the system. The injectors are checked to determine their accuracy. An ATP sample is run to assess the system's response to a positive sample. Furthermore, an empty cuvette is also run to determine the system's response to a negative sample. Software verification and power failure simulations are also performed. Report generation and printing are determined to look for possible problems during data generation.

The Operation Qualification (OQ) verifies that the instrument is operating functionally as a unit. For instance, in the ATP bioluminescence system, OQ verifies and documents that the installed system can perform bioluminescence detection within the specified acceptance criteria. An example of this is shown in Tables 4 and 6. If the response to ATP for a product is between 25% and 200% of the response to broth alone and there is no inhibition/enhancement effect on the bioluminescence reaction, the assay conditions are valid and the system is verified.

The second part of the OQ is to ensure that low-level contamination of a test product by microorganisms can be detected by the system. This is accomplished by challenging the product suspension with different concentrations of pure cultures of microorganisms (Table 5). If the assay is negative, the streak plates from enrichment cultures confirms that the organisms did not grow in the presence of product. If the assay is positive, the streak plates confirm that only the inoculated organism is present and there is no contamination from the preparation of the test.

The Performance Qualification (PQ) demonstrates that the instrument is capable of repeatedly functioning as specified. The PQ is an intensive testing regimen for all products to be tested. Parallel testing is part of this assessment. Both the conventional method and the system should test three consecutive batches. Different concentrations of microorganisms are spiked into the products and samples analyzed using both the rapid and standard method. Some inclusion of naturally contaminated samples is beneficial during this phase (perhaps previously retained samples). Provided that equivalence is shown, then product release can commence using the rapid method system. At this particular stage, written procedures are prepared for operating, training, and maintaining the system. Running a positive and negative ATP sample verifies the system's performance on a daily basis.

According to the PDA technical report, the assay performance must be further evaluated for accuracy, precision, specificity, detection limit, detection limit of quantitation, linearity (if quantitation is desired), ruggedness, and robustness. Accuracy is whether the rapid method generates the same results as the standard method. In the ATP bioluminescence system, a positive sample with the system is confirmed by microbial growth on agar media. Specificity covers the ability of the method to detect all types of microorganisms such as bacteria, yeast, and mold. The ATP bioluminescence system, flow cytometry, PCR, and impedance have been shown to detect all different types of microorganisms (Table 11).

Linearity studies have been conducted to demonstrate the equivalence of ATP bioluminescence and flow cytometry to detect different microorganisms within a given range compared to standard plate counts (Table 11). The limit of detection ascertains the lowest number of microorganisms detected in a sample. This applies to quantitative and qualitative methods. For instance, for flow cytometry and ATP bioluminescence, studies have shown detection limits of 1 CFU/mL for water samples. Range is the interval between the upper and lower limit of detection or quantitation with a degree of accuracy and precision. Precision covers the repeatability of an assay when performed in duplicates for different samples. Ruggedness is defined as how reproducible the assays are when performed under different conditions such as different

TABLE 11 Criteria for Validating Rapid Methods in Pharmaceutical Laboratories as per PDA Technical Report (42)

	Accuracy	Precision	Specificity	Detection limit	Linearity
ATP bioluminescence					
Quantitative	X	X	X	X	X
Qualitative	X	X	X	NA	NA
Flow cytometry	X	X	X	X	X
Impedance	X	X	X	X	X
Immunoassays	X	X	X	NA	NA
PCR assays	X	X	X	NA	NA
Direct viable counts	X	X	X	X	X

NA = Not applicable.
From Ref. 42.

laboratories, analysts, instruments, reagents, and days. Robustness deals with the issue of how reliable the assay is by small changes in assay parameters such as media reagents, incubation time, and so forth. Limits of quantitation, linearity, and range are not applicable to qualitative analysis. However, they must be ascertained when immediate quantitation of microorganisms is performed. Table 11 summarizes the data obtained from scientific studies where some of the recommended criteria and guidelines are ascertained and fulfilled. The PDA guidelines provide a solid foundation for the validation and assessment of rapid microbiological procedures.

23.9. CONCLUSION

On the basis of published scientific studies and conferences, there are several available new technologies that can replace standard microbiological methods. Rapid methods have proved to be effective, reliable, sensitive, and equivalent to standard microbiological assays. However, application must be based on the needs of a given company and in a case-by-case basis. For instance, in some situations microbial enumeration is required; in others, the presence or absence of microorganisms results in rapid quality analysis (Table 12). Microbial contamination is a sporadic event in pharmaceutical environments, so rapid screening of batches using alternative microbiological testing provides a rapid release for approximately 99% of samples tested. When microbial contamination is found, rapid methods such as immunoassays or PCR technology can analyze the sample for the presence of pathogenic microorganisms using high throughput screening (Table 12). However, quantitative systems to date do not have high throughput screening capabilities.

As demonstrated by published scientific reports, validation studies showing equivalency between compendial and rapid methods must be per-

TABLE 12 Comparison of Rapid Methods

Method	Sensitivity (cells/mL)	Detection time (hr)	High throughput	Quantitation ATP
Qualitative	10^4	24–48	Yes	No
Quantitative	1	24	No	Yes
PCR	10^5	24–30	Yes	No
Flow cytometry	1	2	No	Yes
Impedance	10^6	24–30	No	Yes
Immunoassays	10^4	24–27	Yes	No
Direct viable counts	1	24	No	Yes

formed before implementation. Some of the rapid technologies are more accurate than standard microbiological methods (Table 12). For example, enumeration and detection of bacteria that did not grow on standard media will create a situation where changes in specifications will be required. However, changes in specifications can be documented if there is a significant advantage in the use of a rapid method. Several terms such as microbial viability will be redefined as per specific data supporting the changes, indicating that a microorganism can be viable but not able to grow in enrichment media. For instance, in flow cytometry, DVC, and PCR studies, several microbial species have been found to be predominant members of the microbial community but have not been isolated or detected using standard methods (7,9,24). However, this should not discourage the use of these technologies but on the contrary create an environment where their use will develop additional information where process validation and control can be significantly improved.

Future optimization of pharmaceutical manufacturing and quality control requires faster microbiological analysis than standard conventional methods. Rapid methods identify microbial contamination with detection times ranging from 90 min to 30 hr allowing the monitoring of critical control points, reducing losses, and optimizing resources (Table 12). In the 21st century, with advances in computer sciences, automation, combinatorial chemistry, genomics, and medicine, quality control microbiology requires faster turnover times, higher resolution, and sensitivity without compromising efficacy. Rapid technologies enhance the ability of a quality control system for risk assessment and process control.

REFERENCES

1. Casey, W.; Muth, H.; Kirby, J.; Allen, P. Use of nonselective preenrichment media for the recovery of enteric bacteria from pharmaceutical products. Pharm. Technol. 1998, 22, 114–117.
2. Palmieri, M.J.; Carito, S.L.; Meyer, J. Comparison of rapid NFT and API 20E with conventional methods for identification of gram-negative nonfermentative bacilli from pharmaceutical and cosmetics. Appl. Environ. Microbiol. 1988, 54, 2838–3241.
3. Hyde, W. Origin of bacteria in the clean room and their growth requirements. PDA J. Sci. Technol. 1998, 52, 154–164.
4. Whyte, W.; Nive, L.; Bell, N.D. Microbial growth in small-volume pharmaceuticals. J. Parenteral Sci. Technol. 1989, 43, 208–212.
5. United States Pharmacopeial Convention. Sterility tests. In U.S. Pharmacopoeia; United States Pharmacopeial Convention: Rockville, Maryland, 2000; Vol. 24, 1818–1823.

6. Jimenez, L. Rapid methods for the microbiological surveillance of pharmaceuticals. PDA J. Pharm. Sci. Technol. 2001, *55*, 278–285.
7. Kawai, M.; Matsutera, E.; Kanda, H.; Yamaguchi, N.; Tani, K.; Nasu, M. 16S ribosomal DNA-based analysis of bacterial diversity in purified water used in pharmaceutical manufacturing processes by PCR and denaturing gradient gel electrophoresis. Appl. Environ. Microbiol. 2002, *68*, 699–704.
8. Nagarkar, P.; Ravetkar, S.D.; Watve, M.G. Oligophilic bacteria as tools to monitor aseptic pharmaceutical production units. Appl. Environ. Microbiol. 2001, *67*, 1371–1374.
9. Kawai, M.; Yamaguchi, N.; Nasu, N. Rapid enumeration of physiologically active bacteria in purified water used in the pharmaceutical manufacturing process. J. Appl. Microbiol. 1999, *86*, 496–504.
10. Zani, F.; Minutello, A.; Maggi, L.; Santi, P.; Mazza, P. Evaluation of preservative effectiveness in pharmaceutical products: the use of a wild strain of Pseudomonas cepacia. J. Appl. Microbiol. 1997, *43*, 208–212.
11. Papapetropoulou, M.; Papageorgakopoulou, N. Metabolic and structural changes in Pseudomonas aeruginosa, Achromobacter CDC, and Agrobacterium radiobacter cells injured in parenteral fluids. PDA J. Pharm. Sci. Technol. 1994, *48*, 299–303.
12. Desvignes, A.; Sebastien, F.; Benard, J.; Campion, G. Etude de la contaminacion microbienne de diverses preparations pharmaceutiques. Ann. Pharm. Fr. 1973, *31*, 775–785.
13. Coppi, G.; Genova, R. Controllo della contaminazione microbica in preparati farmaceutici orally e topici col metodo delle membrane filtranti. Il Farmaco 1970, *26*, 224–229.
14. Jimenez, L. Molecular diagnosis of microbial contamination in cosmetic and pharmaceutical products–A review. J. AOAC Int. 2001, *84*, 671–675.
15. Hugo, W.B. Bacteria. In *Pharmaceutical Microbiology*; Hugo, W.B. Russell, A.B., Eds.; 6th Ed.; Blackwell Science: Oxford, 1998; 3–34.
16. Bussey, D.M.; Tsuji, K. Bioluminescence for USP sterility testing of pharmaceutical products. Appl. Environ. Microbiol. 1986, *51*, 349–355.
17. Ignar, R.; English, D.; Jimenez, L. Rapid detection of microbial contamination in Triclosan and High Fluoride dentifrices using an ATP bioluminescence assay. J. of Rapid Methods and Automation in Microbiology 1998, *6*, 51–58.
18. Marino, G.; Maier, C.; Cundell, A.M. A comparison of the MicroCount Digital System to plate count and membrane filtration methods for the enumeration of microorganisms in water for pharmaceutical purposes. PDA J. Pharm. Sci. Technol. 2000, *54*, 172–192.
19. Jimenez, L. Light up the life in microbial contamination. Pharmaceutical Quality and Formulation, August/September 2001, 54–55.
20. Scalici, C.; Smalls, S.; Blumberg, S.; English, D.; Jimenez, L. Comparison of Millipore Digital Total Count System and standard membrane filtration procedure to enumerate microorganisms in water samples from cosmetic/pharmaceutical environments. Journal of Rapid Methodsand and Automation in Microbiology 1998, *7*, 199–209.

21. Hauschild, J. Applying rapid enumeration and identification of microorganisms using the Millipore Microstar™ system. Rapid Methods and Automation in Microbiology for Pharmaceutical, Biotechnology, and Devices Applications, San Juan, Puerto Rico, Feb 1–2, 2001.

22. Reynolds, D.T.; Fricker, C.R. Application of lasers scanning for the rapid and automated detection of bacteria in water samples. J. Appl. Microbiol. 1999, *86*, 785–796.

23. Wallner, G.; Tillmann, D.; Haberer, K. Evaluation of the ChemScan system for rapid microbiological analysis of pharmaceutical water. PDA J. Pharm. Sci. Technol. 1999, *53*, 70–74.

24. Gapp, G.; Guoymard, S.; Nabet, P.; Scouvart, J. Evaluation of the applications of a system for real-time microbial analysis of pharmaceutical water systems. European Journal of Parenteral Sciences 1999, *4*, 131–136.

25. Costanzo, S.P.; Borazjani, R.N.; McCormick, P.J. Validation of the Scan RDI for routine microbiological analysis of process water. PDA J. Pharm. Sci. Technol. 2002, *56*, 206–219.

26. Ramond, B.; Rolland, X.; Planchez, C.; Cornet, P.; Antoni, C.; Drocourt, J.L. Enumeration of total viable microorganisms in an antibiotic raw material using ChemScan solid phase cytometer. PDA J. Pharm. Sci. Technol. 2000, *54*, 320–331.

27. McColgan, J. Rapid detection of bacterial contamination in recombinant mammalian cell culture. Rapid Methods and Automation in Microbiology for Pharmaceutical, Biotechnology, and Devices Applications, San Juan, Puerto Rico, Feb 1–2, 2001.

28. Connolly, P.; Bloomfield, S.F.; Denyer, S.P. The use of impedance for preservative efficacy testing of pharmaceuticals and cosmetics. J. Appl. Bacteriol. 1994, *76*, 68–74.

29. Connolly, P.; Bloomfield, S.F.; Denyer, S.P. A study of the use of rapid methods for preservative efficacy testing of pharmaceuticals and cosmetics. J. Appl. Bacteriol. 1993, *75*, 456–462.

30. Hill, W.E. The polymerase chain reaction: application for the detection of foodborne pathogens. Crit. Rev. Food Sci. Nutr. 1996, *36*, 123–173.

31. Ieven, M.; Goosens, H. Relevance of nucleic acid amplification techniques for diagnosis of respiratory tract infections in the clinical laboratory. Clin. Microbiol. Rev. 1997, *10*, 242–256.

32. Jimenez, L.; Smalls, S.; Scalici, C.; Bosko, Y.; Ignar, R.; English, D. Detection of Salmonella spp. contamination in raw materials and cosmetic/pharmaceutical products using the BAX™ system, a PCR-based assay. Journal of Rapid Methods and Automation in Microbiology 1998, *7*, 67–76.

33. Jimenez, L.; Smalls, S.; Grech, P.; Bosko, Y.; Ignar, R.; English, D. Molecular detection of bacterial indicators in cosmetic/pharmaceuticals and raw materials. Journal of Industrial Microbiology and Biotechnology 1999, *21*, 93–95.

34. Jimenez, L.; Smalls, S.; Ignar, R. Use of PCR analysis for rapid detection of low levels of bacterial and mold contamination in pharmaceutical samples. Journal of Microbiological Methods 2000, *41*, 259–265.

35. Jimenez, L.; Smalls, S. Molecular detection of Burkholderia cepacia in toiletries, cosmetic and pharmaceutical raw materials and finished products. Journal of AOAC International 2000, 83, 963–966.

36. Jimenez, L.; Bosko, Y.; Smalls, S.; Ignar, R.; English, D. Molecular detection of Aspergillus niger contamination in cosmetic/pharmaceutical raw materials and finished products. Journal of Rapid Methods and Automation in Microbiology 1999, 7, 49–56.

37. Jimenez, L.; Ignar, R.; D'Aiello, R.; Grech, P. Use of PCR analysis for sterility testing in pharmaceutical environments. Journal of Rapid Methods and Automation in Microbiology 2000, 8, 11–20.

38. Jimenez, L.; Smalls, S.; Scalici, C.; Bosko, Y.; Ignar, R. PCR detection of Salmonella typhimurium in pharmaceutical raw materials and products contaminated with a mixed bacterial culture using the BAX™ system. PDA Journal of Pharmaceutical Sciences and Technology 2001, 55, 286–289.

39. Jimenez, L. Simultaneous PCR detection of bacteria and mold DNA sequence in pharmaceutical samples by using a gradient thermocycler. Journal of Rapid Methods and Automation in Microbiology 2001, 9, 263–270.

40. English, D.; Scalici, C.; Hamilton, J.; Jimenez, L. Evaluation of the TECRA™ visual immunoassay for detecting Staphylococcus aureus in cosmetic/pharmaceutical raw materials and finished products. Journal of Rapid Methods and Automation in Microbiology 1999, 7, 193–203.

41. Hughes, D.; Dailianis, A.; Hill, L. An immunoassay method for rapid detection of Staphylococcus aureus in cosmetics, pharmaceutical products, and raw materials. Journal of AOAC International 1999, 82, 1171–1174.

42. PDA Technical Report Number 33. Evaluation, validation, and implementation of new microbiological testing methods. J. Parent. Sci. Tech. 2000, 54 (3).

Index